Collaborative Construction Procurement and Improved Value

Collaborative Construction Procurement and Improved Value

David Mosey
Centre of Construction Law & Dispute Resolution
Dickson Poon School of Law
King's College London

WILEY Blackwell

This edition first published 2019
© 2019 John Wiley & Sons Ltd

The right of David Mosey to be identified as the author of this work has been asserted in accordance with law.

Registered Offices
John Wiley & Sons, Inc., 111 River Street, Hoboken, NJ 07030, USA
John Wiley & Sons Ltd, The Atrium, Southern Gate, Chichester, West Sussex, PO19 8SQ, UK

Editorial Office
9600 Garsington Road, Oxford, OX4 2DQ, UK

For details of our global editorial offices, customer services, and more information about Wiley products visit us at www.wiley.com.

Wiley also publishes its books in a variety of electronic formats and by print-on-demand. Some content that appears in standard print versions of this book may not be available in other formats.

Library of Congress Cataloging-in-Publication Data

Names: Mosey, David, 1954- author.
Title: Collaborative construction procurement and improved value / David Mosey.
Description: Hoboken, NJ : Wiley-Blackwell, 2019. | Includes bibliographical
   references and index. |
Identifiers: LCCN 2019000081 (print) | LCCN 2019004042 (ebook) | ISBN
   9781119151920 (Adobe PDF) | ISBN 9781119151937 (ePub) | ISBN 9781119151913
   (hardback)
Subjects: LCSH: Buildings–Specifications–Great Britain. | Construction
   contracts–Great Britain. | Building materials–Purchasing–Great
   Britain–Case studies. | Building information modeling–Great
   Britain–Case studies. | Government purchasing–Great Britain. | Public
   contracts–Great Britain.
Classification: LCC TH425 (ebook) | LCC TH425 .M64 2019 (print) | DDC
   354.6/42530941–dc23
LC record available at https://lccn.loc.gov/2019000081

Cover Design: Wiley
Cover Image: © Association of Consultant Architects Ltd.

Set in 10/12pt WarnockPro by SPi Global, Chennai, India

10 9 8 7 6 5 4 3 2 1

Contents

Reviews

Sir Rupert Jackson PC, former Lord Justice of Appeal:
'A successful procurement exercise or construction project is one in which all participants work together collaboratively to achieve a common end. That is not easy to achieve because the participants each have their own commercial interests and reputations to protect. I have long believed that the mere inclusion of platitudes that "the parties will work together in good faith" adds little to the implied term of co-operation, and a series of recent cases have shown that such wording seldom avails the parties when a dispute erupts.

The present book goes far beyond platitudes. It explores new ways of working and new contractual structures which can actually bring about collaborative working. It demonstrates how the use of BIM can facilitate the ready sharing of information between members of the team. It explains how the team members can benefit from the creation and development of a project alliance. The research and case studies set out in this book will offer practical guidance to all who are working in the construction sector.'

Professor Phillip Capper, Head of International Arbitration, White & Case:
'Improvement of risk management in international construction projects is vital, and Professor Mosey's challenging book shows how this can be achieved through team integration and early contractor involvement. The focus has to widen from the typical client/main contractor preoccupation, and this penetrative study shows how supply chain collaboration is vital.

Beginning from the lasting initiatives inspired by Sir Michael Latham in 1994, Professor Mosey explores how team selection and contractual commitments can help build a collaborative culture. We see how collaborative procurement is connected to contracts, and also to BIM and other digital technology. Empirical evidence is grounded in case studies and King's College London research led by Professor Mosey, which is enhanced by leading co-authors' analyses from other major economies. Collaborative construction procurement demonstrably can reduce disputes, and this book enables a global readership to see which forms of contract can achieve better construction project performance.'

Ann Bentley, Construction Leadership Council Board Member, Rider Levett Bucknall Global Board Member and author of 'Procuring for Value':
'As any harassed parent knows, telling restive children to "play-nicely" is no guarantee that they will. Collaboration is much the same, and a broad expression of collaborative

intent is no guarantee of collaborative behaviour: it requires knowledge, structure and commitment. With this comprehensive and far-reaching analysis, taking us from the birth of collaborative contracts to their relevance and use around the world, David Mosey and his King's College team go a very long way to filling important knowledge gaps.

This book should be recommended reading for anyone considering undertaking a construction project, and compulsory reading for their advisers. I commend David and his team for this work and the contributions that it will make to improving the way construction is procured and delivered.'

Mark Farmer, CEO Cast Consultancy and author of 'Modernise or Die':
'There is a crucial need to adopt an integrated procurement model in order to deliver projects more efficiently, for example through increasing "pre-manufactured value" by moving processes from the final site into controlled manufacturing environments. I commend this book whose international co-authors have collated an excellent global reference point, demonstrating how organising projects differently can create better outcomes for all parties.

The recommended procurement and contract systems are shown to achieve better aligned interests by harnessing learning and relationships from project to project and by using value-based selection and remuneration techniques. Unless you can deliver specific value-adding expertise through integrated working behaviours, the construction world will become an increasingly difficult place to make money and survive. Reading this publication is a vital part of future-proofing yourself.'

Professor John Uff CBE QC:
'This seminal work brings together the fruits of studies and writings spanning many years and encompassing many projects throughout the world under a variety of legal systems. The need for collaboration in the construction process has been a constant theme in the search for procedures and systems which can harness the expertise and energies of parties with divergent commercial interests while avoiding disputes.

Procurement is the point at which collaboration begins, with the choice of project alliancing for a single enterprise or a framework or other longer-term arrangement bringing wider opportunities for collaboration. These extended relationships are supported by the authors' work in developing the FAC-1 and TAC-1 models for which impressive case studies are described. The key to success is seen as the development of personal relationships, enhanced by digital technology including BIM, shared knowledge and appropriate motivation.'

Shelagh Grant, Chief Executive, Housing Forum:
'David Mosey's extensive knowledge of the construction industry, and his well thought through solutions to delivering the best possible outcomes, come over strongly in this work. Many examples are given of the collaborative links and early interactions that help achieve good quality and good value in difficult and complex situations.

The elements of successful collaboration are clearly laid out with particular emphasis on the selection of and relationships between team members. The application of digital technology is shown to work in particular alliance with this approach.'

Matthew Bell, Senior Lecturer and Co-Director of Studies for Construction Law, Melbourne Law School:

'For many in the construction industry, collaborative procurement is the holy grail. This new text by Professor David Mosey and leading practitioners from around the world provides a uniquely-valuable road map in pursuit of that goal. It not only explains the benefits of collaborative ways of working, it also helps industry professionals and their lawyers navigate the potential pitfalls by compiling a critical assessment of experience to date.

The text harnesses lessons learned and the value of technological innovations such as BIM. In this way, it provides both a "how to" and "why to" manual for realising the potential of collaborative construction procurement as we enter the third decade of the 21ˢᵗ century.'

Nick Barrett, Editor, Construction Law:

'Anyone viewing a typical construction project sees the impressive collaboration that brings designs, people, machinery and materials together in the one place, but they may not see the dangerous divisions that still exist in construction's procurement and contractual underpinnings.

This book's authors show how a new focus on collaborative procurement can treat many of the industry's ills. Evidence has been gathered internationally, not just from the UK, that collaborative approaches can make a major difference to outcomes.

The need for a new industry strategy has never been greater, particularly after the Grenfell Tower disaster and the Carillion collapse. Collaborative procurement approaches that can be easily adopted are detailed in these pages, with a diversity of case studies that should convince even the sceptical.'

Jason Russell, Executive Director, Highways, Transport & Environment, Surrey County Council:

'As a Local Government Director, I am being challenged as never before to reduce costs whilst improving outcomes for our communities.

This timely book demonstrates that bringing together the wider supply chain at the right time, with clear outcomes and underpinned by effective processes, can deliver significant benefits. It provides a practical guide, built on the experience of many projects that have delivered proven results over a number of years, and it is essential reading for anyone interested in getting better value from their construction projects.'

Dr David Hancock, Construction Director, Infrastructure and Projects Authority, Cabinet Office:

'Since the success of Terminal 5 Heathrow, I have been a great supporter of collaborative approaches and ECI for complex construction projects. This book recognises that collaboration may not be a universal panacea, and it sets out the arguments and opportunities that need to be debated prior to making procurement decisions. Where those opportunities outweigh the risks, it provides the foundations both contractually and behaviourally to ensure the best chance of success, with real examples from industry.

This is a book that will benefit both the novice and the expert, providing a high-level overview and a dive into details for the practitioner to implement, without bias to a single contract type and with guidance on Alliance Contract forms for those who wish to realise their benefits.'

Kevin Murray, Deputy Director - Construction & Property, Government Property Agency:
'This book provides comprehensive evidence that lays waste to the myth that collaboration does not need contractual provisions, commitment and accountability.'

Mark Bew MBE, Chair of PCSG Consultancy and former Chair of the UK Government BIM Task Group:
'The only moving parts in procurement are the client, the supply chain and a contract defining the relationships between them, yet surprising complexities soon emerge in aligning their different agendas, incentives and required outcomes. This is when poor information often creates the shadows, fears and misunderstandings that are the enemies of aligned goals and collaboration.

This book has drawn on the author's vast experience of creating contracts which acknowledge these challenges, and which seek to use existing data and management techniques to create clear, open processes that work effectively. Any client or supplier should read and understand the techniques described in this book, as they remove waste and friction and can be applied to all client-supplier relationships in ways that provide significant benefits.'

Dr Kristen MacAskill, Course Director, Construction Engineering Masters Programme, Department of Engineering, University of Cambridge:
'This book presents a wide-ranging discussion on what underpins successful partnering in the construction industry, supported by evidence of achievements in both UK and global case studies. While recognising that the form of contract alone will not deliver project success, David Mosey demonstrates how contract arrangements can help facilitate more collaborative attitudes and behaviours.

The chapters explore the benefits of collaborative arrangements in construction contracts, potential shortfalls in the execution of those arrangements and solutions to manage risks. David outlines not just the contractual mechanisms but also the wider people and process considerations required for real benefits to be delivered. Many angles are explored and it is possible to pick up this book and start on the questions that resonate most with the reader.'

Don Ward, Chief Executive, Constructing Excellence:
'Many people have worked to implement the recommendations of Latham and Egan for construction reform, but few can match David Mosey's first-hand experience and expertise in delivering the approaches which he promotes in this book with characteristic clarity and skill. He has probably worked on more collaborative projects than anyone else in the UK construction industry in the last two decades. He can literally point to billions of pounds worth of projects which he directly influenced and helped on a journey to implement better collaboration, using contracts and procurement routes as key enabling tools.

Consequently, David has had more success and gained extensive first-hand knowledge of what works and what doesn't, and the plentiful case studies throughout this book illustrate this so very well. I have been honoured to work alongside him, including in the trial projects programme on which he draws heavily, and I hope this book will provide many more people with access to his thinking, approaches and practical advice. I hope you find David's experience and expertise as valuable as I have done, and that he convinces you to implement collaborative procurement just as energetically.'

Preface

Let's dig tunnels.
Let's build bridges.
Let's get close like clouds of midges.
What was under Mr. Brunel's hat?
His love letters
And his sandwiches.
Let us cross that big divide.
Let us go and coincide
And with space between deducted
Let us mind what's been constructed.

You provide the motion and I'll start the debate
You provide the provender and I'll supply the napkin
    and the plate.
Let's combine this life of mine with your own slender
    fate.
Let me elaborate.
Let's be thick as thieves can be.
Let's thicken up the ice and then entice the world to
    skate.
You be narrow. I'll be straight.
You be weight and I'll be volume.
Let's make a pair of zeros
make a bigger figure of eight.
Let's collaborate.

With kind permission of John Hegley, an inspiring collaborator
'*Let us Play*', taken from '*Peace Love and Potatoes*', Profile Books 2012

Acknowledgements

Firstly, I offer heartfelt thanks to my co-authors Howard W. Ashcraft, Dr Wolfgang Breyer, Professor Paula Gerber, Professor Stefan Leupertz, Marko Misko, Mariana Miraglia, Matatias Parente, Dr Alexandre Aroeira Salles, Adriana Spassova and Professor Sara Valaguzza. Their contributions to this book are fascinating and their work in the field of collaborative procurement is an inspiration to anyone exploring what can be done to integrate theory and practice across different jurisdictions.

In addition, I offer my gratitude and respect to friends and colleagues at King's College London, Trowers & Hamlins and the Association of Consultant Architects, and to all the practitioners, academics, project teams and research groups who have helped me understand how collaborative construction procurement can deliver improved economic and social value. I am also very grateful to Jo Howard for her thoughtful proofreading of this book and to Dr Paul Sayer and all at Wiley for their patient support.

I owe a particular debt to my father Victor Alfred Mosey and to four other remarkable men for their wisdom and guidance whenever I embarked on new endeavours, namely Anthony Gosselin Trower, H.E. Yousuf Ahmed Al-Shirawi, Sir Michael Latham and Professor David D. Caron.

This book is dedicated to my wife Cécile.

February 2019

David Mosey
King's College London

About the Authors

Professor David Mosey PhD is Director of the King's College London Centre of Construction Law and Dispute Resolution and is former Head of Projects and Construction at Trowers & Hamlins solicitors where he worked for over 30 years as a specialist construction lawyer. He has advised on a wide variety of construction and engineering projects in the UK and internationally, with a particular focus on collaborative construction procurement.

David has been described in the Chambers Guide to the Legal Profession as a 'partnering guru' who 'gives something to the industry'. He is principal author of the PPC2000 suite of partnering contract forms, and led the research and drafting of the FAC-1 Framework Alliance Contract and the TAC-1 Term Alliance Contract.

David was a lead mentor for the UK Government Trial Projects programme, testing the savings and other benefits achieved through the combination of early contractor involvement, collaborative working and Building Information Modelling. He won the UK Constructing Excellence Achievers Award in 2009, and in October 2015 was awarded the Association of Consultant Architects medal for services to the architectural profession. For three years David was a Senior Visiting Fellow at Melbourne Law School, and he is currently a Fellow of the Cambridge University Engineering Masters Programme. He is also a member of the Comitato Scientifico of the Milan Centre of Construction Law & Management.

Howard W. Ashcraft has led in the development and use of Integrated Project Delivery and Building Information Modelling in the United States, Canada and abroad. Over the past decade, his team has structured over 130 pure IPD projects and worked on many highly integrated projects. He co-authored the AIACC's *Integrated Project Delivery: A Working Definition*, the AIA's *IPD Guide, Integrating Project Delivery* (Wiley 2017) and *Integrated Project Delivery: An Action Guide for Leaders* (Pankow Foundation 2018), and he has chaired subcommittees for the National Building Information Modelling Standard (NBIMS).

A partner in the San Francisco law firm of Hanson Bridgett, Howard is an elected Fellow of the American College of Construction Lawyers, an Honorary Fellow of the Canadian College of Construction Lawyers, a Fellow of the American Bar Foundation and an Honorary Member of AIA California Council. In addition to his professional practice, he serves as an Adjunct Professor of Civil and Environmental Engineering at Stanford University.

Dr Wolfgang Breyer is a specialist construction lawyer advising contractors, engineers and employers on major projects such as wind parks, hydropower stations, road and rail, offshore energy and power plants in the Middle East, Asia, Africa and Europe. His areas of expertise include international and German construction contracts, construction risk management advice, and international construction arbitration, litigation, adjudication and alternative dispute resolution.

In May 2015, Dr Breyer led the creation of the International Construction Law Association, bringing together families of law from around the globe in order to inform and debate comparatively on construction law issues. He is a well-known speaker and writer on construction law, and is also creator and Practice Director of the Masters' Programme in International Construction: Law and Practice at the University of Stuttgart.

Professor Paula Gerber has been a lawyer for over 25 years. She spent five years working as a construction lawyer in London and five years in Los Angeles, before returning to Australia where she became a partner in a leading Melbourne law firm. Paula moved from private practice to academia and is now a Professor at the Monash University Law School. She is an internationally recognised expert in construction law, in particular in the areas of dispute avoidance processes (DAPs) and dispute resolution processes, particularly ADR.

Professor Gerber is the lead author of *Best Practice in Construction Disputes: Avoidance, Management and Resolution* (LexisNexis 2013), which was a finalist in the prestigious Centenary Book Award. She is also the author of numerous journal articles and book chapters.

Professor Stefan Leupertz is a former Judge of the German Supreme Court and is known as one of the leading legal authorities on construction law in Germany. During his tenure as a judge, he was assigned to the Seventh Civil Division, responsible for hearing construction, contract and architectural law disputes. In 2012 Professor Leupertz resigned as a judge and commenced his own practice Leupertz Baukonfliktmanagement with the focus purely on construction law – in particular, dispute avoidance and dispute resolution – acting as an arbitrator, adjudicator and legal expert. Professor Leupertz is highly respected in academia as co-editor of the German construction law journal, *Baurecht*, and as editor/author of numerous publications on German construction law. He sits on boards that include the Arbitral Tribunal Estate Law Germany, the German Baugerichtstag e.V, the Anglo-German Construction Law Platform and the International Construction Law Association.

Mariana Miraglia is a partner of Aroeira Salles Advogados and acts for domestic and overseas clients in relation to major infrastructure projects in Brazil. Her areas of expertise include the legal management of construction contracts, procurement, civil investigations, public project auditing, compliance and regulatory law. Mariana's international experience includes advice in respect of cross-border contracts executed in Brazil, preparing for complex arbitration proceedings in New York and advising on an infrastructure project in Africa.

Marko Misko has been involved in the delivery of major Australian infrastructure projects for almost 30 years, across a wide variety of sectors that include transport, defence, health, education, foreign aid, housing, energy and natural resources, and commercial property development.

Marko is committed to driving greater collaboration within the construction industry through procurement methods, project delivery models, risk allocation and standard form contracts. To that end, Marko has worked with all Australian forms of collaborative contracting including managing contractor, project alliancing, delivery partner model, early contractor involvement and most recently integrated project delivery (IPD). Marko has spent the past five years assisting the Australian Department of Defence in finalising its third generation suite of IPD standard form delivery models, designed to facilitate maximum collaboration among all key project contractors/stakeholders and to make optimal use of Building Information Modelling.

Matatias Parente is a lawyer at Aroeira Salles Advogados, advising clients on all aspects of the delivery of construction and engineering projects, from procurement and the drafting and negotiation of contracts, to contract administration and the avoidance and resolution of disputes. He has advised clients operating across a range of sectors including shipbuilding, oil and gas, energy and mining, also assisting them on regulatory compliance matters.

Dr Alexandre Aroeira Salles is founding partner of the Brazilian law firm Aroeira Salles Advogados and advises on large-scale infrastructure and energy projects. His areas of expertise include construction disputes, procurement, public-private partnerships and concession arrangements, and the legal management of construction and energy contracts. Alexandre is a founding member of the Brazil Infrastructure Institute and a director of the Brazilian Construction Law Institute. He also holds a doctorate from PUC University in São Paulo and is widely published, in particular in the area of construction law.

Alexandre's expertise is recognised internationally. He is on the board of the International Construction Law Association and regularly speaks at high-profile international conferences, including in London, Paris, Beijing, Dubai and São Paulo. Alexandre is independently ranked as a leading lawyer in international legal directories.

Adriana Spassova is a civil engineer with more than 30 years' professional experience as a designer, construction manager and consultant. She has an MSc in Construction Law & Dispute Resolution from Kings' College London and is Vice President of the European Society of Construction Law, Chair of the Bulgarian Society of Construction Law and the only FIDIC Accredited Dispute Adjudicator & Trainer in Bulgaria.

As a partner at EQE Control, Adriana has managed consulting services for the biggest project in Bulgaria, namely the 600 MW TPP AES Galabovo (€1.5 billion), as well as for the 156 MW St. Nikola Wind Park Kavarna, the 176-m Europe Tower Sofia, the Varna Business Park and 6 World Bank projects. In the last three years, she has been a Planning and Contracts Key Expert in the preparation of railway projects with a €1 billion budget.

Adriana has more than 20 years' experience in preparation of FIDIC tender documentation and in acting as an engineer in nuclear safety, waste management and infrastructure. She is also a claims consultant and an expert in arbitration and DAB. She has delivered FIDIC accredited training in Bulgaria, Russia, Ukraine, Kazakhstan, Uzbekistan and Belarus.

Professor Sara Valaguzza is Professor at the University of Milan, where she teaches public-private partnerships, public contracts and environmental law. Her interests focus on strategic regulation, alliancing, construction law, sustainable development, procuring for value and legal BIM. Sara practises as an attorney for Italian and international clients. Her firm is a reputed legal 'boutique' in Italy, assisting public authorities and private companies in many significant construction and urban regeneration projects, such as the Blue and Lilac Line subway construction in Milan and the post-Expo development.

Sara leads an interdisciplinary research group promoting collaborative contracts in public and private fields. Since 2016 she has been Deputy Director of the Milan Centre of Construction Law & Management, and Head of the European Association of Public-Private Partnerships. With more than 40 publications in different languages, her latest book *Procuring for Value – Governare per contratto creare valore attraverso i contratti pubblici* promotes reform of the construction industry by the use of collaborative contracting and BIM to create added value in strategic public procurement.

1

What Is Collaborative Construction Procurement?

1.1 Overview

Collaborative construction procurement comprises a set of processes and relationships through which teams can develop, share and apply information in ways that improve the design, construction and operation of their projects. It supports team selection and team integration, and it offers a fresh approach to legal and cultural issues that can otherwise reduce efficiency and waste valuable resources.

This book explores the delivery of economic and social value through improvements in strategic thinking, team selection, contract integration and the use of digital technology, and it considers how the processes and relationships of collaborative construction procurement can be brought more into the mainstream. It uses analysis, guidance, and case studies to illustrate how collaborative approaches can be adopted successfully by any team in any part of the construction sector. With contributions from six other countries, this book also shows how the models that comprise collaborative construction procurement can operate in a range of common law and civil law jurisdictions.

Many procurement models provide little time or opportunity for consultants, contractors, subcontractors, manufacturers and operators to integrate their work. Instead, these models attempt to fix prices without joint cost analysis and to transfer risks without joint risk management, often encouraging misunderstandings and disputes that lead to cost overruns, delays and defects. This book examines collaborative approaches that can anticipate and avoid the problems created by incomplete data, by fragmented contracts and by the neglect of team integration.

We will review the collaborative bridges that connect and integrate the work of different team members and that translate their aspirations into actions, plus a range of factors that may encourage or obstruct progress. New procurement models will not gain widespread support unless they offer benefits for all parties, and we will examine the ways in which procurement processes, digital technology and collaborative contracts can accommodate the differing aspirations and requirements of all team members.

This chapter summarises recent research and commentaries that identify the need for procurement reform, and notes the links between collaborative procurement, digital technology and contracts. Chapter 2 considers the foundations for collaborative construction procurement and Chapter 3 describes the features of collaborative working through a project alliance. Chapters 4 and 5 consider the greater potential for improved commitments and improved outcomes where team members create strategic alliances through frameworks and other long-term contracts.

Collaborative Construction Procurement and Improved Value, First Edition. David Mosey.
© 2019 John Wiley & Sons Ltd. Published 2019 by John Wiley & Sons Ltd.

Chapters 6–8 consider how collaborative team members are selected, whether collaboration needs a contract at all and, if so, whether a new type of contract is required to fulfil this role. They explore how team selection and contractual commitments can help to build a collaborative culture, and test whether team members can integrate their work by making non-binding declarations. The collaborative and alliance features of standard form project contracts are considered in Chapter 9, and the FAC-1 Framework Alliance Contract and the TAC-1 Term Alliance Contract are explained in Chapters 10 and 11.

Collaborative construction procurement needs to be sustained by personal relationships, and different ways to create and support a collaborative culture are explored in Chapter 12. The potential for digital technology to create new connections between team members, and also to integrate the capital and operational phases of a project, is examined in Chapters 13 and 14, together with the impact of Building Information Modelling ('BIM') on procurement and contracting practices.

The lessons learned from case studies show how economic and social value can be improved through collaborative construction procurement, and these are considered in Chapter 15. The different options available for costing, incentivising and programming a collaborative project or programme of work are considered in Chapters 16 and 17. Collaborative risk management systems are analysed in Chapter 18, and collaborative ways to avoid or resolve disputes are explored in Chapter 19.

Chapters 20–25 have been contributed by leading practitioners in Australia, Brazil, Bulgaria, Germany, Italy and the USA. They describe how new approaches to procurement, contracts, and BIM are adopted in each country, and explore the different challenges arising in common law and civil law jurisdictions.

1.2 What Is Collaborative Construction Procurement?

This book uses the terms 'collaborative construction procurement' and 'collaborative procurement' to describe how projects and programmes of work can be planned and delivered by integrated teams. It examines how collaborative procurement can be supported by early contractor involvement, by digital technology such as BIM and by new contractual structures and processes in order to improve the outputs from construction, engineering and asset management.

Construction projects should always be a team endeavour, yet despite extensive evidence as to the benefits of collaborative working, most procurement models and contracts do not support teamwork but instead focus on the transfer of risk down the supply chain. This traditional defensiveness reminds us that organisations are obliged to protect their own interests, and it is important to examine the extent to which collaborative alternatives provide equivalent or improved legal and commercial protections.

Collaboration among individuals engaged on a project or programme of work is only made possible by integrating the differing needs and commercial priorities of the organisations who employ them. Knowledge is power, and the legal and commercial tests of collaborative construction procurement should include:

- Firstly, whether team members build up shared knowledge at a time when it can be used to improve project outcomes
- Secondly, whether team members use that shared knowledge to improve project outcomes rather than for their individual benefit.

To pass the first test requires integrated and transparent systems, and to pass the second test requires motivation that aligns different commercial interests. The references to case law in this book illustrate how the courts treat a range of issues in the absence of contractual clarity, and why clear contracts are therefore essential to ensure that the features of collaborative procurement are commitments rather than optional extras.

For the construction industry and its clients to get the best out of collaborative working, and for them to avoid creating new barriers in place of the old ones, they should build the bridges through which:

- The agreed objectives of team members are connected to the contracts that support their shared approach to agreed commitments
- The people and organisations who contribute to a project are connected to emerging technology
- The vision of design consultants is connected to the expertise of the contractors, subcontractors and manufacturers who bring those designs to life
- The design and construction of projects are connected to their ongoing operation, repair and maintenance
- The experience of the different sectors and different jurisdictions can provide examples of good practice.

In order to create and maintain commercial and legal bridges, construction clients and the teams with whom they work need to develop integrated relationships, transparent data and a clear view of the issues arising. This book considers the new lenses, as illustrated in Diagram 1, through which collaborative construction procurement can improve data transparency and can develop a clear vision of how team members can work together more efficiently.

We will examine the ways in which collaborative options for team selection, project planning and project delivery can:

- Help a client to decide whether a construction project should be approached as a team endeavour rather than a cascade of risk transfer
- Create new opportunities to ask and answer practical questions as to risk and value
- Avoid making risk and value assumptions before team members are selected
- Establish what involvement, roles and responsibilities it is reasonable and valuable for all team members to accept.

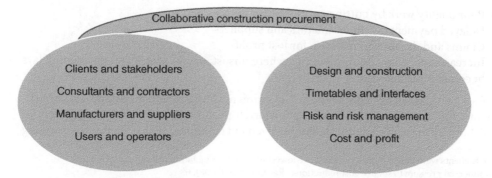

Diagram 1 Team members, data and issues viewed through collaborative construction procurement.

1.3 Why Is Collaborative Procurement Important?

The success of collaborative construction procurement depends on team members making clear what it is they will do together that they would not do alone, going beyond a general sense of joint purpose to agree in detail how they will integrate their roles and responsibilities in a way that can be compatible with their different commercial drivers. Collaborative models have already proven their value, for example in the UK, the USA and Australia, but they need to be presented and explained more clearly and consistently in order to attract wider support.

It should be possible for changes to be made incrementally rather than by demanding that team members leap across a chasm to unfamiliar territory. In addition to exploring the intellectual and legal challenges generated by no blame clauses and shared pain/gain incentives, we will also examine other collaborative procurement systems such as early contractor involvement, Supply Chain Collaboration, joint project governance, shared digital data, long-term strategic appointments and whole life asset management.

The success of collaborative procurement models depends on the people who put them into practice and on the ways that they act, react, communicate and seek consensus. These are the features that comprise a collaborative culture, and it is important to examine the ways that they can be developed and sustained. However, we also need to look critically at assertions that a collaborative culture needs only a general declaration of good faith or that it requires the exclusion or limitation of reasonable legal rights and obligations.

Declarations of collaborative principles can create shared values among team members but do not get us very far if they are not translated into collaborative actions. General declarations are of limited value in helping a team to deal with typical risks and challenges in a new way, and failure to honour a collaborative declaration only increases cynicism. Meanwhile, if collaboration is seen as no more than a set of reluctant compromises through which the parties edge away from opposing positions, it is unlikely to create the foundations for a shared commitment to improved value but instead may create a sense of residual dissatisfaction and mutual suspicion.

The global financial crisis in 2008 placed pressure on emerging collaborative procurement models and tempted clients instead to exploit lowest price bids in a cutthroat marketplace.[1] However, by 2010 the newly-elected UK Coalition Government was concerned that artificially low prices would lead to:

- Poor quality work by cutting corners
- Delayed payment of subcontractors and suppliers
- Claims and disputes to make up for lost profit
- Increased defaults and insolvencies where unsustainable prices and payment delays become intolerable.[2]

As a result, when drafting the 2011 Government Construction Strategy,[3] UK Government Chief Construction Adviser Paul Morrell proposed instead the development of reliable data through early contractor involvement combined with collaborative working

---

1 UK clients were drawn back to single-stage lowest price tendering because it appeared to apply 'commercial pressure to secure cost reductions', Rawlinson, S. (2008), 68.
2 The same concerns, arising after an earlier recession, emerged in Latham, M. (1994).
3 Government Construction Strategy (2011).

and BIM as the cornerstones of more efficient procurement and delivery of construction, engineering and asset management. The 2011 Government Construction Strategy recommended that these approaches should be features of three new procurement models and proposed that they should be tested through a programme of 'Trial Projects'.[4] Early supply chain engagement and collaborative working were also features of BS8534, the British Standard for construction procurement published in 2011.[5]

Recent years have seen a series of reports underlining the need for improved procurement and contracting practices, for example:

- The 2016 Farmer report 'Modernise or Die'[6]
- The 2017 McKinsey report 'Reinventing Construction Through a Productivity Revolution'[7]
- 'Building a Safer Future', the 2018 Hackitt Review of Building Regulations and Fire Safety[8]
- The 2018 Housing Forum report 'Stopping Building Failures, How a Collaborative Approach can Improve Quality and Workmanship'[9]
- The 2018 Construction Leadership Council report 'Procuring for Value'.[10]

McKinsey recommended that poor productivity in the construction sector means that we need to 'rewire the contractual framework'.[11] In the aftermath of the 2017 Grenfell Tower disaster in London, the Hackitt Report suggested that procurement systems need a complete overhaul because 'the primary motivation is to do things as quickly and cheaply as possible rather than to deliver quality homes which are safe for people to live in'.[12] She highlighted a cultural issue in construction procurement which she described as a 'race to the bottom', and emphasised that the reform of current practices should 'lead to a significant increase in productivity'.[13]

The UK best practice body Constructing Excellence has championed collaborative procurement for many years, and in 2011 published guidance that included the following three overriding principles:

- 'Common vision and leadership: an absolute focus on the end purpose based on a clear understanding by all participants of what represents value for the client and end users. Leadership needs to establish this common vision and then constantly relate progress by the project to this vision to reinforce the team's goal'
- 'Collaborative culture and behaviours: collaborative behaviours include teamwork and joint problem solving. Participants demonstrate values such as trust, fairness, openness, no-blame, honesty and transparency'
- 'Collaborative processes and tools: adopting processes and tools which support the development of the collaborative culture and deliver the benefits, such as information

---

4 Government Construction Trial Projects (2012).
5 BS8534:2011.
6 Farmer, M. (2016).
7 McKinsey Global Institute (2017).
8 Building a Safer Future (2018).
9 Housing Forum (2018).
10 Procuring for Value (2018), researched by Ann Bentley, CLC Member and Global Board Director of Rider Levett Bucknall.
11 McKinsey Global Institute (2017), 8.
12 Building a Safer Future (2018), 5. This is considered further in Section 15.8.
13 Building a Safer Future (2018), 5, 8. See also Housing Forum (2018).

collaboration platforms, open book costing, lean and waste elimination, and project bank accounts'.[14]

The interlocking collaborative themes that recur throughout this book include shared vision and leadership, a common understanding of value, a culture of teamwork and joint problem-solving and a range of supporting commercial processes and tools.

1.4 What Research Has Examined Collaborative Construction Procurement?

This book draws on research undertaken by Governments, by construction industry practitioners, and by academics and other commentators, combined with three related research initiatives led by King's College London Centre of Construction Law. The King's research examines:

- The means by which improved value has been delivered using early contractor involvement, collaborative contracts and BIM on Trial Projects, with outputs captured in published case studies[15] and related guidance[16]
- The relationship between procurement, contracts and BIM and their influence on private sector and public sector projects, with outputs captured in a published report[17]
- The potential for proposed new forms of contract to facilitate early contractor involvement, collaborative working and BIM, leading to publication of the standard form 'FAC-1 Framework Alliance Contract'[18] and the 'TAC-1 Term Alliance Contract'.[19]

The research timelines for each of these King's initiatives are set out in Appendix A. The King's research tested the findings of a range of Government and industry reports, including for example:

- The 1962 Emmerson 'Survey of Problems Before the Construction Industries', observing that 'in no other important industry is the responsibility for the design so far removed from the responsibility for production'[20]
- The 1994 Latham Report 'Constructing the Team' and his interim report 'Trust and Money', observing that 'the traditional separation of design and construction has long been a source of controversy'[21]
- The UK Government Construction Strategies 2011 and 2016–2020, both recommending a combination of early contractor involvement, collaborative working and BIM[22]

---

14 CE (2011).
15 The Trial Project case studies are published online by Cabinet Office and Constructing Excellence, and are listed in Appendix C.
16 For example, Two Stage Open Book and Supply Chain Collaboration Guidance (2014).
17 KCL Centre of Construction Law (2016).
18 FAC-1 Framework Alliance Contract (2016).
19 TAC-1 Term Alliance Contract (2016).
20 Emmerson, H. (1962), 9. Banwell picked up this theme and stated that 'those who continue to regard design and construction as separate fields of endeavour are mistaken', Banwell Report (1964), 4, Section 2.6.
21 Latham, M. (1993, 1994).
22 Government Construction Strategy (2011, 2016–2020).

- The 2012 Effectiveness of Frameworks Report, part of the UK Government Procurement/Lean Client Task Group Report,[23] examining the features that influence the success of public sector frameworks
- The 2012 report by the UK Parliamentary All-Party Group for Excellence in the Built Environment, recommending more time for project brief and planning, creation of integrated teams, evaluation using 'balanced scorecards', assessment of capital and operating costs and the adoption of BIM[24]
- The Infrastructure Client Group reports on Alliancing Best Practice in Infrastructure Delivery 2014[25] and Alliancing Code of Practice 2015,[26] recommending how alliances can improve the delivery of infrastructure projects
- Digital Built Britain (2015), setting out a vision for the development of BIM[27]
- The 2016 Infrastructure and Projects Authority Project Initiation Routemap, recommending improved public sector procurement practices.[28]

The King's research was enabled in part by our appointment as lead mentor for the UK Cabinet Office and its Trial Projects Delivery Group, testing the potential of collaborative working, early contractor involvement and BIM through the procurement and delivery model known as 'Two Stage Open Book'.[29] In addition to Two Stage Open Book, the Trial Projects Delivery Group initiated and reviewed Trial Projects examining two other procurement and delivery models known as 'Cost Led Procurement'[30] and 'Integrated Project Insurance'.[31] A summary of the Trial Projects research process is set out in Appendix B, and 10 Trial Project case studies are summarised and reviewed in later chapters.

1.5 What Case Studies Support Collaborative Procurement?

I worked as a trainee solicitor for Anthony Trower,[32] who had on his mantelpiece the Benjamin Disraeli motto 'Never explain, never complain'. This seemed strange advice in a profession whose job includes explaining the solutions that resolve a client's complaints, and it took me a while to understand that it meant we should focus on actions rather than words. There are many justified complaints about the state of construction procurement, and many reports explaining how it should be improved, but we can learn more from those clients and teams who have taken action to find a better approach.

The reports and research initiatives referred to in Sections 1.3 and 1.4 encourage more collaborative procurement but they do not always support their recommendations with

---

23 Procurement/Lean Client Task Group (2012).
24 All-Party Group for Excellence in the Built Environment (2012).
25 Infrastructure Client Group (2014).
26 Infrastructure Client Group (2015).
27 Digital Built Britain (2015).
28 IPA (2014).
29 Two Stage Open Book and Supply Chain Collaboration Guidance (2014).
30 For which the lead mentor was Vaughan Bernand, former Chief Executive of Shepherd Construction and former Chair of Constructing Excellence.
31 For which the lead mentor was Martin Davis, former director of Emcor.
32 Anthony Gosselin ('Cocky') Trower (1921–2006) was a founder member of the Special Air Service. In 1944 he was parachuted into occupied France to work with the French Resistance, so he knew something about the challenges of teambuilding in a difficult environment.

examples of what has been achieved in practice. Empirical evidence can demonstrate why it is worth investing time and effort in new approaches and can provide persuasive examples of how collaborative techniques have been put into practice. Several best practice groups working in the field of collaborative procurement have created valuable case studies, and these include:

- Constructing Excellence[33]
- The Housing Forum[34]
- The Alliance Steering Group[35]
- The Institute for Collaborative Working[36]
- The Infrastructure Client Group.[37]

Drawing in part on the work of these best practice groups, a total of over 50 case studies have been researched for this book, and these are listed in Appendix C. They comprise:

- Trial Project case studies of early contractor involvement and collaborative working (and in some cases BIM), coordinated by Constructing Excellence in collaboration with the relevant project teams
- Case studies of project alliances, framework alliances and term alliances researched by the author for the Alliance Steering Group in collaboration with the relevant project teams
- Alliance case studies researched by the King's College London Centre of Construction Law, in collaboration with the relevant project teams.

These case studies examine collaborative practices that relate to:

- A range of sectors that include education, environment, health, highways, housing, leisure, offices, public buildings, rail and utilities
- The procurement and delivery of individual projects, of frameworks comprising multiple projects and of term call-off contracts, in both the public and private sectors
- A range of construction and engineering projects, delivered through different collaborative procurement models and involving different allocations of roles and responsibilities under a range of standard contract forms.

These case studies illustrate each aspect of collaborative procurement and the agreed actions that can be undertaken in order to deliver improved value. Summaries of the following case studies appear in the following sections:

- Project alliances for housing and engineering projects – Section 3.10
- Framework alliances for housing, offices, custodial facilities and water services – Section 4.10
- Term alliances for housing and highways – Section 5.10

---

33 http://constructingexcellence.org.uk, currently led by Chief Executive Don Ward.
34 www.housingforum.org.uk, currently led by Chief Executive Shelagh Grant.
35 www.allianceforms.co.uk, currently chaired by Shane Hughes of Savills and managed successively by Fiona Griffiths, Alison Low and Shona Broughton.
36 http://www.instituteforcollaborativeworking.com, currently led by Chief Operating Officer David Hawkins.
37 www.ice.org.uk/about-ice/what-we-do/infrastructure-client-group, currently chaired by Dale Evans of Anglian Water.

- Collaborative selection processes for offices and underground rail projects – Section 6.10
- Collaborative construction management for hotel and health projects – Section 7.10
- Joint venture and consortium procurement for local government facilities, housing and schools – Section 8.10
- Use of FAC-1 for programmes of housing, sports, schools, regeneration and public buildings – Section 10.10
- Use of TAC-1 for housing and highways programmes – Section 11.10
- Creation of a collaborative culture on health and highways work programmes – Section 12.10
- Use of BIM on alliance projects for custodial facilities – Section 14.10
- Collaborative costing and incentivisation of housing and schools projects – Section 16.10
- Collaborative time and change management on rail, hospital and harbour projects – Section 17.10
- Collaborative risk management on school and hospital projects – Section 18.10
- Collaborative dispute resolution on housing and office projects – Section 19.10.

In addition to the above summaries, the evidence provided by these and other case studies is quoted and referred to throughout later chapters, and these references are listed against each case study in Appendix C.

1.6 How Is Collaborative Procurement Connected to Digital Technology?

Digital technology enables the rapid creation and sharing of construction project data. Each project needs to connect established production processes with innovations and prototypes on a unique site, and each project relies on the coordination of a diverse network of people, products, services and works. Where digital technology can improve and integrate these activities and the underlying data used by team members, it offers great benefits to the construction industry.

However, digital technology does not offer viable new solutions unless these solutions can be applied by team members who understand their impact and their limitations.

The programming of software for BIM, and for other technologies such as smart contracts, shines a harsh light on vague contract provisions and disjointed procurement practices. It demands increased precision in contractual exchanges and it offers improved ways of capturing the data that supports a more holistic approach to the full project lifecycle. The connections between digital technology and the success of collaborative construction procurement appear in UK Government policy documents such as Digital Built Britain,[38] and the digital disruption of previous norms has fuelled arguments that digital technology creates new barriers to contracts, that it gives rise to completely new contracts and that it does away with the need for conventional contracts altogether.

A King's research initiative analysed emerging BIM practices and related legal issues by reference to 12 case studies and 40 confidential interviews, in order to explore

---

38 Digital Built Britain (2015).

whether the contributions of digital technology to improved value are dependent on new procurement models and new contract terms. With the benefit of grant funding from the Society of Construction Law and the Association of Consultant Architects, the KCL Centre of Construction Law assembled a research group of specialist lawyers and other practitioners to examine over a 24-month period the relationship between procurement, contracts and BIM in the context of common law, statutory and contractual obligations. Details of the research group members, the projects analysed and the specialists interviewed are set out in Appendix D.

Research interviewees described the ways in which BIM enables collaborative working, for example through the use of BIM models to assist joint working by design consultants and through the improved ability to explain design proposals to a client. The interview results were written up and agreed by the research group, a draft research report was shared with delegates at a public conference, and input and comments from delegates were reflected in a final report which was published online in July 2016. The report 'Enabling BIM Through Procurement and Contracts'[39] revealed significant links between procurement, collaborative contracts and the adoption of BIM, and some of the responses from interviewees are summarised in Section 13.10. The research leading to this report also influenced work on the third King's research initiative, namely the development of new alliance forms governing a strategic approach to collaborative contracting.

1.7 How Is Collaborative Procurement Connected to Contracts?

The Trial Projects and other case studies illustrate the role of contracts in supporting collaborative construction procurement. Against this backdrop we will consider in Chapter 7 whether collaborative procurement needs a contract at all, and in Chapter 8 whether the iterative planning and other joint working governed by an alliance contract mean that it should be seen as a new type of contract distinct from typical construction contracts. We will consider the features of classical, neo-classical and relational contracts, and will argue that an alliance contract should be recognised as a new hybrid category and known as an 'enterprise contract'.[40]

Individual projects can be procured under a wide range of standard form contracts, and we will explore in Chapter 9 the extent to which the various standard forms support collaborative construction procurement. For example, publication in 2018 of the NEC4 Alliance Contract[41] invites valuable comparisons with two other multi-party project alliance forms, namely the PPC2000 Project Partnering Contract[42] and the US standard form ConsensusDocs300.[43]

As regards the contracts that reach beyond individual projects and support long-term collaborative relationships, a cross-industry working group collected evidence from UK Government departments and the wider public sector and examined the features of

---

39 KCL Centre of Construction Law (2016).
40 This term was first used in Mosey, D. (2017).
41 NEC4 Alliance Contract (2018).
42 PPC2000 (2013).
43 ConsensusDocs 300 (2016).

frameworks, publishing their 'Effectiveness of Frameworks' report in 2012.[44] Further research and analysis were undertaken by the UK Local Government Association and National Association of Construction Frameworks in 2016, and both reports concluded that the benefits of effective frameworks are vital to the public sector and the construction industry.[45] These reports provided examples of bespoke collaborative contracts created by a range of public sector clients. Other UK clients who have developed their own long-term collaborative contracts include those who led Trial Project teams and also the utilities, rail and highways clients who formed the UK Infrastructure Client Group.

Four of the Trial Projects were supported by a bespoke, multi-party framework alliance, namely Project Horizon, Cookham Wood, North Wales Prison and SCMG, and nearly all other Trial Projects were supported by bespoke, two-party collaborative frameworks. King's analysis of the Trial Projects, and of other examples of successful collaborative procurement, revealed that frequently the success of a project team resulted in part from preparatory work undertaken pursuant to an overarching collaborative framework.

However, the reports and case studies also revealed how each collaborative framework contract was drafted from scratch, creating an additional cost and time burden for its members and restricting the scope to establish new norms of collaborative working. None of these bespoke strategic contracts have been made publicly available for wider use. This led King's to conclude that further research was justified into the potential for new standard form strategic collaborative contracts that could capture the features of successful bespoke frameworks and alliances.

The King's research established that new standard form alliance contracts are needed in order to overcome serious obstacles to the progress of collaborative working. For example:

- Without clear contractual systems the construction industry can lose patience with ideas before they are embedded in working practices, shifting their energy to investigate new fads[46] in preference to the development of innovations that have been proven to work
- If the provisions of alliance contracts are not accessible in a way that enables them to be well understood, enthusiasts may rely instead on alluring headlines and symbolic declarations, in the hope that these offer shortcuts that replace more rigorous contractual understanding[47]
- Without standard form alliance contracts, there is the risk that only those clients with sufficient resources and commercial influence to create bespoke alliance forms are able to adopt collaborative practices, which impedes the sharing of knowledge and the recognition of more widely applicable rules.

These concerns led the King's research team to consider how collaborative contracts could be made more widely accessible by creating new standard forms that avoid the costs and inconsistencies of successive bespoke contracts. King's research included

---

44 Effectiveness of Frameworks (2012).
45 LGA/NACF (2016).
46 Uff, J. (2018), 177, on the risk of 'another failed fad'.
47 For example, the comments on collaborative working and contracts in Section 7.1 and examples of reliance on non-binding agreements in Section 7.2.

a year of industry consultation with 120 organisations in 14 jurisdictions. Responses revealed a demand for new standard form alliance contracts and strong support for the new draft forms that King's circulated for comment. The full list of consultees is set out in Appendix E. Feedback from the consultation process and consequent amendments to the draft contracts are illustrated in Appendix F.

The final contract forms were published by the Association of Consultant Architects as the FAC-1 Framework Alliance Contract and the TAC-1 Term Alliance Contract in 2016 and these forms have since been adopted on a wide variety of procurements.[48] The features of the new forms are explored in Chapters 4, 5, 10 and 11 and are illustrated in other chapters, and guidance on completing FAC-1 is set out in Appendix G.

1.8 Who Was Sir Michael Latham?

The work of my friend and mentor Sir Michael Latham[49] has provided lasting benefits to the construction industry and its clients. His 1994 Government/industry review reflected meticulous research and a commitment to fair business practices.[50] It remains relevant and influential.

Sir Michael's work led directly to statutory adjudication and payment rights that were enacted and implied in construction contracts within two years following publication of his report. However, Sir Michael also made other important recommendations arising from the research he conducted in 143 meetings with leading client and industry bodies, including far-sighted proposals that relate to procurement, design management and contract drafting. While we might assume that these recommendations have become standard business practice, in fact the progress made has been variable.

To improve the quality of bid submissions, Latham recommended 'a two stage tender process for more complex and substantial projects',[51] arguably paving the way for the endorsement of early contractor involvement in the UK 2011 and 2016 Government Construction Strategies. Other ways to improve the procedures and criteria governing team selection are considered in Chapter 6. Latham also sought to save money for bidders through, for example, pooling the costs of ground investigations undertaken where one consultant is retained by all bidders, and through reimbursement of certain tender costs on large and expensive schemes.[52] Efforts to avoid wasted money on bid costs remain rare, for example the commitment by Transport for London to repay certain costs incurred by unsuccessful bidders in creating proposals for the Bank Station Capacity Upgrade.[53]

Latham foresaw BIM in the form of 'advanced computer aided design' or 'virtual reality' by which 'all aspects of the design, manufacture, assembly and use of the product can…be presented in one entity'.[54] His reference to 'use' emphasises the need to adopt a

---

48 Case studies appear in Sections 10.10 and 11.10. Regular updates as to the adoption of FAC-1 and TAC-1 are posted under 'News and Users' on www.allianceforms.co.uk.

49 Sir Michael Latham (1942–2017) was author of the seminal 1994 report 'Constructing the Team', which followed his interim 1993 report 'Trust and Money'.

50 Latham, M. (1994).

51 Latham, M. (1994), Section 6.32.3.b.

52 Latham, M. (1994), Section 6.32.3.c and d.

53 As considered in Section 6.10 and as regards tender costs Section 15.3.

54 Latham, M. (1994), Section 4.8.

whole life approach yet it remains doubtful whether BIM in practice has advanced much beyond the use of digital data in the capital phase of a project. We still need to integrate operation, maintenance and repair into our procurement models in a way that harvests the full potential of BIM, and examples of how this can be approached are considered in Sections 5.7, 13.6 and 14.8.

Latham stated that 'Best practice is about partnering, collaborative working and stripping out of the equation at the earliest possible stage those costs which add no value', and he recommended that this should extend not only to 'first tier contractors' but also to 'specialist contractors'.[55] The engagement of specialist subcontractors and manufacturers during the preconstruction phase, using the Supply Chain Collaboration systems considered in Sections 2.9, 2.10, 3.6, 4.6 and 5.6, provides new ways to achieve 'integration of the work of designers and specialists'[56] and to eliminate the 'fuzzy edges' that Latham described as giving rise to many claims and disputes. Interestingly, Latham also recommended that 'Subcontractors should undertake that, in the spirit of teamwork, they will coordinate their activities effectively with each other',[57] foreseeing supply chain alliances of the type trialled by Kier Highways Services and considered in Section 11.10.[58]

Latham supported alliances and partnering but assessed them objectively through a commercial lens. He recognised the benefits of a 'formal partnering agreement' founded on 'a relationship of trust, to achieve specific primary objectives by maximising the effectiveness of each participant's resources and expertise' and 'not limited to a particular project'.[59] However, he also emphasised that partners must be sought 'through a competitive tendering process' and that partnering arrangements 'should include mutually agreed and measurable targets for productivity improvements'.[60]

In assessing the features of a 'modern contract', Latham underlined the benefits of:

- Commitment to teamwork and fair dealing with all parties, including subcontractors, specialists and suppliers
- The power of shared financial motivation
- The need to spell out work stages through milestones and activity schedules.[61]

The Latham recommendations have stood the test of time, and they inform the ways in which clients and their teams can establish robust commercial foundations for collaborative construction procurement.

1.9 How Can Collaborative Procurement Reflect ISO 44001?

A successful team needs the benefit of ideas from many contributors, and a collaborative approach by one organisation does not have much effect unless robust links are

---

55 NAO (2005), 1.
56 Latham, M. (1994), Section 4.3.
57 Latham, M. (1994), Section 6.41.6.
58 http://constructingexcellence.org.uk/wp-content/uploads/2015/12/Trial-Projects-Horizon-Case-Study-Second-Year-Update_Final.pdf.
59 Latham, M. (1994), Section 6.43.
60 Latham, M. (1994), Section 6.47.
61 Latham, M. (1994), Sections 5.17 and 5.18.

built up with other organisations who are willing to adopt the same approach. In 2017 publication of 'ISO 44001' created an international standard for collaborative business relationship management[62] designed to help businesses manage their collaborative relationships on several different levels:

- 'A single application (including operating unit, operating division, single project or programme, mergers and acquisitions)
- An individual relationship (including one-to-one relationships, alliance, partnership, business customers, joint venture)
- Multiple identified relationships (including multiple partner alliances, consortia, joint ventures, networks, extended enterprise arrangements and end-to-end supply chains)
- Full application organisation-wide for all identified relationship types'.[63]

ISO 44001 is not restricted to the construction sector and is designed for use as a business tool in any part of both the public and private sectors. It is reported that multinational organisations have already recognised its benefits in terms of a 20% reduction in operating costs and 15% savings through supply chain aggregation.[64] ISO 44001 recommends that 'the partners shall establish and agree a formal foundation for joint working, including contractual frameworks or agreements, roles, responsibilities and ethical principles'.[65] However, it does not offer specific recommendations as to where the parties can find the procurement models, technological support and contracts through which collaborative inter-organisational options can be brought to life. These are the missing links that this book will attempt to provide.

ISO 44001 links collaborative working to contractual systems, and proposes that:

- 'Contract terms shall be reviewed to determine clarity of purpose, encourage appropriate behaviour and identify the potential impacts on or conflict with the aims of collaborative working
- All performance requirements and measurement methods should be mutually agreed to ensure clarity
- Risk and reward models, issue management, exit strategy, knowledge transfer and sustainability should be considered when developing an agreement'.[66]

In seeking improved value, ISO 44001 proposes that a collaborative team needs to:

- 'Define what "value" means to the collaborative partners
- Provide a mechanism for the capture of innovation and ideas for improvement
- Provide a method for performing analysis and evaluation of ideas and innovations against relevant criteria…
- Establish a method for reviewing the success or failure of value creation initiatives and record lessons learned for future use'.[67]

In later chapters we will explore the ways in which collaborative construction procurement models and contracts can enable adoption of the ISO 44001 recommendations.[68]

---

62 ISO 44001: 2017.
63 ISO 44001 Section 1.
64 Hawkins, D. (2017), (xiv) as part of a detailed commentary on ISO44001.
65 ISO 44001 Section 8.6.2.1.
66 ISO 44001 Section 8.6.10.
67 ISO 44001 Section 8.7.2.
68 For example, the Connect Plus Sustainable Business Culture Model 'is now accredited in accordance with BS1000 Collaborative business relationships', Connect Plus Trial Project case study, 3.

We will consider why the adoption of collaborative working in the UK construction sector has often appeared less methodical than ISO 44001 suggests, sometimes appearing to offer an idealistic parallel universe of non-contractual engagement. We will explore whether more teams would adopt the collaborative approaches described in ISO 44001 if they could be shown how to develop collaborative systems with greater certainty and without the sense that they are required to give up reasonable legal and contractual rights.

1.10 What Should Collaborative Procurement Provide?

Collaborative construction procurement should provide the timely build-up and exchange of accurate data between team members in order to integrate their work, align their objectives and create improved project outcomes. Despite 60 years of persuasive reports, recommendations and examples, collaborative construction procurement practices are still not widely accepted or well understood. Implementation is often only skin deep because 'beneath the surface you find many so-called partners still seek to avoid or exploit risk to maximise their own profits, rather than find ways to share risk and collaborate genuinely so that all can profit'.[69]

The advice given by Professor Phillip Capper[70] to new students on the King's College London MSc in Construction Law, even though these students are experienced architects, engineers, lawyers and surveyors, is that the course will change the way they think. In order for collaborative construction procurement to reach the mainstream, there are ways in which the industry and its clients can change the way they think about what procurement, contracts and management processes have to offer.

The current performance of the construction industry is far from ideal, and it has been suggested that 'if construction labour productivity were to catch up with the progress made by other sectors over the past 20 years... this could increase the construction industry's value added by $1.6 trillion a year'.[71] The construction industry and its clients also need new ways to reduce the enormous amounts of time and money spent on acrimonious claims and disputes,[72] and continuing to adopt the same deeply ingrained, traditional approaches to procurement is unlikely to provide the answers we need.[73] Changes to procurement models and contracts can provide direction and support for collaborative working, can reconcile the different interests, assumptions and commercial objectives of the parties, and can improve the productivity of all team members.

In considering the means by which productivity can be improved, by which risks can be reduced and by which disputes can be avoided, there are a host of complex technical, financial, behavioural and logistical issues that construction teams encounter. We

---

69 Never Waste a Good Crisis (2009), 8.
70 Head of International Arbitration at White & Case who is described as 'the doyen of construction law'. (Chambers UK 2013).
71 McKinsey Global Institute (2017) (Executive Summary).
72 'A third of those who responded to our survey experienced at least one dispute in the preceding 12 months. The prognosis for the industry is not great, with nearly two in five telling us that the number of disputes is increasing', NBS (2018), 3.
73 'When you think you're through changing, you're through'. Latham M. in Association of Consultant Architects (2010), 1.

will consider in later chapters some of the ways in which these issues can be dealt with differently. Rather than assuming that collaborative construction procurement is the answer to every problem, we will look at how it affects specific situations arising on live projects.

This book sets out to engage with the cynics as well as with the enthusiasts. However, while a healthy amount of cynicism can help to avoid wasted time or wishful thinking, an excessive dose can be toxic. The construction industry and its clients should not undermine proposals for collaborative procurement by assuming that they are not actually new, or are naïve, or are too risky, or disguise someone's hidden agenda. While a 2018 survey concluded that disputes are currently 'a part of doing business in the UK construction sector', it proposed that solutions lay in the wider adoption of 'legally explicit' collaborative working, combined with a better understanding of BIM and other technologies.[74]

In order to build up a clear understanding of what collaborative construction procurement can offer, both to the enthusiasts and to the cynics, we should see it as a range of proactive and pragmatic approaches that improve on traditional options. Collaborative procurement should give us clear ways to ensure that more accurate information is exchanged among a wider range of contributors to design, manufacture, construction and operation, working together earlier in the planning and development of the project when there is time and opportunity for optimum, cost-effective solutions to be agreed. It should set out the steps and techniques by which early contractor involvement, BIM, collaborative working and long-term relationships can have a demonstrable effect on better value, reduced risks and avoidance of disputes.

The demands of a global marketplace challenge the assumption that different legal systems demand different approaches to collaborative construction procurement. We live in an era when new procurement and contracting practices should be accessible to users regardless of geographical boundaries, and in a world where digital technologies such as BIM have been adopted by construction clients, consultants and contractors in many countries. Alliances and a range of comparable collaborative models have been developed and adopted in Australia[75] and the USA,[76] and in Chapters 20–25 leading international academics and practitioners explain the factors that influence the approach to collaborative construction procurement in three European jurisdictions (Bulgaria, Germany and Italy) and three other jurisdictions (Australia, Brazil and the USA).

Although language barriers, differing laws and differing economic circumstances may appear to obstruct collaborative working as a basis for shared transnational procurement models and contract forms, the King's research has shown through wide consultation on FAC-1, and through its early acceptance in civil law and common law jurisdictions, how transnational challenges can be overcome.[77] King's has worked closely with lawyers and academics in Brazil, Bulgaria, Germany and Italy in their

---

74 NBS (2018), 3 and 24.

75 $32 billion total value of alliance projects were delivered in New South Wales, Victoria, Queensland, and Western Australia between 2004 and 2009 – Department of Treasury and Finance, Victoria (2009) 158.

76 The US model known as 'Integrated Project Delivery' has many alliance characteristics.

77 As illustrated in the schedules of required amendments which appear in Sections 21.5 (Brazil), 22.5 (Bulgaria), and 24.5 (Italy).

adaptations of FAC-1 (and in Germany of PPC2000) for translation and use in their respective jurisdictions.[78] Clarifying the systems of collaborative procurement through new standard form contracts underlines how these systems can transcend national boundaries and can help to create new opportunities in a global marketplace.

---

78 As described in Chapters 21–24.

2

What Are the Foundations for Collaborative Construction Procurement?

2.1 Overview

There is increasing recognition of the ways in which collaborative construction procurement can support 'a structured management approach to facilitate team-working across contractual boundaries',[79] yet those who work in the construction industry point out that spontaneous collaborative working occurs every day on every construction project without new contractual structures. So, is collaboration no more than reasonable behaviour by competent individuals?

While there is widespread concern at the complexity and cost of construction disputes, collaborative working is often described as a way to avoid reliance on contracts, as if the contracts themselves are the cause of disputes. Meanwhile, traditional construction procurement models and contracts are still based on the assumption that divergent commercial interests cannot be integrated or even reconciled, and this continues to fuel fragmented and defensive structures. The complexity and cost of construction dispute resolution shows how this fragmentation and defensiveness do not support effective pricing, programming, change management, quality control, safety or sustainability.

Constructing Excellence provide detailed guidance on collaborative working and they also propose six critical success factors:

- Early involvement
- Selection by value
- Aligned commercial interests
- Common processes and tools
- Performance measurement
- Long-term relationships.[80]

In all the case studies examined as part of the research supporting this book, the collaborative teams improved the timing and structure of their team selection process and adopted new contractual relationships. There are four interconnected themes that have been critical to their success:

---

79 Smith et al. (2006), 144.
80 Constructing Excellence Hymn-Sheet (2015).

Collaborative Construction Procurement and Improved Value, First Edition. David Mosey.
© 2019 John Wiley & Sons Ltd. Published 2019 by John Wiley & Sons Ltd.

- The ways in which collaborative working can be established and supported at all stages in a project or programme of work, from team selection through to project completion and operation, while avoiding disputes and delivering improved economic and social value
- The procurement and delivery models through which early contractor involvement can create a foundation for collaborative working, while remaining grounded in the commercial common sense on which safe and intelligent working practices depend
- The ways in which the development and exchange of data, for example through digital technology such as Building Information Modelling (BIM), can support collaborative working
- The ways in which contracts can support collaborative working on individual projects and on multiple projects and term appointments.

The behaviour and attitudes that contribute to a collaborative culture are less likely to be sustained without a corresponding commercial model that can be understood and adopted by all team members. Awareness of the need for collaboration in project procurement and delivery does not on its own overcome innate caution, scepticism or silent resistance to change. This chapter will attempt to define more clearly what we mean by collaborative construction procurement, and will break down some of its components by reference to contracts, technology, selection processes and activities.

Collaborative construction procurement can be accessible to any client and supply chain, and is not an exotic approach that is appropriate only for enthusiasts or for very sophisticated teams. In the high-risk industry of construction, the cautious, conservative and cynical people all take their share of the personal responsibilities required for safe and successful project delivery, and they need to be persuaded rather than bypassed if new collaborative norms are to be established. This chapter will examine the foundations for collaborative procurement in terms of new working relationships, the timing of team members' appointments and the joint activities that can improve the value delivered while implementing a project.

2.2   What Is Different About Collaborative Construction Procurement?

Collaborative procurement in the UK has been explored extensively through 'partnering'[81] as an approach to single projects and longer-term working relationships. The UK National Audit Office suggested that organisations adopting a partnering approach should:

- 'Work in a positive no-blame whole team environment
- Provide early warning to each other of any matters than could affect the achievement of the project objectives

---

81 'A structured management approach designed to promote collaborative working between contracting parties', whether applied to a single project or a series of projects, with the aim of achieving continuous improvement from one project to the next, NAO (2005), 6.

- Use common information systems and work on an open-book basis including showing the elements of contingency and risk allowances added to costs, prices and timing of all future work
- Have incentives for delivery – based around pain/gain share arrangements'.[82]

In 1998, a UK initiative led by Sir John Egan[83] identified successful partnering practices in other industries, particularly manufacturing, and promoted these as a model for the construction sector. Two years later Egan launched the PPC2000 Project Partnering Contract,[84] which emerged from industry engagement led by the Construction Industry Council Partnering Task Force[85] and was published by the Association of Consultant Architects. Partnering became a significant plank of UK Government construction policy, promoted by the Office of Government Commerce in their 2007 'Achieving Excellence in Construction' guidance,[86] who recommended that public sector clients should 'introduce partnering into all property and construction projects'.[87]

More recently, the collaborative system known as 'alliancing' has achieved successes in Australia[88] and has attracted increasing attention in the UK. This has influenced development of the FAC-1 Framework Alliance Contract, the TAC-1 Term Alliance Contract and the NEC4 Alliance Contract.

The terms 'partnering' and 'alliancing' imply a set of predefined techniques and activities but are vague and potentially confusing if used without detailed explanation. The words alone do not provide shorthand for specific collaborative practices and do not provide us with a clear understanding of what mutual commitments are being made. We need to look behind the headlines in order examine on their merits the detailed relationship structures, techniques and activities that these headlines are intended to represent.

Whether described as partnering or alliancing or using any other term of art, collaborative procurement can govern a single project or a longer-term relationship. In all cases 'the key is the promotion and practice of collaborative working and the use of techniques aimed at encouraging co-operation between the parties'.[89] The approaches adopted to collaborative procurement 'can vary considerably across a wide range, from nothing more than a signature of a statement of goals to, at the other end of the spectrum, full integration of management systems, open-book accounting and sharing of other information and mutual incentivisation'.[90]

Collaborative procurement is competing with a range of more embedded commercial models, contracts and attitudes that treat conflict as the unavoidable consequence of differing commercial interests, yet research suggests 'that traditional models of the construction process are increasingly unable to fulfil new demands of clients,

---

82 NAO (2005), 5.
83 Egan, J. (1998).
84 The first edition of PPC2000 (2013).
85 CIC (2002).
86 OGC (2007), Procurement Guides 1–11, 05.
87 OGC (2007), Procurement Guide, 01, 16.
88 As considered in Chapter 20. The US collaborative procurement model known as 'Integrated Project Delivery' is considered in Chapter 25 and in other chapters that examine features of the ConsensusDocs (2016) contract forms.
89 Baker, E. (2007), 344.
90 Baker, E. (2007), 345.

necessitating a search for alternative approaches'.[91] A more collaborative approach to a construction project can be framed as three different processes that are not mutually exclusive:

- A construction process which draws upon synergy and ways to maximise the effectiveness of each project team member's resources
- A management process based on strategic planning so as to improve efficiency
- A non-contractual governance process moving towards trust and informal arrangements for 'putting the handshake back into doing business'.[92]

It is argued in later chapters that these processes, and many of the recommendations of ISO 44001, can only be effective if they are developed into mutual commitments which are supported by binding contracts. Commentators have recognised that successful collaborative construction procurement 'works by making careful plans at the start of projects and then relentlessly putting them into effect'.[93]

Collaborative working can be established and maintained using procurement and contract systems which all team members can understand and follow in a consistent way. The principles underlying collaborative procurement have been described as:

- *Strategy*. Developing the client's objectives and how consultants, contractors and specialists can meet them on the basis of feedback
- *Membership*. Identifying the firms that will need to be involved to ensure all necessary skills are developed and available
- *Equity*. Ensuring everyone is rewarded for their work on the basis of fair prices and fair profits
- *Integration*. Improving the way the firms involved work together by using cooperation and building trust
- *Benchmarks*. Setting measured targets that lead to continuous improvement in performance from project to project
- *Project processes*. Establishing standards and procedures that embody best practice based on process engineering
- *Feedback*. Capturing lessons learned from projects to guide the development of strategy.[94]

Some of the above principles can be applied to a single project, and illustrate that collaborative working is not entirely dependent on a workstream of multiple projects. Other principles can only be applied if there is a workstream of multiple projects. For example, benchmarks that enable continuous improvement from project to project, and feedback capturing lessons learned from project to project, can only be achieved through agreed connections between the team members engaged on successive projects.

---

91 Barlow et al. (1997), 5.
92 Barlow et al. (1997), 6.
93 Bennett & Pearce (2006), 83.
94 Bennett & Jayes (1998), 4.

2.3 How Are Construction Contracts Formed?

A contract can be described as 'an agreement giving rise to obligations which are enforced or recognised by law'.[95] Contracts are the currency of commerce in so far as they enable parties to record and clarify their rights and responsibilities, to assign tasks, obligations and risks, and to form the foundation for relationships. Contracts also create the rights and procedures by which the parties can deal with disputes, hopefully without destroying their working relationships, plus the exit routes necessary to bring relationships to an end.

Through a construction contract 'the law firstly creates a space within which the parties can plan the exchange, making due provision for future contingencies (the planning function), and secondly provides a set of sanctions aimed at inducing performance of the agreed obligations (the incentive function)'.[96] R.J. Smith identifies three roles for a construction contract:

- To set out rights, responsibilities and procedures
- To identify, assign and transfer risk
- To act as a 'planning tool' so that there are 'fewer surprises and dilemmas during construction'.[97]

Traditionally, construction contracts have focused on the first two roles,[98] describing responsibilities and transferring risks between the contracting parties, with significant emphasis on the consequences of failure.[99] The natural reaction to this type of contract is the reinforcement of all available protections because a contracting party will perceive that it can only gain at the expense of another party.

The creation of a construction contract requires the amalgamation of contract conditions with the technical, commercial, and managerial documents that set out the agreed brief, proposals, prices, programme and related procedures. The contract conditions describe important procedures such as those governing payment, change and project completion, but are often seen as unhelpful and not complementary to the other contract documents, something to be kept in the drawer until a dispute arises.

Construction project relationships are governed by standard form contracts or by bespoke contracts or by exchanges of correspondence or by no written contracts at all. The completeness of these contract relationships is also linked to the amount of design, cost and risk information that is available when team selections are made and when contracts are created. Consequently, contract provisions and processes are linked to procurement processes insofar as they govern the completion and integration of additional

---

95 Treitel (2011), para 1-001.
96 Arrighetti et al. (1997), 173.
97 Smith, R.J. (1995), 41 and 42.
98 N.J. Smith described a construction contract as an essential means to 'formalise a set of risks, rules and relationships into one set of words which will govern all dealings between the parties while carrying out that contract', Smith, N.J. (2002), 178.
99 Barlow et al. observed that 'Traditional models view the construction process as the purchase of a product, governed by legal contracts' with minimal uncertainty as to what product is required. They noted that such a contract seeks to include provisions whereby 'any uncertainty in the means by which it is implemented is passed onto contractors or subcontractors as risk', Barlow et al. (1997), 5.

design, cost, programming and risk information and insofar as they deal with errors and inconsistencies in that information.

Construction contract relationships are formed by the selection of consultants and contractors (and of subcontractors and suppliers) through competitive or negotiated procedures which are considered in Chapter 6. The creation of a construction contract under a traditional procurement model often focuses primarily on the allocation of design and construction responsibilities through:

- A 'traditional' allocation of design liability to one or more consultants and construction liability to a main contractor[100]
- A 'turnkey' or 'design and build' allocation of all or most design and construction liability to a main contractor[101]
- A 'construction management' allocation of subdivided liability for design and works packages to one or more consultants and to multiple specialist contractors.[102]

Collaborative procurement can support any of these allocations and can also challenge the following unhelpful, self-imposed limitations:

- Most construction contracts are designed to come into effect just before construction is due to commence,[103] which leaves it too late for this type of contract to be a planning tool. In order for contracts to have a planning role in collaborative procurement, they should describe the systems by which the planning takes place
- There is too much focus on the contracts between a client and main contractor and little attention is given to the contracts between a client and its consultants and those between a main contractor and its subcontractors and suppliers. All these contracts should fit together and influence each other in supporting collaborative procurement
- Most construction contracts cover only one project, and do not describe how the parties can develop their relationships over multiple projects and apply lessons learned from earlier projects so as to improve value on subsequent projects. Contracts that support collaborative procurement can describe systems for achieving improved results by connecting multiple projects
- The construction contracts that receive most attention govern only capital expenditure on design and construction, and there is little analysis of contracts governing a project's ongoing operation, repair and maintenance. Capital and operational contracts should be considered as parts of an integrated procurement model in order to improve the efficiency, safety and value that can be achieved over the full project lifecycle.

Construction contracts typically contain detailed machinery that is designed to deal with complex and changing circumstances, with provisions that help the parties respond to variable progress and payments, to the instruction of changes, to unforeseeable matters that give rise to time and money claims, to the enforcement of warranties and to

---

100 Traditional procurement is known as 'design-bid-build' in the USA and some other jurisdictions.
101 Design and build is known as 'design-build' in the USA and some other jurisdictions.
102 Construction management is also known as 'construction management at risk' in the USA and some other jurisdictions.
103 For example, JCT SBC/Q (2016) and JCT Design and Build (2016) unless used with JCT PCSA (2016); also FIDIC (2017); NEC4 unless used with Option X22.

the insurance of risks. All these provisions aim to preserve the parties' relationships, but when disputes arise:

- 'Doctrines of impossibility of performance, frustration, and mistake are used...to relieve parties'
- 'More covert techniques such as interpretation or manipulations of offer and acceptance and rules governing conditions are also available'
- 'As general proposition these doctrines aim not at continuing the contractual relations but at picking up the pieces of broken contracts and allocating them between the parties on some basis deemed equitable'.[104]

2.4 What Is Different About a Collaborative Construction Contract?

A collaborative construction contract can contain the same machinery as any other construction contract, but it should also contain features such as the following recommended by Latham:

- A specific duty for all parties to deal fairly with each other
- Firm duties of teamwork with shared financial motivation
- Clearly defined roles and responsibilities in a fully integrated set of documents
- Taking all possible steps to avoid conflict
- Incentives for exceptional performance.[105]

These recommendations reflect emerging variances in the features of modern contractual relations, including the ways that they treat:

- Commencement, duration and termination
- Measurement and specificity
- Planning
- Sharing or dividing benefits and burdens
- Interdependence, future cooperation and solidarity
- Personal relations among, and numbers of, participants
- Power.[106]

Contracts are the shared currency of all commercial relationships, and the principles of collaborative procurement need to be converted into this currency so that team members can use and understand them. The transactions governed by contracts comprise:

- A 'bargaining' transaction, governed by a contract of a voluntary nature
- A 'rationing' transaction, governed by a contract exercising authority
- A 'managerial' transaction, governed by a contract for coordination.[107]

---

104 Macneil, I.R. (1978), 875.
105 Latham, M. (1994), 37.
106 Macneil, I.R. (1981), 1025.
107 Commons, J. (1934), 34.

A collaborative construction contract can provide for bargaining transactions by which a team agrees the effect of joint activities, rationing transactions by which these activities are instructed and managerial transactions by which the work of the different team members is integrated. It has been suggested that 'a good contract should be in essence a handbook for performance',[108] and a collaborative contract should be handbook for the performance, management and integration of agreed activities.

Collaborative construction contracts should be created at a time when they can support the planning of projects and programmes of work and when they can help to reduce the risk of surprises and dilemmas by integrating the work of team members. In order to coordinate and motivate the work of collaborative team members, commentators recognise that construction contracts should govern not only actions and payments, but also rules and procedures for planning and decision-making by reference to mutual expectations as to the parties' behaviour.[109]

We will test in later chapters the ways in which collaborative procurement can benefit from detailed contractual machinery, including the creation of a multi-party contract. We will also consider the effects of combining collaborative working with contracts that provide for exemptions from liability or the sharing of liability for the consequences of any party's negligent acts or omissions.

2.5 What Is an Alliance?

An alliance has been described as:

- 'A long-term arrangement which offers opportunities for benefits to be gained by coordinated action and cost-sharing over a number of projects or an on-going programme'[110]
- 'A collaborative and integrated team brought together from across the supply chain. The team share a set of common goals aligned with customer and client outcomes and work under common incentives'.[111]

Alliances are not unique to the construction industry and describe a range of agreements between two or more parties, working in any sector, who agree to pursue a set of agreed objectives. They have their origins in the UK oil and gas sector when the CRINE initiative (Cost Reduction Initiative for the New Era)[112] attempted to generate urgent capital cost savings through collaborative working across multiple companies. The potential value of alliances in public/private sector cooperation is illustrated in the

---

108 Smith, R.J. (1995), 42.
109 'These agreements may encompass the sort of actions each is to take, any payments that might flow from one to another, the rules and procedures they will use to decide matters in the future, and the behaviour that each might expect from the others', Milgrom & Roberts (1992), 127.
110 Baker, E. (2007), 345.
111 Infrastructure Client Group (2015), Section 1.
112 CRINE (1994).

King's College London Cultural Partnerships Report[113] and in the NHS alliance model for joint work by social care commissioners and a range of providers.[114]

In Australia it has been noted that 'there is a plethora of selection guidelines on the use of the alliance delivery method that are inconsistent, confusing, do not reflect current practice and are not focused on optimizing value for money…A consistent approach across jurisdictions would improve the procurement selection strategy and buying power, and ensure consistency in government engagement with industry'.[115] King's research distinguished between the following types of alliance contract:

- A 'project alliance contract' where a team is brought together to deliver a single project
- A 'framework alliance contract' where a long-term relationship is linked to the award of contracts for a number of projects, so that a team can use lessons learned on one project to improve the delivery of other projects
- A 'term alliance contract' where a long-term relationship is linked to orders placed for agreed tasks so that a team can use lessons learned on earlier tasks to improve the delivery of later tasks.

These distinctions reflect Australian research which notes that 'a project alliance is generally formed for a single project, after which the team is usually disbanded', and that 'a program alliance incorporates multiple projects under an alliance framework, where the specific number, scope and duration of projects may be unknown and the same alliance participants are potentially delivering all projects'.[116] Different approaches to alliances in a range of jurisdictions are considered in Sections 20.1 (Australia), 21.1 (Brazil), 22.1 (Bulgaria) 23.2 (Germany), 24.1(Italy), and 25.1 (USA).

There is a general understanding that 'partnering and alliance contracting both involve the development of relationships between parties to a construction or engineering project which, in broad terms, involve the subordination (to some extent) of self-interest to the greater good of the satisfactory completion of the project'.[117] The word 'alliance' is often used to describe a model that creates 'a no blame – no dispute

---

113 King's College London (2015).
114 The NHS England (2015) Standard Contract Template Alliance Agreement has many features in common with FAC-1, for example:

- A multi-party structure connecting multiple clients and providers and their separate service contracts (clause 11)
- Created as an overarching agreement alongside and in conjunction with individual Services Contracts (clauses 2.5, 2.6, 2.7)
- Agreed alliance objectives (clause 5.1) and agreed performance indicators and risk/reward mechanisms (clause 12 and Schedules 3 and 4)
- Commitment to 'effective collaborative processes' (clause 7.1)
- Consensus-based decisions through governance by an alliance leadership team (clauses 6.1 and 8)
- Mutual intellectual property licences and confidentiality provisions (clauses 13 and 14)
- Early warning and alternative dispute resolution steps through the alliance management team, alliance leadership team and independent facilitation (clause 34 and Schedule 7).

115 Department of Treasury and Finance, Victoria (2009), 153.
116 Department of Treasury and Finance, Victoria (2009), 157.
117 Bailey, J. (2016), 1.69, 37.

approach' under which intentional or wilful default is 'the only direct route to legal process'.[118]

The avoidance of disputes is one objective of collaborative construction procurement, and to share risks and liabilities in an equitable manner is at the heart of an alliance business model. However, a 'no blame'[119] clause is only one way of achieving these objectives and it should be assessed objectively by reference to:

- Other principles of collaborative procurement such as transparency, consistency and mutual reliance
- Other business models for costing, incentivising and managing an alliance.[120]

A contract governing an alliance should attempt to answer the following questions:

- How is the alliance created, between which members and can additional members be added?
- Why is the alliance created, what are the measures and targets for its success and how is it brought to an end if it does not succeed?
- How is each stage of the agreed scope of works, services and supplies authorised, in what stages and to which alliance members?
- What will alliance members do together or individually in order to improve economic and social value, by means of what contributions and by what deadlines?
- How will alliance members be rewarded for their work?
- How will alliance members reach decisions, manage risks and avoid disputes?

It can be argued that an effective alliance contract should be a multi-party contract entered into directly between all alliance members and that separate two-party contracts can impede collaboration, can obstruct the sharing of project data and can encourage a 'divide and rule' project culture.[121] The collaborative procurement process leading to creation of an alliance contract, and the management of the contract itself, can support each stage of the ISO 44001 lifecycle:

- *Operational awareness.* Vision, values, leadership and objectives
- *Knowledge.* Strategy, outcomes, business case and Implementation Plan
- *Internal assessment.* Policies, people skills and collaborative maturity
- *Partner selection.* Capability, roles and responsibilities
- *Working together.* Governance, management systems and processes
- *Value creation.* Continual improvement processes
- *Staying together.* Team management, monitoring, measurement and behaviour
- *Exit strategy activation.* Disengagement triggers and processes.[122]

---

118 Infrastructure Client Group (2015), Section 3.4.
119 For example under NEC4 Alliance Contract (2018) clauses 80, 81 and 94 as considered in Sections 7.7 and 18.7.
120 As explored in Chapters 16–18.
121 The implications of multi-party contracts are considered in Section 7.9.
122 ISO 44001 Section 8.

2.6   How Does BIM Affect Collaborative Procurement?

BIM[123] is an example of digital technology that is rapidly evolving and that is increasingly available to construction project teams. It has been described as:

- 'The management of information through the whole life cycle of a built asset, from initial design all the way through to construction, maintaining and finally de-commissioning, through the use of digital modelling'[124]
- 'A way of working that facilitates early contractor involvement, underpinned by the digital technologies which unlock more efficient methods of designing, creating and maintaining our assets'.[125]

While collaborative working can enhance human interactions throughout a construction project, BIM can enhance technological interactions through the creation of improved data as to design, cost, time and operation, all of which can be set out by reference to visible three-dimensional models. BIM increases the scope and speed of data exchanges, and this underlines procurement questions as to who provides what data, when it is best provided and how it is used and relied upon. These questions are brought into sharper focus because BIM enables, and arguably depends upon, more integration and collaborative working among team members.

BIM offers us a clearer view of the mutual dependencies between the activities of team members. However, BIM can only support these dependencies if the team members agree to share design, cost, and time data not only in a digital form but also in the levels of detail required and at the times when this data will be most useful to the project. BIM provides new systems to link the work of team members and to facilitate communication between them. It can ensure that data captured during the design and construction of a project provides the basis for its safe and efficient operation, and that the same data provides the basis for improvement on later projects. BIM also provides the axis around which other technologies, such as smart contracts, can improve the efficiency of supply chain transactions.

In a 2015 UK National BIM Survey, 57% of the respondents agreed that BIM is 'all about real time collaboration', and 31% regarded lack of collaboration as one of the main barriers to BIM.[126] We will explore in Chapters 13 and 14 the steps that need to be taken if it is accepted that:

- 'Industry-wide adoption of digitisation through media such as BIM is predicated on collaboration

---

123 For various definitions of BIM see:

BS EN ISO 19650 (2019)

ICE – http://www.ice-conferences.com/ice-bim-2014/what-is-bim

NBS – https://www.thenbs.com/knowledge/what-is-building-information-modelling-bim

RICS – http://www.rics.org/uk/knowledge/glossary/bim-intro.

124 BSI – http://www.bsigroup.com/en-GB/Building-Information-Modelling-BIM.

125 Government Construction Strategy (2016–2020), Section 22.

126 https://www.thenbs.com/knowledge/nbs-national-bim-report-2015.

- The BIM model sits at the heart of any project and only functions fully if traditional design and construction barriers are broken down by multi-party liaison and working'.[127]

2.7 What Is Early Contractor Involvement?

Early contractor involvement is a feature of collaborative procurement through which, by an early appointment in advance of start on site, 'the contractor has the opportunity to contribute his knowledge of the market and his experience, and the employer has the benefit of an open book system, enabling him to assess whether the contractor is offering a fair price and good value'.[128] Burke observed that the ability of the parties to influence project outcomes, including reduction of cost, creation of additional value, improvement of performance, and flexibility to incorporate changes, is much higher in the earlier conceptual and design stages of the project.[129] By the time that manufacture, delivery and construction are underway, the opportunities to reduce cost, to improve value, to reduce risk or to improve efficiency have reduced significantly.

Early contractor involvement should not be confused with a two-stage tendering system by which bidding contractors are expected to offer speculative design contributions as part of a short-listing or negotiation process, with no clarity as to whether their efforts will lead to appointment or other reward. By contrast, early contractor involvement should be governed by a conditional appointment setting out the joint processes by which additional information is completed to a level of detail sufficient for the parties then to agree that the project should proceed to construction. Early contractor involvement should also describe the processes by which the team move from the preconstruction phase to the construction phase with minimum negotiation.

Early contractor involvement does not impose any specific allocation of design and construction responsibilities among the consultants, main contractor, specialist contractors, manufacturers and operators. Instead, it ensures that main contract and subcontract appointments are made early enough to secure the maximum contributions from each team member, not as an optional extra but as an important component of mainstream risk management. For example, early contractor involvement can provide for the systematic joint analysis and validation of the design contributions and risks that a design consultant, contractor or subcontractor is being asked to warrant. Without this joint analysis and validation, a design warranty is not reliable.

A collaborative contract governing early contractor involvement can set out in detail the steps by which it aligns the differing commercial interests of team members, including for example:

- The extent to which an early invitation to tender enables main contractor bidders to assimilate less detailed project information and to propose their own improvements, plus guidance as to how these proposals are included in the criteria for appointment

---

127 Farmer, M. (2016), 33.
128 Barber & Jackson (2010), 171.
129 Burke, R. (2002), 31.

- Whether cost transparency can be achieved through agreement of main contractor profit, overheads and preconstruction phase costs in a way that enables a more accurate build-up of other costs
- Whether subcontractor, manufacturer and supplier appointments can be finalised prior to start on site in order to improve cost certainty, while also attracting these parties' commitment to improved value
- How client, consultant and main contractor activities such as design reviews, supply chain tenders and joint risk management activities can be programmed in a way that does not delay start on site
- Whether client and consultant risk assessments can be aligned with those of the main contractor, subcontractors and suppliers, and how joint risk management actions can be agreed and implemented without delaying start on site.

It has been suggested that early contractor involvement may create only an unenforceable agreement to agree future prices, and this is a risk if the preconstruction phase processes are not clearly spelled out in a contract. A simple agreement to negotiate can be invalid[130] and, where no work has been done, a 'price to be agreed' will not usually create a binding contract.[131] However, where a preconstruction contract is created, the processes governing the build-up of cost data can be binding contractual commitments.[132] Arguments that this contractual machinery is inadequate will 'exert minimal attraction',[133] and arguments that contractual machinery has broken down are difficult to prove.[134] The legal status of a conditional contract is considered further in Section 8.3.

2.8 What Are Two Stage Open Book, Cost Led Procurement and Integrated Project Insurance?

It has been suggested that 'partnering and alliancing are in the nature of additional elements that can be added to a form of procurement, as opposed to being distinct forms of procurement in their own right'.[135] However, the UK Government have described the following as new procurement models designed to achieve savings in capital and operational costs plus other efficiencies using a combination of early contractor involvement, collaborative working and BIM:

- 'Two Stage Open Book', comprising the use of preconstruction phase conditional appointments of the team as a means to encourage proposals for cost savings and

---

130 A 'bare' agreement to negotiate in *Walford v Miles* [1992] 2 AC 128 (HL) was seen as too uncertain to enforce.
131 As considered in *Foley v Classique Coaches* [1934] 2 KB 1 (CA) and *Courtney & Fairbairn v Tolaini Bros (Hotels)* [1975] 1 WLR 297 (CA). However, where there has been complete or partial performance, an agreement to agree a price can create an obligation to pay a reasonable price *Alstom Signalling v Jarvis Facilities* [2004] EWHC 1232 (TCC).
132 In *Tramtrack Croydon v London Bus Services* [2007] EWHC 107 (Comm), it was agreed that 'the parties shall in good faith agree (acting reasonably) the financial arrangements', and failure to agree was referred to expert determination.
133 As considered in *Queensland Electricity v New Hope Collieries* [1984] UKPC 39, [1989] 1 Lloyd's Rep 205 at 210.
134 As considered in *Secretary of State for Defence v Turner Estate Solutions* [2015] EWHC 1150 (TCC).
135 Bailey, J. (2016), 1.69, 37.

Consultants and Tier 1 Contractor selected on Profit/Fees/Overheads and designs/other proposals (costed as appropriate) against Client brief and Project Budget	Integrated Team develop designs and costs, obtain savings/improved value within Project Budget while finalising work/supply packages with Tier 2/3 Subcontractors and Suppliers	Authority to proceed to construct for fixed price or target cost within Project Budget and Client brief
Brief and team engagement	Design/cost development	Decision to build

Diagram 2 Summary of Two Stage Open Book.

other improved value, within a stated budget, after selection but prior to construction phase appointments[136]

- 'Cost Led Procurement', comprising the use of a mini-competition (usually under an existing framework) as a means to encourage speculative proposals for savings and improved value, within stated cost ceiling, prior to team selection and appointments[137]
- 'Integrated Project Insurance' ('IPI') comprising the use of project insurance without recourse, including cover for design problems and cost overruns combined with restricted contractual remedies, as a means to encourage a more collaborative culture and to obtain additional proposals for savings and improved value.[138]

Two Stage Open Book may appear more familiar than the other two new models but it benefits from a clear structure as illustrated in Diagram 2.[139] Its particular features are set out in published guidance and include client leadership and governance, a collaborative culture in place of adversarial behaviours, use of a conditional contract or equivalent provisions in a framework and use of a preconstruction phase timetable.[140]

Cost Led Procurement is particularly suited to large programmes of work involving projects that form part of a framework where bidders are familiar with client expectations and can propose improvements based on their experience of earlier projects.[141] Guidance for this model states:

- 'The client engages supply chain teams (preferably from a framework agreement) early on in the project to participate in a competition against each other on a particular scheme at the earliest possible moment in the project

136 https://www.gov.uk/government/publications/two-stage-open-book and Two Stage Open Book and Supply Chain Collaboration Guidance (2014).
137 https://www.gov.uk/government/publications/cost-led-procurement and Cost Led Procurement Guidance (2014).
138 https://www.gov.uk/government/publications/integrated-project-insurance and Integrated Project Insurance Guidance (2014).
139 Two Stage Open Book and Supply Chain Collaboration Guidance (2014), 3.
140 Two Stage Open Book and Supply Chain Collaboration Guidance (2014), Section 7.
141 As illustrated in the Rye Harbour Trial Project case study summarised in Section 17.10.

- A two-stage process with two supply chain teams is taken forward to refine their proposal (based on client feedback) and subsequently submit a final proposal that is acceptable to the client'.[142]

IPI uses a bespoke form of contract and 'no blame' clauses which, together with the new insurance policy, relieve team members from the adverse consequences of certain acts and omissions, including design errors and cost overruns.[143] The IPI bespoke contract was not made available while its Trial Projects continued, although basic IPI contract features are summarised in a review by University of Reading of the Dudley College Trial Project.[144]

In 2016 a second Government Construction Strategy reported that:

- 'New models of construction procurement were trialled under GCS 2011–2015 to explore the potential to drive better value and affordability in the procurement process
- The new models include the principles of early supplier engagement, transparency of cost and collaborative working to deliver a value for money outcome
- Alongside the potential for efficiencies, the models can support improved relationships across clients and the supply chain, increased supply chain innovation, and reduced risk'.[145]

The three recommended procurement models are not mutually exclusive, and Diagram 3 indicates how they can be combined.

For example, the Ministry of Justice uses a Two Stage Open Book procurement model under which an integrated project team works together during the preconstruction phase. However, the members of each project team are selected from overarching framework alliances by a process that can include submission of ideas for improved value in line with Cost Led Procurement. In addition to its impact on the Trial Project case studies, an approach equivalent to Two Stage Open Book also features in many of the other case studies considered in this book. Guidance describes how it can combine early contractor involvement with the work of a collaborative project team in the following stages:

- 'The Client invites Consultants and Tier 1 Contractors to bid for a project or programme of work in response to a project brief, concept design and Project Budget cost ceiling
- The Client selects the Consultants and Tier 1 Contractor (and any early approved Tier 2/3 Subcontractors and Suppliers) on the basis of their appropriate skills, approach, capacity, capability, stability, experience, and the strength of their supply chain
- The winning team works up a set of design and other proposals on the basis of Open Book costs that meet the Client's stated outcomes and cost parameters, going to the market for Tier 2/3 Subcontractors and Suppliers (to the extent not previously selected) in relation to specific work/supply packages, and appointing them according to processes, costs, and criteria that are fully transparent to the Client

---

142 Cost Led Procurement Guidance (2014).
143 As considered in Sections 7.7, 18.7, and 18.8.
144 University of Reading (2018), 97–99.
145 Government Construction Strategy (2016–2020).

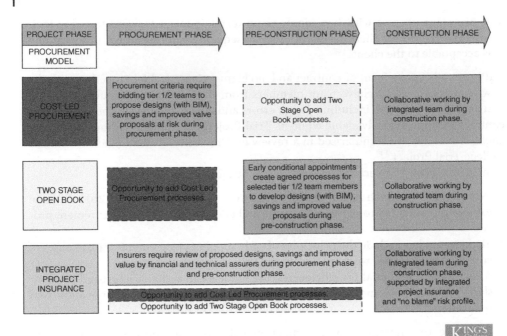

Diagram 3 Two Stage Open Book, Cost Led Procurement and Integrated Project Insurance.

- Early engagement of the Tier 1 Contractor and Tier 2/3 Subcontractors and Suppliers under collaborative and transparent processes enables the Client to develop greater confidence in the Integrated Team members, and in the information and innovation they offer, before they are authorised to commence the Construction Phase of the project
- The Client authorises start on site when the Integrated Team have built up designs in accordance with the project brief, and have finalised a fixed price or target cost within the Client's Project Budget comprising the Tier 1 Contractor's fees/profit/overheads and the Open Book costs of each work/supply package
- Joint working during the Preconstruction Phase encourages all Integrated Team members to share knowledge and focus on driving out unnecessary cost and risk rather than on commercial negotiation of price and scope, also to focus on innovation rather than adverse traditional practices such as quoting the cheapest price and then relying on recovery of additional amounts during construction'.[146]

2.9 What Is Supply Chain Collaboration?

There is often significant unexplored value that can be contributed by sub-contractors, suppliers, manufacturers and operators, whether they are engaged by the client or by other supply chain members. So as to unlock this potential, consultants need

---

146 Two Stage Open Book and Supply Chain Collaboration Guidance (2014), 12, 13.

procurement processes that enable them to work alongside the specialists who will implement their designs. UK guidance suggests that:

- 'Successful alliances are about integrating as much of the value chain as is practical
- Effective integration of the supply chain into the alliance gives genuine opportunity for meaningful early contractor involvement
- Early involvement provides a greater degree of influence over the release of value'.[147]

Early case studies recorded how 'the partnership approach to the majority of the supply chain was one of the primary changes to the project's procurement arrangements. Stakeholders have commented that the project was successful in creating a completely open and non-confrontational environment that allowed for regional works contractors, sub-contractors and suppliers to work with the Department as a single integrated team'.[148]

ISO 44001 notes the following 'risks and opportunities which result either directly or through supply chains or extended enterprises associated with potential collaborative working partners':

- Supply chain and/or extended enterprise performance
- Existing supply chain collaborations
- Combined procurement opportunities
- Relationship interdependencies
- Appropriate capabilities
- Sustainability and vulnerability
- Reputational risks
- Business continuity impacts
- Management of potential conflicts between supply chains
- Equitable collaborative terms.[149]

However, commentators have observed that under current procurement systems:

- 'The lack of integration across the supply chain, manifested in a wide-scale use of sub-contracting and tiered transactional interfaces, is commonplace'
- 'This has created significant non value add costs in the supply chain through multiple on-costing, downward and often inappropriate risk transfer'
- 'This leads to an industry that tends to be cost-focused rather than value-focused'.[150]

There are different ways in which to create closer integration across the supply chain. For example, a client itself can undertake the direct selection and appointment of specialist supply chain members. These can be provisional appointments which are later novated to a main contractor, or they can be complete appointments where the client does not include a main contractor in its team. In either case a public sector client would be required to make these selections pursuant to a formal competitive process so as not to limit open competition.[151]

---

147 Infrastructure Client Group (2014), 19.
148 NAO (2008), 23.
149 ISO 44001 Section 8.3.5.
150 Farmer, M. (2016), 17.
151 Under the Public Contracts Regulations 2015, and as considered further in Chapter 6.

Alternative ways to improve supply chain integration and to create new opportunities for improved value were analysed as part of the research undertaken by King's when supporting and assessing the Trial Projects. The resultant system, known as 'Supply Chain Collaboration', represents probably the most important innovation to emerge from the King's research. It is a process of collaborative procurement that enables a client (including a public sector client) to explore improved arrangements with subcontracted supply chain members through a sequence of agreed activities led by one or more main contractors.

Although subcontracted supply chain members can contribute to the costs and qualitative proposals that are submitted and assessed when a main contractor is selected, it may be possible to improve on these proposals by reconsidering and agreeing new ways to engage with these supply chain members. For example, carving out a clear period of time and an agreed process can:

- Enable the main contractor and its preselected subcontractors, suppliers, manufacturers and operators to engage with the client and consultants in order to check whether they have the same understanding of designs and of the ways in which those designs will be completed and constructed
- Create opportunities for the main contractor to agree improved working arrangements with its subcontractors, suppliers, manufacturers and operators that lead to improved prices and proposals
- Create opportunities for the main contractor and client to consider whether improved value and improved performance can be offered by alternative subcontractors, suppliers, manufacturers and operators.

Unlocking the potential of Supply Chain Collaboration requires a well-planned approach using techniques that go much further than simply prescribing back to back subcontracts. Construction contracts may require that subcontract arrangements reflect main contract terms, for example as to prompt payment.[152] So far so good, but these provisions alone do not create an open working relationship that connects the client and its consultants to the specialist trades and manufacturers who will undertake crucial aspects of many construction projects.

The basic structure of Supply Chain Collaboration is illustrated in Diagram 4 and the recommended sequence of activities is set out in Section 2.10. This new system was pioneered on two Trial Projects by Surrey County Council Highways Department through a term alliance[153] and by Hackney Homes and Homes for Haringey[154] through a framework alliance. Their work and the improved value that they achieved demonstrate how clients and other alliance members can agree greater involvement of subcontractors and suppliers that increases the scope for improvements in working practices, sources of supply, costs, quality, safety, sustainability and social value.

Public sector clients can reconcile Supply Chain Collaboration with the constraints of public procurement regulations[155] as a means to explore with the main contractor, after its selection, whether local or regional businesses offer better value than the main contractor's intended supply chain members. This system enables clients and their teams to focus on these questions by agreement with main contractors in a manner that is

---

152 For example, JCT SBC/Q (2016) clause 3.9 and NEC4 clause 26.
153 Project Horizon Trial Project case study summarised in Section 5.10.
154 SCMG Trial Project case study summarised in Section 4.10.
155 As considered in Section 6 8.

Client(s), main contractor(s) and consultants agree scope for achieving savings/improved value through improved mutual commitments with subcontractors, manufacturers, suppliers and operators	Timetabled sub-contract business cases and tenders are led by main contractor(s) to obtain new proposals from subcontractors, manufacturers, suppliers and operators	Improved mutual commitments and consequent cost savings and other improved value are agreed and recorded
Scoping of Supply Chain Collaboration	Supply Chain Collaboration process	Improved commitments and improved value

Diagram 4 Summary of Supply Chain Collaboration.

compliant with public procurement requirements and that enhances the opportunities for local and regional businesses to prove their worth and to win additional work.

Supply Chain Collaboration has wide-ranging potential as a joint process by which alliance members can improve value. There are already examples of how this process can be undertaken by different combinations of alliance members:

- By a client working directly with trade contractors and suppliers[156]
- By a client working with one main contractor and two subcontractors and suppliers[157]
- By two clients working with six main contractors and 30 types of subcontractors and suppliers[158]
- By a main contractor working with four types of subcontractors and suppliers.[159]

2.10   What Are the Supply Chain Collaboration Activities?

Supply Chain Collaboration should be clearly structured using the sequence of activities set out below.[160]

Client(s) identify the relevant main contractor(s) with projects/programmes of work under new or existing frameworks, alliances or long-term contracts that are compatible/appropriate for Supply Chain Collaboration

Workshop of client(s), main contractor(s) and consultant(s) to explain the processes and opportunities created by Supply Chain Collaboration, namely working with subcontractors, manufacturers and suppliers to improve mutual commitments and secure savings and other improved value

Proposals from main contractor(s) with details of current subcontractors, manufacturers and suppliers and current costs/terms/qualitative and other benefits

---

156 For example, the ORTUS Learning and Events Centre project summarised in Section 7.10, the Jarvis Hotels project summarised in Section 7.10 and the Futures Housing Group framework alliance summarised in Section 10.10.
157 Project Horizon Trial Project case study.
158 SCMG Trial Project case study summarised in Section 4.10.
159 Kier Highways Services supply chain alliance case study summarised in Section 11.10.
160 Two Stage Open Book and Supply Chain Collaboration Guidance (2014), 25.

Client(s) evaluate proposals from main contractors and analyse the current subcontractor/manufacturer/supplier relationships for commonality and potential improved mutual commitments, savings and other improved value in specific work/supply packages

Client(s) agree standardised specifications for relevant work/supply packages with main contractor(s)

Workshop of client(s), main contractor(s), consultant(s) and current subcontractors/manufacturers/suppliers to explain the Supply Chain Collaboration processes and opportunities

Main contractor(s) create and distribute enquiry document to subcontractors/manufacturers/suppliers describing new client/main contractor commitments and inviting new proposals for work/supply packages

Main contractor(s), with client(s), evaluate initial proposals and invite successful subcontractors/manufacturers/suppliers to interviews and discussions

Main contractor(s), with client(s), conduct interviews and discussions with selected subcontractors/manufacturers/suppliers to finalise improved mutual commitments, savings and other improved value in each work/supply package

Main contractor(s) and selected subcontractors/manufacturers/suppliers confirm improved mutual commitments, savings and other improved value in each work/supply package

3

How Does a Project Alliance Operate?

3.1 Overview

Project team members collaborate at a personal level, irrespective of their contractual links, because reasonable, professional people know that they depend on each other. However, the organisations for whom those people work each have their individual objectives and priorities which include the necessity of making a profit or at least avoiding a loss. When the objectives and priorities of different team members diverge, then the reasonable behaviour of individuals can come under strain. In order to anticipate and avoid the breakdown of working relationships, project team members can benefit from the creation and development of a project alliance.

Through the commitment of organisations as well as individuals to a collaborative approach, a project alliance can increase the scope to resolve differing perceptions and to seek improvement and resolution in matters of potential legal liability. However, the successful integration of a project team requires early consideration of the vertical, horizontal and linear links that support mutual commitments and early agreement of systems for planning and coordination.

Collaborative relationships and processes that underpin the work of a project alliance include the agreement by all team members of:

- Shared objectives, success measures, targets and incentives
- Consensus-based decision-making
- Systems for team members to build up, share and manage their design, time, cost and operational data as the basis for agreeing optimum solutions
- Agreed activities by team members that are designed to improve value
- Clear timeframes and deadlines for team members' agreed activities and for other team members' responses and approvals
- Systems for joint management of risks and the agreed avoidance and resolution of disputes.

This chapter explores how a project team can work collaboratively as a project alliance, applying approaches that overcome the potential divergence of their interests and navigating around problems in ways that all team members can see are in their collective best interests.

Collaborative Construction Procurement and Improved Value, First Edition. David Mosey.
© 2019 John Wiley & Sons Ltd. Published 2019 by John Wiley & Sons Ltd.

3.2 What Is a Collaborative Project Team?

Each construction project involves a team of people working together to achieve successful completion and each project should also involve a team of organisations working together, so what is different about a collaborative project team?

Collaborative working is defined by the relationships created between team members and by the ways in which the team members work together. The complex links between the organisations comprising a project supply chain have different dimensions, and a collaborative team creates ways of cementing these links so that they operate more effectively:

- Lateral links where, for example, an architect works alongside a structural engineer, a services engineer, a cost consultant, other consultants and one or more contractors in building up a costed set of designs
- Vertical links where, for example, a contractor delegates design contributions and the implementation of those designs to a supply chain of subconsultants, subcontractors, suppliers, manufacturers and operators
- More lateral links where, for example, supply chain members work alongside each other on site
- Linear links where, for example, a design consultant hands over to a contractor and its supply chain specialists and these contractors in turn hand over to an operator.

Latham emphasised that 'there is no benefit if a company claims "we are always partnering" but then carries on with the same business practices, exploiting any problem incurred by another party or delaying payment to its supply chain'.[161] Performance by project team members of their agreed obligations depends on the coordination of their different interests and the reconciliation of their different motivations, and it has been observed that:

- 'The coordination problem is to determine what things should be done, how they should be accomplished and who should do what'
- 'The motivation problem is to ensure that the various individuals involved in these processes willingly do their parts in the whole undertaking, both reporting information accurately to allow the right plan to be devised and acting as they are supposed to act to carry out the plan'.[162]

Before creating a collaborative project team, a client should consider:

- Its needs, objectives and budget in relation to the proposed project
- How and when to create concept designs and a description of performance outputs that reflect its needs and objectives
- How and when to create an outline cost plan that reflect its needs and objectives

---

161 Association of Consultant Architects (2010), 1.
162 Milgrom & Roberts (1992), 126.

- How and when to create the other documents used for selection of collaborative team members
- Whether it wishes to create a project alliance.[163]

3.3 How Can a Collaborative Team Form a Project Alliance?

Applying the tests in Section 2.5, the following questions should be answered in order to create an effective project alliance:

- How is the alliance created, between which members and can additional members be added?
  A project alliance can be created using a multi-party contract in order to integrate the roles of the client, consultants, contractors and other members and can include provision for adding alliance members as and when they are selected.
- Why is the alliance created, what are the measures and targets for its success, and how is it brought to an end if it does not succeed?
  A project alliance can be created for the design, construction and operation of a single project, with agreed measures and targets linked to the success of that project, and with agreed procedures for performance review and for ending the alliance (or the role of any member) if it does not achieve agreed targets.
- How is each stage of the agreed scope of works, services and supplies authorised, in what stages and to which alliance members?
  A project alliance can map out the roles and responsibilities of each alliance member at each preconstruction, construction and post-construction stage of the project, and there is no need for an alliance to blur the lines between who does what.
- What will alliance members do together or individually in order to improve economic and social value, by means of what contributions and by what deadlines?
  A project alliance can describe systems for improving value through Supply Chain Collaboration and other agreed activities, with a timetable governing the deadlines and interfaces of each alliance member's contributions.
- How will alliance members be rewarded for their work?
  A project alliance can incentivise performance through payments related to results, for example through alliance members sharing agreed cost savings.
- How will the alliance members reach decisions, manage risks and avoid disputes?
  A project alliance needs a decision-making forum with agreed terms of reference for those matters requiring consensus among alliance members, including joint risk management and non-adversarial dispute resolution.

It is important that alliance members include the client and all key contributors to the design, construction and management of the project, and that formation of the alliance

---

163 There is a range of guidance as to the ways in which a client should explore its options in advance of deciding what project it needs and whether it needs a project at all, for example Infrastructure and Projects Authority (2014).

does not delay or omit the membership of the main contractor, key subcontractors and preferably the future operator. PPC2000, the US ConsensusDocs 300 and the NEC4 Alliance Contract all create multi-party project alliance contracts, and under each of these forms there is no minimum or maximum number of members of a project alliance.

Not all members need to join a project alliance at the same time. The processes of early contractor involvement and Supply Chain Collaboration discussed in Sections 3.5 and 3.6 show how a project alliance may add new members at different stages. For example, the client's needs and objectives need to be made clear in the documents used for selection of the alliance members, and certain alliance members can be appointed at an earlier stage than others in order to prepare those documents. For these purposes the first alliance members are likely to be:

- A design consultant who creates concept designs and a description of performance outputs
- A cost consultant who creates an early outline cost plan.

A project alliance contract should describe the machinery for adding new members, and for replacing any member who drops out of the alliance. For example, PPC2000 provides a standard form 'Joining Agreement' by which all team members agree to the additional or replacement member and its proposed role.[164] ConsensusDocs 300 provides for the team to comprise the 'Owner, Design Professional and Constructor' and then for any additional consultant, subconsultant, or subcontractor to become a 'Joining Party' by signing a 'Joining Agreement'.[165] The NEC4 Alliance Contract states that 'the Alliance Board may add a Partner to the Alliance' by instruction to the appointed 'Alliance Manager'.[166]

3.4 How Does a Project Alliance Operate?

The members of a project alliance should:

- Start working together at a time when their collaborative relationships can be developed and when project value can be improved
- Describe clearly the ways in which they will implement value-adding processes
- Establish relationships that are fair, open and consistent and that connect alliance members to each other, to the wider supply chain and to other stakeholders
- Create timetables and decision-making processes that support consensus and that enable efficient progress.

A practical question facing any alliance is how and to what extent each alliance member should deploy its expertise either individually within the scope of a clear role or collectively with other alliance members where each party's contributions cannot be separately identified. For example, should an architect or engineer create designs for review and input by other alliance members or should it create those designs jointly with

---

164 PPC2000 (2013) clauses 10.2 and 26.10 and Appendix 2.
165 ConsensusDocs 300 (2016) Articles 2.2, 21.30, 21.31, and 21.39.
166 NEC4 Alliance Contract (2018) clause 21.8. The roles of the Alliance Board and the Alliance Manager are considered in Sections 12.4 and 12.7.

other alliance members? Even where alliance members are closely integrated through co-location, they need to consider the legal and practical risks of any confusion as to who has prepared each design or other document and who is responsible for each aspect of manufacture, delivery and construction.

Some critics of collaborative procurement allege that an alliance risks the creation of a murky contractual soup in which it is not possible to identify the party responsible for each contribution, and this is a risk that should be avoided. The following points should be borne in mind:

- It is possible for a project alliance contract to make clear the interfaces between specific activities undertaken by each alliance member, for example in PPC2000 where an alliance member 'who prepares or contributes to any one or more Partnering Documents shall be responsible for the consequences of any error or omission in, or any discrepancy between, such Partnering Documents or its contributions to them, except to the extent of its reliance (if stated in such Partnering Documents) on any contribution or information provided by any one or more other Partnering Team members'[167]
- Even where there is contractual emphasis on collective working, in practice the alliance members will fulfil distinct roles. For example, the NEC4 Alliance Contract 'Implementation Plan' describes 'the roles and responsibilities of the members of the Alliance'[168] and its guidance states that 'whilst the Alliance is contractually responsible for Providing the Works, this can only be done effectively if each member of the Alliance understands what every other member's role and responsibilities are'[169]
- If the contributions of alliance members cannot be identified, then they cannot be supported by professional indemnity insurance and the alliance members will depend on a new type of insurance cover, as considered in Section 18.8.

There are differing options governing the extent to which alliance contracts envisage that alliance members will work collectively as if they formed a single virtual organisation.[170] For example:

- PPC2000 states that alliance members 'shall work together and individually in the spirit of trust, fairness and cooperation for the benefit of the Project, within the scope of their agreed roles, expertise and responsibilities as stated in the Partnering Documents'[171]
- The NEC4 Alliance Contract states that 'The members of the alliance…work collectively to support the delivery of the contract on a best for project basis'[172]

PPC2000 seeks to establish a commitment to work collaboratively 'together' but at the same time limits this commitment by reference to the interfaces established between the alliance members working 'individually'. The PPC2000 wording can be criticised as ambiguous, and its interpretation depends on agreement of details as to:

---

167 PPC2000 (2013) clause 2.4.
168 NEC4 Alliance Contract (2018) clause 11.2(20).
169 NEC4 *preparing an alliance contract* (2018), 57.
170 The concept of a virtual organisation is considered in Section 7.6.
171 PPC2000 (2013) clause 1.3.
172 NEC4 Alliance Contract (2018), clause 20.1.

- How and when the alliance members will work together, for example in 'Partnering Team' meetings, in 'Core Group' meetings, in 'Design Team' meetings and in the joint 'Value Engineering', 'Value Management' and 'Risk Management' exercises organised by the 'Client Representative'[173]
- How and when the alliance members will work individually, according to the agreed roles, expertise and responsibilities stated and integrated in the client's 'Project Brief', the main contractor's 'Project Proposals' and the various 'Consultant Services Schedules'.[174]

The NEC4 Alliance Contract wording merges the alliance members' roles more closely, and illustrates the ways in which 'all members of the alliance…share in the performance of the alliance as a whole as opposed to their own individual performance'.[175] For example, it provides that 'the Alliance carries out the design necessary to Provide the Works',[176] whereas PPC2000 breaks this obligation down among alliance members and provides that 'the Lead Designer and the other Design Team members shall develop the design and process of the Project…'.[177]

Other features of the NEC4 Alliance Contract and PPC2000 are explored in Sections 9.8 and 9.9 and in other sections of this and later chapters. Case studies illustrating how PPC2000 operates as a project alliance contract in practice are summarised in Section 3.10 and in later chapters.[178] At the time writing, there are no case studies of how the NEC4 Alliance Contract operates in practice, although it was announced at the time of its publication that an early user would be Yorkshire Water Services.[179]

A project alliance contract may incorporate a 'no blame' clause which limits or combines the individual liabilities of alliance members, for example in relation to any matter other than an intentional act or omission or a wilful default.[180] The rationale for these clauses is that alliance members will offer stronger contributions in support of each other's work if they have no fear of legal liability. The impact of no blame clauses is considered in Sections 7.7, 16.8 and 18.7, including the central question of whether they operate as intended.

In addition to clarifying the collective and individual rights and obligations of alliance members, a project alliance contract should:

- State its agreed objectives and targets[181]
- Create direct connections between alliance members so that they are aware of each other's roles, so that they have a shared communication system and so that the timing and impact of their work are integrated through clearly agreed interfaces[182]

---

173 PPC2000 (2013) clauses 3.5, 3.8, 8.13 and 5.1(iii). The Client Representative fulfils the role of alliance manager as considered in Section 12.4.
174 PPC2000 (2013) clauses 1.3 and 2.4.
175 Introducing the new NEC4 Alliance Contract (2018).
176 NEC4 Alliance Contract (2018), clause 23.1.
177 PPC2000 (2013) clause 8.1.
178 See also the cross-references to PPC2000 in the projects listed in Appendix C, and the additional PPC2000 case studies in Association of Consultant Architects (2010) and at www.ppc2000.co.uk.
179 Introducing the new NEC4 Alliance Contract (2018).
180 For example, NEC4 Alliance Contract (2018) clauses 80, 81 and 82.
181 For example, NEC4 Alliance Contract (2018) clause 53 and PPC2000 (2013) clause 4.
182 For example, both the NEC4 Alliance Contract and PPC2000 create a multi-party structure and describe systems for communications, timing, integration and interfaces.

- Create a forum for alliance members to review and agree innovations and other change proposals that are not already agreed when the alliance was created[183]
- Create a forum for team members to share early warning of problems and to agree corrective actions[184]
- Describe the incentives that motivate alliance members to fulfil their agreed objectives.[185]

By these means, project alliance members can ensure that their contract fulfils the functions of 'a planning and incentive device' which links the agreed objectives of the project team to the means of achieving them and to the behaviour required from the alliance members.[186]

3.5   How Can a Project Alliance Use Early Contractor Involvement?

A project alliance team can use early contractor involvement to support any allocation of design and construction responsibilities and any design team structure, including:

- Where the design team leader is a lead consultant or a main contractor throughout the project[187]
- Where a consultant lead designer hands over at some point to a main contractor who completes the designs, for example where design consultants are novated to become subconsultants of the main contractor[188]
- Where there is no main contractor appointed to take overall responsibility for construction but instead there are multiple appointments by the client of specialists and trade contractors.[189]

Latham commented that early contractor involvement 'motivates the team to drive down cost in a systematic way. It also helps identify carbon reduction and energy efficiency measures, as well as opportunities for employment and skills during the conditional preconstruction phase. These can be properly costed and jointly assessed with key subcontractors and manufacturers at a time when all team members have the same objective, namely to finalise a brief within budget so that work can proceed on site'.[190]

The UK Institution of Civil Engineers suggests that the early involvement of contractors and suppliers will often:

- Facilitate the early establishment of an integrated team comprising the client, its advisers and all designers, contractors and other supply chain members
- Establish a longer-term relationship that supports good performance

---

183 For example, NEC4 Alliance Contract (2018) clause 21 and PPC2000 (2013) clauses 3.5, 3.6, 12.10 and 23.4.
184 For example, NEC4 Alliance Contract (2018) clauses 15 and 95 and PPC2000 (2013) clauses 3.7 and 27.3.
185 For example, NEC4 Alliance Contract (2018) clause 53 and PPC2000 (2013) clause 13.
186 Arrighetti et al. (1997), 171.
187 For example, the Hampstead Heath Ponds case study summarised in Section 3.10.
188 For example, the Bath and North East Somerset Council case study summarised in Section 12.10.
189 For example, the ORTUS and the Jarvis Hotels case studies summarised in Section 7.10.
190 Association of Consultant Architects (2010), 1.

- Ensure common understanding of the requirements, outcomes and key outputs
- Maximise the time available for construction planning in terms of phasing, resources, interfaces and health and safety
- Enable contractor input to design development and facilitate collaborative design
- Allow time for the contractor, designers, and wider supply chain to develop innovative and better value solutions, including off-site construction opportunities
- Manage risks to accelerate delivery, reduce costs and secure health and safety
- Remove procurement from the critical path.[191]

Early contractor involvement is integral to the two-stage structure of PPC2000, which describes:

- The conditional preconstruction phase appointment of all team members and the preconditions to the construction phase proceeding[192]
- The processes of preconstruction phase design developed by the design team, coordinated by a lead designer and approved by a core group[193]
- The processes for appointing subcontractors and suppliers by approval of business cases or by mini-competitions organised by the main contractor, with all proposals subject to core group consultation[194]
- The build-up of the cost data that leads to agreement of an agreed maximum price[195] that cannot exceed the previously agreed budget.[196]

The NEC4 Alliance Contract treats early contractor involvement only as an option rather than as an integral feature and does not set out the preconstruction phase processes in detail.[197] Its Option X22 leaves the build-up of agreed designs, cost, timescales and risk treatment to be set out in other contract documents and describes only:

- The appointment of alliance members to the preconstruction phase and the submission of a proposal for the construction phase, to be prepared by an alliance manager and to be approved by an alliance board[198] in accordance with a procedure to be stated in the contract 'Scope'[199]
- The requirement for this proposal to forecast its effect on the Scope, the project budget, the 'Performance Table'[200] and the project completion date.

If the project alliance team members cannot finalise a mutually acceptable basis for proceeding to construction, then the preconstruction phase project alliance contract should allow them appropriate leeway to withdraw. For example:

- Arup in their report for OGC observed that PPC2000 'is based on a two-stage tendering process whereby time and cost data is developed incrementally and reported

---

191 ICE Guidance Note (2018).
192 PPC2000 (2013) clause 14.
193 PPC2000 (2013) clause 8.3, with defined terms as stated in Appendix 1.
194 PPC2000 (2013) clause 10.
195 PPC2000 (2013) clause 12.
196 PPC2000 (2013) clauses 8.7 and 12.3.
197 NEC4 Option X22 is also considered in Sections 9.7 and 9.9.
198 NEC4 Alliance Contract (2018) Option X22.2. The role of the Alliance Manager and the Alliance Board are considered in Sections 12.4 and 12.7.
199 The 'Scope' specifies or describes the works and any constraints on how the alliance provides the works, NEC4 Alliance Contract (2018) clause 11.2(29).
200 The 'Performance Table' is considered further in Section 9.9.

on an open-book basis. This means that there can be a focus on value at all material points and the contract can still enable the parties to withdraw if the value profile is not satisfactory'[201]

- The PPC2000 clause permitting withdrawal does not state whether the client can then appoint other parties to complete the construction phase,[202] and it will be important to clarify what rights the client has to use the designs and other documents prepared by team members
- The NEC4 Alliance Contract includes an option for its alliance board not to accept the 'Stage Two' construction phase proposals and to issue a notice that Stage Two will not proceed[203]
- The NEC4 Alliance Contract makes clear that in these circumstances the client can 'appoint another person or organisation to complete Stage Two and can use any material, information, or design which the Alliance has provided'.[204]

3.6   How Can a Project Alliance Use Supply Chain Collaboration?

A critical process in building a project alliance is the early appointment of supply chain members by a main contractor, or by the client in the absence of a main contractor, so that they can contribute to design, programming and risk management. Supply chain contributions can include proposals for improved quality and improved programming, and also for cost savings and reduced risk, if early planning creates the time and opportunity for supply chain members to validate, challenge and improve on assumptions made by the client, the design consultants and the main contractor.

The contributions of subcontractors and suppliers can be built into the preconstruction phase planning and can be made:

- Speculatively prior to selection, if potential supply chain members recognise that this gives them a better chance of being appointed by an alliance main contractor who is already in place
- Speculatively during selection, if potential supply chain members recognise that proposals are part of the preconstruction phase selection process for supply chain members and are submitted for consideration by the main contractor with client and consultant participation
- Under early conditional preconstruction phase appointments of supply chain members who are appointed back to back with the early appointment of the main contractor.

The Two Stage Open Book procurement model envisages the development of an integrated team through which the main contractor and the subcontractors and suppliers are appointed early in order to make the maximum contributions to improved design, costing, risk management and programming of the work/supply packages comprising a

201 Arup (2008), 37.
202 PPC2000 (2013) clause 26.1.
203 NEC4 Alliance Contract (2018) Option X22.3.
204 NEC4 Alliance Contract (2018) Option X22.9.

project or programme or work.[205] The Integrated Project Insurance procurement model also provides for the early conditional appointment of team members that include the main contractor and key subcontractors and suppliers.[206] The Cost Led Procurement model places more reliance on speculative early contributions from main contractors and their supply chain members in advance of their appointment and as part of their selection usually under a pre-existing framework contract.[207]

Collaborative contracts can make clear how team members agree to extend collaborative procurement to all supply chain members, for example through:

- Selection according to agreed cost and quality criteria that are consistent, where appropriate, with the collaborative procurement model[208]
- Subcontracts and supply contracts that are fair and that are consistent with the contracts between other team members[209]
- Supply Chain Collaboration that enables the detailed costs of individual works and supply packages to be examined and agreed by all project alliance team members and that enables development and completion of designs in ways that reduce risk for clients, consultants and contractors.

Historically, clients have been reluctant to take an active role in selection of a main contractor's supply chain members in case they inadvertently take additional legal responsibility for their performance and the quality of their work. Supply Chain Collaboration should be led by the main contractor, in order to avoid client involvement in the selection of supply chain members being categorised as 'nomination' or 'naming' and thereby potentially diminishing the main contractor's responsibility for those parties.[210]

Clients may also be concerned as to time spent in extra meetings with subcontractors and may anticipate direct claims in respect of late contractor payments. One client explained that it would have no time or appetite for Supply Chain Collaboration, regardless of its possible benefits, and summed up its view with the mixed metaphor that 'dealing with subcontractors is like herding cats so why should I buy a dog and bark myself?'[211] This was not an example of the hands-on client leadership that I was hoping for.

Details of the participation by the client and consultants in reviewing documents and attending meetings should be agreed in advance, making it clear that client and consultant involvement in these processes does not detract from the fact that supply chain selection and review are led by the main contractor and give rise to subcontract appointments. The Supply Chain Collaboration process should follow the steps set out in Section 2.10, using an agreed timetable so as to avoid delays.

---

205 Two Stage Open Book and Supply Chain Collaboration Guidance (2014).
206 Integrated Project Insurance Guidance (2014).
207 Cost Led Procurement Guidance (2014).
208 For example, PPC2000 (2013) clause 10.8 provides for all documentation relating to selection of supply chain members to 'encourage their maximum contribution to and participation in an integrated design, supply and construction process'.
209 For example, PPC2000 (2013) clause 10.1 provides for all subcontracts and supply contracts to 'establish, wherever possible, partnering relationships complementary to those described in the Partnering Contract'.
210 Provisions for client nomination of sub-contractors appeared in JCT standard form contracts until they were omitted in the 2005 editions. Provisions for 'Named Specialists' appear in JCT 2016 contracts, for example in Supplemental Provision 9 in JCT SBC/Q (2016).
211 Seminar on FAC-1 provided by the author for public sector clients, October 2017.

Main contractors will be concerned to ensure that Supply Chain Collaboration does not lead to subcontract appointments that prejudice existing good supply chain relationships or that lead to appointment of supply chain members whose work the main contractor is unwilling to warrant. Any prospective supply chain members need to be agreed by the main contractor, the client and the other alliance members, so that there is no risk of the main contractor being obliged to appoint a party that it has not approved.

However, in concluding a consensus-based alliance system of Supply Chain Collaboration, it is possible that there will be different views as to which team members should be selected. It is, therefore, important to embed Supply Chain Collaboration from the beginning of the procurement process and to include clear contractual machinery that describes when and how:

- A shortlist of prospective supply chain members will be agreed
- Prospective supply chain members will be briefed and invited to put forward proposals
- Supply chain members will be selected and appointed.

No standard form project contracts provide for Supply Chain Collaboration except PPC2000 which sets out procedures for the development and review of business cases and tenders by which alliance members can build up a full picture of agreed costs while securing important contributions from supply chain members who may or may not become members of the alliance team.[212]

The NEC4 Alliance Contract omits any Supply Chain Collaboration procedures and makes a binary distinction between specialists who are inside the alliance and subcontractors and suppliers who are not,[213] with only a rudimentary approval process for the latter.[214] This omission is compounded by the treatment of early contractor involvement only as an option[215] and by the omission of pre-construction phase processes that could describe the costing of subcontract work.[216]

3.7 How Can a Project Alliance Measure and Reward Performance?

A project alliance should state the agreed objectives of its members and the success measures and targets by which achievement of these objectives is measured. The principal objectives of a project alliance are likely to be completion of the project in accordance with the client's brief and within the client's budget, but these can be combined with additional objectives linked to collaborative working and improved value. For example, the objectives under the PPC2000 contract include:

- 'Trust, fairness, mutual cooperation, dedication to agreed common goals and an understanding of each other's expectations and values

---

212 PPC2000 (2013) clause 10.
213 'If inside, they become members of the Alliance as Partners, in the same way as first tier suppliers and are not Subcontractors. If outside the Alliance, they are contracted in the traditional way to one or more members of the Alliance...' NEC4 *managing an alliance contract* (2018), 1.
214 NEC4 Alliance Contract (2018) clause 26.
215 As considered in Section 3.5.
216 As considered in Section 16.5.

- Finalisation of the required designs, timetables, prices and supply chain for the Project
- Innovation, improved efficiency, cost-effectiveness, lean production and improved Sustainability
- Completion of the Project within the agreed time and price and to the agreed quality
- Measurable continuous improvement by reference to the Targets…and the KPIs
- Commitment to people including staff and Users'.[217]

Project alliance objectives need to be distilled into success measures and targets, so that a collaborative contract can state what level of performance will determine incentives such as shared pain and gain.[218] The contract also needs to make clear what scope there is for falling short of the agreed objectives at all and whether team members may receive additional rewards for meeting agreed targets in respect of the work they specialise in and for which they are already being paid. For example, the performance criteria on the ORTUS multi-party project alliance were devised on the basis that 'the client wanted all criteria to be wholly achieved but added a notional weighting to assist decision-making (and also to provide the basis of an incentivisation mechanism for some team members):

- Health and Safety 25% (25% = No reportable accidents)
- Programme 15% (15% = Project completion date May 2013)
- Quality 30% (Comprising 20% = Design Brief met; 5% = Quality Expectation met; 2.5% = Zero defects at project completion; 2.5% = Minor Defects Liability Period defects only)
- Cost 20% (Comprising 15% = Build costs not exceeding budget; 2.5% = Fees and other planning/development costs not exceeding budget; 2.5% = Lifecycle cost forecast (net present value ex build) not exceeding target)
- Public Relations 10% (Comprising 7.5% = Good public relations maintained; 2.5% = Less than five complaints to the client).

The above criteria were shared with all team members, and were accompanied by statements as the 'vision and ambitions for the project'.[219] Regular progress reviews should be part of alliance governance whereby:

- 'Processes will be in place to maintain both team integration and partner alignment
- Teams will review their own performance and how they are using integration to deliver value
- The connection with partners will be maintained through a structured and formal review process that focuses on alignment, partner contribution to the alliance and relationships'.[220]

A project alliance should also make clear the ways in which its members can earn agreed incentives. While a framework alliance or term alliance can reward performance with additional work, this is not an option on a project alliance unless it is subdivided into successive project phases. Project alliance incentives are more likely to be financial, for example through payments by way of:

---

217 PPC2000 (2013) clause 4.1.
218 For example, as in PPC2000 (2013) clauses 4.2, 13.2 and 13.3 and Appendix 8.
219 ACA case study at www.ppc2000.co.uk as summarised in Section 7.10.
220 Infrastructure Client Group (2014), 20.

- Shared savings achieved against a client budget or cost plan that has been set early in the preconstruction phase, by reference to alliance activities that are undertaken during the preconstruction phase before the budget or cost plan is converted into a fixed price
- Shared savings that are achieved against a budget or cost plan or target price that is not converted into a fixed price, by reference to alliance activities that are undertaken at any time during the life of the project.

Incentives are considered further in Sections 16.8 and 16.9.

3.8 How Can a Project Alliance Improve Value and Outcomes?

Project alliance teams should expect that their agreed collaborative relationships, interfaces and activities will create measurable improvements in economic and social value. The range of improved value that can be achieved through a project alliance is considered in Chapter 15, and this covers improved cost and certainty and cost savings, improved quality, local and regional business opportunities, new employment and training initiatives, improved safety and environmental benefits. The case studies in Section 3.10 and in later chapters describe how a project alliance can improve value for all alliance members. It can also improve project outcomes by:

- Avoiding misunderstandings through improved communication, as considered in Chapter 12
- Avoiding final account escalation by building up accurate agreed costs and incentivising savings, as considered in Chapter 16
- Avoiding delays by creating and managing integrated timetables and programmes, as considered in Chapter 17
- Reducing risks by joint investigations and other early risk management, as considered in Chapter 18
- Using collaborative governance to avoid the wasted costs and distractions caused by disputes, as considered in Chapter 19.

The activities to which systematic early contractor involvement can add value have been identified as including:

- 'Better teamwork – critical to the success of any project
- Better programming – shorter project delivery times, or better fit to client constraints
- Better design and specification – better buildability and more effective sourcing
- Better care of the environment – less waste and damage, better public perceptions
- Better budgeting – greater sensitivity to the market, specialist knowledge
- Better management of risk and value – participation in risk and value management'.[221]

---

221 CIRIA (1998), 14. A 1975 UK industry report identified the following circumstances where contractor design contributions would be most effective: 'Projects where some form of proprietary building system is being used; serial contracts where the contractor can provide feedback to the design team from experience on an earlier contract; where the method of construction to be used is new or complex; where the construction methods and plant employed are central to finding the most economical design solutions; and where time is of such priority that the construction programme must be compressed', NEDO (1975), Sections 7.24 and 72.

The types of cost savings that can be agreed through collaborative procurement were itemised as follows on the Archbishop Beck Trial Project:

- 'De-bundling certain elements of precast to avoid additional layers of profit and overheads, wastage and to achieve more efficient coordination generated savings of approximately £45 000
- Extensive rounds of value engineering in respect of mechanical and electrical designs, with the preferred specialist subcontractor A&B Engineering and M&E consultant Mouchel contributed to a contract saving of £258 688
- Removal of suspended ceilings and adding additional acoustic treatments produced a saving of £21 068
- Recycling of waste materials for re-use on site, combined with redesign of external landscaping, resulted in a saving of £55 164
- Installation of under-floor heating in the ground floor slab to allow earlier installation and rationalisation of the heating system produced a saving of £31 234
- Acquisition and re-use of surplus bleacher seating from another school produced a contract saving of £18 564'.[222]

3.9 How Can a Project Alliance Deal with Problems?

Problems that can arise on a project alliance include the following 'four major road blocks to success':[223]

- 'A shift in business conditions: if conditions change and the project is behind schedule, with unanticipated technical problems and cost overruns, the strategy within each organisation may change and even revert to an "us versus them" attitude'
  However, a shift in business conditions is not necessarily a reason to revert to an adversarial position. For example, if there is a contractual commitment to share cost information, as considered in Chapter 16, the implications of the shift can be examined by all parties in seeking to agree a solution[224]
- 'Uneven levels of commitment: unevenness of commitment often develops from the basic differences between organisations'
  Uneven levels of commitment can be tested and ironed out during the preconstruction phase, and can also be monitored and managed during the construction phase, if the project alliance team members clarify the activities and support by which they will create and sustain a collaborative culture, as considered in Chapter 12
- 'Lack of momentum: a partnership requires nurturing and development throughout the life of the project. The representatives from each side must constantly work to maintain the health of the partnership'
  Lack of momentum can be addressed by agreement of deadlines for completion of key activities by each alliance member during the preconstruction phase and the construction phase, as considered in Chapter 17. Avoiding delays also depends on

---

222 Archbishop Beck Trial Project case study summarised in Section 16.10.
223 Smith, N.J. (2002), 304.
224 The North Wales Prison Trial Project case study summarised in Section 14.10 describes how a collaborative team can respond to a shift in market conditions.

personal leadership and on the use of an early warning system through which any concerns are notified to the other team members, as considered in Chapters 12 and 19.

- 'Failure to share information: partnering requires timely communication of information and the maintenance of open and direct lines of communication among all members of the partnering team. The failure to share information is most likely to arise when team members revert to past practices'

Failure to share information can be addressed through agreement of a communication system that sets out the scope of requirements for shared information, as considered in Chapter 12, including for example the extent of agreed cost data as explored in Chapter 16. This system can also make clear the agreed constraints in respect of confidential information and the intellectual property rights attaching to information that is shared, as explored in relation to Building Information Modelling (BIM) in Section 14.4.

A project alliance team can use effective management, co-location, training, independent advice and other collaborative systems as ways to maintain the commitment of team members to solving problems rather than exploiting them, and these are explored in Chapter 12. They can also use contractual joint risk management to mitigate the impact of problems as explored in Chapter 18. The techniques described in these chapters are designed to enable earlier identification of risk issues combined with joint decision-making and agreed actions.

Where a team member breaches a project alliance contract then, subject to any agreed exclusions or limitations of liability such as those available through no blame clauses,[225] the remedies are the same as under any other contract. However, as to the actions that team members take in response to a breach, a project alliance contract should incorporate provisions for non-adversarial dispute avoidance and resolution, as considered in Chapter 19.

For example, a project alliance can attempt to resolve a potential dispute through referring it to a representative forum with agreed terms of reference, such as an 'Alliance Board' under the NEC4 Alliance Contract or a 'Core Group' under PPC2000, through which alliance members can seek a common understanding of the problem and can seek to agree an acceptable solution.[226] The basis for agreed resolution needs to be unanimous agreement of all those present, firstly to encourage attendance and secondly to avoid fear of being in a minority.[227] In 18 years of PPC2000 project alliance contracts being used on many hundreds of projects, the Core Group as a forum for early warning and agreed dispute resolution has helped to ensure that only two disputes were referred to the courts.[228]

Where collaborative construction procurement does not produce the expected results, a project alliance contract should establish any agreed flexibility for adjusting

---

225 As explored in Sections 7.7 and 18.7.
226 As considered in Sections 12.7 and 19.7.
227 As considered in Section 12.7.
228 The two known cases are *TSG Building Services v South Anglia Housing* [2013] EWHC 1151 (TCC) as considered in Section 7.5 and *Willmott Dixon Housing v Newlon Housing Trust* [2013] EWHC 798 (TCC) as considered in Section 19.9.

team members' expectations.[229] If matters arise that exceed the agreed parameters of this flexibility, there needs to be a system for abandoning the project and agreeing the consequent rights and entitlements of the team members.

UK guidance suggests that:

- 'Integration and the right behaviours are enabled by knowing the arrangement for the closedown of the alliance
- The commercial model needs to be very clear about the separation of the parties, the remaining liabilities and how staff will be affected
- This gives both individuals and organisations clarity and so provides the confidence that allows them to commit to the greater integration required to establish the alliance'.[230]

3.10 Project Alliance Case Studies

The following case studies describe work led by public sector clients who created project alliances using early contractor involvement and collaborative working under a multi-party contract. The first is London Borough of Hackney who procured the Rogate House refurbishment project on the Nightingale Estate in 1999 using a bespoke prototype of the contract which was later published as PPC2000.[231] This multi-party project alliance contract provided the basis for the team to undertake joint development of designs and joint risk management, with early appointment of subcontractors including concrete specialists and asbestos specialists to analyse design solutions and methods of working. In addition, the entire team agreed the construction phase timetable in advance on the basis of resident consultation that established an agreed decanting programme.

Completion of previous comparable works using a traditional procurement approach had required 115 weeks to refurbish 108 flats, whereas the Rogate House integrated team worked at approximately double the speed and took 90 weeks to refurbish 192 flats. Hackney have gone on to use PPC2000 as their preferred contract form under successive frameworks culminating in the Supply Chain Management Group (SCMG) Trial Project described in Section 4.10.

The second case study was led by a very different public body, namely the City of London Corporation, who have adopted PPC2000, and the related TPC2005 term alliance contract, on a wide range of different projects including schools, offices, listed court buildings, housing and (in the case study below) complex engineering.[232]

---

229 Williamson envisaged the need to adjust conflicting positions of self-interest by means of 'cooperative adaptation' through 'conscious, deliberate and purposeful efforts to craft adaptive internal coordinating mechanisms', Williamson, O.E. (1993), 48.
230 Infrastructure Client Group (2014), 32.
231 The client's willingness to make this decision depended on advice from Noel Foley as client lead and Thelma Stober as the client's in-house legal adviser. It also depended on the other named team members being willing to trust in the robustness of new relationships and systems that the project alliance contract created.
232 Other City of London Corporation project alliance case studies are summarised in Sections 6.10 and 18.10.

This case study describes a range of risk management and design solutions developed jointly by City of London Corporation as client, the Superintendent of Hampstead Heath as operator, Atkins as comprehensive engineering, landscaping and ecology consultant, BAM as main contractor and BAM's specialist subcontractors over the following timeline:

- November 2013 – March 2014: Atkins advised on feasibility
- March 2014: PPC2000 Project Partnering Agreement signed, including BAM, Atkins and Capita
- March 2014 – March 2015: ECI period, BAM carried out site investigation and tree clearance
- July 2014: planning approval granted
- July 2014 – November 2014: Regular stakeholder meetings
- November 2014: Judicial review
- December 2014: Detailed costing
- March 2015 – October 2016: PPC2000 Commencement Agreement, construction phase completed.

Rogate House, London Borough of Hackney

The £13 million PPC refurbishment of Rogate House included specialist concrete works and the cut-through of an existing building to form two separate buildings undertaken by a project alliance comprising London Borough of Hackney (client), Wates Construction (main contractor), Leonard Stace (client representative), Abbey Holford Rowe (architect), Babtie Allott (structural engineer), Lomax Consulting Engineers (services engineer).

Hackney wanted an alternative procurement approach that recognised the complexities of a major refurbishment project and enabled it to achieve time and cost certainty. For this purpose, it commissioned the first local government use of the PPC2000 partnering contract for conclusion with its main contractor and consultants.

The alliance team established the joint basis for development of designs and risk management, with early appointment of subcontractors including concrete specialists and asbestos specialists to analyse design solutions and methods of working. In addition, the entire team agreed the construction phase timetable in advance on the basis of resident consultation to establish an agreed decanting programme.

Multi-skilled specialists were selected for internal works, in preference to traditional trades, so as to require fewer visits to each flat. This was instrumental in obtaining support from residents that was not evident on the earlier similar Alma House project.

The collaborative approach to design also allowed agreement of aesthetic improvements such as external metal balconies which were designed and installed in collaboration with the balcony supplier for a cost less than that incurred at Alma House. Another example of design collaboration involved the main contractor recommending metal refuse chutes as a cheaper and simpler alternative to the Alma House precast concrete chutes. Completion of works at Alma House had taken 115 weeks to refurbish 108 flats whereas at Rogate House it took 90 weeks to refurbish 192 flats. The Rogate House team worked at approximately double the speed.

(Continued)

Noel Foley, the London Borough of Hackney project leader, commented: 'The pre-construction phase agreement must be part of the business case review or gateway management process as projects proceed to site. This could then act as a check on rushing to site.' Leonard Stace added: 'We needed concrete specialists on board early in case we had to rejig the programme: they would know whether these changes were achievable. A close-knit team with a common objective…was the only way this approach could work.'

Association of Consultant Architects (2010), 35.

Hampstead Heath Ponds

This £15 million PPC2000 project alliance comprised improvements to the Hampstead Chain and Highgate Chain of Dams, a statutory reservoir and part of the ponds infrastructure covering seven sites at Hampstead Heath. The highly complex and diverse work required consultation with local user groups and statutory authorities.

The alliance team comprised City of London Corporation (client and later PPC2000 Client Representative), the Superintendent of Hampstead Heath (member of core group), Atkins (comprehensive engineering, landscaping, and ecology consultancy), Capita (cost consultant and initially Client Representative), BAM Nuttall (BAM) (main contractor) and Shane Hughes of Savills (independent adviser).

Joint design solutions agreed and implemented by the project alliance members included:

- BAM proposed precast concrete for key elements such as the headwalls and piling as the use of wet concrete is extremely difficult in a pond environment. A wall to be placed on a dam in order to raise it was originally planned using in situ concrete. BAM suggested instead a sheet pile wall provided by a specialist subcontractor, an improved engineering solution that significantly reduced the amount of concrete required
- BAM suggested a prefabricated steel frame structure which had been a provisional sum in the tender. The whole team, including the local end-user stakeholders, visited the proposed subcontractor in Yorkshire, which proved a highly valuable exercise. Design team meetings were then held at the subcontractor's offices, which assisted the flow of the design and implementation process. The prefabricated solution was very important because the existing deck could no longer be used, and a solution had to be found which did not require a significant suspension of access to the pool. The prefabricated solution meant that this element of the project was efficiently concluded within six months
- BAM and Atkins worked together on effective tree loss minimisation, following enabling works carried out by BAM.

Risk management workshops took place throughout the project. The Risk Register stated responsibilities, for example splitting gas main risk permanent works (City of London) and gas main risk temporary works (BAM). Examples of joint risk management included:

- It had been agreed that clay from the site would be used as part of the dam construction works but ground investigation suggested that the top layers of clay would be weathered and therefore not fit for this use. The parties agreed to assume a usable level of 1 m and then agreed a 50% pain/gain share to the extent that the usable clay was above or below that level. This avoided re-measurement. The agreement worked well on one clay bed but proved more difficult to apply on another where there the 50% split proved an over-simplification
- BAM attended early stakeholder meetings, crucial in explaining to the many interested parties how the impact of the works would be minimised. Introducing key individuals such as the plant drivers at an early stage was an approach carried forward to the construction phase so as to ensure the continual building of relationships
- BAM reversed the proposed order of the pond works, so that the riskiest pond should be worked on first, in order to create 'float' for later works. This significantly de-risked the project programme.

Reported for the KCL Centre of Construction Law by City of London Corporation and by Shane Hughes of Savills as independent adviser.

4

How Does a Framework Alliance Operate?

4.1 Overview

Longer-term collaboration on multiple projects can 'not only foster mutual trust, but also facilitate the sharing of knowledge and information to generate innovation and value for the parties to the relationship'.[233] There is greater scope for improved value to be achieved on multiple projects because they attract increased personal commitment and investment, because team members can plan with a clearer understanding of potential additional work, and because team members can be expected to learn from project to project.

While collaborative procurement can improve commercial value on a single project, this may have limited potential because:

- 'Personnel from different organisations in the project chains are unfamiliar with each other and there may only be a limited level of trust'[234]
- Procurement systems based on single projects provide 'little opportunity for achieving incremental improvements in efficiency and/or effectiveness'.[235]

Multiple projects can be procured through a framework alliance which sets out systems for awarding each project, systems for improving value and measures that establish an ongoing pipeline of work based on agreed levels of performance. A framework alliance can also ensure that all parties are aware of each other's roles and that their respective contract terms are consistent. These features can motivate the mutual trust necessary for successful joint working, through 'a horizontal agreement between the respective partners [which] capture[s] the principles within the commercial model, particularly those that jointly incentivise performance and create collaboration'.[236]

In 2012 a cross-industry working group collected evidence from UK Government departments and the wider public sector, and reported that benefits from the use of effective frameworks include:

- Delivery of sustainable efficiency savings

233 Colledge, B. (2005), 32.
234 NEDC (1991), 34. 'The long-term nature of partnering means that parties on all sides are familiar with the project requirements and the level of trust which has been built up over a series of projects is there from day one on the next project'.
235 Egan, J. (1998).
236 Infrastructure Client Group (2015), Section 3.4.

- Reduction in consultancy and construction costs
- Delivery of projects closer to target cost and time
- Reduction of disputes, claims and litigation
- High client satisfaction rates
- High proportion of value of work undertaken by small and medium-sized enterprises
- High proportion of local labour and subcontractors
- High take-up of government initiatives such as fair payment and apprenticeships
- High proportion of construction, demolition and excavation waste diverted from landfill
- Good health and safety performance against national average
- Acting as a key enabler to integration of the supply team.[237]

Specific frameworks considered in the 2012 Effectiveness of Frameworks Report included the Ministry of Justice (MoJ) framework alliance summarised in Section 4.10, and also successful frameworks created by:

- Department for Education, who created 'regular contractor forums in which issues are raised and discussed', whose performance indicators included 'SME engagement, apprenticeships, waste and carbon measures' and whose framework generated 'a 9.5% reduction in outturn costs…when compared with single procurements'[238]
- Department of Health, who encouraged early engagement of the supply chain 'to increase quality of design, engage key stakeholders, ensure cost robustness, minimise risk and increase certainty of delivery in time and budget', and who established 'a solid governance structure that involves suppliers and clients in development of the framework'[239]
- Environment Agency who 'have adopted a partnering approach…and work collaboratively in the delivery of all contracts called off'.[240]

This chapter considers the characteristics of frameworks and how they can be structured and developed more effectively as framework alliances in order to create a strong and flexible basis for collaborative construction procurement.

4.2 What Is a Framework?

Frameworks are an established means for the delivery of multiple projects and are recognised in UK public sector procurement regulations as 'an agreement between one or more contracting authorities and one or more economic operators, the purpose of which is to establish the terms governing contracts to be awarded during a given period, in particular with regard to price and, where appropriate, the quantity envisaged'.[241] A framework contract governs a long-term relationship with a contractor, consultant, supplier or service provider and sets out the procedure for the award of specific work under supplementary contractual commitments. It should describe:

---

237 Procurement/Lean Client Task Group (2012), 83.
238 Procurement/Lean Client Task Group (2012), 98.
239 Procurement/Lean Client Task Group (2012), 97.
240 Procurement/Lean Client Task Group (2012), 97.
241 Public Contracts Regulations 2015, Reg 33(2).

- The systems by which successive projects will be planned, designed and built
- The investments and rewards expected by the parties
- The agreed measures of the parties' performance
- The machinery for the parties to review and improve on their original underlying commercial calculations and expectations.

A framework provides commercial motivation for the parties to behave less opportunistically because they are aware of the additional projects on offer. Even the most cynical parties may calculate:

- 'I shall be better off in the longer term if I continue my relationship with you instead of terminating it; and I also estimate that you similarly have calculated that you will be better off if the relationship continues'[242]
- 'Concern with one's reputation may be an effective check on ex post opportunism, overcoming temptations to renege or renegotiate'.[243]

Longer-term commitments enable the parties to agree plans for delivering work linked to plans for improving performance so that they can jointly 'monitor delivery', 'review performance', 'take corrective action', 'implement and maintain procedures for measuring...effectiveness', and 'establish a process of continual improvement'.[244]

Trial Project case studies have illustrated the benefits of collaborative frameworks, including the Supply Chain Management Group (SCMG) housing framework alliance which created systems that are 'easily replicable by any local authorities or housing associations under new procurements and also under current frameworks and long-term contracts that contain processes for continuous improvement'.[245] The Trial Projects results led the Southern Construction Framework to use Two Stage Open Book as the basis for their four-year framework awarded in April 2016.[246]

However, despite their potential as vehicles for integrating teams and improving value, framework contracts are sometimes used cynically as shortcuts to market and as a means to attract lower prices by exaggerating the potential pipeline of work. Without confidence in collaborative working relationships and in the conditions governing the award of future work, the expectations that are created when awarding a framework can deteriorate into mutual indifference, disillusionment and distrust.

4.3 How Does a Framework Contract Operate?

Until 2005 there were no published forms of framework contract at all, but then two came along at once:

- The JCT Framework Agreement, now part of the JCT 2016 suite
- The NEC Framework Contract, now part of the NEC4 suite.

---

242 Campbell & Harris (2005), 23, 24.
243 Milgrom & Roberts (1992), 139.
244 ISO 44001 Section 8.6.6.
245 SCMG Trial Project cases study.
246 The Southern Construction Framework reported procurement of a total of £857 million in the first year for 27 public authorities across 48 projects, with framework users including central and local government, plus further and higher education establishments. Southern Construction Framework Report (2016).

The JCT Framework Agreement sets out a system for call-off of particular 'Tasks' by the issue of an 'Enquiry', the submission of a response and the creation of an 'Order' in the form of an 'Underlying Contract' for each project.[247] The NEC4 Framework Contract does not offer details of the required call-off procedures, and provides only spaces for insertion of 'the scope', 'the selection procedure' and 'the quotation procedure'.[248]

The JCT Framework Agreement requires that a main contractor should endeavour to achieve closer involvement from members of its supply chain in matters such as 'design development', 'project planning', 'risk assessment' and 'value engineering'.[249] This encourages supply chain contributions to design, planning and risk but is likely to have limited effect in the absence of any detailed machinery for joint working with supply chain members to achieve improved value.

The JCT Framework Agreement includes clauses by which:

- 'The Parties will continually impress upon all personnel involved with the Tasks their keen desire to work with each other and with all other Project Participants in an open, co-operative and collaborative manner and in a spirit of mutual trust and respect with a view to achieving the Framework Objectives'[250]
- The parties will share information and give early warning,[251] although these expectations may be optimistic if they are not integrated with the JCT 'Underlying Contracts'
- The parties will 'promptly volunteer and share such knowledge or information' as would be of value to the other party in the performance of underlying project contracts,[252] but with no provision for any reward or for any recognition of the intellectual property rights to which shared knowledge or information would be subject
- The parties will undertake early 'collaborative risk analysis',[253] with provision in the original 2005 version that amended Underlying Contracts could reflect the effects of joint risk analysis, although with this link was removed in a revised edition two years later.[254]

Although it encourages the collaborative potential of frameworks, the JCT Framework Agreement omits:

- Any timeframes or procedures through which the closer involvement of supply chain members will be achieved
- Any agreed rewards for the main contractor or the members of its supply chain in return for providing added value
- Any explanation as to how the main contractor or its supply chain members will be appointed on specific projects in ways that would enable their added value to be incorporated into the JCT Underlying Contracts.

The NEC4 Framework Contract has no collaborative procurement provisions, in contrast with the commitment to collaborative systems in other NEC4 forms and despite

---

247 JCT Framework Agreement (2016) clause 4.
248 NEC4 Framework Contract (2017) Contract Data Part One – Data provided by the Client. The NEC4 Framework Guide (2017) explains how each section of the NEC4 Framework Contract can be completed.
249 JCT Framework Agreement (2016) clause 10.2.
250 JCT Framework Agreement (2016) clause 9.
251 JCT Framework Agreement (2016) clauses 11, 17, 19.
252 JCT Framework Agreement (2016) clause 11.
253 JCT Framework Agreement (2016) clause 14.
254 JCT Framework Agreement (2005) and JCT Framework Agreement (2007) clause 14.

the greater potential for a collaborative approach when working with the same team over successive projects. The NEC4 Framework Contract does not include clauses that describe agreed objectives, success measures, incentives, improved value, supply chain involvement or joint risk management. It contains no provisions for any processes that could improve value, measure performance or reward success, and it is silent on risk management, non-adversarial governance and alternative dispute resolution.

There are provisions in the NEC4 Framework Contract for:

- The award of a 'Work Order' namely an NEC4 construction contract or professional services contract[255]
- The issue of a 'Time Charge Order' to govern early advice, comprising 'a contract to provide advice on a proposed Work Order on a time charge basis'.[256]

This suggests the basis on which the main contractor could undertake preconstruction phase activities in order to assist in finalising details of the proposed project, although the NEC4 Framework Agreement does not provide guidance as to the preconstruction processes that might be covered by a Time Charge Order.[257]

Neither contract form encourages the parties to invest in a long-term relationship. They both permit early termination without a requirement to demonstrate breach or failure to meet agreed objectives, and this does not offer the stable basis for the investment and commitment that framework contracts are intended to attract:

- The JCT Framework Agreement provides for termination by at least one month's notice which may be given at any time after the first 12 months.[258]
- The NEC4 Framework Contract allows either party to 'terminate their obligations under the contract at any time by notifying the other Party'.[259]

The statutory definition of a framework contract as 'an agreement made between one or more contracting authorities and one or more economic operators'[260] opens the door to a multi-party approach. However, the JCT and NEC forms are designed to be entered into between only two parties, with no provision for strategic integration between multiple supply chain members or between multiple clients. The results achieved under more collaborative, bespoke framework contracts led to the development of a new type of standard form.

4.4 What Is a Framework Alliance?

Research by the UK National Association of Construction Frameworks identified the importance of frameworks that have the collaborative features of an alliance. They found that: 'significant savings, benefits and other efficiencies in construction can be achieved

---

255 NEC4 Framework Contract (2017) clause 23.4.
256 NEC4 Framework Contract (2017) clauses 11.2 (4) and 22.1.
257 NEC4 Framework Guide (2017), Preparing and managing a framework contract, 5, suggests that a Time Charge Order is usually governed by an NEC4 Professional Services Short Form Contract.
258 JCT Framework Agreement (2016) clause 22.2.
259 NEC4 Framework Contract (2017) clause 90.1.
260 Public Contracts Regulations 2015, Reg. 33(2).

by effective frameworks through the longer-term arrangements, non-adversarial relationships, common incentives, integrated teams and objective assessment of performance associated with such frameworks'.[261] In order to fulfil this potential and in order to build greater industry credibility, a framework can be established as a framework alliance, using a multi-party contract that creates mutual commitments to:

- Shared objectives, success measures, targets and incentives, assessed jointly by alliance members through consensus-based decision-making and other means to create and sustain an alliance culture
- Systems for alliance members to build up and manage shared design, time, cost and operational data as the basis for agreeing optimum solutions
- Activities to be undertaken by alliance members that are designed to improve value
- Clear timeframes and deadlines for alliance members' activities, responses and approvals
- Systems for joint management of risks and the agreed avoidance and resolution of disputes.

A framework alliance contract should make clear the purposes of the framework programme, through provision for objectives and success measures agreed by all alliance members. These provisions should recognise not only the objectives of the alliance and the framework programme, but also the objectives of each alliance member in relation to the alliance and the framework programme.

A framework alliance contract functions as an umbrella contract, as illustrated in Diagram 5, that can integrate multiple project contracts entered into by a wide range of members and governing a wide range of work. For example, a framework alliance could be created by:

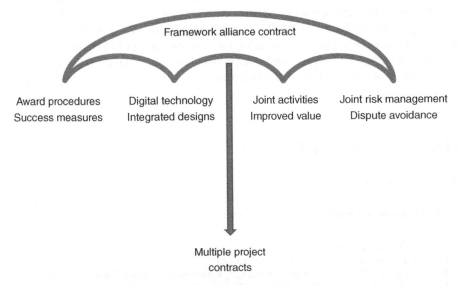

Diagram 5 Integration through a framework alliance contract.

---

261 LGA/NACF (2016).

- One or more clients seeking to award project contracts in respect of a single complex project or in respect of one or more compatible work programmes, who can benefit from integrating the standards and working practices governing their projects or the different parts of their single project
- One or more consultants and/or contractors competing for the award of project contracts, who agree to seek improved value that will enhance the outcomes from all their projects
- One or more project teams who agree to integrate the work of their subcontracted supply chain members so as to meet overarching shared objectives.

A framework alliance can strengthen the detailed preconstruction phase planning for each project, so that individual construction contracts for specific projects are created only once preconstruction phase processes have been satisfactorily completed.[262] For example, a framework alliance contract can act as a preconstruction phase contract for each project, setting out how joint design development, joint price and supply chain development, joint risk management and joint agreement of each construction phase programme are to be undertaken.

For example, the SCMG Trial Project framework alliance created systems under which 'the project teams have a clear process for exchanging information on a collaborative basis at an early stage, with participants in early contractor involvement meetings working together to agree solutions that promote the best method of delivering the project. Often such discussions are led by the tier 1 contractor (with tier 2/3 support), so as to utilise experience from recent similar projects and to offer clear and well considered methods for the efficient delivery of the works'.[263]

The potential use of a framework alliance in a range of jurisdictions, and the extent to which similar models are already used in those jurisdictions, are explored in Sections 20.4 (Australia), 21.4 (Brazil), 22.4 (Bulgaria), 23.3 (Germany), 24.4 (Italy) and 25.4 (USA).

4.5 How Does a Framework Alliance Award Work?

Compliance with a framework alliance contract requires all members to follow agreed procedures and to enter into project contracts for works, services or supplies on pre-agreed terms according to a legally binding set of procedural commitments.[264] However, there is a risk that a loosely worded framework alliance contract may be unenforceable as a mere agreement to negotiate.[265] Hence, it is important that the machinery leading to the award of individual project contracts is clearly set out.

It is fundamental to framework alliance relationships that members understand not only the prospective scope and nature of the work covered by a framework alliance but also how that work will be awarded. Framework alliance contracts create conditional arrangements under which a stated procedure will lead to the award of project

---

262 Macneil described how a strong relationship between the parties acts as a starting point for moving into the substantive planning required for particular projects, Macneil, I. R. (1981), 1029.
263 SCMG Trial Project case study.
264 The consequences of breach are considered in Section 4.9.
265 Interestingly, JCT Framework Agreement (2005) was published with binding and non-binding options.

contracts governing works, services or supplies. Framework alliance contracts do not usually govern the provision of the works or services or supplies themselves, with the limited exception of any early works or services or supplies ordered in advance of a project contract award.[266]

A framework alliance contract should state whether there is any agreed minimum value or type of project contracts to be awarded to any framework alliance member, so that members have a clear picture as to the level of certainty in the flow of work.[267] It should also state whether any exclusivity is granted to alliance members,[268] and the basis for adjustment of that exclusivity according to achievement of agreed targets.[269]

A framework alliance contract can set out the procedures giving rise to the award of work:

- By direct allocation of work to any one or more alliance members
- Through a competition in which two or more alliance members submit prices and proposals in response to a project brief according to agreed rules and evaluation criteria set out in the framework alliance contract.[270]

A framework alliance contract should explain how a client will make the decisions that lead to the conditional or unconditional award of each project contract, including how it will decide if and when each procedure will be used. It should set out:

- The agreed procedures and timescales for the issue of each project brief, for the submission of prices and other project proposals and for all steps leading to award of project contracts, including the method, rules and criteria for evaluation of prices and other project proposals and the required format for their submission
- The model documents that will be applied to each project, including cross-reference to project contract conditions that will govern the projects awarded
- The procurement model used for each project, including the sources and timing of all contributions to design, supply chain engagement, costing, programming and risk management
- The required approach to Building Information Modelling (BIM) as appropriate.

4.6   How Can a Framework Alliance Use Supply Chain Collaboration?

A framework alliance contract should create the conditions most likely to meet the client's requirements and to achieve the agreed objectives. Therefore, it is important to build into a framework alliance contract the commitment of all parties to implement specific activities designed to improve value. The failure of a framework alliance is a client failure too, not least because re-procurement involves significant costs for the client as well as the bidders.

---

266 For example, FAC-1 Framework Alliance Contract (2016) clause 7 and Appendix 3.
267 For example, in FAC-1 Framework Alliance Contract (2016) clause 5.6.
268 For example, in FAC-1 Framework Alliance Contract (2016) clause 5.7.
269 For example, FAC-1 Framework Alliance Contract (2016) Schedule 1 Part 2.
270 As considered in Sections 6.6 and 6.8, public sector framework procedures are subject to the Public Contracts Regulations 2015, Regs. 33(7), (8), (9), (10) and (11).

Over the life of a framework alliance, cost savings and other improved value can be achieved through Supply Chain Collaboration by revisiting the original design, cost, risk and programme assumptions made by any one or more main contractor. This creates greater scope for improved value because one or more clients can work jointly with one or more main contractors to test and review:

- Whether improved mutual commitments with subcontracted supply chain members can be achieved through activities that generate additional integration, information and innovation
- Whether these improved mutual commitments and activities can secure savings and other improved value in specific work or service or supply packages, in specific projects and in the programme of work as a whole.

Supply Chain Collaboration can also enable framework alliance members to earn greater rewards because:

- Contractors, consultants and subcontracted supply chain members are incentivised by the prospect of a continued pipeline of work
- Clients can expect and measure continuous improvement
- Alliance members can capture lessons learned and pass them on from project to project
- Joint commitment to the new processes can be replicated and can become standard business practice.

For example, under the SCMG Trial Project framework alliance:

- 'After Tier 1 contractors are conditionally engaged, they work up project proposals in conjunction with the clients on an open-book basis and (with SCMG support) go to the market for Tier 2 and Tier 3 selecting them on fully transparent criteria'
- 'Over the course of the programme the clients gain insight and understanding as to the implications and value of client interventions, choices and decisions'
- 'This assists Hackney Homes and Homes for Haringey in becoming intelligent clients and encourages Tier 1 contractors to share knowledge through the supply chain, focusing on driving out unnecessary cost and risk rather than on any commercial "gaming" over price and the scope of works included'.[271]

Contractors on the Property Services Cluster Trial Project framework formed 'a Cluster Delivery Team which fully collaborates to share resources, information and supply chains. The Team aims to establish common designs, elements and components and, subsequently, common supply chains. This is allowing subcontract appointments to be made across the whole Tranche, seeking to drive efficiencies through aggregation'.[272]

Similarly, on the Ministry of Justice framework alliance, 'Over the course of successive projects, framework contractors build up designs and solutions that they know will meet MoJ specifications, and are able to transfer these from one project type to another'.[273]

---

271 SCMG Trial Project case study.
272 Property Services Cluster Trial Project case study summarised in Section 8.10.
273 Cookham Wood Trial Project case study summarised in Section 14.10.

4.7 How Can a Framework Alliance Measure and Reward Performance?

A framework alliance can motivate conscious adaptation from short-term self-interested behaviour to a long-term cooperative strategy, taking into account the following components:

- Analysis of the benefit of the relationship and assessment of the value that other parties place on the relationship
- The incentive to continue the relationship combined with the disincentive against terminating the relationship
- The expectation of an undefined share in the joint gains that can be generated by the relationship, provided that such share is expected to be greater than any gain achievable by the party proceeding alone
- The risk of losing specific investments made in the relationship and the potential cost of developing and investing in a new alternative relationship.[274]

A framework alliance should agree its joint objectives and the individual objectives of its members, plus the measures of its success and the targets that alliance members seek to meet.[275] The objectives can clarify both the shared objectives of the alliance and those of individual alliance members. They form the basis for the success measures and targets, for seeking improved value in accordance with the agreed alliance activities, and for rewarding success by means of agreed incentives.[276]

The agreed success measures are the performance indicators that determine whether the framework alliance members have achieved the agreed objectives. Targets for each success measure should be objectively measurable, stating the method of recording relevant data, the alliance member responsible for measuring that data and the system for reporting to the other alliance members. For example, the Connect Plus DBFO highways framework established a 'Balanced Scorecard' process which 'enables Connect Plus to understand and measure progress towards its declared objectives of:

- Creating and maintaining a group of directors and facilitators empowered with the skills and behaviours to support and lead the cultural change and role model collaboration
- Delivery of a whole life approach
- Minimising the impact of maintenance works
- Maintaining project facilities
- Enhancing knowledge of project facilities
- Respect for the environment
- Reduced risk'.[277]

A framework alliance can link success in meeting its agreed targets to incentives that reward alliance members for achieving agreed objectives.[278] For example, the prospect

---

274 Campbell & Harris (2005), 23.
275 For example, FAC-1 Framework Alliance Contract (2016) Schedule 1 Part 1 and Part 2.
276 For example, FAC-1 Framework Alliance Contract (2016) Schedule 1 Part 2 and Part 3.
277 Connect Plus Trial Project case study summarised in Section 12.10.
278 Milgrom & Roberts described the need 'to structure a relationship and set common expectations, and … establish mechanisms that will be used to make decisions and allocate costs and benefits', Milgrom & Roberts (1992), 132.

of successive projects is a major incentive for consultants and contractors, and a system of performance measurement provides a commercial basis for the client to award additional projects to the same alliance member or members.[279] The following examples of incentives are set out in the FAC-1 Framework Alliance Contract:

- 'Additional payments including shares of savings achieved through Supply Chain Collaboration and other Alliance Activities'
- 'Adjustment of any exclusivity in the award of Project Contracts'
- 'Extension of the scope of the Framework Programme and/or the duration of the Framework Alliance Contract'.[280]

4.8 How Can a Framework Alliance Improve Value?

The evidence of improved value achieved by multi-party framework alliance contracts is substantial. For example, on a £240 million programme of work, the Whitefriars Housing Group and two main contractors 'set up a joint framework agreement to govern a five-year programme of housing work, which included a model two-stage project contract providing for preconstruction phase processes governing completion of the client brief, price framework and construction programme as preconditions to start on site'.[281] They achieved agreed cost savings of 10% and agreed time savings of 20% plus new local jobs and training opportunities for 38 people in the first year and a total of over 200 during the programme as a whole.

The Eden Project built their innovative Education and Resource Centre (the 'Core') plus a number of smaller projects using a 'multi-party framework agreement that contained provisions for joint preconstruction phase design development, pricing, risk management and programming prior to award of each construction phase contract'.[282] The bespoke multi-party framework was linked to the NEC Option C project contract form and integrated the work of Grimshaw as lead designer, Buro Happold as services engineer, Anthony Hunt Associates as civil and structural engineer, Haskoll as architect, Land Use as landscape architect, Scott Wilson Kirkpatrick as design manager and supervisor, Davis Langdon as cost consultant and project manager and Sir Robert McAlpine and Alfred McAlpine as joint venture main contractor. Their collaboration overcame significant challenges caused by unstable ground conditions and delivered an award-winning project.[283]

The Department for Work and Pensions (DWP) and Land Securities Trillium (LST) appointed 14 main contractors and a range of subcontractors and suppliers under a multi-party framework alliance contract to govern a national programme of government office refurbishment, achieving overall savings of 24.8% and winning a health and safety award. The alliance members agreed a common set of preconstruction phase processes that were applied to each of a series of 969 projects during a three-year period. Corresponding frameworks were created between the clients and key subcontractors and

---

279 JCT Framework Agreement (2016) clause 21 refers to measurement of performance against agreed key performance indicators, but does not link that measured performance to the award of future projects.
280 FAC-1 Framework Alliance Contract (2016) guidance note in Schedule 1 Part 3.
281 Mosey, D. (2009), Case Study, 251 as summarised in Section 4.10.
282 Mosey, D. (2009), Case Study, 257.
283 Mosey, D. (2009), Case Study, 257. The quality of the Core project is considered in Section 15.4.

suppliers under which works and supplies were called off by individual main contractors on their respective projects. Operation of the integrated relationships and processes set out in the framework alliance and the underlying supply chain frameworks enabled the team to deliver 'a final cost in 2006 of £737 million against a 2003 forecast of £981 million'.[284]

Other examples of successful framework alliances include:

- A £1 billion housing refurbishment and new build programme procured by Glasgow Housing Association with 24 main contractors and 27 consultants, where Savills as independent adviser reported that 'the ability to get the constructors on board early, and to involve them in the design, programming and scoping decisions before works began, undoubtedly saved the client – and the constructors – hundreds of thousands of pounds'[285]
- A £70 million schools programme procured by London Borough of Harrow with Kier, where the client reported that 'the complete programme has been delivered on time to provide a great boost to education in Harrow. Relations between team members and the schools' head teachers have remained positive in what could have been stressful circumstances'.[286]

Nearly all the UK Government Trial Projects[287] were procured using bespoke collaborative frameworks, including the Property Services Cluster programme, the M25 Connect Plus programme, the Archbishop Beck School project and the Rye Harbour project. In each case the strategic relationships established under the framework contracts contributed to the improved economic and social value that was achieved by the Trial Project teams. For example, the Archbishop Beck School project used the wider SCAPE framework to ensure that the team 'benefitted from lessons learned on the previous Notre Dame School project. It also contributed to the City-wide initiative led by Liverpool City Council for the engagement of local Tier 2/3 Subcontractors and Suppliers and improvement of local employment and skills commitments'.[288]

Three Trial Projects were supported by bespoke multi-party framework alliances. Firstly, the UK Ministry of Justice multi-party framework alliance was analysed in the 2012 Effectiveness of Frameworks Report, and was credited with achieving reduced operating costs of £10 million, reduced burden on industry of £30 million and procurement risk mitigation of £2 million.[289] The Cookham Wood Trial Project case study showed how early engagement of the supply chain under the Ministry of Justice framework alliance contract supported the project team in achieving agreed cost savings of 20%, in reducing their construction programme from 50 to 44 weeks (with a consequent saving of £85 000 in time-related site overheads) and in a range of

---

284 Mosey, D. (2009), Case Study, 254 and Association of Consultant Architects (2010), 29, as summarised in Section 4.10.

285 Association of Consultant Architects (2010), 17–18.

286 Association of Consultant Architects (2010), 22.

287 The only Trial Projects not procured under collaborative frameworks were Project Horizon procured under a term alliance and summarised in Section 5.10, and Dudley College procured under a project alliance and summarised in Section 18.10.

288 Archbishop Beck Trial Project case study.

289 Effectiveness of Frameworks (2012), 103–4, summarised in Section 4.10.

agreed design improvements.[290] The North Wales Prison Trial Project team used the same framework alliance contract to enable delivery of improved value such as:

- 26% agreed post-tender cost savings
- A reduced footprint for the entry building and energy centre using lessons learned from the earlier Oakwood prison alliance project
- Adoption of an alternative lighting solution previously used on alliance projects in Scotland
- Extensive use of local and regional subcontractors and suppliers plus high numbers of apprenticeships and the creation of new local employment opportunities.[291]

On a third Trial Project, the Supply Chain Management Group ('SCMG') procured a two client, £200 million multi-party housing framework alliance which used Supply Chain Collaboration to create new opportunities and benefits for local residents and local businesses. The SCMG framework alliance achieved post-tender agreed 'savings averaging 14%' plus:

- 'Reduced risks, costs savings and time savings through accelerated constructor/supply chain briefing'
- 'Subcontractor/supplier innovations in proposed new materials and development of specifications' and 'Exchange of best practice between specialist competitors'
- 'Improved repair and maintenance' and 'more sustainable solutions'
- 'Development of opportunities for local tier 2/3 sub-contractors and suppliers… across 30 different disciplines'
- 'Additional employment and skills opportunities'
- Lower bid costs for other alliance members, 'specifically £719 per £1 million of turnover (under SCMG) as against £4,808 per £1 million of turnover (under the comparable traditional bid)'.[292]

The categories of improved economic and social value that can be achieved through the collaborative procurement of framework alliances, and the ways in which they can create benefits for all team members, are considered in Chapter 15.

4.9 How Can a Framework Alliance Deal with Problems?

Any dispute suggests the failure of an alliance and demands urgent attention, and a framework alliance contract should include provisions that enable the all iance members to manage risks and to avoid or resolve disputes effectively. The 2012 Effectiveness of Frameworks Report examined evidence taken from a range of Government frameworks and noted the following factors as major risks that can undermine the effectiveness of frameworks:

- 'Framework agreements that are not driven by demonstrable business need
- Framework agreements that are not designed to effectively deliver the business needs of potential clients

---

290 Cookham Wood Trial Project case study summarised in Section 14.10.
291 North Wales Prison Trial Project case study summarised in Section 14.10.
292 SCMG Trial Project case study summarised in Section 4.10.

- Framework agreements that are merely used as short cuts to market rather than a means of sustainable effective delivery
- Public sector clients engaging advisers/consultants who are not familiar with or committed to collaborative partnering processes and who promote lowest cost tendering.
- Frameworks perceived as an opportunity to generate income, sovereignty and job protective behaviours
- Frameworks perceived as a quick route to market (OJEU avoidance)
- Less expert clients believing that lowest cost tendering will deliver best value
- Less expert clients not understanding that more complex schemes may benefit from retaining some risk by the client'.[293]

A framework alliance can mitigate the impact of problems and disputes by collaborative governance and by active risk management through, for example, early identification of risk issues combined with joint decision-making and agreed actions. Alliance decision-making and risk management are explored in Chapters 12 and 18.

It is possible that an alliance member will breach the terms of a framework alliance, and this will give rise to any of the remedies for breach that are stated in the framework alliance contract or otherwise available under the applicable governing law. For example, if a UK client breaches a framework alliance contract in relation to the award of a project contract, it is arguable that another alliance member could suffer loss consequent on its unrealised expectation of work. If it can be established that this expectation is more than speculative, there will be grounds for recovery of losses,[294] generally calculated on a percentage basis according to the expectation and prospect of work being awarded.[295]

It is also possible that alliance members will not achieve their agreed objectives or targets. It is therefore important that the framework alliance contract sets out an agreed exit route, preferably after a review process that allows the alliance members to explore whether other courses of action are preferable. Clear procedures governing an open and orderly exit process should not adversely affect the parties' investment in and commitment to a long-term relationship. For example, the FAC-1 Framework Alliance Contract provides for:

- Termination of an alliance member's appointment in the event of it failing to remedy a notified breach, but only after review by a core group of agreed individuals seeking consensus on any alternative options[296]
- Termination of an alliance member's appointment in the event of it failing meet specific, agreed performance targets, but only after the core group has attempted to agree remedial actions necessary to achieve the required targets.[297]

A framework alliance contract should include provisions that enable the alliance members to avoid or resolve disputes effectively. The options for effective avoidance

---

293 Effectiveness of Frameworks (2012), 94.
294 *Allied Maples v Simmons & Simmons* [1995] EWCA Civ. 17, [1995] 1 WLR 1602.
295 For example, one of four framework contractors will be assumed to have a 25% chance of winning the relevant work.
296 FAC-1 Framework Alliance Contract (2016) clause 14.4.
297 FAC-1 Framework Alliance Contract (2016) clause 14.2.

and resolution of alliance disputes are considered in Chapter 19. For example, FAC-1 provides for non-adversarial dispute avoidance and dispute resolution by means of:

- Early warning as soon as an alliance member is aware of any dispute and its consideration by a core group, possibly supported by an independent adviser[298]
- Options to refer a dispute to conciliation or a dispute board.[299]

4.10   Framework Alliance Case Studies

The improved value created through four of the following case studies has been considered in Section 4.8. They comprise the £240 million housing programme procured by Whitefriars Housing Group, the £737 million office programme procured by Department for Work and Pensions with Land Securities Trillium, the £1.2 billion Ministry of Justice prison alliance frameworks and the £200 million SCMG Hackney and Haringey alliance frameworks.

Kevin Murray[300] led the work by the UK Ministry of Justice in pioneering the use of multi-party framework alliances that were designed to integrate the strategic appointments of their consultants and contractors, to motivate improved performance and to encourage the adoption of new technologies. These framework alliances enabled the Ministry of Justice to combine systematic early contractor involvement, collaborative working and BIM on all their projects.[301] The Ministry of Justice framework alliances and its Cookham Wood Trial Project were explored in my inaugural lecture for King's College London, 'Putting builders into prison and getting them out'.[302]

The fifth case study in this section describes a new business model that reflects the commitment of Anglian Water to integrated, collaborative working and to the development of long-term supply chain relationships. The Anglian '@one' strategic alliance created relationships aligned directly with Anglian Water's customer outcomes, as defined through a process of engagement and consultation with customers. Partners were selected against their capability to deliver these outcomes and were incentivised to deliver improvements against historic baseline performance.

Anglian and its main partners were brought together in a multi-party alliance which operates as a virtual organisation in its own right, a model that is considered in Section 7.6. The alliance team acts collectively as an alliance 'integrator', developing strategies for how the programme should be delivered and driving improvement initiatives. The partners that make up the alliance, along with Anglian Water, generate their entitlement to profit as if they were shareholders in a separate company, by outperforming historic benchmarks for delivery of outcomes.

---

298 FAC-1 Framework Alliance Contract (2016) clause 15.1, and as considered in Sections 19.7 and 19.8.
299 FAC-1 Framework Alliance Contract (2016) clause 15.2, and as considered in Section 19.9.
300 Since appointed as Deputy Director of Construction and Property, Government Property Agency, where he has led the use of FAC-1 as the basis for a range of consultant and contractor framework alliances.
301 As illustrated by the Cookham Wood and North Wales Prison Trial Project case studies and the HMP Bure case study referred to in Section 17.8.
302 This can be viewed, with an introduction by Professor David D. Caron, on YouTube at https://www .youtube.com/watch?v=khsgOT_GZzk.

The alliance manages the wider supply chain, with longer-term frameworks that develop more effective relationships and secure earlier involvement of suppliers. As with the main integrator, framework suppliers generate a return by delivering value against historic baselines, not by delivering work or providing hours. The Anglian strategic alliance is linked to a digital transformation programme, as considered in Section 14.8.

Whitefriars Housing Group Framework Alliance

Whitefriars Housing Group set up a strategic partnering arrangement with windows specialists Graham Holmes and Anglian Windows in late 2000. This programme was expanded to comprise a total of £230 million of refurbishment work undertaken by Wates Construction and Lovell.

A three-way framework alliance contract between Whitefriars, Wates and Lovell awarded annual programmes of work according to available client funding, contractor performance on previous work and main contractor capacity for further work. When the framework alliance was set up, Whitefriars did not have sufficient funding to cover its entire programme, and efficiencies were essential in order to reduce anticipated costs. Lovell had obtained cheaper prices from its long-term kitchen supplier, and Wates agreed to utilise the same kitchen supplier, with all consequent savings reverting to the client.

In addition, establishment of a steady volume of work enabled both main contractors to operate using a stable workforce and to increase their efficiency on site, for example reducing the turnaround time for the installation of new kitchens from three to two weeks per flat. Reduced time on site achieved savings in preliminaries costs of £2 million, all reverting to the client. The resultant efficiencies reduced a five-year programme to four years and saved the client **10%** of its £240 million expected cost. The client, with both constructors and in partnership with Mowlem, established the Whitefriars Housing Plus Agency which secured training opportunities for 38 people in the first year and a total of over 200 during the programme as a whole.

Association of Consultant Architects (2010), 46.

Department for Work and Pensions/Land Securities Trillium Job Centre Plus Offices

This was a fast-track programme of office adaptations undertaken through a framework alliance set up jointly by a public sector client and a private sector client, using standard designs, materials and equipment adapted to a wide variety of different buildings. The joint objective of Department for Work and Pensions (DWP) and Land Securities Trillium (LST) was to create an efficient contract structure to enable a quick start-up on site, utilising model PPC2000 processes and contract documents to streamline a nationwide programme. They subdivided England, Wales and Scotland into districts, and a contractor was appointed to undertake works in each district. DWP and LST wanted to ensure that there was a cross-pollination between districts, which was initiated through the use of a single multi-party framework alliance contract between all 14 contractors and the joint clients. DWP and LST also wanted to create a fully integrated supply chain to support the roll-out programme, with specialist framework agreements negotiated in parallel with key subcontractors and suppliers. Strict timetables were agreed to govern both the pre-construction and construction phases of each project.

The framework alliance used cost reimbursement combined with incentives that linked cost savings to the future award of work, and the final cost of £737 million against a forecast of £981 million achieved savings of **24.8%**. DWP and LST won the Building Magazine 'Integrating the Supply Chain' Award 2004; the LABC 'Services Award for Integrated Site Safety' 2004; the OGC Government Opportunities Award for 'Public Procurement Excellence' 2003; and the Building Magazine 'Health and Safety Award' 2003.

Association of Consultant Architects (2010), 28.

Ministry of Justice New Build Alliance Frameworks

[Extracts from Effectiveness of Frameworks (2012), 103, 104, regarding the Ministry of Justice £1.2 billion strategic new build and refurbishment alliances created in 2012, and regarding the previous Ministry of Justice alliances that delivered 10 000 new or refurbished prison places between 2004 and 2012.]

'Ministry of Justice (MoJ) uses three Alliance Frameworks for the delivery of newbuild and refurbishment projects (above £150k) on the MoJ estate in England and Wales. There is no upper value threshold. The alliances are as follows:

- *New build construction*. Awarded in September 2004 (maximum term of 11 years)
- *Refurbishment*. Awarded in February 2005 (maximum term of seven years)
- *Consultancy*. Awarded in November 2003 (maximum term of 11 years).

There are eight suppliers appointed to the New Build Alliance, eight suppliers appointed to the refurbishment Alliance (four of these are also on the New Build Alliance) and 14 principal suppliers appointed to the consultancy Alliance.

There are more than 200 suppliers registered in the supply chains of the Alliance suppliers. The MoJ has the ability to influence the supply chain to incorporate local suppliers and SMEs. The MoJ manages the Alliance and contracts directly with the suppliers as projects are called off. Individual project appointments are made via mini-competitions or by direct appointment.

There is a standard two-stage selection process that enables the MoJ to comply with all relevant procurement legislation and enables early contractor involvement in projects. Selections are based on project specific criteria in line with the principles of the Alliance and are based on quality and cost criteria. A first stage selection process (to identify the preferred supplier) can be completed within as little as three weeks, but normally takes six to eight weeks (saving at least six to nine months in comparison with tendering each project via an OJEU process).

Early engagement of the supply chain is encouraged by the two-stage approach. This serves to gain contractor and supply chain input into design, ensure cost robustness and appropriate risk management strategies for all projects. This increases the certainty of delivery on time and budget and the quality of the overall product. The framework

(Continued)

enables the department to react quickly to emerging procurement requirements. It is estimated that risk has been mitigated by over £2 million over the frameworks' operation.

The MoJ uses the PPC2000 standard form of Partnering Contract that has minimal amendments. A standardised suite of processes and contract templates are used to ensure consistency and ease of use by the project team. An Agreed Maximum Price for each project is agreed between the Alliance Supplier and the MoJ. Key performance information and cost analysis data is collected for all schemes and is made available to all schemes and the Cabinet Office.

The Alliance has a solid governance structure through a 'Strategic Core Group' comprising representatives from the MoJ and the Alliance suppliers. Information on the delivery pipeline and updates on the MoJ ways of working, challenges, initiatives, etc. are discussed as part of 'Strategic Core Group' meetings… A 'Core Group' comprising representatives from the MoJ and the Alliance suppliers deal with any issues that may arise on projects as part of a defined structured hierarchy for project governance applicable to each project.

The major benefits of operating the Alliance Framework include reduced procurement costs estimated at £10 million, reduced burden on industry tendering of around £30 million and procurement risk mitigation (as stated above) of about £2 million. This suggests a total framework operation cost saving in the order of **£42 million** to industry and the department.

Hackney Homes and Homes for Haringey (SCMG) Housing Framework Alliance

Hackney Homes and Homes for Haringey (together SCMG) used Two Stage Open Book to deliver their £200 million housing framework alliance, creating a multi-client, multi-contractor framework alliance team working with main contractors Mulalley, Keepmoat, Mansell, Lakehouse, Lovell and Wates, and with a wide range of SME subcontractors and manufacturers under a standardised system of costing and long-term engagement that created major savings and significant qualitative benefits.

The agreed procurement route combined Two Stage Open Book with programmed systems of Supply Chain Collaboration. Main contractors were selected to join multi-party framework alliance contracts which incorporated contractual processes for joint working with suppliers and subcontractors. Through these systems the SCMG clients, main contractors and subcontractors, suppliers and manufacturers including Veka, Bauder, Sovereign and Birchcroft worked collaboratively as integrated teams to deliver cost savings, improved employment and skills outputs, extended warranties and more sustainable solutions. The full range of subcontractors and suppliers brought into the SCMG system covered 30 disciplines.

Cost savings achieved included **16.5%** by Hackney on its 2010 framework alliance procurement, plus further savings averaging **14%** achieved by Homes for Haringey and Hackney Homes by agreement with other alliance members and with subcontractors, suppliers and manufacturers through the application of the SCMG processes from 2010 to 2013.

Examples of the qualitative benefits achieved as a result of early involvement of all supply chain members include:

- Transparent and shared development of standard specifications and basket rates, leading to more efficient pricing and better value
- Reduced risks, cost savings, and time savings through accelerated constructor/supply chain briefing process
- Subcontractor/supplier innovations in proposed new materials and development of specifications, such as future-proofing green roofs at no additional cost and upgrading windows from Grade C to Grade A at no additional cost
- Improved quality control through manufacturer attendance on site
- Exchange of best practice between specialist competitors
- Availability of extended warranties above industry standards, managed by suppliers/installers, such as windows warranted for 30 years
- Improved repairs and maintenance through, for example, self-cleaning glass on high-rise blocks
- Time savings, such as through quicker build-up of prices leading to earlier start on site and reduced client/consultant time/costs
- Additional employment and skills opportunities for individuals, for example 46 new apprenticeships over the first 18 months of the Hackney programme.

SCMG Trial Project case study.

Anglian Water@ One Alliance

Anglian Water provides water services to an area stretching from the Thames Estuary to the Humber. It provides 1.2 billion litres of water a day, to 6 million customers, through 112 000 km of pipe and 1257 water treatment sites. Anglian Water is responsible for maintaining and improving this network, delivering high quality drinking water to their customers and recycling the resulting waste water. Anglian Water's programme is defined in five year Asset Management Periods (AMPs), typically investing £4.5 billion in each AMP, covering replacement and refurbishment of above- and below-ground assets. Historically, in line with infrastructure generally, projects were delivered in a largely transactional manner, with Anglian tendering works and selecting the most economically advantageous proposal. AMP 3, while successfully delivering the required outputs through a partnering approach, was felt to be less effective than it could have been. With AMP4 requiring a further focus on efficiency, effective solutions and customer service, Anglian Water decided to shift to a different delivery model, developing both its capability as an asset owner and more effective relationships with its partners. A review of best practice across different sectors and an identification of the underpinning best practice characteristics, led to Anglian adopting a strategy based on a more integrated and collaborative working and the development of long-term supply chain relationships.

These relationships were aligned directly with Anglian Water's customer outcomes, which had already been defined through a process of engagement and consultation with

(Continued)

customers. Partners were selected against their capability to deliver these outcomes and incentivised to deliver improvements against historic baseline performance. Anglian and the main partners were brought together in an alliance. This alliance, as an integrated and collaborative organisation, was engaged at outcome level, not at project or scope, giving partners and the wider supply chain the opportunity to develop more innovative solutions and to challenge standards.

The alliance team is the integrator, developing strategies for how the programme should be delivered and driving improvement initiatives. The partners that make up the alliance, along with Anglian Water, are shareholders that generate a return by outperforming historic benchmarks for delivery of outcomes.

All parts of the alliance work collaboratively, taking a best for task approach to the development of integrated teams. The alliance manages the wider supply chain, with a longer-term framework used to develop more effective relationships and secure earlier involvement of the right suppliers. As with the main integrator, framework suppliers generate a return by delivering value against historic baselines, not by delivering work or providing hours.

An example of the alliance acting as integrator was in the development of product-based delivery. The alliance was able to shift from the historic project approach, recognising the opportunity to translate repeatability within the programme into standard products and components and to use a 'product catalogue' as the starting point for delivering the programme. This created significant value when compared to a previous approach that amplified variance and redesign at all levels, including unique project led solutions and multiple variations of critical components.

This was allied with a digital transformation strategy which has seen the alliance design and build everything virtually, including rehearsing and optimising construction in virtual rehearsal suites before going to site. Not only has this shifted delivery from construction to assembly, it has provided health and safety benefits through off-site construction of products. Digital transformation has also led to more effective engagement with users and operators, with greater involvement in the virtual development of solutions improving operability and operator buy-in. The progress of digital rehearsal demonstrates the value in delivering through integrated teams, where all the influential parts of the wider supply chain are involved in optioneering and solution development.

The alliance has established a strong track record, delivering significant improvements in cost, carbon and time. Anglian Water's future plans commit to further development of aligned and collaborative relationships with its supply chain partners.

Reported to KCL Centre of Construction Law by Dale Evans and Chris Candlish of the Anglian @One alliance.

5

How Does a Term Alliance Operate?

5.1 Overview

Operation, repair and maintenance are fundamental to the use, safety and financial viability of all completed projects. The capability of the construction industry to integrate these types of works and services with capital projects has been demonstrated through the ongoing asset management that forms part of many public private partnerships.[303] It is therefore surprising that capital project works often remain disconnected from post-completion operation, repair and maintenance, and that little attention is given to the scope for the latter to provide a basis for innovation and collaborative working.

Term contracts govern long-term commitments to undertake a range of activities such as:

- Responsive activities with varying call out times according to the urgency of the problem, for example in terms of its impact on health and safety
- Cyclical activities such as decoration of a building or servicing of equipment
- Planned activities such as improvements and upgrades.

The potential duration of a term contract offers scope for the development of systems that recognise agreed objectives, for the parties to agree joint and individual activities that improve the prospect of achieving those objectives and for the measurement of success according to agreed targets. Where this collaborative approach is adopted, a term contract can be developed into a term alliance.

Term alliances can motivate the parties by identifying their common commercial goals and objectives, including the prospect of a continued course of business based on accepted levels of performance and pricing. Continued workflow is a major incentive for collaborative working, and a transparent system of performance measurement under a term alliance enables the client to justify the continued issue of orders.

There are specific challenges in the performance of operation, repair and maintenance activities that justify collaborative joint risk management under a term alliance, for example a constrained environment in which 'construction companies are forced to work on tight, often occupied sites where it is difficult to anticipate what complications they may uncover, and where it is hard to work at scale and with a high degree of standardization'.[304]

---

303 PFI projects and other public/private partnerships are considered in Section 8.9.
304 McKinsey Global Institute (2017), 43.

Collaborative Construction Procurement and Improved Value, First Edition. David Mosey.
© 2019 John Wiley & Sons Ltd. Published 2019 by John Wiley & Sons Ltd.

Term alliances also have an important role to play in supporting a collaborative procurement model for whole life asset management, creating integrated systems through which 'whole life asset performance will shift focus from the costs of construction to the costs of a building across its life cycle, particularly its use of energy'.[305]

This chapter will consider the characteristics of a term contract, the features of a term alliance and the ways in which a term alliance contract can deal with some of the most pressing issues that challenge the safety and efficiency of construction procurement.

5.2   How Does a Term Contract Operate?

A term contract creates a system for ordering any type of works, services or supplies. In the context of the built environment, a term contract often governs the repair, maintenance and operation of a completed project. It can also govern the call-off of goods, materials, equipment and any other capital project components manufactured off-site. The required activities vary widely in complexity and in scope, and operational providers will be aware that clients are seeking to minimise their scope and costs.

The basic pricing model for a term contract is often a schedule of rates, either inclusive gross rates or net rates with separate pricing of profit and overheads.[306] These rates are likely to be subject to indexation during the life of a term contract.[307] Alternative pricing models may provide for yearly or monthly lump sums, or for lump sum unit prices, all of which may be periodically adjustable and all of which are usually designed to reduce the administration of payments for individual orders. To calculate lump sum prices under a term contract, particularly those relating to responsive activities, requires bidders to base their expected pipeline of work, and their related risk assessments, on the client's historic data and/or predictions as to the relevant volumes and types of work. The reliability of this historic data may depend in part on reports provided to the client by the outgoing provider.

Although term contracts typically provide for no guarantee of continuous work and govern activities that will be undertaken in the stages and sequences that the client instructs in its orders, they include implied collaborative obligations on the part of the client:

- Not to hinder or prevent the provider from carrying out its obligations in a regular/orderly manner, subject to any express power of the client to regulate timing and continuity
- To take all steps in its power reasonably necessary to enable the provider to carry out its obligations in a regular/orderly manner, again subject to any express power of the client to regulate
- To provide or arrange for such full, correct and coordinated information as is known to the client or reasonably should have been known, in such a manner and at such times as are reasonably necessary to enable the provider to carry out its obligations.[308]

---

305 BEIS (2018), 7.
306 As considered in Section 16.7 together with other term alliance pricing options.
307 For example, under JCT MTC (2016) clause 5.6 (Rates-Fluctuations) and TAC-1 Term Alliance Contract (2016) clause 4.2 (Fluctuations).
308 *Scottish Power v Kvaerner Construction* [1999] SLT 721.

Cyclical and responsive works of the type included in term contracts are often undertaken in occupied buildings or while other infrastructure is in active use. These constraints suggest the need for an alliance approach that takes account of the interests of stakeholders. For example, it is important that term contracts provide expressly for non-exclusive possession and limited access so as not to place the client in breach of implied obligations to the provider.[309]

A significant feature in the procurement and contracting for operation, repair and maintenance is the point at which a term contractor or other provider assumes responsibility under its contract, either following handover from the capital works team or from a previous provider, for which purpose:

- There are usually no implied links or representations made between the capital works team and the operation, repair and maintenance contractor, and these links have to be built into the capital works procurement as well as the operation, repair and maintenance procurement
- Handover from one term provider to another will occur at a single moment in time[310] and will benefit from contract terms setting out a preparatory process so as to ensure that the new provider is ready to assume its responsibilities.

Provision of a continuous service requires the prior mobilisation of staff, equipment and supply chain members together with the handover of employees from any previous provider,[311] the handover or interface of all relevant IT systems and a practical process of familiarisation with the relevant assets and their occupants. The importance of this continuity was underlined by the Hackitt Report in its focus on the need for a 'golden thread of good quality information',[312] yet most term contract forms do not deal with these mobilisation and handover interfaces at all.[313]

5.3 What Are the Term Contract Forms?

There are a range of UK published forms of term contract:

- The Infrastructure Conditions of Contract (ICC) 2011 Term Version[314]
- The Joint Contracts Tribunal (JCT) 2016 Measured Term Contract (MTC)[315]
- The NEC4 Term Service Contract (TSC)[316]

---

309 The licence granted to a contractor which implies that is it entitled to unlimited access to a site is a serious obligation on the part of a client and is often neglected. This implied obligation was considered in, for example, *Hounslow London Borough Council v Twickenham Gardens Development* [1971] Ch 233.
310 For example, midnight on 31 December.
311 For example, under the UK Transfer of Undertakings (Protection of Employment) Regulations 2006 and related pensions provisions.
312 Building a Safer Future (2018), 35.
313 An exception is TAC-1 Term Alliance Contract (2016), clause 14.9, which governs handover from one provider to another.
314 ICC Term Version (2011).
315 JCT MTC (2016).
316 NEC4 TSC (2017). There is also an NEC4 (2017) Design Build and Operate Contract, published for the first time in 2017, which covers the provision of works and related services.

- The TPC2005 Term Partnering Contract[317]
- The TAC-1 Term Alliance Contract.[318]

There is no FIDIC standard form term contract.

The ICC Term Version Guidance Notes suggest that it 'should be suitable for planned and reactive maintenance or refurbishment work as well as for new work and emergency works'.[319] However, certain of its provisions appear to be suitable only for large and complex tasks, such as those clauses governing ground conditions,[320] task final accounts,[321] liquidated damages,[322] and certification of task completion.[323]

The JCT 2016 Measured Term Contract includes the following provisions:

- Submission of provider proposals for sustainable development and environmental considerations, but with no agreed share of benefits and with no arrangements for supply chain members to participate[324]
- Performance indicators and monitoring, but with no provider incentives[325]
- Submission of provider proposals for cost savings and value improvements, but with no agreed share of benefits and with no arrangements for supply chain members to participate[326]
- Collaboration in terms that follow broadly the terms of other JCT 2016 contracts.[327]

The NEC4 Term Service Contract includes the following provisions:

- The options of fixed prices or target costs or cost reimbursement[328]
- Joint risk management through 'early warning meetings' equivalent to the systems in other NEC4 contracts[329]
- 'Key Performance Indicators (KPIs)' linked to an 'Incentive Schedule'[330]
- Collaboration in terms that follow broadly the terms of other NEC4 contracts[331]
- Submission of provider proposals for reduced operating and maintenance costs but with no arrangements for supply chain members to participate.[332]

An order under a term contract may require a task to be completed by the provider alone and could cover all the design, supply and construction activities that the provider may consider the task requires. Under English law, the acceptance of these combined responsibilities can a imply an obligation on the part of a provider that goes beyond

---

317 TPC2005 (2013).
318 TAC-1 Term Alliance Contract (2016).
319 ICC Term Version Guidance Notes (2011), 1.
320 ICC Term Version (2011) clause 12.
321 ICC Term Version (2011) clause 60(4).
322 ICC Term Version (2011) clause 47.
323 ICC Term Version (2011) clause 48.
324 JCT MTC (2016) Supplemental Provisions 4.
325 JCT MTC (2016) Supplemental Provisions 5.
326 JCT MTC (2016) Supplemental Provisions 3.
327 As considered in Section 9.6.
328 NEC4 TSC (2017) Options A, C and E.
329 NEC4 TSC (2017) clause 15, as considered in Sections 18.4 and 19.6.
330 NEC4 TSC (2017) Option X20.
331 As considered in Section 9.7.
332 NEC4 TSC (2017) Option X21.

the usual test of its reasonable skill and care[333] and that instead obliges the provider to produce a result that is fit for its intended purpose.[334]

In these circumstances, clear wording is required in a term contract if it is intended to reduce this implied duty to one of reasonable skill and care. This clarification can be an important limitation on the provider's liability, particularly if the relevant design activities are supported by professional indemnity insurance where policy terms can provide that a fitness for purpose warranty invalidates the cover. This clarification appears in the TPC2005 and TAC-1 contracts considered in Section 5.4[335] but, surprisingly, not in the NEC4 Term Service Contract.[336] Term contracts can govern design activities that require the support of professional indemnity insurance, for example where these design activities relate to the provision of modular components or other off-site manufacture.

5.4 What Is a Term Alliance?

A term alliance should make clear the purposes of the term programme, through provision for objectives and success measures agreed by all alliance members. These provisions should recognise not only the objectives of the alliance and the term programme, but also the objectives of each alliance member in relation to the alliance and the term programme. A term alliance should provide for:

- Shared objectives, success measures, targets and incentives
- Systems through which members build up and manage shared design, time, cost and operational data
- Activities through which members can improve economic or social value
- Clear timeframes and deadlines for alliance members' activities, responses and approvals
- Systems for joint management of risks and the agreed avoidance and resolution of disputes.

The ICC 2011, JCT 2016 and NEC4 term contracts do not set out to create or support a term alliance as they omit all or most of these features. The standard form term alliance

---

333 Under the Supply of Goods and Services Act 1982, Section 13, where a supplier of services is acting in the course of a business, there is an implied term that the supplier will carry out the services with reasonable care and skill. This is an objective test consistent with the findings in *Bolam v Friern Hospital Management Committee* [1957] 1 WLR 582 (QB) and *Greaves v Baynham Meikle* [1975] 1 WLR 1095 (CA).

334 Reasonable fitness for their intended purpose is an implied term under the Supply of Goods and Services Act 1982, Sections 4(5) and 4(6), in respect of goods transferred in the course of a business where the required purpose for which they are provided is made clear, except in circumstances where the client does not rely, or it is unreasonable for him to rely, on the skill or judgement of the provider. Construction responsibilities that combine design, supply and construction can be treated as goods pursuant to, for example, the findings in *Viking Grain v TH White Installations* [1985] 33 BLR 103 (QB) and *Tesco v Costain* [2003] EWHC 1487 (TCC).

335 For example, the TAC-1 Term Alliance Contract clause 10.1 provides that all team members have a duty of 'reasonable skill and care appropriate to their respective roles, expertise and responsibilities as stated in the Term Documents'.

336 Most NEC4 contracts include an option at NEC4 X15 to reduce the duty of care to reasonable skill and care, but this option is missing in NEC4 TSC (2017).

contracts are the TAC-1 Term Alliance Contract and its predecessor the TPC2005 Term Partnering Contact. TPC2005 has been used successfully on a wide range of term programmes including planned, cyclical and responsive works and/or services and/or supplies to support housing stock, offices, schools, highways and hotels.[337] TPC2005 and TAC-1 can govern the appointment of consultants, main contractors, subcontractors, manufacturers, suppliers and operators.

The selection and mobilisation processes for the members of a term alliance provide opportunities to test their collaborative commitments and to search for improved value. In its tender for a £27 million term alliance contract for responsive and void repairs to 7000 properties, AmicusHorizon Housing Group (now Optivo):

- 'Focussed on shared service delivery, innovation and communication. In this way, strategic partnering issues were clearly addressed, such as pain and gain incentive models, clear dispute resolution procedures and a partnering timetable'
- Arranged for 'integration of the TPC form of contract with the tender documentation to ensure appropriate evaluation of tenders'.[338]

Marian Burke, who led a term alliance procurement at AmicusHorizon, reported that 'it was vital that we worked against a formalised partnering contract. By electing to use a TPC contract from the beginning we could also use this to shape our tender documentation. It ensured that we did not miss critical elements, such as the development and use of risk registers, as they are an integral part of the contract'.[339]

A term alliance contract should describe the agreed basis for determining the success of the alliance and the term programme in achieving its stated objectives, and for measuring the performance of the alliance members. It should state the targets set for each success measure, including the method of recording relevant data, the alliance member responsible for measuring against that data and the system for reporting to the other alliance members.

Because it describes a series of tasks, a term alliance contract can link its agreed targets to a wider range of incentives, for example:

- Additional payments including shares of savings achieved through alliance activities
- Adjustment of the scope of the term programme, or of any exclusivity granted to the provider
- Extension of the duration of the term alliance contract.[340]

It is important for providers to understand how objectives and success measures affect the future award of work, with clarity as to which targets are so important that a failure to meet them will require urgent action and may ultimately determine whether a term alliance appointment may be terminated.[341]

---

337 Association of Consultant Architects (2010).
338 Association of Consultant Architects (2010), 9.
339 Association of Consultant Architects (2010), 9.
340 TAC-1 Term Alliance Contract (2016) guidance note in Schedule 1 Part 3.
341 For example, as set out in TAC-1 Term Alliance Contract (2016) Schedule 1 and clause 14.2.

5.5   How Does a Term Alliance Award Work?

A term alliance contract should provide clarity as to the ways in which work will be awarded, by means of:

- The preparatory processes that are preconditions to the award of work[342]
- The procedure governing the award of work[343]
- The order form by which work is instructed[344]
- Whether or not the client offers any exclusivity or minimum value of work.[345]

Orders may be in electronic or paper form, and a term alliance contract should set out:

- The agreed form of communication that constitutes an order
- How the parties provide evidence of the authority of signatory/issuer of the order.

The contractual significance of the order form is underlined by the fact that a term contract usually creates no contractual obligations to perform tasks or pay money until an order is issued. An order needs to contain all the information in relation to a particular task, such as its specification, price and deadline for completion. If this is not possible, the parties need to consider whether it is appropriate to issue a task order based on incomplete information without risking the uncertainty and unenforceability of an agreement to agree.[346]

Where early contractor involvement is enabled by a two-stage order form under a term alliance contract, it should set out a binding step by step process building on pre-agreed contract details through which the parties complete relevant task designs, supply chain details, prices and timetable before authorising implementation of the relevant task. Any uncertainty in a two-stage order under a public sector term contract could inadvertently convert it into a framework contract and place it subject to additional regulatory constraints such as a maximum four-year duration.[347]

The variable nature of responsive and cyclical works may create a challenge in terms of how to prescribe appropriate times for commencement and completion of different types of task. Clients can set specific timeframes for individual task types, but this is appropriate only where the nature of the tasks is capable of being commoditised in this way. An alternative is to establish task times in each order form, but this will mean either that the provider is expected to accept unilaterally expressed requirements of the client or that it is able to reject them and leave this key issue as a matter to be agreed.

---

342 For example, TPC2005 (2013) clause 6.1 and TAC-1 Term Alliance Contract (2016) clause 5.1.

343 For example, TPC2005 (2013) clause 6.2 and Appendix 6, and TAC-1 Term Alliance Contract (2016) clauses 5.2 and 5.3 and Schedule 4.

344 For example, TAC-1 Term Alliance Contract (2016) Appendix 3. None of the ICC 2011, JCT 2016 or NEC4 term contract forms include a pro forma order form for calling off tasks.

345 TPC2005 (2013) clause 6.11 and TAC-1 Term Alliance Contract (2016), clauses 5.6 and 5.7.

346 As considered in Sections 2.7 and 11.4.

347 Public Contract Regulations 2015, Reg. 33(3).

5.6 How Can a Term Alliance Use Supply Chain Collaboration?

Over the life of a term alliance, cost savings and other improved value can be achieved through Supply Chain Collaboration by joint processes which revisit the original design, cost, risk and time assumptions made by a provider so as to test and review:

- Whether improved mutual commitments can be achieved, in conjunction with activities that generate additional integration, information and innovation
- Whether these improved mutual commitments and activities can secure savings and other improved value in specific work, service and supply packages and in the term programme as a whole.

Supply Chain Collaboration requires agreement of planned processes, supported by contractual commitments and timescales agreed with each supply chain member. For example, on the Project Horizon highways term alliance 'Surrey County Council has worked successfully with its tier 1 contractor Kier and tier 2/tier 3 supply chain members Aggregate Industries and Marshall Surfacing to achieve substantial savings, improved quality and an integrated team culture'.[348]

Over the duration of a term alliance contract, Supply Chain Collaboration can create increased rewards for alliance members where:

- Providers and their supply chain members are incentivised by the prospect of a continued pipeline of work
- The pipeline of work increases the prospect of alliance members achieving and exceeding success targets
- Alliance members capture lessons learned and pass them on from task to task
- Clients measure and reward continuous improvement
- Joint commitment to new processes is replicated and becomes business as usual.

It is also possible to build into a term alliance a collaborative mobilisation procedure under which, after contract award, the alliance members go through a number of joint preparatory steps leading up to the date and time when the new provider takes over responsibility for the agreed works, services or supplies. During this period the alliance members undertake agreed alliance activities that can include early Supply Chain Collaboration designed to ensure that each of them, and any outgoing capital works team or outgoing operation, repair and maintenance provider, have done all that is necessary for the issue and implementation of orders under the new term alliance contract.[349]

---

348 Project Horizon Trial Project case study: 'Surrey's vision is to demonstrate that the same techniques as those used on Project Horizon can achieve comparable savings and benefits for other local authorities through creation of a Supply Chain Collaboration Toolkit pursuant to the Highways Maintenance Efficiency Programme'. This was duly published as the HMEP Supply Chain Collaboration Toolkit (2014).
349 For example, TPC2005 (2013) clause 6.1 and TAC-1 Term Alliance Contract (2016) clause 5.1 provide for the satisfaction of agreed preconditions before the term programme is implemented and state that no orders will be issued until these preconditions have been satisfied.

5.7 How Can a Term Alliance Improve Asset Management?

Where a term alliance governs a combination of planned, responsive and cyclical tasks, it is particularly valuable as a procurement model for long-term asset management. By adopting an integrated and informed approach to asset management:

- Capital improvements can reduce responsive and cyclical costs
- Repair and maintenance can postpone or reduce the risk of failure of project components and can avoid or delay the need for additional capital expenditure.

Most UK procurement and delivery models are still based on the construction phase reaching completion and focus on when and how designers and contractors are released from liability. These capital project procurement and delivery models do not provide for repair, maintenance, and operation and rarely incentivise the efficiency of these ongoing activities. As the capital project teams step away, their view seems to be 'après moi le deluge', possibly literally. As to whether clients adopt procurement structures that enable and require a whole life approach to asset management, commentators have observed that 'there are many public sector procurement departments who fail to take account of both capital and revenue expenditure. This frustrates the main purpose of public sector procurement – to appoint the most economically advantageous bid over the lifetime of the project'.[350]

Successful asset management depends on creating interfaces between the capital works team and the operation, repair and maintenance team. These interfaces give rise to a number of provisions that should appear in capital works contracts and should be mirrored in all related term contracts:

- Suitable intellectual property rights licences in respect of Building Information Modelling (BIM) models and other design documents that enable the operation, repair and maintenance team to access and use all available asset data
- A clear interface between the defects liability obligations of the capital works team and commencement of the obligations of the operation, repair and maintenance team, with clarification as to who responds to notification of a problem and at whose expense
- An understanding of all exclusions and limitations in the liability of the design and construction team, including all specialist subcontractors, suppliers and manufacturers, so that it is clear where the operation, repair and maintenance team must step in to avoid leaving any gaps in the service
- Availability to the operation, repair and maintenance team of information regarding plant and equipment warranties, including the terms and conditions of those warranties, so that the operation, repair and maintenance team do not invalidate them through any act or omission
- A clear understanding of the specific obligations of the operation, repair and maintenance team in relation to warranted plant and equipment
- Clarity as to the effect on the liability of the capital works team, including subcontractors, suppliers and manufacturers, in the event of an error or omission by the operation, repair and maintenance team.

---

350 Never Waste a Good Crisis (2009), 21.

Term alliance case studies illustrate how collaborative procurement can improve the repair, maintenance and operation of completed buildings. For example, the Welwyn Hatfield approach to collaborative procurement was recognised in the Institute of Building Management Awards, and their 'many examples of good practice' attracted praise from the Council's District Auditor.[351] They created a term alliance with Mears for the repair of 9500 homes, developing a new culture under which:

- 'The shake-up in performance had clear financial advantages beyond the social benefits of housing more people and cutting the time they spend in temporary accommodation'
- 'For example, the capital cost of increasing housing capacity (by building more houses instead of shortening the time properties are empty) would have been about £10 million'
- 'Customer satisfaction, independently measured by MORI and monitored by a tenants' panel, showed solid improvements in landlord service, value for money and quality of repairs and maintenance'
- 'Viewings accompanied by Mears increased the number of tenants accepting the first property offered from 30% to 80%'
- 'Void turnaround time more than halved, leading to quicker lettings and increased rent receipts'
- 'Under Open-book pricing the annual cost increase for maintenance jobs ran below inflation'.[352]

Trial Projects such as those led by Surrey County Council,[353] Hackney Homes and Homes for Haringey[354] have combined capital and responsive works, with direct client/subcontractor relationships giving rise to extended warranties. In addition, the potential of BIM is increasingly focused on supporting a whole life model,[355] for example through the UK Government Soft Landings initiative.[356]

More imaginative approaches to the pricing of term alliances can be crafted where the same provider is also responsible for capital works and is incentivised to complete the capital works in a way that reduces ongoing responsive and cyclical expenditure. Appropriate incentives could be, for example, a share of savings and/or the award of additional capital works, in each case linked to anticipated reductions in operation, repair and maintenance costs.

5.8   How Can a Term Alliance Improve Value?

As the client, provider, supply chain, and stakeholders all build greater confidence and become familiar with each other's expectations and methods of working under a term

---

351 Association of Consultant Architects (2010), 45. This successful collaborative procurement was led by Peter Sharman, a member of the UK Alliance Steering Group.
352 'The reduced void turnaround time has enabled more families to move in and the tenants really like the viewings which are much friendlier', Peter Sharman, then Deputy Chief Housing Officer, Welwyn Hatfield Council, Association of Consultant Architects (2010), 45.
353 Project Horizon Trial Project case study.
354 SCMG Trial Project case study.
355 Digital Built Britain (2015) and as considered in Section 13.6.
356 As considered in Section 13.6.

alliance, it is likely that new information will become available that creates opportunities for improved efficiency. Where incentives can be offered such as an extended term or greater exclusivity or additional remuneration, the alliance members can calculate the benefit of making efforts to identify these opportunities and to share them with other members:

- NEC4 TSC provides several options for incentives and performance-related payments but also for deductions[357]
- TPC2005 provides for agreed shared savings, shared added value, pain/gain shares, links between payment and achievement of KPI targets, and performance-related benefits for employees[358]
- TAC-1 provides incentives that can include additional payments such as shares of agreed savings, adjustment to the scope or exclusivity of the term programme and extension of the duration of the term alliance contract linked to the measurement of improved value by reference to agreed Targets and Success Measures.[359]

Term alliances offer the potential stability of a long-term term relationship with provision for works, service and supplies to be called off under a straightforward system. Their potential to encourage improved value has been demonstrated on a wide range of programmes. For example:

- On a £200 million Royal Borough of Greenwich housing term alliance:
  - 'Performance for the first full year showed immediate improvements with time to re-let vacant properties down by 23% from 43 to 33 days, average time for non-urgent repairs down from 18 to 11 days, complaints down by 50% and post-job satisfaction scores of 95%
  - "Open-book" accounting and supply chain management led to savings to such an extent that the client agreed to extend the scope of work to compensate for falling contract values'.[360]
- Maidstone Housing Trust created a £20 million term alliance programme for improvement of existing housing stock by VINCI Facilities Building Solutions and Symphony Group, reporting that:
  - 'Cost and time efficiencies were key to the client so that savings could be reinvested in the programme, but the works were complex and a robust structure was needed if these efficiencies were to be achieved
  - Working closely with the client team, VINCI Facilities identified alterations that could be made to the properties to meet design specifications and minimum storage requirements
  - They also found alternative materials to those specified by the client that matched longevity and durability and gave rise to time and cost savings, sufficient for an additional four properties to be upgraded
  - The team was tasked with reducing cost and time. Moving away from the traditional schedule of rates and costs per property, a mechanism of three fixed costs

---

357 NEC4 TSC (2017) Option C, Option X12.4 and Option X17 'Low service damages' and Option X20.
358 TPC2005 (2013) clause 2.6 and Appendix 3 Part 2.
359 TAC-1 Term Alliance Contract (2016) clauses 2.3 and 5.7 plus Schedule 1 Part 2 and Part 3.
360 Association of Consultant Architects (2010), 19, as summarised in Section 5.10.

was developed that covered all types of property, facilitating greater transparency and planning, and significantly reducing overheads'.[361]

Implementation of Two Stage Open Book and Supply Chain Collaboration, pursuant to the agreed processes set out in term alliance contracts, has enabled:

- Agreement by a term alliance comprising Surrey County Council, Kier, Marshall Surfacing, and Aggregate Industries of savings averaging 12% against previously tendered rates, sustained over five years through to completion of the programme of work in 2018 and combined with extended warranties, local business opportunities and measures to improve sustainability[362]
- Agreement by a term alliance comprising Kier, Surrey and a range of subcontractor supply chain members of savings averaging 8% against previously tendered rates, combined additional apprenticeships and other measures to improve social value.[363]

The full range of improved economic and social value that can be achieved through term alliances and other collaborative procurement models is considered in more detail in Chapter 15.

5.9   How Can a Term Alliance Deal with Problems?

It is possible that the term alliance members will not achieve their agreed objectives or targets. It is therefore important that the term alliance contract sets out an agreed exit route, preferably after a review process that allows the alliance members to explore whether other courses of action are preferable. Clear procedures governing an open and orderly exit process should not adversely affect the parties' investment in and commitment to a long-term relationship.

For example, TAC-1 provides for:

- Termination of an alliance member's appointment in the event of of it failing to remedy a notified breach, but only after review by a core group who are required to consider any alternative options[364]
- Termination of an alliance member's appointment in the event of it failing to meet specific, agreed performance targets, but only after the core group has attempted to agree remedial actions necessary to achieve the required targets.[365]

However, if the alliance members agree to a contract clause permitting termination without default,[366] then this right of termination cannot be constrained on the basis that it was not exercised in good faith.[367] For example, a right to terminate at will has been upheld notwithstanding TPC2005 clause 1.1 establishing a 'spirit of trust, fairness

---

361 Association of Consultant Architects (2010), 32.
362 Case study summarised in Section 5.10 and additional data in Section 11.5.
363 Case study summarised in Section 11.10.
364 As set out in TAC-1 Term Alliance Contract (2016) clause 14.4.
365 TAC-1 Term Alliance Contract (2016), clause 14.2.
366 For example, by adding this right under TAC-1 clause 14.1.
367 The impact of implied and express good faith is considered in Sections 7.4 and 7.5.

and mutual co-operation' and notwithstanding the parties' commitment 'in all matters governed by the Partnering Contract to act reasonably'.[368]

The sources of problems and disputes under a term contract are less likely to derive from uncertain or incomplete information, because a contractual system based on the issue of orders is predicated on an agreed brief and agreed prices against which orders can be called off. That said, it is possible that errors or omissions in the brief can emerge over successive orders and also that the orders themselves may lack the necessary detail or accuracy. Where problems under a term contract give rise to claims for additional time or money, the standard forms treat these as follows:

- NEC4 TSC provides for 18 compensation events[369] compared to 14 under the NEC3 edition, plus related time and money claims,[370] although it removes the previous seven compensation events that were provided for under the NEC3 edition at Option X19.10 by reference to individual task orders
- JCT MTC (2016) provides for an extension of time if the provider is unable to complete a task for 'reasons beyond his control (including compliance with any instruction of the Contract Administrator that does not arise from the Contractor's default)'[371] but is silent on additional cost
- TPC2005 and TAC-1 provide for an extension of time in respect of any matter beyond the reasonable control of the provider and any of its supply chain members unless restricted to any listed matters.[372] Both forms provide an option to agree payment for matters giving rise to an extension of time but limit these payments to, for example, 'amounts properly incurred in respect of unavoidable work and expenditure and time-based Site Overheads, but not in any event any additional Profit or Central Office Overheads or any loss of profit on other activities or any other additional payment of any kind'.[373] This leaves it for the parties to make express provision for additional payment entitlements dependent on whether this is practical or appropriate in the context of the type of operation repair and maintenance activities.

Any dispute suggests the failure of an alliance and demands urgent attention. Early warning of a potential problem or dispute to a consensus-based governance group can avert the need for a more formal dispute resolution procedure. A term alliance can mitigate the impact of problems and disputes by active risk management through, for example, early identification of risk issues combined with joint decision-making and agreed actions.[374]

It is possible that an alliance member will breach the terms of a term alliance, and this will give rise to any of the remedies for breach that are stated in the term alliance contract or otherwise available under the applicable governing laws. However, under

---

368 *TSG Building Services v South Anglia Housing* [2013] EWHC 1151 (TCC) was decided by reference to the TPC2005 form of contract. This case is considered further in Section 7.5.
369 NEC4 TSC (2017) clause 60.1.
370 NEC4 TSC (2017) clauses 62 and 63.
371 JCT MTC (2016) clause 2.10.
372 TAC-1 Term Alliance Contract (2016) clause 9.6 and TPC2005 (2013) clause 8.4.
373 TAC-1 Term Alliance Contract (2016) clause 9.7 with similar wording in TPC2005 (2013), clause 8.5.
374 Alliance decision-making and risk management are explored in Chapters 12 and 18.

English law, there is no additional implied obligation of 'trust and confidence' arising by reason of a long-term relationship under a term contract such as a highway maintenance contract.[375]

A term alliance contract should include provisions that enable the alliance members to avoid or resolve disputes effectively. The options for effective avoidance and resolution of alliance disputes are considered in Chapter 19. For example, TAC-1 provides for non-adversarial dispute resolution by means of:

- Early warning as soon as an alliance member is aware of any dispute and its consideration by a core group, possibly supported by an independent adviser[376]
- Options to refer a dispute to conciliation or a dispute board.[377]

5.10   Term Alliance Case Studies

The following case studies show how term alliances can integrate teams and improve value on construction and engineering works and services. The Royal Borough of Greenwich, working in a term alliance with their direct service organisation and with Kier, agreed to trial a bespoke prototype of the contract form later published as TPC2005. Greenwich awarded four ten-year contracts, each with a total value of approximately £50 million, and reported how performance showed immediate improvements. Greenwich identified the key changes under their term alliance as improved communication and joint working. Their use of open-book accounting and supply chain management generated significant savings, and the client extended the scope of work so as to compensate for the fact that improved efficiency in repairs had led to falling contract expenditure.

Surrey County Council originally trialled a TPC2005 prototype but with limited success. Their problems were attributable in part to the use of a cost reimbursement pricing model[378] for statutory repair obligations, resulting in excessive provider resources being diverted to these statutory repairs and away from potential capital improvements. The Surrey team then adopted the published version of TPC2005 with far more success, and developed in agreement with the provider and its supply chain members a four-way term alliance pursuing Supply Chain Collaboration. Surrey reported on the cumulative Supply Chain Collaboration savings and other improved value delivered by alliance members over a five-year period,[379] and also worked with Kier on a £54 million contractor-led supply chain alliance.[380] Work on the Surrey term alliance was led by Jason Russell[381] and achieved extensive recognition, for example through:

- A case study published by the UK Infrastructure Client Group[382]

---

375 *Bedfordshire County Council v Fitzpatrick* [1998] 62 Con LR 64 (TCC).

376 TAC-1 Term Alliance Contract (2016), clause 15.1, and as considered in Sections 12.9, 19.6, 19.7 and 19.8.

377 TAC-1 Term Alliance Contract (2016), clause 15.2 and as considered in Section 19.9.

378 Section 16.4 considers the use of cost reimbursement.

379 As summarised in Section 11.5.

380 Case study summarised in Section 11.10.

381 Executive Director of Highways, Transport, and the Environment, Surrey County Council.

382 Infrastructure Client Group (2015).

- A case study in the HM Treasury Procurement Routemap[383]
- The model for a Supply Chain Collaboration Toolkit published by the Highways Maintenance Efficiency Programme[384]
- A principal client case study in the P13 Blueprint, which recognised how Surrey developed the roll-out of their alliance 'beyond road maintenance to the whole capital maintenance portfolio' and then 'extended their supplier engagement to build a supply chain alliance'.[385]

Royal Borough of Greenwich Repairs and Maintenance Alliance

£200 million maintenance contracts were undertaken by a direct service organisation/private sector term alliance including Royal Borough of Greenwich as client, Greenwich Building Services and Kier Building Maintenance as providers, with Cameron Consulting and Trowers & Hamlins as partnering advisers.

Greenwich Council reorganised property maintenance into four geographic areas covering 26 000 tenants and 3000 leaseholders. Greenwich awarded three ten-year TPC term contracts to Greenwich Building Services and the fourth to Kier Building Maintenance, each with a total value of approximately £50 million.

Performance for the first full year showed immediate improvements with time to re-let vacant properties down by 23% from 43 to 33 days, average time for non-urgent repairs down from 18 to 11 days, complaints down by 50% and post-job satisfaction scores of 95%.

Improved communication and joint working were key changes. Open-book accounting and supply chain management led to savings to such an extent that the client agreed to extend the scope of work to compensate for falling contract values.

The TPC Core Group and Partnering Team structures promoted communication which ensured the right people were dealing with issues at appropriate levels. The Early Warning and Core Group systems encouraged collective resolution of problems, and Greenwich Council suffered no claims under its TPC contracts.

Contract extensions and assessments of achievement were based on performance according to Greenwich's key performance indicators against targets agreed in advance with all stakeholders.

Tyron Stalberg, then Greenwich Housing Services Manager, commented: 'We knew we needed to adopt partnering, but the Council was nervous about how we would make it work. When we discovered TPC made partnering contractual, not just bolted on, we felt this was what we needed. You can't opt out! What we didn't realise then was how powerful a vehicle for partnering it really is'.

Association of Consultant Architects (2010), 19.

---

383 IPA (2014), 22.
384 HMEP Supply Chain Collaboration Toolkit (2014).
385 P13 Blueprint (2018). The P13 Blueprint suggests that Surrey 'utilised learning from others through the Alliancing Code of Practice', albeit that this code was published in 2015 and Surrey's procurement strategy was created in 2011.

Surrey County Council Project Horizon Term Alliance

Surrey County Council set up a term alliance with main contractor Kier and supply chain members Aggregate Industries and Marshall Surfacing, achieving substantial savings, improved quality, a range of social value and an integrated team culture in the delivery of capital highways repairs and improvement works across Surrey. Kier, with support from Surrey County Council, ran an agreed process to select subcontractors and suppliers and to create an integrated team for a five-year £100 million programme of capital highways and repair works. This enabled early collaboration with the selected tier 2 and tier 3 supply chain members so as to maximise savings, added value and joint identification of opportunities, including longer-term, larger-scale opportunities in return for savings and added value committed to at all levels in the supply chain.

Two Stage Open Book and Supply Chain Collaboration enabled and supported a culture of collaborative working at all levels of the supply chain as well as the creation of integrated project teams with better defined roles for individuals employed by the client, main contractor and other supply chain members. The parties entered into a four-way alliance contract to establish supplementary arrangements for joint working on Project Horizon.

Surrey achieved savings of **16%** at the point of selecting Kier in 2011, against the prices previously paid for comparable works. Implementation of Two Stage Open Book and Supply Chain Collaboration pursuant to the agreed TPC2005 alliance processes led to additional agreed savings in excess of **12%** that were sustained over a period of five years. These cost savings were agreed in exchange for:

- Visibility and continuity of pipeline of work through larger scale, longer term work offered to Kier by Surrey and in turn to Marshall Surfacing and Aggregate Industries **(5%)**
- Advance planning of work on each annual cycle **(2%)**
- Prompt payment of Marshall Surfacing and Aggregate Industries by Kier **(1%)**
- Closer involvement in the design and planning of individual tasks within the programme **(2%)**
- Availability of storage facilities in depots **(2%)**.

Qualitative benefits comprised:

- Improved whole life value, including agreement of an extended 10 year warranty for material and pavement design
- Improved quality control through joint risk assessments and integrated team agreement of appropriate surface treatments and monitoring work on site
- Improved apprentice commitments by Kier, Marshall Surfacing and Aggregate Industries, supplementing employment and skills commitments already agreed by Surrey and Kier under an Employment and Skills Plan in accordance with the CITB Client-Based Approach
- Lean programming of individual tasks leading to time savings
- Innovation through collaborative working, for example to increase recycling and reduce landfill.

Analysis of work on site showed no major remedial work required, no major health and safety incidents, and additional improvements to drainage systems and footways as part of agreed design solutions. Surrey received over 100 complimentary letters from residents and Council members, having never received any before.

Project Horizon Trial Project case study.

6

How Are Collaborative Team Members Selected?

6.1 Overview

The process of selecting team members can raise expectations, arising from the client's brief and the bidders' promises, but failure to meet those expectations causes disillusionment and leads to disputes. The procedures that govern selection of team members must be treated seriously as they create their own implied or express contractual obligations.[386] In addition to the implied procedural contract governing any tender process, a framework contract or framework alliance contract creates express procedural rights and obligations. Public sector selection procedures are also subject to regulatory requirements that are designed to ensure non-discrimination, equal treatment, proportionality and transparency.[387]

The team selection process should make a balanced assessment of cost, quality and other measurable proposals and capabilities but instead it often focuses on prices in a way that does not deliver reliable information or good value. A disproportionate focus on selecting team members according to lowest price, particularly where costs are not analysed jointly, can lead to the neglect of other important criteria.

Clients can be attracted by single-stage, lowest price selection of contractors because it suggests a clear commitment by a bidder at the point of selection to accept unconditional responsibility for delivering its works or services or supplies for a fixed amount of money. This model is intended to fix a competitive price that can be relied on but, when applied to construction works, commentators note that:

- 'As a practical matter, things seldom work out this way'[388]
- 'If consultants and contractors are selected on lowest price, there is a risk that lowest cost procurement drives the minimum standard of material and finish to meet the specification'.[389]

---

386 The UK courts recognise implied contractual obligations under any tender process, as illustrated in *Blackpool and Fylde Aero Club v Blackpool Borough Council* [1990] EWCA Civ 13, [1990] 3 All ER 25 and *Harmon CFEM Facades (UK) v House of Commons* [1999] EWHC 199 (TCC), [1999] 67 Con LR 1.

387 Currently, in the UK, these are the Public Contracts Regulations 2015, which may or may not be reformed after the Brexit process is concluded. While often criticised as bureaucratic, the Public Contracts Regulations 2015 provide a reasonably balanced approach and, compared to previous regulations, go a long way towards enabling public sector clients to evaluate quality as well as cost.

388 Smith, R.J. (1995), 44. Smith recognised the appeal of apparent cost certainty obtained through fixed price quotes, but the inherent risks led him to describe this philosophy as a 'legalised gambling' approach to contracting.

389 Housing Forum (2018), 9.

Collaborative Construction Procurement and Improved Value, First Edition. David Mosey.
© 2019 John Wiley & Sons Ltd. Published 2019 by John Wiley & Sons Ltd.

The extensive use of single-stage selection owes less to its track record in achieving efficiencies and avoiding claims and disputes than to cautious advice regarding available alternatives. Sometimes this caution may be encouraged by 'professionals with a vested interest in old ways of working' but with the result that clients are 'all too often…sadly disappointed as they discover that claims, delays, defects and disputes make this an expensive and ineffective approach'.[390]

A two-stage selection process can give rise to provisional appointments governing preparatory activities during which team members' tender proposals and commitments can be tested, and often improved upon, before the full implementation of the project is approved. The UK Joint Contracts Tribunal recognised that early selection 'increases the scope for value engineering, through early contractor involvement, team work and fixed (rather than estimated) subcontractor pricing, and…reduces the scope for claims that result from inaccurate or inadequate designs or specification. With the design and procurement processes being in part concurrent, it may also save time'.[391]

The UK Government expressed its commitment in 2018 to 'improve the lifetime performance of buildings, through better procurement', and for this purpose to:

- 'Develop an industry wide definition of value which takes into account more than capital cost'
- 'Produce a universally applicable methodology for procurement and promote common and consistent standards across industry'.[392]

This chapter will explore the criteria by which team members can be selected, the differences between single-stage and two-stage selection processes, how selection operates under a framework alliance and how the selection process itself can include alliance activities.

6.2   What Are the Problems with Single Stage Selection?

Single-stage, lowest price selection can give rise to problems where a client is provided with inaccurate fixed prices based on incomplete or inaccurate information. A client and a main contractor will both be at risk if fixed prices offered by a bidder are based on in-house estimates or on inaccurate prices received from subcontractors and suppliers who do not have the time and data necessary for fixed quotations.[393] Two of my MSc students who work for large contractors once explained that, when preparing fixed prices for typical tender submissions, they do not usually obtain any supporting prices from potential subcontractors. Instead, they rely on their own estimators to calculate what they expect subcontract packages to cost, and then obtain actual subcontractor prices only after they been awarded the project. Quoting fixed prices to a client in this way creates the illusion of cost certainty and increases the risk of later claims and disputes if the underlying estimates prove to be insufficient.

---

390 Bennett & Pearce (2006), 7.
391 JCT Tendering (2017), Section 16. The JCT see this as a suitable procurement route for complex projects but in practice its advantages have been proven on a wider range of projects where a client uses early contractor involvement and a collaborative approach.
392 BEIS (2018), 13.
393 Burke, R. (2002), 85, as to the risks of single-stage pricing.

In order to provide an accurate price in a single-stage tender, each bidding contractor would in theory need to present the client's proposed requirements to each of its subcontractors, suppliers and manufacturers so as to obtain subdivided fixed price quotes prior to submitting its own fixed price quote to the client. The time and cost of conducting these additional tender procedures in a thorough manner can be prohibitive for bidding contractors and for their prospective supply chain members.[394] Instead, as a matter of commercial necessity, bidders often make judgements as to the level of detail and accuracy required in responses to enquiry documents according to the importance of each project element, and with allowances by which they hope to cover the risk of inaccurate pricing.

It has been noted that 'clients tend to fixate on lowest initial tendered price and this is often perpetuated by their advisors, who, in a traditional procurement model, are implicitly employed (at least partly) to manage a fixed and adversarial transactional interface between clients and industry. The cost-based procurement model often hinders the ability to focus on value, outcomes or performance if appropriate weightings are not made.'[395] Other critics of single-stage selection have noted that it:

- Can provide an incomplete assessment of the costs and capabilities of prospective team members because it depends on a process of 'private information' where the principal is not informed of certain important characteristics of bids at the time when the contract is awarded, for example the bidders' actual underlying costs[396]
- Can increase the risk of later claims and disputes because selection 'on a take-it-or-leave-it basis' by 'short and sharp consent' and, without joint examination of all relevant issues, denies the opportunity for mutual planning and creates 'a process heavily laden with conflict'[397]
- Does not achieve the price certainty it claims and, in terms of predictability between contract prices and final prices, has proven the least likely to provide predictable cost results, with only 56% of projects completed within plus or minus 5% of the contract price.[398]

Problems in the construction industry by way of unpredictable outturn costs, delays and defects can often be traced to a single-stage, fixed price selection process by which the contractor is expected to assess a correct market price for a project that it has not previously built on a site in respect of which there is little information, adopting a design which may still be incomplete or subject to revision and using a labour force and supply chain not yet recruited.[399] Single-stage, fixed price bidding by reference to incomplete information can lead bidders to add arbitrary contingencies or premiums to their quoted prices, with the following possible consequences:

---

394 The bidder costs of single-stage procurement can be more than six times higher than those of a two-stage framework alliance procurement, as illustrated in Two Stage Open Book and Supply Chain Collaboration Guidance (2014), 19.
395 Farmer, D. (2016), 24.
396 Milgrom & Roberts (1992), 129.
397 Macneil, I.R. (1974), 770, 771, 777.
398 NEDO (1975), Section 5.5, 43.
399 Burke, R. (2002), 237.

- Inflation of the prices quoted to cover perceived risks, with difficulty for the client in then challenging the pricing of a bidder's risk assessments after selection if the client wishes to negotiate reduced risk contingencies
- A windfall by way of additional profit for the successful bidder, and therefore wasted money for the client, if the contingency is higher than necessary to cover the bidder's actual costs
- A loss for the successful bidder, resulting in pressure to make additional claims on the client, if the contingency is insufficient to cover the bidder's actual costs.

These failings are not necessarily the result of deliberate tactics and instead may be attributable to the lack of time and available information, because:

- 'Under traditional single stage tendering arrangements the opportunity to plan for the construction stage is restricted
- [Bidders] will do enough preparatory work to be successful at tender but are unlikely to be able to understand fully all aspects of the project or have sufficient time to identify and consider how to manage the potential risks to the project'.[400]

It is argued that new approaches are required because:

- For contractors 'successfully to take part in the preparation of a project will mean that, in some cases, changes will have to be made in the time-honoured procedures under which contracts are let'[401]
- 'There is broad agreement across the industry that early supplier engagement is advantageous, and that innovation starts with the supply chain' but 'perversely, the procurement process and behaviours adopted tend on balance to dis-incentivise innovation and focus on lowest cost'[402]
- 'The low bid syndrome can be recognized as a major determinant behind the adversarial behaviour leading to failure…while sequential involvement of the parties does not allow mutual exchange of information and collaboration for the benefit of the project'.[403]

6.3 How Can Early Contractor Selection Enable Collaborative Procurement?

When considering new options for appointing team members, we can learn from the way that clients often develop their relationships with design consultants. The selection of consultants is in effect made under a conditional appointment which progresses in multiple stages during the preconstruction phase of a project according to approval of the consultant's designs and other proposals. The authorisation of successive design

---

400 JCT CE Guide (2016), Section 37. This is known as 'information impactedness', which 'refers to the limited knowledge of the parties to a transaction which denies them the ability to make correct purchasing or selling decisions', and as a result of which 'opportunism…occurs whenever firms take advantage of the information impactedness of the co-transactor', Gruneberg & Hughes (2006), 18.
401 Banwell Report (1964), 6, Section 2.11.
402 London Underground (2014), 11.
403 Lahdenpera, P. (2017), Present knowledge.

stages provides the client with a system of control which can validate the original selection. This familiar approach to client/consultant relationships can also be applied to conditional contractor appointments during the preconstruction phase of a project.

Early contractor selection can be combined with robust competitive processes, commencing with selection of the main contractor through 'a preliminary competition based on an outline, in which the offers of selected firms are considered in the light of such factors as management and plant capacity, and the basis of their labour rates, prices and overheads'.[404] Under a second-stage procedure 'the chosen contractor works as a member of the team, while details are developed and bills of quantities drawn up, and at the end of this time submits a more detailed price which if satisfactory becomes the formal contract sum'.[405]

This two-stage system can provide 'competition in a new sense' while at the same time enabling 'the contractor to join the team at a time which is precluded by existing procedures'.[406] It was recommended as early as 1964 because it was seen to offer 'undeniable advantages for the client in solving some of the failures in communication and understanding between designers and contractors and contractors and subcontractors which have hampered the industry in recent years'.[407] The Two Stage Open Book procurement and delivery model develops the 1964 recommendations into a full two-stage selection system that enables a client to select a main contractor early and thereby to obtain and approve in stages the additional data that will form the basis for award of a construction phase contract.

Early contractor selection also reduces the overall number of subcontract tendering exercises conducted in the marketplace and the costs and time that these incur. Instead of prospective subcontractors, manufacturers, suppliers, and operators each wasting time and money bidding to one or more of a number of main contractors who are themselves bidders, and somehow having to recover that wasted cost from the clients on whose projects their bids are successful, they will be in the position of bidding to a preselected main contractor. This significantly increases their chances of success and their likely commitment to proposing safe and reliable solutions, as well as ideas for improved value, both during and after the subcontract tender process.

Main contractors have their own preferred supply chains which they use regularly and which they appoint on competitive terms under their own systems. These subcontractors and suppliers know the main contractor procurement teams and are familiar with their ways of working. The Trial Project case studies have shown that Two Stage Open Book selection does not undermine the potential efficiency of contractors using their established relationships. Instead it tests the value of these relationships and at the same time exposes prospective supply chain members to direct contact with the client and consultants, in ways that have been demonstrated to:

- Provide additional time for joint reviews of designs, programmes and risk assessments
- Enable the client and consultants to agree improvements on the brief and proposals that formed the basis of the main contractor's selection
- Enable more accurate costing of service, supply and work packages and the build-up of cost data that is available to the client and consultants as well as the main contractor

---

404 Banwell Report (1964), 10, Section 3.14.
405 Banwell Report (1964), 10, Section 3.13.
406 Banwell Report (1964), 10, Section 3.15.
407 Banwell Report (1964), 10, Section 3.15.

- Reinforce integration among team members through their joint activities and shared knowledge.[408]

Two Stage Open Book guidance suggests that:

- 'At the point of selection of the Consultants and Tier 1 Contractor, Two Stage Open Book provides the basis for a transparent competitive process in respect of their fees/profit/overheads, and any other components of the project for which it is appropriate to test costing, such as risk contingencies and the provisional cost of particular proposals submitted
- Evaluation of fees/profit/overheads and such other costs needs to be balanced appropriately against evaluation of qualitative proposals and the proven ability of the Consultants and Tier 1 Contractor to deliver the project/programme within the Project Budget cost ceiling
- At the point of selection of Tier 2/3 Subcontractors and Suppliers, Two Stage Open Book provides the basis for further transparent competition based on accurate costing and additional qualitative proposals'.[409]

The UK Government noted that 'the Two Stage Open Book model reduces industry bidding costs, enables faster mobilisation and provides the opportunity for Clients to work earlier with a single Integrated Team testing design, cost and risk issues ahead of start on site following full project award at the end of the second stage'.[410]

However, it is also recognised that a two-stage selection procedure:

- Calls for 'a greater input from the client or his advisers than simple competition'[411]
- Can 'place considerable demands on contractors', for example in preparing qualitative proposals designed to meet the client's evaluation criteria[412]
- Needs to be 'carefully thought out, to balance the expenditure of resources against the benefit to be obtained'.[413]

6.4   How Can Early Contractor Selection Create More Accurate Prices?

Concerns may be expressed that two-stage selection allows the appointed main contractor to inflate its prices towards the end of the first stage. Case studies show how management of the preconstruction process can contain any opportunistic attempts to inflate costs, as illustrated on the North Wales Prison Trial Project and on the Bermondsey Academy project where, for example, it was necessary to 'keep a close watch on the development of the preconstruction risk register, so as to ensure that risk contingencies allocated to particular items were subject to scrutiny and agreed actions'.[414] In practice

---

408 As illustrated in the SCMG and Project Horizon Trial Project case studies, and by the Kier contractor-led supply chain alliance summarised in Section 11.10.
409 Two Stage Open Book and Supply Chain Collaboration Guidance (2014), 36.
410 Two Stage Open Book and Supply Chain Collaboration Guidance (2014), 2.
411 NEDO (1975), Section 5.42, 50.
412 CIRIA (1998), 8.
413 CIRIA (1998), 8.
414 Mosey, D. (2009), 236. Risk registers are considered in Chapter 18.

two-stage selection can produce the most predictable results in terms of contract prices corresponding to final prices:

- 'If the selection process has been properly managed and documented (for example, by an elemental cost plan)
- If a reliable basis of pricing has been established by the client's cost advisers and
- If there are no significant changes in the client's brief or design concept'.[415]

In addition, a model by which the parties can build up additional shared information as to underlying costs can reduce 'informational asymmetries'[416] and will help the parties to focus on accuracy through 'incentive-efficient mechanisms'.[417] A survey of predictability between contract prices and final prices, analysed according to the method of main contractor selection, concluded that two-stage selection was the most likely to produce predictable results, namely 82% of projects successful within plus or minus 5% of the contract price.[418] It has also been suggested that clients who use more enlightened procurement and risk management procedures have not only encountered fewer delays and disputes but have also 'obtained more competitive bids'.[419]

The means by which two-stage selection can achieve improved cost certainty and agreed cost savings are explored by reference to a range of cost models in Chapter 16 and include:

- The separate agreement of profit and overheads
- The use of a budget as a cost ceiling
- The use of fixed prices or target costs according to the features of the project
- The use of appropriate incentives to manage costs.

Two-stage selection can create a fixed price ahead of start on site but not at the point of main contractor selection because 'the first stage of the two-stage process involves competitive tenders, but that is often inevitably restricted to preliminaries, overheads, contractor's profit and those elements of the design that are sufficiently advanced to allow proper pricing'.[420] Certain elements of the project costs can be established at the point of main contractor selection where bidders are given enough information to be able to provide accurate calculation of costs.

Evaluation of fixed prices can form part of the criteria applied to those aspects of a project where designs are already well understood by the client and its consultants, for example where they have been successfully used on previous projects forming part of a framework alliance. Inclusion of fixed price elements can also be made easier by the use of designs contained in Building Information Modelling (BIM) libraries.[421] While it is therefore possible to combine two-stage selection with obtaining fixed cost proposals for some parts of a project, this mixed economy approach requires caution and careful analysis so as to ensure that it does not lead the team back into the pricing problems inherent in single-stage selection.

---

415 NEDO (1975), Section 5.42, 50.
416 Milgrom & Roberts (1992), 140.
417 Milgrom & Roberts (1992),143.
418 NEDO (1975), Section 5.5, 43.
419 Smith, R.J. (1995), 44, 45.
420 JCT Tendering (2017), Section 17.
421 As considered in Chapters 13 and 14.

For the remaining costs, early contractor involvement enables the client and cost consultant to work with the design consultants, the main contractor and its prospective supply chain members to establish detailed costs more accurately and to identify opportunities for agreement of cost savings in conjunction with other improved value. It is important to set out in a two-stage contract the detailed machinery for agreeing costs and prices that is governed by the early provisional appointment of alliance members.[422]

6.5 How Are Collaborative Team Members Selected?

It is possible that a collaborative team will be selected by an informal process, in reliance on relationships that have been established through work on previous projects. A competitive selection process may appear bureaucratic, and a confident private sector client may be able to short-circuit the selection process using direct negotiation. However, Australian research has noted that:

- 'The complex nature of alliances can result in Owners being exposed to serious asymmetry of information, commercial capability and capacity in their engagement with Non-Owner Participants'
- 'The exposure of Owners can be increased when there is no price competition as there has not been the "traditional" competitive tension which can alleviate such asymmetry'.[423]

It is therefore worth exploring competitive selection procedures that are compatible with collaborative procurement. These should start with an understanding of the marketplace in terms of, for example, its appetite, capacity and capability.[424] A process of market consultation through informal early market engagement will often reveal important information that influences the client's approach to its formal selection of team members.[425] For public sector clients the publication of a Prior Information Notice[426] can alert potential bidders to an imminent procurement and can be used to initiate market consultation.

A successful selection process also depends on the quality of the client's brief, which should set out the maximum information as to the client's business needs and all relevant external factors, including:

- 'The initial goals and objectives of the project, signed off by the client as the definition of the business need to be met'
- 'All project specific requirements and constraints that may be pertinent'
- 'Any time and budgetary constraints'.[427]

---

422 If agreed pricing machinery breaks down, there are instances where the court has substituted its own machinery to determine a fair and reasonable price if it finds that the machinery is a subsidiary and non-essential part of the contract, *Sudbrooke Trading v Eggleton* [1983] 1 AC 444 (HL).
423 Department of Treasury and Finance (2009), 155.
424 IPA (2014), 26, 27.
425 Market consultation is permitted by Public Contracts Regulations 2015, Reg. 40, and resultant 'advice may be used in the planning and conduct of the procurement procedure'.
426 As provided for in Public Contracts Regulations 2015, Reg. 48.
427 Selecting the Team (2005), 6, 7.

Guidance on selecting contractors by value recognised the need for clients and their advisers to invest the time and money necessary:

- 'Thoroughly to work through and prioritise what they are seeking to gain from a project
- To set projects up to enable contractors to contribute the maximum value
- To identify relevant criteria for their selection
- To gather information to enable these criteria to be applied'.[428]

The sequence of the above guidance is important. In order to run a selection process other than on the basis of price, the client needs to establish other criteria that can be assessed objectively.[429]

The evaluation criteria for the members of a collaborative team can include assessment of each organisation's commitment to collaborative working, including for example:

- 'Collaborative profile and experience
- Cultural compatibility
- Customer relationship management
- Supplier relationship management
- Stakeholder implications'.[430]

Evaluation criteria need to be detailed and measurable rather than vague commitments to collaborate. A client's general expression of its requirements for collaborative working can easily be played back as the very thing a bidder is doing already. When the bid has been won, collaborative working is then interpreted differently according to whatever each team member wants it to be, often to reflect their existing internal systems whether or not fully understood or agreed by other team members. Each party can claim that it wants to work collaboratively but that other team members have let it down, supposedly then justifying a reversion to narrow self-interest as a commercial necessity when challenges arise.

For these reasons, robust evaluation criteria should look closely at a prospective team member's:

- 'Objectives
- Requirements
- Expectations
- Risks and risk management approach'.[431]

A University of Reading report described how evaluation of team members for the Dudley College Trial Project included behaviours such as:

- 'Ability to work in a spirit of mutual trust
- Ability to work with a "no-blame mindset"

---

428 CIRIA (1998), 8.
429 CIRIA concluded that selecting contractors by value under a two-stage approach results in better teamwork, programming, design and specification, care of the environment, budgeting and management of risk and value-CIRIA (1998), 14.
430 ISO 44001 Section 8.4.6.
431 ISO 44001 Section 8.4.6.

- Ability to understand/appreciate perspective of others and adapt behaviour appropriately
- Mutual respect between differing disciplines and personalities'.[432]

The Reading report explained how 'following post-tender interviews, a behavioural workshop was held with bidding parties to validate the assessments made by the Client Advisory Team from previous assessment activities'.[433] This level of behavioural analysis of individuals can be revealing but is a demanding and arguably subjective process and, in the absence of contractual constraints, it cannot prevent those individuals leaving an organisation after its appointment.[434]

The members of a collaborative team can be selected individually or they can be selected together as members of a complete team. Bidders comprising a team can demonstrate how they are already integrated and arguably can move more quickly to the creation of improved value.[435] However, the need to create an integrated team is another piece of work for bidders to undertake after an invitation has gone out and before bids are returned, and this should lead clients to increase the time allowed for bid preparation.

UK Cabinet Office research found that 'the best projects [we saw] and the best private sector clients put time into getting the right team. They assessed the quality of the individuals, their ability to work together and their experience'.[436] The Cabinet Office were concerned that public sector clients frequently put together teams with undue emphasis on lowest price or expediency. They recommended, for example, that interviewing the individuals who will actually work on the project should be normal practice.[437]

Interviews are an important feature of collaborative procurement, providing 'the opportunity to compare the applicant's creative approaches to the design process, as well as their interpretation and understanding of project implementation' plus 'an important insight into each applicant's management style and communications abilities'.[438] For example, on the ORTUS project alliance:

- 'New prospective alliance members were introduced to the project through a careful selection process by way of tenders and interviews
- All consultants and constructors were invited to attend workshops in small groups hosted by the partnering adviser [who] explained the contract carefully to each party and highlighted the key differences between the alliance and typical building contracts
- Queries were raised and discussed and each alliance member was made aware that the partnering adviser would be available as a resource to them should they deem it necessary
- These workshops were not only informative but helped bond relationships between partners and reinforce the partnership culture'.[439]

---

432 University of Reading (2018), 52, 53.
433 University of Reading (2018), 55.
434 Constraints on individuals leaving a collaborative team are considered in Section 12.2.
435 See for example the Tottenham Court Road case study in Association of Consultant Architects (2010), 42–3, summarised in Section 6.10.
436 Efficiency Unit (1995), 253, 76.
437 Efficiency Unit (1995), Section 254, 76.
438 Selecting the Team (2005), 13.
439 Case study summarised in Section 7.10.

6.6   How Do Framework Selection Procedures Operate?

The procedures by which members win a place on a framework or framework alliance are subject to the same evaluation considerations as those explored in the remainder of this chapter. The procedures governing the award of work under a framework or framework alliance, once its members have been selected, should be set out in the framework contract itself.[440] It is expected that selection to undertake a specific project will require the provision of further details in relation to that project and, where necessary, these details can include more precisely formulated terms.[441]

For example, the Ministry of Justice framework alliance provided for project appointments to be made by mini-competitions or by direct appointment under 'a standard two-stage selection process that enables the MoJ to comply with all relevant procurement legislation and enables early contractor involvement in projects', and it was noted that:

- 'Early engagement of the supply chain is encouraged by the two-stage approach
- This serves to gain contractor and supply chain input into design, ensure cost robustness and appropriate risk management strategies for all projects
- This increases the certainty of delivery on time and budget and the quality of the overall product
- The framework enables the department to react quickly to emerging procurement requirements'.[442]

The systems for selection for specific projects are set out as mutual commitments in the framework contract or framework alliance contract. For example, the Environment Agency Rye Harbour Trial Project team used a 'three-stage selection process that enables the EA to comply with all relevant procurement legislation and enables early contractor involvement',[443] and this approach was identified as having the features of Cost Led Procurement. The Rye Harbour Trial Project case study reported that 'One of the biggest benefits of Cost Led Procurement in this respect was the ability for the Environment Agency to streamline the upfront processes involved in the procurement of this project, enabling them to move forward very quickly'.[444]

The selection of team members for one or more projects under a well-structured framework or framework alliance can save significant time:

- On the Property Services Cluster Trial Project 'the constructor partners, Osborne, Miller and Mansell (now Balfour Beatty) were jointly appointed from the .... framework through a mini-competition for all of the work included in the Tranche. This process took seven weeks, allowing for full and early contractor involvement'[445]
- The Supply Chain Management Group (SCMG) framework alliance reported
  - 'Reduced risks, cost savings and time savings through an accelerated constructor/supply chain briefing process'

---

440 For example, under the Public Contracts Regulations 2015, Reg. 33(11)(d).
441 For example, under the Public Contracts Regulations 2015, Reg. 33(11).
442 Effectiveness of Frameworks (2012), Case Study, 103.
443 Effectiveness of Frameworks (2012), Case Study, 97.
444 Rye Harbour Trial Project case study summarised in Section 17.10.
445 Property Services Cluster Trial Project case study summarised in Section 8.10.

- 'Time savings, such as through quicker build-up of prices leading to earlier start on site and reduced client/consultant time/costs'
- 'I don't know any other London borough that has managed to carry out so much work in so short a time'.[446]
- Under the Ministry of Justice framework alliance 'A first stage selection process (to identify the preferred supplier) can be completed within as little as 3 weeks, but normally takes 6 to 8 weeks (saving at least 6 to 9 months in comparison with tendering each project via an OJEU process).'[447]

The award of a framework contract enables a public sector client not to have to run a further public procurement process for the award of contracts pursuant to that framework.[448] However, public sector clients need to obtain a minimum amount of cost information at the point of selection of the members of a framework, as considered in Section 16.6. Public sector clients are also subject to other constraints on their selection procedures, as summarised in Section 6.8 and below:

- The process for deciding which framework award procedure will apply should be set out as part of the framework alliance award procedures using objective criteria[449]
- The regulations governing award of contracts prohibit 'substantial modifications to the terms laid down'[450]
- Although framework alliance contracts may provide for the later addition of new members, the clients are limited to those who are 'clearly identified for that purpose in the call for competition or the invitation to confirm interest',[451] and appointments are limited to 'those economic operators party to the framework agreement as concluded'[452]
- For the award of contracts under a framework concluded with one party (the 'economic operator'), the client may make direct awards without further competition and 'may consult the economic operator…in writing, requesting it to supplement its tender as necessary'[453]
- For the award of contracts under a framework concluded with more than one party which sets out 'all the terms governing the provision of the works, services and supplies concerned' the client may make awards without further competition according to 'objective conditions for determining which of the economic operators…shall perform them' or may make a selection by 'reopening competition amongst the economic operators which are party to the framework agreement'[454]
- For the award of contracts under a framework concluded with more than one party where 'the terms governing the provision of the works, services and supplies concerned are not laid down in the framework agreement', the client may only award

---

446 SCMG Trial Project case study.
447 Effectiveness of Frameworks (2012), Case Study, 103.
448 Public Contracts Regulations 2015, Reg. 33.
449 Public Contracts Regulations 2015, Reg. 33(8)(b).
450 Public Contracts Regulations 2015, Reg. 33(6).
451 Public Contracts Regulations 2015, Reg. 33(5).
452 Public Contracts Regulations 2015, Reg. 33(5).
453 Public Contracts Regulations 2015, Reg. 33(7)(b).
454 Public Contracts Regulations 2015, Reg. 33(8)(a).

contracts by 'reopening competition amongst the economic operators which are party to the framework agreement'.[455]

6.7   How Can Evaluation Criteria Balance Cost and Quality?

The UK Office of Government Commerce included among its critical factors for success:

- 'Award of contract on the basis of best value for money over the whole life of the facility, not just lowest tender price'
- 'An integrated process in which design, construction, operation and maintenance are considered as a whole'
- 'Procurement and contract strategies that ensure the provision of an integrated project team'.[456]

Alliance selection criteria can assess a wide range of capabilities and proposals, for example:

- 'Technical knowledge and skills – experience in engineering specialist elements; appropriate design capacity
- A number of management skills:…managing time…managing cost…managing value…managing quality…managing risk…managing health and safety
- Effective internal organisation – clear communications; sound administration; empowered staff
- Collaborative culture – record of "partnering"; positive lead from the top; client focus
- Appropriate human resources – qualified and enthusiastic personnel available to do the job
- Supply chain management – sound dealing with subcontractors/suppliers; established relationships
- Financial resources – sound balances and cash flow; reliable references
- Generally – a sound, relevant and demonstrable track record'.[457]

Evaluation by reference to these criteria is more demanding for the client and its consultants than a straightforward comparison of prices. However, the bidders' qualitative submissions provide valuable information that should assist the client in making the right choice and that should provide stronger foundations for commercial relationships. For example:

- On the Hampstead Heath Ponds project alliance, the basis for team selection was 80% quality and 20% price[458]
- On the Tottenham Court Road project alliance, 'Early selection of the team members was made on the basis of:
  - Tendered profit, overheads and fees
  - Identifiable costs in respect of site overheads/preliminaries and risk contingencies
  - Full costing of pre-designed items

---

455 Public Contracts Regulations 2015, Reg. 33(8)(c).
456 OGC (2007), 1.
457 CIRIA (1998), 15.
458 As summarised in Section 3.10.

- Demonstration by the team members of their ability to deliver the project in accordance with the client's brief and within the pre-determined project budget'.[459]
- On the Bleak Hill School project alliance, the main contractor was selected on criteria comprising weighted proposals for design, supply chain, programming, quality control, construction process, cost control, health and safety (CDM), and communications. The selection process 'examined 8 method statements as well as the experience, aptitude and commitment of the contractors with a view to the added value they could contribute to the project'.[460]

An examination of collaborative procurement in the housing sector noted that 'On housing programmes, the financial elements of the bid will include the construction costs, overheads and profits, costs of staff transferring as a result of TUPE (particularly on repairs and maintenance programmes), and the cost of any social value proposals including apprenticeship opportunities. However, there are other financial elements that can be evaluated including discount cost savings over the lifetime of the contract and life-cycle costs'.[461]

A persistent concern is that any financial evaluation will inevitably dominate a selection procedure and will tempt bidders to undercut each other regardless of other criteria.[462] It has been suggested that, unless quality accounts for at least 75% of the evaluation criteria, it is likely that lowest cost will be the ultimate deciding factor.[463] Overcoming this problem requires the client and its advisers to make clear their priorities in a way that bidders respond to, for example by evaluating quality first and then evaluating cost, taking the benchmark price from the highest quality bid. Specialist commentators have suggested other evaluation models 'that seek to protect the contracting authority and the bidders from an unrealistic pricing risk':

- 'The optimum pricing model in which the contracting authority sets out the optimum price which it considers appropriate for the contract, based on market research. The tenderer is then incentivised to make the effort to reach the optimum price without undercutting it. The tenderer closest to the optimum price receives the highest mark. This should protect against abnormally low bids but arguably curbs the potential for truly innovative approaches'[464]
- 'The fixed price model where the contracting authority fixes the price for the contract and then undertakes a value for money evaluation on the non-price element of the contract's delivery, such as the quality and experience of the team, choice of materials, health and safety standards, liaison with residents, or environmental and social aspects of the project. By fixing the price and considering alternative value for money proposals, the contracting authority will again be neutralising the effect of any abnormally low bids on the overall evaluation'.[465]

---

459 Case study summarised in Section 6.10.

460 Iain Beaton, Assistant Chief Executive, St. Helens MBC in SCALA (2000). Bleak Hill School was a £2.5 million project using a PPC2000 prototype.

461 Housing Forum (2018), Saunders, K. and Maqbool, A., Case Study 4, 18.

462 'Used in the wrong way (with the wrong price/quality split or the wrong sub-criteria), price can still become and overriding factor in selecting the preferred bidder and, consequently, quality is compromised', Housing Forum (2018), Saunders, K. and Maqbool, A., Case Study 4, 18.

463 Rebecca Rees of Trowers & Hamlins, FAC-1 Masterclass 6 October 2018.

464 Rees, R. (2016).

465 Housing Forum (2018), Saunders, K. and Maqbool, A., Case Study 4, 18.

6.8   Can Early Contractor Selection Comply with Public Procurement?

Selection by a public sector client pursuant to the UK Public Contracts Regulations[466] is not restricted to price comparisons, but it is required to be on the basis of the 'most economically advantageous tender'.[467] The most economically advantageous tender should 'be identified on the basis of the price or cost, using a cost-effectiveness approach, such as life-cycle costing', and 'may include the best price-quality ratio, which shall be assessed on the basis of criteria, such as qualitative, environmental and/or social aspects, linked to the subject matter of the public contract in question'.[468]

Selection according to the most economically advantageous tender can include evaluation criteria consistent with the objectives of improving both economic and social value, for example permitting the client to evaluate 'quality, including technical merit, aesthetic and functional characteristics, accessibility, design for all users, social, environmental and innovative characteristics'.[469] Qualitative evaluation may be undertaken alongside cost evaluation as part of a price-quality ratio or may be the only basis for evaluation where 'the cost element may [also] take the form of a fixed price or cost on the basis of which economic operators will compete on quality criteria only'.[470]

There is no requirement under the Public Contracts Regulations that every element of a project brief or of a sample project brief must be fully priced prior to selection.[471] There is also no prohibition of the early conditional selection of a contractor to undertake preconstruction phase activities, nor a requirement that evaluation criteria for the most economically advantageous tender should include a fixed price for the project.[472]

The Public Contracts Regulations describe a range of selection procedures for individual projects and for frameworks which include:

- The 'Open Procedure' which comprises a single-stage invitation for any party to bid with no negotiation[473]
- The 'Restricted Procedure' which comprises a pre-qualification stage before a shortlist of parties are invited to bid, again with no negotiation[474]

---

466 The Public Contracts Regulations 2015 require and regulate specific competitive procedures in respect of the award by public sector clients (known as contracting authorities), of contracts for works, services and supplies of a value over stated thresholds. These regulations implement the European Public Contracts Directive 2014/24/EU and, at the time of writing, there is no indication as to how these regulations may change following Brexit.
467 Public Contracts Regulations 2015, Reg. 67(1).
468 Public Contracts Regulations 2015 Reg. 67(2).
469 Public Contracts Regulations 2015, Reg. 67(3)(a).
470 Public Contracts Regulations 2015, Reg. 67(4).
471 If this was the case then no public sector project could use the long-established system of 'provisional sums' for elements of a project that it is agreed will be finalised and costed during the construction phase, for example under JCT SBC/Q (2016) clause 5.2.
472 Public Contracts Regulations 2015, Reg. 67.
473 Public Contracts Regulations 2015, Reg. 27.
474 Public Contracts Regulations 2015, Reg. 28.

- The 'Competitive Procedure with Negotiation' and the 'Competitive Dialogue Procedure', each of which comprises a pre-qualification stage before shortlisted parties are invited to bid and to enter into structured negotiations or dialogue.[475]

The restrictions on negotiation have caused concern where clients see early contractor involvement, for example under Two Stage Open Book or Supply Chain Collaboration, as deferring completion of full pricing in a way that involves negotiation. This concern is not new, and in 1964 Banwell attributed the hesitation of public sector clients regarding early contractor involvement to fears that 'to do so would be contrary to established notions of public accountability'.[476] However, the selection process for early contractor appointments can avoid negotiation and can comply with the Public Contracts Regulations if public sector clients ensure that they:

- 'Request sufficient pricing information during the competition to select the contractor for the purpose of Stage One
- Include a prescriptive and clear pricing mechanism in the contract, which will be applied in order to determine the price for the purpose of Stage Two (i.e. the price of the construction works)'.[477]

In order to be compliant, the second-stage pricing mechanism forming part of Two Stage Open Book or Supply Chain Collaboration should:

- Be a contractual process led by the main contractor
- Not increase the main contractor's quoted profit, overheads and risk contingencies
- Not affect the agreed responsibility of the main contractor for delivering the project within the budget stated by the client prior to the main contractor's selection.

Contractual processes for second-stage pricing are set out in PPC2000 and NEC4 Option X22 and these are considered in Sections 16.3 and 16.5. Both forms create two-stage conditional contracts, and for early contractor selection to comply with the Public Contracts Regulations depends on adopting this two-stage contract structure. By comparison, the JCT Pre-Construction Services Agreement (PCSA)[478] does not set out second-stage pricing processes, and is also vulnerable to breach of public procurement rules because it treats the PCSA as a separate contract distinct from the JCT 2016 construction phase contract: this may cause a problem unless both contracts are awarded to the same contractor simultaneously following a single competitive process.[479] If the JCT preconstruction phase appointment is awarded separately from the construction phase contract, it is arguable that two separate selection exercises should be undertaken, one for preconstruction phase services and another for the construction phase works.[480]

A two-stage selection process that is compliant with the Public Contracts Regulations could adopt the following sequence:

---

475 Public Contracts Regulations 2015, Regs. 29 and 30, as considered further in Section 6.9 and as illustrated in the Bank Station case study summarised at Section 6.10.
476 Banwell Report (1964), 10, Section 3.13.
477 Barber & Jackson (2010), 183.
478 JCT PCSA (2016).
479 Barber & Jackson (2010), 181.
480 The vulnerability of the JCT structure as an agreement to agree is considered in Section 8.2.

- Selection of team members under whichever of the Open, Restricted, Competitive Procedure with Negotiation or Competitive Dialogue Procedure reflects the complexity of the project or, in the case of a framework, the programme of works
- Inclusion in the invitation to tender of all available specifications and other data as part of the project brief or, in the case of a framework, the brief for one or more sample projects and overarching requirements that will apply to all projects
- Invitation for bidders to submit their proposed profit, overheads and risk contingencies, plus fixed or maximum prices for those elements of the brief for the project or, in the case of a framework, sample projects that are capable of pricing
- Invitation for bidders also to submit qualitative proposals
- Evaluation of those prices and proposals according to a matrix and weightings selected by the client and its advisers in establishing the most economically advantageous tender
- Specification in the contract of the process by which the remainder of the prices will be determined for the project, or in the case of a framework, each project.

The devil is in the detail of the tender documents and, if in doubt, public sector clients should seek advice as to the regulatory compliance of any selection procedure. For example, a recent Counsel's opinion explored the compliance of a 'requirement for bidders to commit (as part of the Alliance) to agree a Target Cost that is within the Investment Target for the overall project' where 'cost effectiveness will be evaluated by requiring bidders individually to commit to a "not to exceed" value for their Lot'. The answer provided by Counsel, after reading the tender documents, was that 'the approach to the Investment Target is akin to fixing the price and evaluating only on quality, which is permitted by the Regulations'.[481]

6.9   Can Collaborative Working Occur During Selection?

ISO 44001 recommends that 'during the evaluation of collaborative business opportunities, organisations shall consider the risks and opportunities which result either directly or through supply chains or extended enterprises associated with potential collaborative working partners'.[482] These can include:

- 'Supply chain and/or extended enterprise performance
- Existing supply chain collaborations
- Combined procurement opportunities
- Relationship interdependencies
- Appropriate capabilities
- Sustainability and vulnerability
- Reputational risks
- Business continuity impacts
- Management of potential conflicts between supply chains
- Equitable collaborative terms'.[483]

---

481 University of Reading (2018), 60. See also Public Contracts Regulations 2015, Reg. 67(4) which provides that the cost element 'may take the form of a fixed price or cost on the basis of which economic operators will compete on quality criteria alone'.
482 ISO 44001 Section 8.3.5.
483 ISO 44001 Section 8.3.5.

It is possible to combine selection processes with development of improved value, but an alliance is not created until its members are selected and their objectives, roles and rewards can be agreed and integrated. For so long as selection is still in progress the principal motivation of bidders is to win, and this can limit:

- Bidders' willingness to invest in developing designs and other new ideas
- Bidders' willingness to share these ideas
- The possibility of constructive discussions and exchanges with the client and other alliance members.

The selection process can include a requirement for bidders to submit proposals for improved value at their own risk and with no guarantee of being appointed, and this can be an important part of qualitative selection if bidders are willing to invest in speculative activities. For example, the Cost Led Procurement model envisages that bidders will be members of a framework that motivates their speculative proposals with the reasonable prospect of a pipeline of work.[484]

Selection using the public sector Competitive Dialogue Procedure or Competitive Procedure with Negotiation processes can include a period of structured negotiation or dialogue with bidders that is designed to interrogate bid proposals, to develop a better understanding of what is offered and to improve the content of offers in line with the established award criteria. For example, the Competitive Procedure with Negotiation includes the following steps:

- Establishment of technical specification and award criteria
- Publication of call for competition
- Pre-qualification of bidders to establish a shortlist
- Preparation and submission of tender proposals
- Negotiation with bidders
- Evaluation of tender proposals and reduction to two bidders
- Preparation and submission and evaluation of final offers.[485]

The SCMG and Project Horizon procurements both used a Competitive Dialogue Procedure but in each case the clients relied on post-award Supply Chain Collaboration rather than the pre-award dialogue process as the means to develop and agree improved value.[486]

When asked to submit speculative proposals for improved value, bidders may be concerned that their intellectual property rights are at risk and that the client may use their ideas even if they are unsuccessful. It is difficult to guard against this risk unless intellectual property rights are expressly protected in the tender documents. Some clients have offered to pay for the unsuccessful bidders' work in order that they are then entitled to release all or some of this work to the successful team for use on the project. This approach is illustrated in the Bank Station case study in Section 6.10.

---

484 Cost Led Procurement Guidance (2014).
485 Public Contracts Regulations 2015, Reg. 29.
486 SCMG and Project Horizon Trial Project case studies.

6.10   Collaborative Selection Case Studies

A constant theme in evaluation of prospective members of a collaborative team is that it should not focus excessively on comparative prices, and that qualitative evaluation is a more demanding and more rewarding process. The Tottenham Court Road case study describes how a team can be selected early in the life of a project on the basis of:

- Tendered profit, overheads and fees
- Identifiable costs in respect of site overheads/preliminaries and risk contingencies
- Full costing of pre-designed items
- Demonstration by the team members of their ability to deliver the project in accordance with the client's brief and within the pre-determined project budget.

The Bank Station case study describes how a Competitive Dialogue Procedure can use evaluation criteria by which bidders demonstrate how they will meet the client's stated objectives. The Bank Station selection process also involved the client agreeing to reimburse unsuccessful bidders for costs incurred in developing certain of their qualitative proposals, by means of a process governing:

- Confidential engagement of bidding teams for six months to review scheme information and work up their proposals for improvement of the client's business case
- Validation of bidders' ideas which are taken forward into the formal tender
- Evaluation, due diligence and contract award
- Purchase of losing bidders' innovations.

Implementation of the Bank Station project has gone on to adopt collaborative working whereby 'the Tier 1 contractor and owner project staff work in an integrated way in a joint project office with joint management'.[487]

Tottenham Court Road, London, Office Development

The Tottenham Court Road, London, office development was a major commercial, mixed-use development to provide new retail, office and residential buildings and was undertaken by a project alliance comprising client City of London Corporation, main contractor Costain, client representative Mouchel, cost consultant EC Harris, design consultants Watkins Grey International, Campbell Reith and Roger Preston & Partners.

City of London Corporation selected its design consultants and main contractor at the same time and utilised the conditional preconstruction phase appointment under PPC2000 as a basis for the team to establish firstly the feasibility of the project and then its detailed designs. Early selection of the team members was made on the basis of:

- Tendered profit, overheads and fees
- Identifiable costs in respect of site overheads/preliminaries and risk contingencies
- Full costing of pre-designed items

(Continued)

---

487 P13 Blueprint (2018), 19.

- Demonstration by the team members of their ability to deliver the project in accordance with the client's brief and within the pre-determined project budget.

The team members had to deliver a new state of the art office building within the confines of an existing nineteenth century façade. The façade had to be protected while the new office building was built inside it, and then the two were joined together with retail units below. The team reported that the project benefited from:

- A fully integrated multi-party design and construction team
- Joint feasibility study work under a first stage integrated contract timetable to establish viability of project and of innovative designs
- Joint planning and detailed design under a second stage integrated contract timetable
- Successful completion of a major commercial project on two sites in Central London under a third stage integrated contract timetable.

Peter Snowdon, Projects Director, City of London Corporation, reported: 'Using PPC2000 encourages a high level of collaboration between all team members which was certainly the case on this project. This collaborative philosophy enabled the Core Group to resolve the sometimes difficult issues referred to it quickly'.

Association of Consultant Architects (2010), 42–3.

Bank Station Capacity Upgrade

The selection process for the Bank Station Capacity Upgrade 2015 was a public sector Competitive Dialogue Procedure which provided a rare example of how alliance activities can be undertaken as part of a public procurement process in a way that leads to improved value offered by bidders and paid for by the client, and how this can be reconciled with public sector procurement constraints. London Underground ('LU') as client were keen to make a decision based on bidder proposals for achieving the business objective of 'increasing the number of passengers that can travel through the station at peak times, alleviating the impact of forecast passenger growth'. They structured their selection process as follows:

- Stage 1 pre-qualification of contractor teams
- Stage 2 Competitive Dialogue Procedure around a draft set of tender documents, with confidential engagement of bidding teams for six months to 'review the scheme information, including business case, and work up their proposals to improve this business case'
- Validation of bidders' ideas by the client, which are taken forward into the stage 3 formal tender for a target price design and build contract
- Evaluation, due diligence and contract award by LU including purchase of losing bidder innovations.

LU created 'a legally binding agreement between LU and each of the bidders and set out the obligations of either party in respect of confidentiality, intellectual property, and other commercially sensitive information'. Bidders' ideas were not shared with other bidders but 'bidder teams are paid for participating in the confidential engagement stage and unsuccessful bidders are paid for the subsequent use of their innovative ideas which are complementary to those of the winning bidder'. This was in line with Latham's recommendation referred to in Section 1.8 that, in order to improve the quality of bid submissions, certain tender costs should be reimbursed on large and expensive schemes. LU claimed that this approach 'demonstrated to the market that it is able to maintain confidentiality of bidders' ideas during a negotiated dialogue tender process, and behave equitably and fairly, reinforcing the principle that the bidders are entitled to make a reasonable profit'.

The Bank Station competitive dialogue was a demanding process, and 'around 150 meetings and 350 Requests for Information were handled during the dialogue'. An independent observer ('IOs') was appointed who 'provided written statements during and after the completion of the dialogue confirming the neutrality and objectivity of the engagement. Bidders conversed independently with the IOs and confirmed their satisfaction to the IOs at the end of the dialogue'.

In the context of capturing qualitative proposals as part of the contract awarded to the successful bidder, it is interesting to note LU's recognition that 'With hindsight it would have been preferable to constrain the format of the bidders' responses in the ITT and receive a uniformly structured return that could be readily bound into the contract documents'.

The benefits of inviting innovative proposals as part of the dialogue, while securing their confidentiality and reimbursing unsuccessful bidders, were evaluated by LU as including:

- '£61 155 000 (9.8%) reduction in the Estimated Final Cost to £563 812 000'
- '£148 625 000 (19.2%) increase in Journey Time Social Benefit over the 60-year life of the project'
- Five weeks reduced closure duration of the Northern Line, calculated as providing '£35 884 000 (52.9%) saving in social disbenefit'
- 'A more effective Step-Free Access solution to platform'
- 'A more efficient fire and evacuation strategy'.

Quotes are from London Underground (2014), reviewed for KCL Centre of Construction Law by Dr. Simon Addyman (who led the Bank Station procurement process on behalf of LU).

7

Does Collaborative Procurement Need a Contract?

7.1 Overview

Jason Russell, when briefing bidders for the Surrey Highways alliance,[488] described the TPC2005 standard form term partnering contract as 'intuitive, like my iPhone'.[489] A collaborative contract needs to be intuitive if it is to function as a practical routemap through collaborative processes rather than as an adversarial refuge of last resort.

To put it another way, 'it is the function of the law of obligations to smooth away inequalities and chance, and thereby to make it possible for the values that are inherent in humanity to become effective in their proper proportions'.[490] However, contracts do not bring out the best in people or organisations if they are created with the mindset that one party can only do well at the expense of another. Whether contracts are created by a tender process or by negotiation, their finalisation becomes an adversarial process if it is based on passing the maximum risks from one party to another and back again. The adversarial formation and enforcement of contracts can inhibit and undermine the sharing of objectives, efforts and data through collaborative working.

Some commentators suggest that collaborative procurement can be undertaken with no contract at all, and that excessive attention to contractual matters can undermine the working culture required for successful partnering because 'negotiating the terms of a formal contract tends to destroy partnering attitudes [because] working to rules and procedures defined in a standard form of contract inhibits partnering behaviour',[491] and that 'the mere suggestion or introduction of contracts may signal distrust of another party's intentions, thereby disrupting the process of trust development'.[492] For example, one approach adopted in the USA has used only 'a non-binding partnering charter or letter of intent that includes a management strategy with specific partnering principles'.[493]

Other commentators express concern at 'those who simplistically believe that collaboration based on trust alone, without an effective hierarchy of control in the relationship, can achieve improvements in construction outcomes'.[494] They suggest that collaborative

---

488 As summarised in Section 5.10.
489 At a bidders' conference during the procurement process leading to award of the term contract that underpinned the Project Horizon Trial Project.
490 Kohler, J. (1909). 'The philosophy of law', 135–136.
491 Bennett & Pearce (2006), 41.
492 Malhotra & Lumineau (2011), 983.
493 Tvarno, C. (2015), 307.
494 Cox & Townsend (1998), 333.

Collaborative Construction Procurement and Improved Value, First Edition. David Mosey.
© 2019 John Wiley & Sons Ltd. Published 2019 by John Wiley & Sons Ltd.

contracts can 'provide a foundation for "systems trust" by formalising shared expectations and assumptions of what constitutes accepted behaviour'.[495] When created in a way that supports collaboration 'contracts provide the means by which the parties can coordinate their expectations and efforts. As a result, common knowledge structures and routinized interactions emerge that make it easier for the parties to communicate their ability to meet each other's needs. The process of coordination can thus facilitate competence-based trust development'.[496]

Creative thinking, helpful actions and fair decisions are enabled and motivated by complex human factors, many of which have nothing to do with procurement processes and contract relationships. When people work together, they hope to find a chemistry that tolerates some idiosyncratic features, some unexpected developments, some space for trial and error. For example, although the following are not suitable standards for building quality, they tell us something important about people:

- 'Out of the crooked timber of humanity, no straight thing was ever made'[497]
- 'There is a crack in everything, that's how the light gets in'.[498]

With these observations in mind, some commentators are concerned that written processes and contracts may suck the air out of collaborative relationships and may inhibit innovation. Creating and maintaining the right balance between clear rules and spontaneous collaboration is a delicate responsibility in the hands of the people who lead and manage a team. However, the processes and contracts that put the right people in the right place with the right information at the right time can only improve their chances of success.

An underlying construction contract is formed every time a specification is agreed in return for a price, and suggestions that collaborative procurement can fall outside this contract need further analysis. The alternatives are as follows:

- Not to try to express collaborative commitments in a contract and instead to undertake collaborative working through a separate non-binding or informal arrangement, as explored in Sections 7.2 and 7.3
- To create a contract which is collaborative because it relies on implied or express good faith or creates a virtual organisation or excludes the right to make claims, as explored in Sections 7.4–7.7
- To create a contract which is collaborative because it contains new processes and working relationships that seek to improve and reward the value obtained from all parties' contributions, as explored in Sections 7.8 and 7.9 and in Chapter 8.

7.2 What Is the Impact of a Non-binding Agreement?

It has been suggested that, in a collaborative relationship, contracts are optional because there are some 'business transactions when a contract is not necessarily required'.[499]

---

495 Arrighetti et al. (1997), 175.
496 Malhotra & Lumineau (2011), 983.
497 Immanuel Kant (1724–1804), Idea for a Universal History with a Cosmopolitan Purpose.
498 Leonard Cohen (1934–2016), Anthem.
499 Cox & Thompson (1998), 83.

In the absence of a contract, the arrangements governing collaborative procurement can be based on unwritten understandings developed at meetings and in ad hoc communications or can be recorded only in non-binding documents. This suggests that collaborative procurement is a parallel universe where relationships and activities do not require the same level of clarity and commitment as in other commercial dealings. Yet the same reputational, technical and financial risks are at stake in a collaborative project as in any construction project, and even without a written contract the law will frequently intervene in a manner outside the parties' control in order to imply contractual, regulatory and other obligations.

The possibility of working on a collaborative project without a contract was considered in 1998 by Egan, who concluded that contracts frequently had a negative effect on the success of projects and who stated that 'if the relationship between a constructor and employer is soundly based and the parties recognise their mutual interdependence, then formal contract documents should gradually become obsolete'.[500] This statement was extensively quoted but was not demonstrated by supporting evidence in the Egan Report. Also, the idea that a collaborative understanding would be implied into a commercial relationship if it was not written down was soon called into question in the English courts, when one party attempted unsuccessfully to establish the existence of a contractual relationship through a course of dealing but in the absence of clear written terms.[501]

An unwritten approach to collaborative working invites the following risks:

- Misunderstandings, because anyone who has disputed the minutes of a meeting will know that recollections and interpretations of verbal exchanges always seem to differ and can lead to delays, frustrations and disputes
- Inefficiencies because, without a written contract, the team members need to rely on minutes of meetings, correspondence and other means to remind each other what they are doing
- Bad faith, because, without a clear understanding of agreed activities and deadlines, one party can deny what was agreed and can deprive another of its expected entitlements
- Excessive caution, because lack of clarity gives rise to a lack of confidence which may make the parties hesitate to share their ideas and to offer improved value.[502]

One trend in collaborative working has been a 'non-binding' protocol or charter.[503] This is an informal agreement describing a collaborative relationship in terms that are not intended to be legally binding, although the non-binding status of a protocol of this type cannot be guaranteed and has limited effect on a range of implied legal liabilities. The advisers who create protocols and charters are often not lawyers and may genuinely believe that words such as 'subject to contract' or 'non-binding' create a cloak of no

---

500 Egan, J. (1998), 33.
501 *Baird Textile Holdings v Marks & Spencer* [2001] EWCA Civ 274. where the Court of Appeal found that there was no evidence of an intention to be legally bound.
502 Mosey, D. (2009), 134.
503 For example, JCT Practice Note 4 set out a specimen 'non-binding partnering charter for single project' which is a blank form except for the statement that 'The team agree to work together on [the project] to produce a completed project to meet agreed client needs and meet agreed quality standards within agreed budget/price and agreed programme'.

liability. However, the views of the courts have shown that anyone who creates a protocol or charter believing its obligations to be optional should be encouraged to 'please step away from the contractual vehicle'.

For example, it may come as a surprise that, despite use of the words 'subject to contract', informal commitments have been treated as legally binding where it can be construed that a contract exists by reason of agreement of all essential terms. The courts will not lightly waive a 'subject to contract' provision but 'all will depend on the circumstances...'.[504]

The courts have also made clear that specifying the 'non-binding' status of a protocol or charter may not prevent them from taking it into account if necessary for them to establish the parties' substantive contractual rights. This lack of certainty can lead to unpredictable results in the event of a dispute. In the UK case of *Birse Construction v St David* a supposedly non-binding charter described how the parties would 'produce an exceptional quality development within the agreed time frame, at least cost, enhancing our reputations through mutual trust and cooperation'. A contract was never signed and, 10 months after signature of the charter, the client gave notice that it would deduct liquidated damages for delay. The contractor left the site, the client alleged abandonment of the project, the contractor claimed there was no abandonment because there was no contract, and it became clear that the aspirations set out in the charter would not be fulfilled.

In seeking to untangle this mess, HH Judge Humphrey Lloyd stated that a non-binding charter could be taken into account if necessary by the court in establishing that the parties intended to be contractually bound to each other and as a measure of the agreed standards of their conduct. The judge observed that 'the terms of that document, though clearly not legally binding, are important for they were clearly intended to provide the standards by which the parties were to conduct themselves and against which their conduct and attitudes were to be measured'.[505]

The Construction Industry Council offered this fresh perspective on the role of protocols and charters in the formation of collaborative relationships:

- 'While it is recognised that partnering charters have served a valuable role, the time is right to see a fully integrated approach, so that the relationships and processes required for effective partnering are not at odds with the contractual roles and relationships of partnering team members'[506]
- 'For the avoidance of doubt what we are talking about is a legally binding contract and not a non-legally binding charter or any equivalent'.[507]

---

504 In *RTS Flexible Systems v Molkerei Alois Muller GmbH & Co KG* [2010] UKSC 14, [2010] BLR 337 a letter of intent was issued for automated dairy equipment, noting the wish to proceed '*as set out in the offer*' subject to finalising price and completion date. There was no stated limit on cost or scope under the letter of intent and there was reference to an amended MF/1 contract to be concluded within 4 weeks. After 3 months of negotiation the latest draft said that the contract will not be effective until executed, but the Supreme Court held that a contract already existed by agreement of all essential terms despite lapse of the letter of intent.
505 *Birse Construction v St David (No 1)* [2000] 1 BLR 57 (CA).
506 CIC (2002), 12.
507 CIC (2002), 12.

Another trend in collaborative relationships was to rely on an underlying traditional contract.[508] As a result, some collaborative project teams adopted a twin-track contractual structure, whereby the parties signed a protocol or charter as a basis to pursue collaborative working but also entered into a traditional construction contract that they could fall back on if their collaborative relationship did not provide them with the desired results.

However carefully a collaborative protocol is prepared, there is a risk that it will not dovetail with the contract that it overlays and amends. To assume that collaborative contracts can be 'mimicked using traditional contracts that are interlinked through an addendum or rider' creates serious risks because 'in practice it is very difficult to rationalise the different terms among different contracts' and 'inconsistency among the contracts is almost guaranteed'.[509]

7.3 What Is the Impact of a Letter of Intent?

In the absence of a contract that governs collaborative procurement through early contractor involvement, a client may seek to establish preconstruction phase commitments by an informal agreement such as a 'letter of intent'. A letter of intent is a bespoke document not provided for in any standard form construction contract, and as a result its structure, content and impact vary widely. Depending on its terms, a letter of intent can have any of the following contractual effects:

- Creating no obligations at all but only a description, without commitment, of what is expected to happen in the future
- Creating a preconstruction contract governing limited activities pending the unconditional award of the construction phase contract[510]
- Creating an informal, but nevertheless unconditional, award of the entire construction phase contract.[511]

The effects of a letter of intent are unpredictable, yet they continue to be widely used in place of a contract governing early contractor involvement. This may be attributable to 'part of the folklore of the construction industry that there exists a mythical beast, the letter of intent, the legal effect of which, if acted upon, is that it entitles the contractor to payment for what he does, but does not expose him to any risk because it imposes no contractual obligations on him'.[512]

A letter of intent can have unexpected consequences, for example where a consultant was liable for failing to conclude a construction contract on a project which was completed using only a series of letters of intent, but where the client was unable to recover

---

508 Barlow et al. (1997), 34.
509 Fischer et al. (2017), 357.
510 As illustrated in *Turiff Construction v Regalia Knitting Mills* [1971] 9 BLR 20 (QB) and in *Mowlem v Stena Line Ports* [2004] EWHC 2206 (TCC).
511 As illustrated in *Mirant Asia-Pacific Construction (Hong Kong) v Ove Arup* [2004] EWHC 1750 (TCC) and in *RTS Flexible Systems v Molkerei Muller* [2010] UKSC 14, [2010] BLR 337.
512 HH Justice Seymour in *Tesco Stores v Costain Construction* [2003] EWHC 1487 (TCC). See also *Haden Young v Laing O'Rourke Midlands* [2008] EWHC 1016 (TCC).

liquidated damages for delay in accordance with the rights that would have arisen under formal contract terms.[513]

In order to operate as a collaborative basis for early contractor involvement, a letter of intent needs to adopt the characteristics of a preconstruction phase contract, including provisions that deal with:

- The timing of preconstruction phase activities
- Relationships with consultants, for example interfaces with design consultants for the purpose of agreeing value engineering and buildability
- The effect of preconstruction phase activities on any quoted fixed prices
- Any limits on the contractor's authority to incur costs, and any payment arrangements during the preconstruction phase
- The timing and procedure for finalising the construction phase contract, particularly if the position starts to change significantly from the documents or terms referred to in the letter of intent
- A procedure to agree the finalisation of detailed designs, the prices of each element of the works, the date of start on site and the date for completion and any other key issues left outstanding as at the date of the letter of intent.

Problems can arise under a letter of intent where it leaves important matters for later agreement, as those matters may be unenforceable as an agreement to agree.[514] While a letter of intent may be triggered by the need to agree limited early services, supplies or works in advance of a commitment to proceed with the whole project, there is a risk if these commitments create uncertainty as to the contract terms that govern them and the remainder of the project.

Alternatives to a letter of intent that allow early commitments to be based on agreed contract terms are the preconstruction agreements governing early activities under the JCT PCSA (2016), NEC4 Option X22 and PPC2000. The Arup report described the PPC2000 approach as 'a well thought out method of allowing works to be carried out whilst the documentation for the project is being developed. Providing a document to commence the project which is coordinated with the main contract processes and that prompts the parties to continue with developing the main contract documentation is superior to a stand-alone letter of intent'.[515]

7.4 What Is the Impact of Implied Good Faith?

It is often suggested that good faith is the key to successful collaborative procurement, and it is important to analyse what commitments the parties hope to obtain

---

513 *Trustees of Ampleforth Abbey Trust v Turner & Townsend PM* [2012] EWHC 2137 (TCC) where it was held there was no absolute obligation on the consultant to ensure the contract was signed but the consultant was liable for failure to take steps reasonably expected of a competent project manager for the purpose of finalising the contract: 'The execution of a contract is to be seen not as a mere aspiration but rather as fundamental'.

514 As also referred to in Section 2.7. In *Walford v Miles* (1992) 2 AC 128 (HL) Lord Ackner stated that: 'The reason why an agreement to negotiate, like an agreement to agree, is unenforceable is simply because it lacks the necessary certainty'. Lord Ackner also queried: 'How can courts be expected to decide whether, subjectively, a proper reason exists for the termination of negotiations?'

515 Arup (2008), 47. The PPC2000 Pre-Possession Agreement was renamed as the 'Pre-Construction Agreement', PPC2000 (2013), Appendix 3, Part 1.

from agreeing to act in good faith or from expecting this duty to be implied. When setting good faith alongside more specific alliance rights and obligations, is it only an aspiration or does it stand alone as an express or implied measure of contractual performance?

The term 'good faith' has been described by the English courts as an obligation 'to adhere to the spirit of the contract, to observe reasonable commercial standards of fair dealing, to be faithful to the agreed common purpose and to act consistently with the justified expectations of the parties'.[516] However, this definition leaves us no wiser as to:

- What is the 'spirit of the contract' and how can the judiciary discover and enforce it?
- What are 'reasonable commercial standards' and where do you find them?
- How far can a court go to discover and protect the 'justified expectations of the parties' and to reconcile these expectations when they are in conflict?

Another approach is to look at the ways that a team member on a construction project is expected to act in order to be doing so 'in good faith', for example:

- Acting honestly[517] in disclosure of all relevant information, either through a duty to warn[518] or a duty to notify a particularly onerous contract provision, for example where a party 'did not do what was necessary to draw this unreasonable and extortionate clause to [the defendants'] attention'.[519] However, it has also been suggested that 'there is no obligation in general to bring difficulties or defects to the attention of a contract partner or prospective contract partner', and that in such circumstances 'silence is golden'[520]
- Acting fairly and reasonably, for example when exercising a discretion under construction contract provisions for payment certification and other decisions by an independent project manager.[521] The contractual basis for such independent decisions is the reasonable judgement of the third party rather than a strict requirement to make an accurate assessment[522]
- Acting so as to take into account another party's interests by not enforcing a contractual right, for example where it is expected that good faith could lead to leniency because 'the partnering ethos…would naturally have led to a sympathetic approach to questions of extension of time and deduction of damages for delay'.[523] However, the English courts have also declared that good faith does not create an obligation to give up a freely negotiated financial advantage where, despite a good faith clause, it

---

516 Vos J. in *CPC Group v Qatari Diar Real Estate Investment Company* [2010] EWHC 1535 (Ch), following Morgan J. in *Berkeley Community Villages v Pullen* [2007] EWHC 1330 (Ch). This definition was qualified by the judge adding 'it might be difficult to understand how, without bad faith, there can be a breach of a duty of good faith, utmost or otherwise'.

517 Good faith as honesty can be taken as a minimum requirement on the basis that 'it is hard to envisage any contract which would not reasonably be understood as requiring honesty in its performance' – Leggatt J. in *Yam Seng Pte v International Trade Corporation* [2013] EWHC 111 (QB).

518 As considered in Sections 18.4 and 19.6.

519 Bingham L.J. in *Interfoto Picture Library v Stiletto Visual Programmes* [1987] EWCA Civ 6.

520 The Court of Appeal *ING Bank v Ros Roca* [2011] EWCA Civ 353.

521 *Sutcliffe v Thackrah* [1974] AC 727 (HL).

522 *Socimer International Bank v Standard Bank* [2008] EWCA Civ 116; [2008] 1 Lloyd's Rep 558.

523 HH Judge Humphrey Lloyd in *Birse Construction v St David* (No 1) [2000] BLR 57 (CA).

was not accepted 'that there is to be routinely implied some positive obligation upon a contracting party to subordinate its own commercial interests to those of the other contracting party'.[524]

The English courts are likely only to imply a contract term that corresponds to 'business common sense'[525] and may resist implying a duty of good faith unless it is 'reasonable and equitable', is 'necessary for business efficacy', 'goes without saying', is 'capable of clear expression' and is 'not in conflict with express terms'.[526] They 'have resisted being drawn into adopting a wide-ranging principle of implied good faith',[527] and their rationale for this caution includes:

- 'The proposition that "good faith" may be used as a fall-back device tellingly shows us why it is tempting but wrong to consider with the advantage of hindsight whether a term should be implied'[528]
- 'There is a real danger that if a general principle of good faith were established it would be invoked as often to undermine as to support the terms in which the parties have reached agreement'.[529]

The English courts recognise good faith as a 'moulding force' but not as a rule of English law,[530] and they prefer an incremental approach through which to 'develop piecemeal solutions in response to demonstrated problems of unfairness'.[531] Some English judgments have suggested that in refusing to accept a 'general obligation of good faith, this jurisdiction would appear to be swimming against the tide',[532] and recently a limited good faith obligation has been implied in a 25-year Private Finance Initiative (PFI) contract for road maintenance. However, even in that case, the judge confirmed that 'a duty of good faith is not usually implied into commercial contracts under English law'.[533] In contrast, good faith is enshrined in the US Uniform Commercial Code requirement that 'every contract or duty within this Act imposes an obligation of good faith in its performance or enforcement'.[534] Other common law jurisdictions have also moved closer to an implied duty of good faith, but they are not consistent in their approaches.[535]

Good faith is a statutory obligation that is implied in all commercial dealings in many civil law systems, and is enshrined, for example, in the civil code under German law.[536]

---

524 *Gold Group Properties v BDW Trading* [2010] EWHC 1632 (TCC).
525 *Rainy Sky S.A. v Kookmin Bank* [2011] UKSC 50 where Lord Clarke held that 'Since the language…is capable of two meanings it is appropriate for the court to have regard to considerations of commercial common sense in resolving the question of what a reasonable person would have understood the parties to have meant'.
526 *Mediterranean Salvage & Towage v Seamar Trading & Commerce (The Reborn)* [2009] EWCA Civ 531.
527 Mosey, D. (2015), 394.
528 HH Judge Humphrey Lloyd in *Francois Abballe v Alstom UK* LTL 7.8.00 (TCC).
529 Moore-Bick L.J. in *Mediterranean Shipping v Cottonex Anstalt* [2016] EWCA Civ 789 at 45.
530 Steyn L.J. in *First Energy v Hungarian International Bank* [1993] 2 Lloyd's Rep 194 (CA).
531 Bingham L.J. in *Interfoto Picture Library v Stiletto Visual Programmes* [1987] EWCA Civ 6.
532 *Yam Seng Pte v International Trade Corporation* [2013] EWHC 111 (QB).
533 *Portsmouth City Council v Ensign Highways* [2015] EWHC 1969 (TCC).
534 US Uniform Commercial Code, Section 1–304.
535 For example, Sir Rupert Jackson assessed the varying treatment of good faith in Australia, Canada, Hong Kong and Singapore in Jackson, R. (2018).
536 Article 242 of the German Civil Code provides that 'An obligor has a duty to perform according to the requirements of good faith, taking customary practice into consideration'.

The statutory implication of a duty of good faith provides judges with another measure of contract interpretation, but this alone does not create sufficient clarity in the parties' relationships and mutual obligations to reduce the number of construction disputes. For example, a German judgment recorded how 'the duty of cooperation requires that in case of disputes [the] arguments, alternatives and counterproposals of the other contractual partner have to be registered and should be part of the exchange of views. Here, the court held that the duty to cooperate was breached which led to a claim for compensation'.[537] There are other cases that illustrate how an implied duty of good faith has been used as a determining factor in a dispute,[538] but it is harder to find empirical evidence of how an implied duty of good faith may have avoided disputes coming to court at all.

As to whether pre-contract commercial negotiations should be conducted in good faith, the English courts have held that to imply 'a duty to carry on negotiations in good faith is inherently repugnant to the adversarial position of the parties', and that each party is 'entitled to pursue his (or her) interests as long as he avoids making misrepresentations'.[539] However, where there is sufficient clarity, an express agreement to act in good faith in negotiations has been treated by the courts as enforceable, dependent on how much of the contract still remains to be agreed.[540] Other instances of implied good faith under English law are considered in the context of joint ventures and other relational contracts in Section 8.2.

Where good faith equates to an implied requirement of honest behaviour, the courts are unlikely to intervene if the parties are equally matched.[541] However, where there is an imbalance, English statute law has used good faith as a measure of required behaviour when protecting the consumer, as a more vulnerable party, from unfair contract terms imposed to their detriment.[542] When the courts have been asked to interpret this wording, it has been stated that 'The requirement of good faith in this context is one of fair and open dealing...which looks to good standards of commercial morality and practice'.[543]

The English courts will look at good faith alongside 'good industry practice' and have defined the latter as including a party 'seeking in good faith to comply with its

---

537 *OLG Brandenburg*, judgment dated 16 March 2011 – 13 U 5/10, one of several examples of how a statutory duty of good faith affects a range of disputes under German law, as considered in Jackson & Fuchs (2015), 406.
538 Jackson & Fuchs (2015).
539 Lord Ackner in *Walford v Miles* [1992] 2 AC 128 (HL).
540 In the case of *Alstom Signalling v Jarvis Facilities* [2004] EWHC 1232 (TCC), a 'pain/gain' arrangement was under negotiation with a view to a subcontract pain/gain share matching the equivalent main contract provisions, and it was held that there was an implied primary obligation to make reasonable endeavours to agree and to negotiate in good faith, and that the court could determine the missing element if no agreement was reached.
541 For example, in *Esso Australia v Southern Pacific* [2005] VSCA 228, it was held that 'If one party to a contract is more shrewd, more cunning and out-manoeuvres the other contracting company who did not suffer a disadvantage and who was not vulnerable, it is difficult to see why the latter should have greater protection than that provided by the law of contract'.
542 The UK Consumer Contracts Regulations 1994 construe an unfair term as one which 'causes a significant imbalance in the parties' rights and obligations under the contract to the detriment of the consumer in a manner or to an extent which is contrary to the requirement of good faith'.
543 *DGFT v First National Bank* [2001] UKHL 5.

contractual obligations and complying with relevant legislation'.[544] Yet a definition of commercial morality remains elusive and this gives rise to challenges when seeking to create a model of ethical conduct for the construction industry.[545] It therefore remains important to create 'cooperative adaptations' as protections against 'the hazards of opportunism'.[546]

Other implied obligations in any English construction contract could be argued to have a similar effect to implied good faith. A general duty to cooperate has been implied by common law for many years.[547] It has also been implied that the parties to a construction contract will act reasonably in all matters related to the project, and 'will not do anything to prevent the other party from performing a contract or to delay him in performing it' and 'will do whatever is necessary in order to enable a contract to be carried out'.[548]

7.5   What Is the Impact of Express Good Faith?

Whatever the courts' view on the meaning of good faith, it appears as an express obligation in many construction contracts. Various versions of this commitment appear in the following standard forms:

- ConsensusDocs 300: 'Each IPD Team member accepts the relationship of mutual trust, good faith, and fair dealings established by this Agreement, and agrees to cooperate and to exercise its skill and judgement to further the interests of the Project'…'The IPD Team will promote harmony and collaboration on the Project'[549]
- Infrastructure Conditions of Contract (ICC) 2014: 'The Contractor, the Employer and the Engineer on his behalf shall each, in the performance of the Contract, collaborate in a spirit of trust and mutual support in the interests of the timely, economic and successful completion of the Works'[550]
- JCT SBC/Q 2016: 'The Parties shall work with each other and with other project team members in a co-operative and collaborative manner, in good faith and in a spirit of trust and respect. To that end, each shall support collaborative behaviour and address behaviour which is not collaborative'[551]

---

544 *Fujitsu v IBM* [2014] EWHC 752 (TCC).
545 A fascinating exercise is recorded in Lavers, A. (2008), which referred to the seven principles of ethical conduct devised by John Uff Q.C. and H.H. Judge Antony Thornton Q.C., namely fair reward, integrity, objectivity, honesty, accountability, fairness and reliability. These principles were stated to apply to 'all professionals working in the construction industry, whatever their original qualification of affiliation'. A questionnaire compared the varying treatment of these principles by reference to the legal systems and contractual practices of eight European jurisdictions.
546 Williamson, O.E. (1979), 251 and Williamson, O.E. (1993), 48.
547 As established in *Mackay v Dick* (1881) 6 App Cas 651 (HL).
548 Vinelott, J. in *London Borough of Merton v Stanley Leach* [1985] 32 BLR 51 (Ch).
549 ConsensusDocs 300 (2016) clause 2.3.
550 ICC (2014) clause 6.1.
551 JCT SBC/Q (2016) clause 1, Schedule 8 (Supplemental Provisions).

- JCT CE: Under an 'Overriding Principle' the parties agree to 'work together with each other and with all other Project Participants in a co-operative and collaborative manner in good faith and in the spirit of trust and respect'[552]
- NEC4: 'The Parties, the Project Manager and the Supervisor shall act in a spirit of mutual trust and co-operation'[553]
- PPC2000: 'The Partnering Team members shall work together and individually in the spirit of trust, fairness and mutual cooperation for the benefit of the Project, within the scope of their agreed roles, expertise and responsibilities as stated in the Partnering Documents'.[554]

These commitments encourage good faith, but they are vague and their legal effect is debatable. For example, the NEC4 wording appears to create obligations for the project manager and supervisor, even though an NEC4 construction contract is made only between the client and the contractor and cannot bind the project manager or supervisor. The JCT SBC/Q 2016 wording arguably goes further than that of the other standard forms, but it is not clear whether the JCT commitment to 'other project members' creates rights for those other project members,[555] nor is it clear what rights arise from the rather sinister JCT obligation to 'address behaviour which is not collaborative'.

The English courts have decided that the use of the above PPC2000 wording in a TPC2005 term partnering contract[556] did not prevent a party relying on a clear contractual right of termination 'at will' by written notice, and that this right could be exercised by a party without the 'good faith' wording creating any obligation for it to provide justification for its decision.[557] The judge did not accept that the good faith wording created an implied overriding duty to preserve a long-term relationship because there was also an express right to terminate.[558]

The above decision provides a valuable clarification without which any NEC4 or PPC2000 contract, and any other contract form that requires good faith or mutual cooperation, could be subject to requirements by one party for the judiciary to second-guess the impact of those general words on every other contractual provision. However, the court might have reached a different conclusion if the relationship had been governed by the JCT CE provision for good faith as an 'Overriding Principle',[559]

---

552 JCT CE (2016) clause 2.1. Although JCT CE was one of the forms recommended by the UK Government for early contractor involvement and collaborative working none of the Trial Project teams adopted it, and there is no other evidence available of how the JCT CE wording has been applied in practice.
553 NEC4 (2017) clause 10.2.
554 PPC2000 (2013) clause 1.3.
555 Pursuant to the Contracts (Rights of Third Parties) Act 1999.
556 TPC2005 (2013) clause 1.3 does not expressly mention good faith but, as in the case of PPC2000 (2013), requires the parties to 'work together and individually in the spirit of trust, fairness and mutual cooperation'.
557 In the case of *TSG Building Services v South Anglia Housing* [2013] EWHC 1151 (TCC) there was an expectation by TSG that a long-term relationship would be preserved despite provision for South Anglia to be entitled to terminate the contract at any time without giving any reason. Mr Justice Akenhead determined that, even if a duty of 'good faith' could be implied in a construction contract, that duty would not override or affect other express rights and obligations such as an entitlement to terminate by notice.
558 Mr Justice Akenhead determined that, even if a duty of 'good faith' could be implied in a construction contract, that duty would not override or affect other express rights and obligations such as an entitlement to terminate by notice.
559 JCT CE (2016) clause 2.1.

as this wording could have the effect of qualifying the enforceability of all other JCT CE contract rights and obligations.

Two UK cases involving NEC contracts show how the English courts continue to interpret good faith wording in different ways:

- In one case it was concluded that, while a party should not exploit another party or take advantage of another's false assumption, 'the spirit of mutual trust and cooperation' does not require a party to act against its own self-interest[560]
- In the other case a party was obliged to reveal cost and time data that was contrary to its interests because to not do so would be 'entirely antipathetic to a spirit of mutual trust and cooperation'.[561]

Term contracts in respect of outsourced services have tested the extent to which good faith creates a basis for collaborative working. For example, a dispute arose under a term contract for the provision of catering and other services to an NHS Trust, which led to the imposition of very harsh financial deductions for apparently minor contract breaches, such as £84 540 for a chocolate mousse being one day out of date. Despite a clause requiring the parties to 'co-operate with each other in good faith', the court limited this duty to specific obligations set out in the clause where the good faith wording appeared, namely 'the efficient transmission of information and instructions'.[562]

7.6   What Is a Virtual Organisation?

Integration between the members of an alliance is sometimes described as the creation of a virtual organisation, as if they are all serving the interests of a single company.[563] A virtual organisation has no legal definition, and team members creating collaborative commitments should define what they mean by a virtual organisation. For example, a clear understanding is needed to ensure that an alliance contract, in stating that it creates a virtual organisation, does not create a business relationship with unintended consequences in terms of profit sharing, joint taxation and joint and several liability.

Alliance members who refer to a virtual organisation might wish their relationships to have the features of a contractual joint venture. Examples of how collaborative contractual systems can support a joint venture are considered in Section 8.8 and in the Kier/Sheffield case study in Section 8.10. Alternatively, alliance members might wish to create and join an actual organisation. For example, the USA AIA C195 series of contracts 'uses a limited liability company to integrate owner, architect and contractor'.[564]

A joint venture may involve the creation of a new organisation such as a distinct corporate entity or partnership, or it may be governed only by the terms of a joint venture agreement between members who remain separate organisations.[565] A contractual joint venture, when combined with the co-location of staff, may bring the

---

560 *Costain v Tarmac Holdings* [2017] EWHC 319 (TCC).
561 *Northern Ireland Housing Executive v Healthy Buildings (Ireland)* [2017] NIQB 43.
562 *Compass Group v Mid-Essex Hospital NHS Trust* [2012] EWHC 781 (QB). Despite the wider wording at the end of that clause linking good faith to each party deriving 'the full benefit of the Contract'.
563 'This anticipatory, commonly held "sense" of the parties may virtually obliterate and present separation as [self-interested] maximisers, thereby making them effectively a single maximiser for many purposes, including, for example, the selection of governance structures', Macneil, I.R. (1981), 1024.
564 Fischer et al. (2017), 383.
565 Joint Venture Guidance (2010), 11–17.

members so close together that they feel like they are running a single organisation, but in legal terms they remain distinct entities who agree to contribute their expertise and resources for specific commercial purposes. In the absence of a separate corporate vehicle in which alliance members become shareholders, any contractual references to a virtual organisation will not create the relationships and duties that are implied among the directors of a single corporate entity.

Even where a duty of good faith is established among team members, each will remain bound by their primary duty to protect the interests of their own organisations and to construe any recognition of other parties' interests in this context. For example, English law struggles to construe a duty of good faith owed between different organisations but states expressly the balance of duties by which a company director:

- Is required in 'good faith' to promote the success of his or her company[566]
- Is also required to have regard to 'the need to foster business relationships with suppliers, customers and others' and 'the desirability of the company maintaining a reputation for high standards of business conduct'.[567]

PPC2000 provides for all alliance members to sign a single contract but distinguishes between the members' different rights and obligations so that these are limited to 'their respective roles, expertise and responsibilities'.[568] It also states that 'nothing in the Partnering Documents shall create, or be construed as creating, a partnership between any of the Partnering Team members'.[569]

By contrast, the multi-party NEC4 Alliance Contract creates collective responsibilities by which the roles of alliance members are intertwined, and its guidance describes the alliance as 'the overall entity undertaking the works'.[570] In addition, while describing certain alliance members as 'Partners', the contract does not make clear that they do not form a partnership. This may benefit from clarification.

Some alliances create collaborative relationships designed to mirror the integration and common purpose that can underpin a single organisation. For example, Anglian Water reported that their 'alliance, as an integrated and collaborative organisation, was engaged at outcome level, not at project or scope, giving partners and the wider supply chain the opportunity to develop more innovative solutions and to challenge standards'. This level of integration was supported by:

- The alliance team itself acting as 'the integrator, developing strategies for how the programme should be delivered and driving improvement initiatives'
- The partners that make up the alliance, along with Anglian Water, being treated as 'shareholders that generate a return by outperforming historic benchmarks for delivery of outcomes'
- The alliance also managing 'the wider supply chain, with a longer-term framework used to develop more effective relationships and secure earlier involvement of the right suppliers', and with framework suppliers motivated to 'generate a return by

---

566 For example, the UK Companies Act 2006, Section 172.1 requires a company director to 'act in a way that he considers, in good faith, would be most likely to promote the success of the company'.
567 Companies Act 2006, Sections 172.1(c) and (e).
568 PPC2000 (2013) clause 22.1.
569 PPC2000 (2013) clause 25.1.
570 NEC4 *managing an alliance contract* (2018), 5 by reference to NEC4 Alliance Contract (2018) clause 11.2(1).

delivering value against historic baselines, not by delivering work or providing hours'.[571]

The P13 Blueprint describes a collaborative approach 'based on enterprise, not on traditional transactional arrangements', integrating the roles of an owner, an investor, suppliers, advisers and an integrator who 'brings in appropriate suppliers and advisors at relevant points within the enterprise as and when they can best add value'.[572] The P13 Blueprint outlines a collaborative 'enterprise' as:

- 'An integrated organisation, aligned and commercially incentivised to deliver better outcomes for customers from infrastructure investment'
- 'Organisations characterised by sophisticated, maturing and typically longer-term relationships between owners, investors, integrators, advisors and suppliers'.[573]

This description of a collaborative enterprise is not the same as other collaborative concepts such as the enterprise planning and enterprise contracts described in Chapter 8. It is not clear whether the P13 Blueprint envisages the enterprise model as a new corporate entity or as a contractual joint venture. However, the first P13 Blueprint case study describes the strategic models for collaborative procurement created by the Surrey Project Horizon Trial Project and the Kier Highways Services supply chain alliance created, both of which combined term alliance contracts and framework alliance contracts with co-location and collaborative mentoring but neither of which expressed their arrangements as a new virtual or actual organisation.[574]

7.7 What Is the Impact of a No Blame Clause?

Some alliance contracts include 'no blame' clauses by which alliance members agree not to take legal action against each other and/or agree to share the costs arising from third-party legal claims.[575] The philosophy behind a no blame clause is that it is possible to cultivate more efficient and innovative practices if collaborative working cannot be undermined by the fear of claims.[576] These clauses, also known as 'no claim' clauses, have strong supporters and mark a significant departure from contract norms. Hence, they deserve analysis of their meaning, scope, enforceability and financial effects.

Under PPC2000 there is an option for team members to adjust their duty of skill and care to each other, for example by each team member being responsible for loss and damage suffered by the client in agreed proportions 'irrespective of the extent of each Partnering Team member's actual contribution to the cause of that loss or damage'.[577] However, in practice, the PPC2000 teams who have agreed to incentives by way of shared

---

571 Case study summary in Section 4.10.
572 P13 Blueprint (2018), 2 and 4.
573 P13 Blueprint (2018), 1.
574 P13 Blueprint (2018), 9, 10 and case study summaries in Sections 5.10 and 11.10.
575 As illustrated in the NEC4 Alliance Contract (2018), in the Australian alliance model described in Chapter 20 and in the US IPD model described in Chapter 25.
576 Infrastructure Client Group (2015) Section 3.4 states that a key requirement of an alliance is 'A no blame – no dispute approach (wilful default being the only direct route to legal process)'.
577 PPC2000 (2013) clause 22.1 option at page (v).

cost savings and cost overruns have not combined these incentives with waivers of legal liability.

The NEC4 Alliance Contract goes further and provides for:

- A 'no claim' clause by which 'any failure by a member of the Alliance to comply with their obligations stated in these conditions of contract does not give rise to any enforceable right or obligation at law except for an event which is a Client's or Partner's liability'[578]
- The definition of the Client's or Partner's liability that includes 'claims and proceedings from Others and compensation and costs payable to Others' which 'are a result of an intentional act or omission to not comply with' the relevant alliance member's obligations under the contract[579]
- The definition of 'Others' as 'people or organisations who are not the Client, the Alliance Manager, a Partner or any employee, Subcontractor or supplier of a member of the Alliance'[580]
- The option to state additional Client's or Partner's liabilities in the Contract Data.[581]

The NEC4 Alliance Contract is the first attempt to capture a no blame clause in a UK standard form but in some respects the above provisions appear confusing, for example as to:

- Why the limitation of liability to an 'intentional act or omission' appears to apply only to claims by third parties ('Others') and not to claims between alliance members (who do not fall within the definition of 'Others'), and whether this will lead to additional liabilities being stated in the Contract Data
- Why alliance members appear to have no legal rights and obligations in respect of the operative contract terms they have agreed that do not fall within the stated Client's or Partner's liabilities, for example as to payment of the agreed shares of Alliance Costs, and whether this will lead to additional liabilities being stated in the Contract Data
- The benefits and risks of providing that alliance members cannot rely on any 'enforceable right or obligation' in respect of their agreed alliance commitments
- Why the guidance states that all other liabilities, agreed to be 'jointly held by the Alliance' as an alliance cost, 'will not extend beyond the final account',[582] and whether it is intended that the alliance members have no liabilities at all following the final account.

The NEC Alliance Contract in its 2017 consultative version contained a different no blame clause linked to 'wilful default'.[583] Whether a no blame clause uses the term 'intentional' or 'wilful', the interpretation of these words may generate the disputes that they are intended to avoid. For example, one judicial interpretation of a clause that used the term 'deliberate default' construed it as meaning that the relevant person knew that their

---

578 NEC4 Alliance Contract (2018) clause 94.
579 NEC4 Alliance Contract (2018) clauses 80.1 and 81.1.
580 NEC4 Alliance Contract (2018) clause 11.2 (23).
581 NEC4 Alliance Contract (2018) clauses 80.1 and 81.1. The Contract Data records agreed details, including those in respect of alliance members' liabilities.
582 NEC4 *managing an alliance contract* (2018), 61 by reference to NEC4 Alliance Contract (2018) clause 82.1.
583 NEC4 Alliance Contract (2018), 2017 consultative version clauses 80 and 81.

act was a default, whereas if the person was merely reckless then their default would not be deliberate.[584] An interpretation of wilful default in the context of Australian alliancing is 'such wanton and reckless act or omission as amounts to a wilful and utter disregard for the harmful and unavoidable consequences thereof…but shall not otherwise include any error of judgement, mistake, act or omission, whether negligent or not, made in good faith…'.[585]

Despite the lack of clear judicial guidance, it is likely that a no blame clause will be interpreted as relieving the team members of individual liability for their negligence. This relief needs to be construed in the context of current concerns as to willingness of the construction industry to take responsibility for its projects,[586] and may lead to arguments that a no blame clause:

- Could reduce the vigilance of alliance members rather than increasing their efficiency and innovation
- Could dilute the responsibilities of alliance members in relation to matters affecting the safety or functionality of a completed project.

The limitations and exclusions of liability created by a no blame clause are often linked to shared pain/gain through agreed contributions to remedying certain types of loss and damage.[587] The impact of managing risk through no blame clauses is considered in Section 18.7.

No blame clauses are sometimes linked to a new type of insurance. For example, because professional indemnity insurance in respect of design liability covers the legal liability of the designer, this type of insurance has no role where no legal liability arises. Instead a no blame clause can be combined with insurance that is unconnected with identification of who is responsible for causing loss or damage. The implications of this type of insurance are considered in Section 18.8.

In the absence of new insurance cover, a no blame clause often provides that problems will be dealt with by sharing costs among all alliance members, including the client, irrespective of which party caused the relevant loss or damage. However, some clients and other team members may question why collaborative procurement obliges them to contribute a share to cost overruns, delays or defects that arise from another alliance member's negligent acts or omissions. This question should be considered through a comparison with the other incentives that support collaborative procurement, in order to establish:

- The value of agreeing a no blame clause and the consequent sharing of costs in order to discourage a blame culture, for example because it prevents alliance members from relying on risk transfer, risk contingencies and other defensive tactics, and because it encourages a collaborative culture by which alliance members are motivated to assist each other in resolving any issue
- Whether, instead, the good effects of sharing of costs and rewards can be achieved through a shared pain/gain incentive of the type considered in Section 16.8 without the addition of a no blame clause

---

584 Mr Justice Edwards-Stuart in *De Beers UK v Atos Origin IT Services* [2010] EWHC 3276 (TCC).
585 Cheung et al. (2006).
586 See, for example, Building a Safer Future (2018) and Section 15.8.
587 As considered in Section 16.8.

- Whether or not it is necessary, therefore, to combine an incentive clause with a no blame clause that prevents alliance members from enforcing their contractual rights, including for example their rights under the incentive clause itself
- Whether the client has behavioural and commercial influence sufficient to ensure that other alliance members perform and collaborate, if and to the extent that a no blame clause prevents any alliance member from enforcing its contractual rights when another alliance member does not perform or collaborate.

7.8 What Is the Impact of a Collaborative Contract?

Collaborative procurement cannot thrive in a parallel universe separate from team members' commercial expectations and commitments. In the absence of a binding contract, collaborative arrangements are likely to be seen as an option overlaying other, less collaborative arrangements for the members to fall back on at any time.[588]

Collaborative activities require coordination and agreed timelines in order to create mutual confidence among team members that each will create and share the data that they all need to deliver the project. Collaborative contracts provide the machinery for coordination and the mutual commitments to agreed activities and timelines.[589] The value of coordination provisions in contracts has been recognised where 'common knowledge structures such as shared language and routinized interactions emerge that make it easier for the parties to communicate their ability to meet each other's needs'.[590] Coordination provisions have been distinguished from control provisions in collaborative contracts and, unlike control provisions, can 'increase the likelihood of continued collaboration after a dispute and that perceptions of competence mitigate this effect'.[591]

The strengths of collaborative contracts have been demonstrated in all the case studies examined in this book. For example, AmicusHorizon and Morrison Facilities Services used a TPC2005 term partnering contract as the means to establish:

- 'Agreement of prices initially on the basis of a schedule of rates with agreed TPC timetable to move to establishment of an "Open-book" approach
- Development and use of a Risk Register
- Development of incentivisation through pain and gain
- Use of alternative dispute resolution procedures
- Description of all key activities in the Partnering Timetable'.[592]

---

588 A survey of organisations involved in partnering noted that 'lack of understanding could cause confusion between partnering arrangements and negotiated contracts. It also led to a lack of involvement of suppliers and users', Housing Forum (2000), 13.
589 These roles are explored in Mosey, D. (2003/1).
590 Malhotra & Lumineau (2011), 983.
591 Malhotra & Lumineau (2011), 993 whose research found 'that contract design affects the degree of trust that exists after a conflict has arisen, and, through this effect, the likelihood of relationship continuance'. They concluded that 'control provisions have a negative effect on the willingness to continue a damaged relationship, and goodwill trust mediates this effect', whereas 'coordination provisions increase the likelihood of continued collaboration after a dispute and that perceptions of competence mitigate this effect'.
592 Association of Consultant Architects (2010), 9.

The UK Construction Industry Council recommended that:

- 'An effective partnering contract should support the full partnering team and aim to deliver an integrated project process. Logically, it should replace any of the existing standard forms'
- 'An effective contract can play a central role in partnering. It sets out the common and agreed rules; it helps define the goals and how to achieve them; it states the agreed mechanism for managing the risks and the rewards; it lays down the guidelines for resolving disputes'.[593]

ISO 44001 recognises the need to:

- 'Ensure a defined process and clear guidelines are in place to capture and manage knowledge creation and sharing between organisations…[which] shall include, where appropriate designated areas which are to be protected from unintended knowledge transfer to collaborative partners'[594]
- 'Establish and agree a formal foundation for joint working, including contractual frameworks or agreements, roles, responsibilities and ethical principles'[595]
- 'Determine clarity of purpose, encourage appropriate behaviour and identify the potential impacts on or conflict with the aims of collaborative working'.[596]

7.9 What Is the Impact of a Multi-Party Contract?

Most construction contracts create only two-party relationships, even in respect of their collaborative provisions.[597] Collateral warranties or third-party rights provisions establish additional direct contractual links between team members,[598] but they are designed only to extend a duty of care rather than to integrate working relationships. Collaborative contracts can comprise an integrated set of two-party contracts,[599] but it is difficult to see how two-party contracts can support all the features of an alliance. If alliance members' interests under contracts are not properly aligned, this 'leaves room for self-interested behaviour to thwart the realisation of efficient plans'.[600]

There is an emerging argument that collaborative procurement generally, and alliances in particular, can create clearer relationships and more integrated processes if expressed in a multi-party contract which establishes:

- The ability of each alliance member to see the roles and responsibilities of the other alliance members so as to check that these correspond to their expectations and so as to ensure that there are no gaps or duplications

---

593 CIC (2002), 12.
594 ISO 44001, Section 8.3.4.
595 ISO 44001, Section 8.6.2.1.
596 ISO 44001, Section 8.6.10.
597 For example, FIDIC (2017), ICC (2014), all JCT 2016 contracts except the JCT CE Project Team Agreement (2016) and all the NEC4 contracts except the NEC4 Alliance Contract (2018).
598 For example, pursuant to the UK Contracts (Rights of Third Parties) Act 1999.
599 For example, NEC4 (2017).
600 Milgrom & Roberts (1992), 129.

- Transparency that can help to create a level playing field that builds trust and confidence because each alliance member can see that it is engaged on the same terms as all other alliance members
- The ability of alliance members to establish and build a system of direct peer group rights and mutual reliance, without routing all contractual concerns via the client as an intermediary
- Simpler contract structures such as direct mutual intellectual property licences, a single integrated set of timeframes and deadlines, and a single governance structure agreed among alliance member representatives.[601]

Where the client enters into separate two-party contracts relating to the same project, there are no direct contractual links that establish peer group rights between, for example, the architect, the structural engineer, the services engineer, the cost consultant, the main contractor and those of its specialist subcontractors who contribute to designs. Each has a contractual route only via the client, or via the main contractor, as regards exercise of their authority, the review of their contributions and the effect of their communications.

For example, a multi-party contract can clarify the collaborative role of a project manager or alliance manager, which should be integrated with its decision-making powers, but which is often blurred where two-party contracts give powers to a third party who does not sign them.[602] For example, JCT SBC/Q (2016) provides that its 'Architect/Contract Administrator' shall:

- Issue instructions regarding provisional sums[603]
- Issue certificates for interim payments within agreed deadlines[604]
- Ascertain amounts of loss and/or expense[605]
- Instruct and value variations[606]
- Certify practical completion.[607]

These powers suggest that the Architect/Contract Administrator has rights and obligations as a party to the JCT contract but, just like the 'Project Manager' and 'Supervisor' under NEC4 or the 'Engineer' under FIDIC (2017), the JCT Architect/Contract Administrator is not itself a party to the same JCT contract as the main contractor. In the absence of a direct multi-party contractual relationship, the liability of a project manager or contract administrator to a main contractor is very limited and its powers are, therefore, not matched by corresponding obligations.[608]

The client, as the only common signatory to a series of two-party contracts, can be asked to make decisions based on conflicting views and conflicting information

601 'The success of PPC2000, and more recently TPC2005, rest on their clear commitment to integration of the partnering team around a single contractual hub', Sir Michael Latham in Association of Consultant Architects (2010), 1.
602 The role of the alliance manager is considered in Section 12.4.
603 JCT SBC/Q (2016) clause 3.16.
604 JCT SBC/Q (2016) clause 4.9.
605 JCT SBC/Q (2016) clause 4.21.4.
606 JCT SBC/Q (2016) clause 5.
607 JCT SBC/Q (2016) clause 2.30.
608 As illustrated in *Pacific Associates v Baxter* [1990] 1 QB 993 and *Costain and Others v Bechtel* [2005] EWHC 1018 (TCC).

provided by different alliance members under the terms of their different contracts. The client becomes inextricably involved in the operation of interfaces between alliance members, even though the client is usually the least well-qualified party to decide who is right. This leads us to consider the pros and cons of direct contract links under a multi-party structure.

Concerns may be expressed that multi-party contracts may confuse the roles and responsibilities of different alliance members, may create additional liability of alliance members for other members' acts and omissions and may be harder to insure as a consequence. Whether any of this is true depends on the drafting of the multi-party contract itself and in practice:

- A multi-party contract can restrict the agreed liability of alliance members to their respective roles, expertise and responsibilities and can exclude each party's responsibility for any error or omission in or discrepancy between any documents, for example to the extent that it is agreed that a party will rely on contributions and information provided by other signatories to the contract[609]
- A multi-party contract can create an integrated set of mutually agreed commitments with maximum clarity as to alliance members' roles and responsibilities, because each alliance member can see, for example, the contractor's project brief and the various consultant services schedules and can see where its own role and responsibilities fit in[610]
- No problems or concerns have been raised by insurers in relation to any of the multi-party contracts described in the case studies referred to in this book.

Alliances have been defined in multi-party terms as:

- A 'multi-party integrator alliance'
  - Under which a client appoints multiple contractors or consultants under consistent terms with an extended supply chain
  - Which creates potential for the integration of tier 2 or tier 3 supply chain members and a transparent basis for the award of varying amounts of work to different contractors or consultants dependent on performance and other agreed criteria
- A 'single integrator multi-stakeholder alliance'
  - Under which a client appoints one contractor and agreed tier 2 or tier 3 subcontractors and suppliers
  - Which creates agreed procedures for the award of work by the contractor to tier 2 or tier 3 subcontractors and suppliers with integrated performance management/incentive arrangements.[611]

Multi-party project alliance contracts are considered further in Sections 3.4, 8.6 and 9.9. Multi-party framework alliances and term alliances are considered in Chapters 4, 5, 10 and 11. There has also been growing interest in the role that a multi-party contract can play in supporting the delivery of projects that use Building Information Modelling (BIM), and this is explored in Section 14.8.

---

609 For example, under PPC2000 (2013) clause 2.4.
610 For example, PPC2000 (2013).
611 Infrastructure Client Group (2014), 5.

7.10   Collaborative Construction Management Case Studies

Alliances and other collaborative contracts are illustrated in all the case studies that appear throughout this book, in most cases using a multi-party structure. The following case studies show how a multi-party alliance contract structure can be extended to specialist contractors and can create a more collaborative approach to the procurement model known as construction management.[612]

The first case study describes a hotel refurbishment project led by Jones Lang LaSalle on behalf of Jarvis Hotels where the team adopted a construction management adaptation of PPC2000.[613] They combined team integration with the programming of a complex refurbishment programme and the management of risks arising from commencing early works on site in an occupied building. Their Core Group comprised representatives of the client and four construction managers, each leading a workstream, resolving concerns and ensuring that work stopped if necessary for customer relations. The team used the PPC2000 contract change management mechanisms to minimise cost and time consequences deriving from a continually evolving work scope.

The second collaborative construction management case study describes a new build learning and events centre for the Maudsley mental health charity, where particular attention was given to the selection and introduction of trade contractors in order to ensure that they understood the ethos and vison of the client. Early engagement of the construction manager ensured that valuable practical considerations were incorporated into the early stage designs, and collaborative procurement ensured efficient project planning and enabled costing by trade contractors based on realistic designs.

Jarvis Hotels, Manchester Piccadilly

Jarvis Hotels created a construction management project alliance for the £17 million refurbishment of the Manchester Piccadilly Ramada hotel, working with specialists County Contractors, Gratte Brothers, Swift Asbestos, Curtis Furniture, Skopos Design, CT Jackson, All Facility Services, and with consultants Jones Lang LaSalle (Project and Development Services), Faber Maunsell, Applied Energy, Baxter Glaysher, Dunbar and Boardman, JWA Architects.

Their innovative, multi-party, construction management (CM) version of the PPC2000 partnering contract allowed package specialist contractors to be managed by sector-specific lead consultants in place of a traditional construction manager. The complexity of the refurbishment programme and the need to commence early works pointed towards a CM approach. It also recognised the client's strong relationship with its existing specialist contractors and suppliers.

A major difference from conventional CM lay in Jarvis' early decision to have four construction managers, each leading a workstream (enabling works, refurbishment works, external works, and lift works) who, together with the client, formed the PPC2000 Core Group.

(Continued)

---

612   As defined in Section 2.3.
613   Created by Chris Paul of Trowers & Hamlins.

Sub-division for tenancies raised challenges, particularly as to the mechanical and electrical upgrade. Also, the need to keep the hotel operating meant that some parts could not be surveyed in advance of start on site. The team used the contractual change mechanisms to minimise cost and time consequences deriving from this continually evolving scope.

Power cuts, floods and false fire alarms caused disruption. The Core Group was used to resolve concerns and to ensure that work stopped if necessary for customer relations. The alliance members reported that their contract contributed to:

- Advance planning of the preconstruction phase through an overarching timetable, and creation of an integrated team through a multi-party contract, with a Core Group of key individuals oiling the wheels through collaborative working
- Workstreams divided into packages, each dependent on a PPC2000 Commencement Agreement authorising start on site issued only when preconditions (e.g. finalisation of surveys, design development and agreement of price and package timetable) were met
- Attention to critical preconstruction activities needed to minimise disruption on site
- Use of the contractual change mechanism to deal with evolving and changing scope of works
- Use of the Core Group to manage unforeseeable events.

Michael Kilduff of Jones Lang LaSalle commented: 'This was never going to be a project for the faint-hearted. The breakthrough that unlocked the project was to find a mechanism where the entire ethos was about finding solutions, yet still bound all parties into key deliverables. I am not aware of anyone coming away from the project without a sense of achievement'.

Association of Consultant Architects (2010), 26–7.

ORTUS Learning and Events Centre, King's College Hospital

The ORTUS project was an award-winning new build learning and events centre for the Maudsley charity, part of the King's Health Partnership which specialises in mental health research and education. The client concluded that a construction management alliance would be the best basis for managing procurement as this would:

a) Enable the client to handpick trade contractors that were not only competitive but also bought into the collaborative ethos of the project
b) Provide an easier mechanism for the brief and design to develop through the duration of the project without exposing the client to the risk of major main contractor loss/expense claims
c) Improve value through savings on main contractor overheads, risk allowances and profit margins
d) Reduce the decision-making hierarchy without a main contractor management team
e) Reduce the risks associated with main contractor insolvency

f) More easily align trade contractors with the client's ethos and project vision through direct dialogue rather than communication via a main contractor

g) Provide a direct and therefore shorter payment route for trade contractors which, in recessionary times, would add to their commercial appetite for the project

h) Improve the client's ability to manage its direct reputational risk in the local community by having direct control over conduct and performance of trade contractors.

Selection was a vital part of the process. New prospective alliance members were introduced to the project through tenders and interviews. The construction manager via the client representative made appointment recommendations to the client. All consultants and contractors were invited to attend workshops in small groups hosted by the partnering adviser, who explained the contract carefully to each party and highlighted the key differences between the alliance and typical building contracts. Queries were raised and discussed, and each alliance member was made aware that the partnering adviser would be available as a resource to them should they deem it necessary. The construction manager and client representative also attended each of these workshops and reminded new alliance members of the vision and ambitions for the project. These workshops not only were informative but helped to build strong relationships between alliance members and to reinforce the partnership culture of the project and the contract.

One of the aspirations of the project was to bring the industry's best practical skills alongside those of the design consultants to get the best value results. Also, in order to meet the client's programme objectives, detailed design and construction drawings needed to be developed during the construction timeframe.

The early engagement of the construction manager ensured that valuable practical considerations were incorporated into the early stage designs. This approach also allowed the project team to plan the procurement process more efficiently and to enable trade contractors to provide costs based on realistically developed designs.

Reported to the ACA and published at www.ppc2000.co.uk.

8

What Types of Contract Support Collaborative Procurement?

8.1 Overview

It is easier to express what we want from collaborative procurement than to explain how we will achieve it. For example, we have seen how a simple declaration provides only a symbolic banner of intended collaboration. It does not give the parties the shared information they need to make progress or provide the alternatives that will stop them sliding towards a dispute when a difficult matter arises. To create a fully functioning system of collaborative procurement requires detailed contract terms, but what types of contract can achieve this?

The object of contractual governance has been stated to be protecting the interests of the respective parties and also, where required, adapting the relationship to changing circumstances.[614] The following types of contract describe the following types of transaction:

- A 'classical contract' describes a complete transaction 'which entails comprehensive contracting whereby all relevant future contingencies pertaining to the supply of a good or service are described and discounted with respect to both likelihood and futurity'.[615] An example is a contract for the sale and purchase of bricks collected from a builders' merchant
- A 'neo-classical contract' describes a more complex transaction where 'not all future contingencies for which adaptations are required can be anticipated at the outset' and where 'the appropriate adaptations will not be evident for many contingencies until the circumstances materialize'.[616] An example is a typical construction contract.

It is recognised that we should not treat 'the construction process as the purchase of a product',[617] and that to attempt to treat a construction contract as a simple and complete transaction puts pressure on the ability of that contract to cope with situations for which it was not designed.[618] A range of important construction phase interactions require procedures to be set out in a neo-classical contract, for example to deal with:

---

614 Williamson, O.E. (1979), 258.
615 Williamson, O.E. (1979), 236.
616 Williamson, O.E. (1979), 237.
617 Barlow et al. (1997), 5.
618 Macneil, I.R. (1974), 815.

Collaborative Construction Procurement and Improved Value, First Edition. David Mosey.
© 2019 John Wiley & Sons Ltd. Published 2019 by John Wiley & Sons Ltd.

- Periodic payments according to interim progress
- Instruction of changes and the agreement of their time and cost impact
- Submission and assessment of claims for additional time and money
- Agreement of aspects of the project that are not yet fully designed or priced and that have to be treated as provisional, with procedures for their resolution during the construction phase.

All these unforeseeable matters require communications with other team members by way of notifications, proposed actions, suggested cost and time for those actions, assessment of the knock-on effects on others, approvals by third parties and approvals to proceed. However, in order to create a collaborative contract, we need to look beyond these arm's length mechanisms and to explore the ways in which a contract can also set out the agreed means to build up and share additional information and can govern new interactions among team members.

This chapter will consider the categorisation of collaborative construction contracts and will suggest that recognition of a new type of contract, known as an 'enterprise contract', may provide a better way to categorise the alliance contracts that underpin and enable collaborative construction procurement.

8.2 What Is a Relational Contract?

As seen in Chapter 7, collaborative aspirations can tempt a team to bypass contracts altogether or to try to neutralise their effect through exclusions of liability. Collaborative procurement has also been linked to a more open-ended model of contractual governance classified as a 'relational contract' where:

- As in the case of a neo-classical contract, adaptations will be required under a relational contract so as to meet future contingencies
- Unlike a neo-classical contract, the parties to a relational contract 'settle for an agreement that frames the relationship' and 'do not agree on detailed plans of action but on goals and objectives'[619]
- Unlike a neo-classical contract, a relational contract reflects only the commencement of the relationship and is followed by 'a complex succession of exercises of choice and agreement'.[620]

In considering the interaction between relational and neo-classical contract models, it is interesting that Macneil chose to cite as an example the construction of complex buildings. He perceived that there exists in such cases the practice 'of having architects, engineers, designers, etc., and the building contractor(s) all work together with the owner from the inception of site location, through building design, and through all other planning, which lasts until the final completion of the project'.[621] Macneil's perception does not reflect the approach adopted on every construction project: instead it appears to describe a project alliance.

Williamson envisaged that relational contracts are more appropriate to long-term contracts, on the basis that the increased costs of setting up a complex governance

---

619 Milgrom & Roberts (1992), 131.
620 Macneil, I.R. (1981), 1041.
621 Macneil, I.R. (1981), 1041, 1042.

structure should only be reserved for 'complex relations', which he defined by reference to the following attributes:

- Uncertainty as to the outcome of the relationship
- The need for the parties to invest in the transactions with a view to ensuring their success
- The number of projects giving rise to frequently recurring transactions.[622]

Relational contracts accurately describe the features of partnerships, joint ventures and other open-ended business relationships where the team members are committed to working together but do not yet know what will be expected of them from third parties. They also appear consistent with some aspects of collaborative procurement, for example where a framework alliance contract lacks clarity as to the extent and type of work that the alliance members will implement.

Relational contracts have also been subject to different treatment of good faith by the English courts, for example where a party to a joint venture agreement issued misleading statements that were inconsistent with the terms of that agreement and the court recognised the possibility of implying a duty of good faith based on evidence as to whether the contract was a 'relational contract'.[623] In this case a contract was seen as relational because it had the following features:

- 'A long-term business relationship
- Investment of substantial resources by both parties
- Implicit expectations of co-operation and loyalty that shape performance obligations in order to give business efficacy to the project
- Implicit expectations of mutual trust and confidence going beyond the avoidance of dishonesty'.[624]

Some commentators see a project alliance contract as relational because it uses 'jointly shared risk and reward to create a system that inherently enables and reinforces collaboration'.[625] However, the open-ended aspect of a relational contract is not consistent with the need for a project alliance contract to set out not only the alliance members' objectives but also clear processes as to how they will achieve them. As a result, there are characteristics of a construction alliance that justify consideration of a different type of contract.

An alliance contract has some features of a relational contract in that it requires incremental choices as part of the mutual planning processes by which the members gather increasing project information. However, there are important distinctions because an alliance contract can provide for these mutual planning processes to be governed by a

---

622 Williamson, O.E. (1979), 239.

623 *Yam Seng Pte v International Trade Corporation* [2013] EWHC 111 (QB) which concerned a joint venture for the marketing of toiletries branded with 'Manchester United'. The decision was based on the contract having 'a high degree of communication, co-operation and predictable performance based on trust and confidence and expectations of loyalty which are not legislated for in the express terms of the contract but are implicit in the parties' understanding and necessary to give business efficacy to the arrangements'. The judgment in *Bristol Groundschool v Intelligent Data Capture* [2014] EWHC 2145 (Ch) also found that a relational contract existed and confirmed the *Yam Seng* judgment.

624 Collins, H. (2016), in light of cases such as *Yam Seng Pte Ltd v International Trade Corporation* [2013] EWHC 111 (QB).

625 Fischer et al. (2017), 47.

timetable and to be progressed through a series of agreed tasks and interactions rather than relying only on an open-ended agreement of common purpose.[626] It has been argued that, in order to operate efficiently, collaborative behaviour 'must be understood as consciously cooperative', and that 'those parties which contract efficiently act cooperatively'.[627] An alliance contract should provide clarity sufficient to show all team members why it is in their commercial interests to act in a manner that is consistent with achieving successful outcomes to the agreed objectives. For an alliance contract to align the commercial interests of its members it should, therefore, not be open to the uncertainties that are inherent in a relational contract.

8.3 What Is a Conditional Contract?

A conditional construction contract can describe project planning processes and can come into effect at an early point when the risks and responsibilities to be assumed by the team members are not fully known. It establishes the means and timescales whereby additional information can be completed that will enable the parties then to agree that a project should proceed to construction. Although an early contract for preconstruction phase contributions can be freestanding, the absence of links to the construction phase contract breaks the continuity of the contractual relationship and requires two separate contracts to be concluded.[628]

But is a conditional contract binding and enforceable? If the contractual relationship is conditional and incomplete during the period until a construction phase contract is concluded, might the preconstruction contract be unenforceable because of its conditionality or incompleteness?[629]

If a conditional preconstruction phase contract is only an agreement to negotiate the design or price or programme for a project, then it will be vulnerable not only to the conflicting positions inherent in negotiation but also to the challenge that it is void for lack of certainty.[630] This is a greater risk if it is not made clear how those processes will lead to commencement of construction.[631] For example, the JCT 2016 Pre-Construction Services Agreement is entirely separate from a JCT 2016 construction phase contract, with no stated machinery by which performance under one contract will lead to creation of the other.[632]

---

626 Macneil, I.R. (1981), 1041 as to an incremental process of agreement using increasing amounts of information.

627 Campbell & Harris (2005), 6.

628 Which can also give rise to a public procurement problem, as considered in Section 6.8.

629 As considered in Section 2.7. Chitty states that an agreement can be unenforceable if it lacks meaning without agreement of further terms, but recognises that enforceable contractual machinery may be the means of achieving the required further agreement, Chitty (2008), 2–131, 216.

630 In *Courtney and Fairbairn v Tolaini Bros (Hotels)* [1975] 1 WLR 297 (CA) Lord Denning M.R. stated '...the price in a building contract is of fundamental importance. It is so essential a term that there is no contract unless the price is agreed or there is an agreed method of ascertaining it, not dependent on the negotiations of the two parties themselves...'.

631 In *Foley v Classique Coaches* [1934] 2 KB 1 (CA), Maugham L.J. stated: 'It is indisputable that unless all material terms of the contract are agreed, there is no binding obligation. An agreement to agree in the future is not a contract; nor is there any contract if a material term is neither settled nor implied by law and the document contains no machinery for ascertaining it'.

632 The operation of JCT (2016) contracts is also considered in Section 9.6.

By contrast, NEC4 Option X22 creates a conditional contract governing 'Stage One' activities and a mechanism by notice from the Project Manager to proceed to 'Stage Two' under the same contract,[633] thereby escaping the trap of an agreement to agree. However, for the construction phase to commence entirely at the option of the Project Manager does not suggest a collaborative decision-making process. This approach may lead to tensions and problems if the main contractor or any other team member considers that the project is not ready to be constructed, for example due to lack of progress in designs or due to the absence of third-party consents.[634] Project alliance forms such as PPC2000, ConsensusDocs300 and the NEC4 Alliance Contract all recognise the need for agreement among alliance members before the construction phase proceeds, but this complicates an assessment of whether the contract is in fact conditional or whether it creates only an agreement to agree to proceed to the construction phase.[635]

Under a conditional contract, the team members can agree in detail the activities and interfaces by which they finalise agreement of the design, price and programme. If they agree to utilise machinery set out in the contract itself in order to arrive at more complete information, then their conditional contract should not be open to challenge as unenforceable.[636] Lord Justice Lloyd observed:

- 'It is sometimes said that the parties must agree on the essential terms and it is only matters of detail which can be left over'
- 'This may be misleading since the word "essential" in that context is ambiguous...[and] there is no legal obstacle which stands in the way of the parties agreeing to be bound now while deferring important matters to be agreed later'.[637]

A conditional contract requires preconditions to be satisfied before there is a right to proceed with additional work. It is not essential that every detail necessary to satisfy the conditions in a conditional contract is determined using contractual machinery. Agreeing certain details by negotiation should not undermine the effectiveness of the conditional contract if the parties can be relied upon to apply a 'standard of reasonableness' or if the relevant matters are of 'subsidiary importance' and do not negate or overturn the intention of the contracting parties to be committed to the other terms that they have agreed.[638] This is an important qualification as it recognises the human element in alliances and the need for complex enterprise interactions between alliance members in order to develop and agree complete information. It also highlights the role of an alliance in supplementing a project management system so as to encourage a culture of reasonable behaviour by the parties, both in applying agreed preconstruction phase contractual machinery and in negotiating subsidiary matters where required.

---

633 NEC4 (2017) Option X22.5.
634 The operation of NEC4 contracts is considered further in Section 9.7.
635 The operation of standard form alliance contract forms are considered further in Section 9.9.
636 In *Alstom Signalling v Jarvis Facilities* [2004] EWHC 1232 (TCC) where Mr Recorder Rees Q.C. held that 'Neither party could thwart the agreement by refusing to negotiate in good faith and/or by refusing to allow an Adjudicator or a TCC Judge to resolve the matter'.
637 *Pagnan v Feed Products* [1987] 2 Lloyd's Rep 601 (CA). The extent to which a contract can be incomplete or conditional is also considered at Sections 2.7 and 7.3.
638 In *Pagnan v Feed Products* [1987] 2 Lloyd's Rep 601 (CA), it was held that failure to agree does not invalidate the contract unless it 'renders the contract as a whole unworkable or void for uncertainty'.

Examples of conditional contracts in the construction sector include:

- A design consultant appointment which requires a client approval in respect of each stage of design development as a precondition before the consultant is authorised to undertake the next stage
- A conditional construction contract by which a client authorises early contractor involvement on a project, and under which the parties satisfy agreed preconditions before the client authorises commencement of the construction phase
- A framework alliance contract by which the parties agree the procedures and success measures that are preconditions to the award of project contracts
- A term alliance contract by which the parties agree the procedures and success measures that are preconditions to the issue of task orders.

In part we can define an alliance contract as a conditional contract. However, although that definition reflects the importance and status of early team appointments, it does not capture the multiple interactions between team members that contribute to an alliance.

8.4 What Is Enterprise Planning?

Macneil noted that 'when goods became more complex and more complexly produced, more extensive planning was required than could be supplied by relying on contemporaneous exchange transactions', so that plans have to be firmed up before production and in anticipation of distribution in order to promise 'the degree of certainty necessary both to exploit and to control change'.[639] He recognised that a contract can have planning functions, and distinguished between:

- 'Transactional planning', which is binding and allocative rather than mutual[640]
- 'Enterprise planning', which may be binding but some or all of which 'is characterised by some degree of tentativeness'.[641]

Alliance contracts can provide for transactional planning where deliverables of an alliance member can be allocated in advance and are independent from the work of other alliance members. They can also recognise enterprise planning where activities are agreed but are dependent on interaction between alliance members by way of successive contributions, comments and approvals. Macneil offered the following enterprise planning techniques for avoiding the completion of details giving rise to the conflicting interests that are inherent in negotiation:

- Merging an allocative issue into enterprise planning as a result of which mutual, non-negotiating activities resolve the issue because the parties pursuing these activities do not perceive the need for negotiation.[642] For example, while a project team may be aware of scope for negotiation of outstanding costs, many elements

---

639 Macneil, I.R. (1974) 765.
640 'Purely allocative planning is a zero-sum game, and hence conflict laden. But planning for the relational enterprise itself…need involve no conflict whatever among the participants…but…[is a game]in which all hope to, and quite normally do, gain', Macneil, I.R. (1974), 778.
641 Macneil, I.R. (1974), 739.
642 Macneil, I.R. (1974), 780.

of those costs can be completed without negotiation by using an agreed system for subcontract tendering after a main contractor has been appointed. An alliance contract can set out a system that describes the agreed activities whereby alliance members develop and agree detailed cost information without negotiation

- Creating a business case for a particular course of action sufficient to demonstrate to all parties the benefits of that business case to the project as a whole, rather than leaving particular team members to haggle over prices or look for alternatives.[643] For example, an alliance contract can provide the system whereby a main contractor builds up a preconstruction business case for the use of an in-house team or a preferred subcontractor, supplier, manufacturer or operator whose work it believes will benefit the project and will be in the interests of all other alliance members. Presentation of a business case gives the main contractor the opportunity to demonstrate the qualitative benefits that justify its proposals, and the relevant costs and the remainder of the business case can be subject to approval by other alliance members.

In Macneil's view, 'lack of measurability makes binding planning difficult to accomplish in the first place and hard to carry out once done, and hence tends to make planning subject to change'.[644] However, enterprise planning activities agreed under an alliance contract can be measurable by reference to specific goals, whether these goals are the readiness to commence work on site under a project alliance or the achievement of strategic objectives under a framework alliance or term alliance.

8.5   What Is an Enterprise Contract?

Neither neo-classical nor relational fully describes the features of an alliance contract whose purpose is to agree activities and timelines, usually among multiple parties, that map out the development of additional data for agreed purposes. While adaptation to meet future contingencies is a feature of a neo-classical contract, the adaptations under an alliance contract do not derive from contingencies that are not foreseeable by the parties but instead reflect the methodical build-up of information through systems of enterprise planning.

An alliance contract creates systems for enterprise planning that, arguably, are a hybrid of neo-classical and relational contracting. For example, an alliance contract should draw alliance members together by setting out the ways in which they agree to explore the potential for shared risk and reward, whereas 'the paradigm of the neoclassical model presupposes sharp division of benefits and burdens between the parties, with one party's benefit equalling the other party's burden'.[645]

Macneil envisaged 'standardised construction contracts' as 'relational agreements containing a great deal of process planning'.[646] However, an alliance contract should only be categorised as 'relational' if its planning processes are uncertain. Where the enterprise planning activities and interactions between alliance members are set out in a detailed and methodical manner, it is arguable that this gives rise to an enterprise contract, a sub-type not expressly recognised by Williamson or Macneil.

---

643 Macneil, I.R. (1974), 780.
644 Macneil, I.R. (1974), 777.
645 Macneil, I.R. (1981), 1046.
646 Macneil, I.R. (1974), 760.

Neo-classical contract systems reflect a 'willingness to recognize conflict between specific planning and subsequent changes in circumstances and to do something about them'.[647] However, an enterprise contract maps out the agreed activities and interactions at each stage of its evolving scope in ways that go beyond the reactive adaptations of a neo-classical contract. In order to achieve timely progress and a clear understanding among its parties, an enterprise contract also specifies the nature and timing of the agreed activities and interactions in a level of detail that goes beyond the open-ended goals and objectives of a relational contract.

In establishing detailed plans of action, an enterprise contract:

- Not only provides for default rules to fill information gaps but also provides processes that incrementally increase the quantity and quality of that information
- Not only deals with contingencies in respect of unknown matters but also sets out processes and interfaces that achieve future expectations and agreed objectives
- Not only uses conditions precedent to be satisfied before proceeding from one stage to the next but also provides systems that give a clearer structure to the activities that enable progress.

An enterprise contract provides for a succession of choices in order to accommodate and utilise increasing information. However, unlike a relational contract, an enterprise contract creates the machinery by which joint activities govern the development of all or most of the increasing information and minimise the role of negotiation in making choices as to how the increased information should be used. Typical features of an enterprise contract include:

- Contractual processes with a timetable that governs the build-up of agreed data, enabling the parties to move from incomplete to complete information
- Contractual processes, approvals and governance systems that avoid or minimise negotiation
- Integration of multiple parties, for example under a multi-party structure or through the work of an agreed integrator.

A simple example of the iterative nature of an enterprise contract is provided by a design consultant appointment, which creates:

- Evolving services linked to designs that are agreed in stages, particularly where these services and designs are integrated with related services and design contributions provided by other consultants
- A process for the creation and submission of design data in successive levels of detail, and for successive interactions with other parties providing contributions, comments and approvals
- Recognition of approved design data as the basis for each next stage of the design services, when that data was not in existence at the start of the previous stage
- Recognition of approved design data as the basis for an appointment for services relating to administration of the construction phase of the project.

The distinctions between a classical or neo-classical contract as described in Section 8.1, a relational contract as described in Section 8.2, and an enterprise contract as described in Sections 8.5–8.7 are illustrated in Diagram 6.

---

647 Macneil, I.R. (1978), 876.

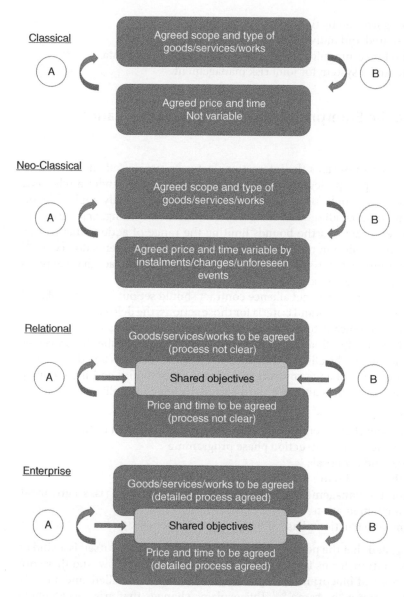

Diagram 6 Features of classical, neo-classical, relational and enterprise contracts.

Under a multi-party alliance contract 'the presence of multiple poles of interest is likely to lead to increased planning of structures and processes at the expense of measured substantive planning of exchange'.[648] The enterprise machinery of an alliance contract can provide processes with agreed rules of engagement that enable more substantive planning by reference to:

- Shared and individual objectives, with agreed success measures and criteria

---

648 Macneil, I.R. (1974), 792.

- Agreed activities governed by timelines and deadlines
- Recognition of shared and individual commercial interests
- A stable medium for communication and the build-up of reliable data
- Recognised risks and a system for joint risk management.

8.6   What Are the Enterprise Features of a Project Alliance Contract?

An enterprise contract prescribes the interactions between project alliance members that govern the build-up of project data. It has been suggested that under a relational contract the parties agree only 'on general provisions that are broadly applicable, on the criteria to be used in deciding what to do when unforeseen contingencies arise, on who has what power to act and the bounds limiting the range of actions that can be taken, and on dispute resolution mechanisms to be used if disagreements do occur'.[649] These general provisions are not sufficient to create an alliance contract that supports the procurement and delivery of a project.

As an enterprise contract, a project alliance contract should set out a detailed plan of agreed actions, the preconditions and criteria for those actions, the delegation of authority to individuals and the means to resolve disagreements. It should also recognise that the alliance members need to finalise missing data, and should describe the processes and techniques that enable the alliance members to complete that missing data. Enterprise features of a project alliance contract include agreed activities and interactions, both between alliance members and with third parties, by which the alliance:

- Builds up design and cost data
- Interacts with supply chain members through Supply Chain Collaboration
- Builds up data to create a construction phase programme
- Completes prices and proposals
- Interacts with third parties to obtain approvals
- Undertakes joint risk management activities including assessment of risks, agreement of actions and implementation
- Agrees when the construction phase of the project will proceed.

It has been suggested that the performance of a design appointment 'itself is a kind of adjustment' where it 'transforms figments of the imagination into a new, and therefore different, reality. A set of blueprints and specifications, however detailed, and a newly built house simply are not the same'.[650] This explains changes that arise accidentally from the development of details, and it touches on one of the reasons for construction disputes.[651] However, it does not acknowledge the potential for deliberate, incremental changes that reflect the development of details pursuant to an enterprise contract.

PPC2000 describes enterprise activities and timescales whereby additional design, cost and supply chain data are completed sufficient for the team members then to agree that the project should proceed to construction.[652] The Arup report described

---

649 Milgrom & Roberts (1992), 131.
650 Macneil, I.R. (1978), 873.
651 As considered in Sections 19.3 and 19.4.
652 PPC2000 (2013) clauses 8, 12, 14 and 15.

PPC2000 as 'a procurement system that provides the processes and mechanisms for planning, procurement and delivery of construction works. The system is based on the application of a number of processes and it is essential that the processes stated are applied'.[653] However, the need to agree the PPC2000 construction phase timetable, before authorising start on site, is an aspect of the alliance relationship that is more relational, in that 'relational planning is the planning of processes for accomplishing substantive planning and exchange in the future'.[654]

Equivalent enterprise activities are set out in ConsensusDocs 300, for example where:

- 'Constructor and Design Professional shall develop consultant, Precon Trade, Subcontractor and Supplier interest in the Project and, in collaboration with other integrated project delivery (IPD) Team members, develop a list of potential consultants, Precon Trades, Subcontractors and Suppliers from whom proposals shall be requested'[655]
- 'Constructor shall contract with Precon Trades and Suppliers during the early Project design to provide preconstruction services and facilitate an integrated, collaborative design process as approved by the Core Group'[656]
- 'The IPD Team shall continually review the Design Documents for clarity, consistency, constructability, and coordination among the construction trades and collaborate with the IPD Team in developing solutions to any identified issues'.[657]

8.7 What Are the Enterprise Features of a Framework Alliance Contract?

Some aspects of a framework alliance contract are relational, for example where incomplete data is available in respect of the briefs for individual projects and where relational planning can govern how some of this data is clarified. However, once project briefs exist, an enterprise contract governing enterprise planning processes more accurately describes the interactions between the framework alliance members in advance of each project commencing on site. The enterprise contract processes of a framework alliance contract can be used to build up the prices and proposals that comprise additional design, cost, and risk data during the selection and mobilisation of the team for each project under the agreed direct award or mini-competition procedures.[658]

An enterprise contract also describes the interactions between framework alliance members in sharing data and working practices, and in capturing and adopting the

---

653 Arup (2008), 37.
654 Macneil, I.R. (1981) 1044.
655 ConsensusDocs 300 (2016) Article 3.2 (Selection and replacement of builders and consultants).
656 ConsensusDocs 300 (2016) Article 3.4 (Participation in collaboration).
657 ConsensusDocs 300 (2016) Article 6.7.3 (IPD Team constructability analyses).
658 By setting out the rules governing the award of a number of project contracts, a framework alliance contract clarifies rights and obligations that may otherwise be implied but open to dispute. For example, in *Harmon CFEM Facades (UK) v House of Commons* [1999] EWHC 199 (TCC), [1999] 67 Con LR 1 it was implied that 'a contract comes into existence whereby the prospective employer implicitly agrees to consider all tenders fairly'.

lessons learned from earlier projects on later projects. Enterprise contract processes can be used to build up more reliable design, cost and risk data as follows:

- Joint reviews by means of exchange of data, with contributions and interactions between alliance members that establish the scope for greater consistency and greater efficiency,[659] and the potential for revised and shared working practices, including reviews after completion of each project that consider how improvements can be made in the later projects[660]
- Joint reviews and exchange of data, through Supply Chain Collaboration and other alliance activities, that explore the scope for supply chain members to offer improved value in exchange for improved commitments by alliance members[661]
- Joint risk management activities including assessment and agreement of actions, timelines and related risks.

The enterprise features of a term alliance contract follow a similar pattern to the above features of a framework alliance contract. Other chapters review the ways in which the enterprise features of a project alliance contract, a framework alliance contract and a term alliance contract address the activities and interactions required for collaborative governance, costing and incentivisation, time and change management, joint risk management and dispute avoidance.

8.8 How Can Collaborative Procurement Support a Joint Venture or Client Consortium?

A joint venture in the construction sector has been described as 'a number of firms collaborating on a project, or a number of projects, with a view to sharing the profits, each firm being paid on the basis of its agreed contributions in kind or in financial terms,'.[662] We have seen in Section 8.2 how a joint venture contract can be treated as a relational contract with an additional implied duty of good faith.

The declared commitment to shared objectives suggests that a joint venture can have the characteristics of a long-term alliance and should operate as an enterprise contract.[663] A joint venture should create a collaborative set of working relationships, and some construction contracts constrain changes in joint venture arrangements without prior consent.[664] Against this backdrop it is surprising that one in four construction joint ventures have been reported to end in dispute.[665] It is therefore worth examining what is needed to ensure that a joint venture is an effective vehicle for collaborative working.

---

659 The prospect of a significant workflow should motivate the parties to agree these improvements in advance of and separate from the award of specific project contracts.

660 Measurement of improved value across the alliance as a whole, linked to its continuation or extension or to other agreed incentives, should motivate the parties to share the feedback necessary to agree these improvements.

661 For example, as set out in FAC-1 Framework Alliance Contract (2016) clause 6 and demonstrated in the SCMG Trial Project.

662 Gruneberg & Hughes (2006), 10.

663 Guidance on public/private joint ventures recognises that they can have the features of partnering and alliances, Joint Venture Guidance (2010), 5.

664 For example, FIDIC (2017) clause 1.14(c).

665 Arcadis Global Construction Disputes Report (2017).

Joint venture contracts illustrate how collaborative contractual relationships can imply particular legal rights and obligations. For example, a contractual joint venture may create fiduciary relationships between its members.[666]

However, the liabilities of joint venture members may be worded in vague and general terms,[667] and problems can arise in a joint venture if the members do not clarify their roles or if they do not share a clear vision of what they wish to achieve.[668] For example:

- Objectives may not be made clear in a way that reconciles the different interests of members
- Governance may not be made clear so as to enable effective decision-making
- Members may not offer balanced contributions in terms of expertise, investment or assets
- Members' roles may not be integrated so as to avoid gaps or duplications
- There may be no agreed exit strategy linked to failure to meet specific targets.

A clear joint venture contract and governance system can help to manage any tensions between the duties owed by a company director to his or her own company and the fiduciary duties owed to a joint venture.[669] A joint venture has to align these differing commercial interests of its members, in the knowledge that:

- 'On the one hand both parties have an incentive to sustain the relationship rather than to permit it to unravel'
- 'On the other hand each party appropriates a separate profit stream and cannot be expected to accede readily to any proposal to adapt the contract'.[670]

A joint venture between a client and a main contractor can create the risk of additional conflicts of interest because the client also has to enforce a contract for works, services and supplies provided by the joint venture organisation in which it is a shareholder. These potential conflicts are more complicated where joint venture members create a real or virtual organisation that combines their agreed contributions in a way that makes it hard to trace who is responsible for each aspect of the joint venture's work.[671]

Potential conflicts of interest can be mitigated through enterprise contract processes under a term alliance contract. For example, Sheffield City Council and Kier Sheffield

---

666 As established in *Donnelly v Weybridge Construction* [2006] EWHC 2678 (TCC). In *Al Nehayan v Kent* [2018] EWHC 333 (Comm), the High Court implied a duty of good faith into an oral joint venture contract. It then found that the claimant's conduct (acting through its representatives) in obtaining the defendant's consent to enter into agreements for demerger and repayment of capital contributions was a breach of the implied duty of good faith.

667 FIDIC 2017 clause 1.14 provides that, if the contractor constitutes a joint venture, all its members shall be deemed to have joint and several liability whatever the joint venture structure, and also requires that the joint venture members must nominate a leader with authority to bind the joint venture and all its members.

668 'A party may provide land, capital, intellectual property, experienced staff, equipment or any other form of asset. Each generally has an expertise or need which is central to the development and success of the new business which they decide to create together. It is also vital that the parties have a 'shared vision' about the objectives for the JV'. Joint Venture Guidance (2010), 4.

669 In *Ross River v Waveley Commercial* [2013] EWCA Civ 910, the Court of Appeal upheld an implied fiduciary duty in a joint venture arrangement, including a fiduciary duty owed by the director of one of the joint venture partners to the other joint venture partner.

670 Williamson, O.E. (1979), 251.

671 Virtual organisations are considered in Section 7.6.

combined a client/contractor joint venture with a term alliance which set out enterprise contract techniques such as:

- Procedures for award of work supported by joint reviews
- Targets for improved value supported by Supply Chain Collaboration
- Joint risk management supported by integrated communications.[672]

A term alliance contract is also being used to clarify enterprise contract activities undertaken by the members of the £103.5 million Avela joint venture between South Liverpool Housing Trust and Penny Lane Builders, governing their repair and maintenance programme and separately their new build housing development.[673]

Another procurement model that can benefit from the use of an enterprise contract is a consortium of clients who together procure a programme of work in order to establish improved value through greater volume and consistency. A client consortium can be created:

- Through a combined procurement such as the Property Services Cluster Trial Project summarised in Section 8.10
- Through a public sector central purchasing body such as those used by Optivo, Crown Commercial Service and LHC in the case studies referred to in Section 10.9 and summarised in Section 10.10
- Through an agreement that links a number of related procurements such as the Supply Chain Management Group (SCMG) Trial Project summarised in Section 4.10 and the National Change Agent programme summarised in Section 8.10.

One area where a client consortium is very important to collaborative procurement is in supporting the flow of work necessary to sustain the off-site manufacture of modular or other prefabricated parts of a project. While there are examples of clients using collaborative working in conjunction with off-site manufacture on a single project,[674] the success of off-site production is dependent on continuous throughput at a factory, and a single project or a single client is unlikely to offer a pipeline of work sufficient to sustain that throughput.

The enterprise contract features of a client consortium agreement can include the procedures for the award of work and planned interactions that support Supply Chain Collaboration, shared working practices and joint risk management. They can also include provisions governing the management of supply and demand among consortium members and supply chain members, through joint planning and joint governance intended to minimise peaks and troughs by establishing and balancing among consortium members the sequence and quantities of projects called off.

An early example of a client consortium established to support off-site manufacture was the Amphion Consortium led by Charlie Adams of Hyde Housing Association. Together they created a progressive model for enterprise contracting but the investment by the main contractor in a new factory was put at risk by client consortium members hesitating to adopt the agreed off-site solutions, particularly when questions were raised

---

672 Case study summarised in Section 8.10.
673 News and Users, www.allianceforms.co.uk.
674 For example, the modular cells on the Cookham Wood Trial Project and the precast elements of the North Wales Prison Trial Project.

by local planning authorities.[675] More recent examples of client consortia establishing a collaborative approach to off-site manufacture include:

- The LHC new build housing programme summarised in Section 10.10
- The Construction Leadership Council (CLC) initiative for modern methods of construction summarised in Section 10.9.

8.9 How Can Collaborative Procurement Support a Public Private Partnership?

Public private partnerships take many forms, such as:

- Private Finance Initiative ('PFI')[676] projects, where the private sector provides funding for the design and construction of an asset and then operates an asset over a long period in return for payment of a fixed or variable charge
- Concessions, which have similar features to other PFI deals except that the private sector operator recovers payments direct from users of the asset
- Outsourcing of services, with or without private sector investment in capital works to create or improve assets related to provision of those services, which can take the form of PFI contracts or long-term service contracts depending on the funding model
- Public/private sector joint ventures as considered in Section 8.8, where the public and private sectors contribute to the joint provision of a service, with or without related capital works to create or improve assets related to provision of those services, under long-term service contracts.

It is recognised that 'the more successful PFI projects have allowed close interaction between building users and a broader range of the supply chain at an early stage in the consideration of designs'.[677] In some sectors it has been suggested that 'public private partnerships…almost always involve a collaborative-delivery method' and that they 'integrate private finance support into the collaboration',[678] but the central question is whether or not the collaborative delivery method involves the client.

For example, those contracts which govern public/private partnerships for the outsourcing of services, without significant private sector capital investment, can have all the features of a term alliance contract in which the client is an active participant. By contrast, public/private partnership contracts based on private capital funding are less likely to have any collaborative features that involve the client. Although it might appear

---

675 As considered in Kaluarachchi & Jones (2007), 1053–1061.
676 The UK Private Finance Initiative ('PFI') was designed to accelerate delivery of assets and to keep the capital costs off the Government's balance sheet but gave rise increasingly to questions around value and long-term public sector debt. The 'Channel Tunnel Fixed Link' was the first PPP in 1988, and PFI started in 1998.
£29.9 billion of PFI/PPP liabilities were on the UK Government balance sheet by 2009/10, with 717 signed contracts in place at end of 2012 when a Government review as to value for money led to the brakes being firmly applied.
677 Be PFI (2003), 4.
678 Water Design-Build Council (2016).

that a PFI contract is a type of relational contract because of its duration,[679] or even that it could operate as a long-term enterprise contract because of its clearly defined purposes, in practice the rigid structure created by the PFI funding model is rarely consistent with a collaborative relationship.

PFI funding is based on the assumption that design, construction and ongoing operational risks are passed by the public sector to the private sector.[680] PFI contracts reflect this approach and usually include a construction contract designed primarily to pass significant design and construction risk down from the private sector provider to its main contractor, with no opportunity for joint risk management with the client in ways that could mitigate those risks under conditional preconstruction appointments.[681] Preconstruction enterprise contract processes are undertaken privately among the members of each bidding team in order to create a robust financial model on the basis of which each prospective private sector provider can submit a unitary charge in its bid. These processes involve joint working between the design consultants, contractor, specialist subcontractors and operators to arrive at reliable figures, deadlines and risk profiles, but are unlikely to involve these details being shared with the client, even during a structured competitive dialogue procedure.[682] Multi-party 'interface agreements'[683] are often used to integrate the contributions of each member of the provider's team but the client is usually not a party to these agreements.

Despite creating a strategic model that integrates capital and operational expenditure from the outset, 'the potential benefits of the PFI process have frequently been lost by the failure to set up integrated teams or by awarding too many one-off contracts which prevents the transfer of learning between projects'[684]. The 2013 voluntary 'Code of Conduct for Operational PFI/PPP schemes'[685] showed how joint working might overcome the constraints of an arm's length PFI model. It encouraged renegotiation of PFI/PPP contracts in a joint public sector and private sector search for efficiency savings and cost reductions, for example by:

- Review of the specification for facilities management services and reduced frequency of non-essential services
- The sale or mothballing of surplus assets
- New ways to maximise third-party income
- Refinancing of project borrowings
- Review of risk transfer, such as change of law, insurance costs, maintenance/service provision costs and lifecycle expectations.[686]

---

679 In *Amey Birmingham Highways v Birmingham City Council* [2018] EWCA Civ. 264 Jackson L.J. commented that 'The contract before the court is a PFI contract intended to run for 25 years. It may therefore be considered as a relational contract'.

680 'In the past Government saw risk transfer as a cornerstone of PFI. Thus, PFI has focused on risk abdication rather than collaborative risk management', Be PFI (2003).

681 For example, SoPC4 (2007).

682 Competitive dialogue is considered in Sections 6.8, 6.9 and 6.10.

683 PFI interface agreements have some features of certain alliance contracts insofar as they overarch a series of two-party contracts with shared objectives and joint timetables.

684 Never Waste a Good Crisis (2009), 14.

685 CIPFA (2013).

686 CIPFA (2013).

Collaborative working among the members of a PFI team, but to the exclusion of the client, is illustrated by the Connect Plus consortium who operate the M25 motorway under a PFI contract and who introduced a range of collaborative processes led by their appointed framework contractors. Their 'Sustainable Business Culture Model' did not involve the client in any collaborative activities or provide a client share of the resultant benefits. Collaborative processes included mutual assessment by all supply chain members under a balanced scorecard, joint analysis of technical and logistical challenges and the development of improved solutions, new designs and the use of alternative materials.[687] The resultant 8% savings were retained by the Connect Plus consortium, although it was argued that:

- Certain qualitative benefits could continue to benefit the client after the end of the PFI contract
- Improved designs and risk management solutions would improve value for clients when applied on other programmes.[688]

In 2012, the UK Government published reforms to the Private Finance Initiative, proposing a new model known as 'PF2'.[689] There has been very little take-up of PF2,[690] but one interesting feature of the reformed model was the proposal that the Government would take a minority equity stake in the special purpose vehicle (SPV) that is responsible for delivery for each project. This was a significant shift away from the arm's length relationship between the public and private sectors that has been typical on many earlier PFI projects, and the change was intended to:

- Encourage greater collaboration between the private and public sectors
- Increase transparency through increased access to SPV information
- Allow the public sector to share in some of the SPV gains.[691]

A combination of private finance with collaborative procurement is being tested in new ways in Italy, where it has been agreed that the FAC-1 Framework Alliance Contract will integrate the objectives and activities of a team undertaking the design, construction and operation of a privately funded campus regeneration project. Alliance members will include the client, the concessionaire, the public company responsible for urban regeneration, the construction manager and the safety coordinator. Stakeholders will include project funders and municipalities.[692]

8.10 Collaborative Joint Venture and Consortium Case Studies

An enterprise contract describes the processes by which multiple parties can use enterprise planning to build up more detailed and reliable shared data, by which they can develop and agree increasingly accurate cost, time and risk parameters and by which

---

687 Connect Plus Trial Project case study summarised in Section 12.10.
688 Connect Plus Trial Project case study.
689 PFI 2 (2012).
690 NAO (2018), 46.
691 PFI 2 (2012), 3 and NAO (2018) 39, 40.
692 As summarised in Section 10.10 and considered in more detail in Chapter 24.

they are enabled and motivated to look for improvements. Identifying the features of an enterprise contract, as distinct from those of a classical, neo-classical or relational contract, helps to make clear the functions of contracts in mapping out and supporting the key actions, reactions and interactions between team members.

All the case studies in this book have features of enterprise contracts, whether in support of a project alliance, a framework alliance or a term alliance. The three case studies below illustrate features of enterprise contracts that supported:

- The client/main contractor term alliance joint venture formed by Sheffield City Council and Kier
- The National Change Agent programme where a group of client consortia integrated their housing refurbishment programmes under multiple collaborative frameworks created with a range of contractors and consultants
- The Property Services Cluster Trial Project, a multi-client, multi-contractor collaborative framework by which Hampshire County Council, Surrey County Council, Reading Borough Council and West Sussex County Council integrated the procurement of a diverse range of school new build, extension and refurbishment projects.

The factors that motivated the Sheffield/Kier joint venture team included:

- Client commitment of their workstream and workforce to the joint venture in return for a capital receipt and profit share
- Client expectations of long-term commitments and improved value generated by Kier's expertise
- Opportunities for the joint venture to generate profits from third-party business
- Access to all joint venture data through open-book costing
- The dependence of continued exclusivity on performance measurement.

Sheffield City Council/Kier Joint Venture Term Alliance

A £65 million per annum TPC joint venture term alliance agreed between Sheffield City Council and Kier Sheffield limited liability partnership (LLP) provided building repair and maintenance services to 55 000 council homes and a range of other council buildings such as libraries, schools and social services centres.

The term alliance transformed Sheffield City Council's repairs and maintenance service through a joint venture with Kier Group. Sheffield was the first local authority to exploit the cost and efficiency savings possible through a limited liability partnership (LLP) combined with a TPC term partnering contract. The joint venture involved the transfer of 1200 staff from Sheffield's direct labour organisation to the LLP which assumed responsibility for the Council's repair and maintenance of housing and other public buildings as well as some modernisation and new construction.

The principal aim of the contract was to achieve best value for the residents and the Council. Decisions implemented via the joint venture and the term alliance with Kier resulted in a major improvement in service together with lower unit costs. The contract was also a vehicle to tackle social inclusion via innovative training and employment projects. The joint venture earned numerous awards including the Local Government Chronicle award for 'Public/Private Partnership of the Year', 'Beacon Status' and the 4Ps 'Innovation Award'.

The term alliance contract included provisions for performance measurement, which after the first three years showed:

- 94% of Tasks completed on time
- 96% of urgent repairs attended to within the permitted time
- tenant satisfaction in housing repairs at 86%
- Integrated communications and review systems assisting in successful implementation
- The TPC 'Open-book' approach to pricing that underpinned the principles of profit share and incentivisation.

Dave Sheridan, then Managing Director, Kier Sheffield LLP, commented: 'Since the formation of the partnership we have made significant progress towards achieving our aim of providing a repairs and maintenance service that is unparalleled throughout the UK'. The '4P's Report recorded: "The Kier Sheffield LLP has been a great achievement and success to date for both Sheffield City Council & Kier Group Plc. This unique partnership is driving forward the enhanced benefits of partnership to invest in and maintain the housing and property of the Council'".

Association of Consultant Architects (2010), 40, and Kier (2005).

National Change Agent Procurement Consortia

The National Change Agent initiative was launched in 2005 and a total of 15 procurement consortia were formed, involving 155 social landlords with over 1.3 million homes under management. Procurement consortia utilised their preferred forms of delivery contract pursuant to multi-client framework alliance contracts, in conjunction with related consortium agreements and long-term second tier supply chain agreements.

Consortia reengineered second tier supply chain arrangements to create greater savings and efficiencies using a two-stage award process for project contracts. The NCA consortia as a whole achieved over a five-year period:

- Efficiency savings totalling **£226 million** from cumulative expenditure of £1.6 billion
- Over 500 apprentices successfully completing NVQ training to levels 2 and 3 and helped into full employment, with 80% retention
- Establishment of numerous SME businesses and social enterprises
- A joint initiative with WRAP to halve waste to landfill
- Completion of improvements to the homes of over 700 000 residents
- Early engagement of contractors and service providers under joint timetables to undertake surveys, value engineering, joint supply chain reviews and full resident consultation
- Clear construction phase timetables establishing key dates and deadlines for activities
- Open-book costing to establish where savings could be achieved and to balance the cost and benefits of added value initiatives such as improved energy efficiency, waste reduction and training and employment outputs
- Effective change management and risk management systems, as a result of which there were no reported disputes by any of the NCA consortium members with any of their

(Continued)

contractors or service providers over the five years of the NCA programme to the date of the report.

Lord Kerslake, then Chief Executive of the Homes and Communities Agency [now Homes England], commented: 'It isn't just the efficiencies that have been achieved. It has also had social benefits as well'.

Association of Consultant Architects (2010), 34.

Property Services Cluster – Education Basic Needs Programme

Hampshire County Council, Surrey County Council, Reading Borough Council and West Sussex County Council created a multi-client schools framework with a total value of £119 million. The contractor partners Osborne, Miller and Mansell (now Balfour Beatty) were jointly appointed through a mini-competition and formed a Cluster Delivery Team (CDT) which collaborated to share resources, information and supply chains. The team aimed to establish common designs, elements and components and, subsequently, common supply chains.

The team used bespoke preconstruction agreements for the preconstruction phase of each project and standard form JCT and NEC contracts for the construction phase of each project. All contractors jointly engaged with existing and potential tier 2 and tier 3 subcontractor and supplier partners to identify the pipeline of opportunities available allowing those subcontractors and suppliers to contribute to cost efficiencies and provide added value through early engagement.

The team agreed cost savings of **7%** and additional benefits that included:
Effective engagement of the tier 1–3 supply chains by and through the CDT:

- Nearly 200 Memoranda of Understanding (MOU) signed with tier 2 and 3 subcontractors and suppliers
- Supply chain delivering discounts for multiple projects
- Collaborating early with specialists selected through the MOU supply chain process
- Taking a 'one team' approach on the projects with all parties from the client team through all tiers of the supply chain.

Greater and more accurate market intelligence through sharing knowledge (cost trend data and avoiding supply chain overload):

- A co-ordinated approach to procurement by the CDT
- Developing a tranche-wide programme for key trades to identify, plan and manage peaks in procurement and construction activities
- Continued communication to identify the impact of change and delay to procurement and delivery activities
- 24 key manufacturers offering discounts across their product ranges.

High levels of Stakeholder Satisfaction across the Programme:

- Higher iESE Gateway 3 and Gateway 4 scores (Westfield School awarded 100% by client project team at Gateway 3 for both Supply Chain and Project Management)

- Satisfied end-users where school projects handed over to date
- Two-stage process: Better alignment of outcomes with the originally planned brief, budget and programme.

Delivering opportunities for skills training and new apprenticeships:

- A co-ordinated approach by the CDT delivering employment and skills plans for groups of projects in the same localities.

Property Services Cluster Trial Project case study.

9

What Standard Form Contracts Support Collaborative Procurement?

9.1 Overview

Construction contracts are often imposed by one party on another as part of the documents that the second party is invited to price, with little or no scope for suggesting and agreeing amendments. This is seen as a way of avoiding the cost, delay and inconsistencies that could result if a contract is separately negotiated with each consultant, contractor, subcontractor and supplier. The unilateral imposition of contracts is more palatable if they comprise standard forms that are already well known, and if these forms are recognised to be fair and are consistent with the contracts created for other team members. The Royal Institute of British Architects recommends that, in order to get 'the best possible outcomes', a client should 'choose a standard form of building contract and consultant appointment that promotes collaboration, integration and direct communication with your design team'.[693]

This chapter reviews the role of standard form construction contracts in collaborative procurement by measuring the following UK standard forms against a range of collaborative features identified by Sir Michael Latham:

- FIDIC 'Red Book' 2017 Conditions of Contract for Construction Second Edition and FIDIC 'White Book' 2017 Client/Consultant Model Services Agreement
- ICC 2014 Infrastructure Conditions of Contract With Quantities Version and ICC Partnering Addendum 2011
- JCT SBC/Q 2016 Standard Building Contract, JCT 2016 Pre-Construction Services Agreement and JCT 2016 Consultancy Agreement (Public Sector)
- NEC4 Engineering and Construction Contract June 2017
- PPC2000 Standard Form of Contract for Project Partnering Amended 2013.

We have considered the JCT and NEC framework contracts in Section 4.3, and we will look at the FAC-1 Framework Alliance Contract in Chapter 10. We have considered the JCT, ICC and NEC term contract forms in Section 5.3, and we will look at the TAC-1 Term Alliance Contract and its TPC2005 predecessor in Chapter 11.

Taking into account the conclusions drawn in Chapter 8 as to the role of a contract in supporting a project alliance, this chapter also considers which standard form project

---

693 RIBA Ten Principles (2017).

Collaborative Construction Procurement and Improved Value, First Edition. David Mosey.
© 2019 John Wiley & Sons Ltd. Published 2019 by John Wiley & Sons Ltd.

contracts have the characteristics that would make them suitable as alliance contracts, comparing PPC2000, the JCT 2016 Constructing Excellence Contract and the NEC4 Alliance Contract. Further analysis of these standard form project contracts also appears in other chapters, for example:

- What is the role of a core group or alliance board? (Section 12.7)
- What is the role of an independent adviser? (Section 12.9)
- How is Building Information Modelling (BIM) treated in construction contracts? (Section 14.7)
- How can collaborative procurement achieve a fixed price? (Section 16.3)
- How do target costs and cost reimbursement operate? (Section 16.4)
- How do shared pain/gain incentives operate? (Section 16.8)
- How are programmes and timetables treated in construction contracts? (Section 17.4)
- How do construction contracts treat risk management? (Section 18.4)
- How can collaborative procurement avoid or resolve disputes? (Sections 19.5–19.9).

9.2   What Is the Role of Standard Form Contracts?

Standard form construction contracts are used extensively in the UK. Certain forms have been encouraged by the UK Government although none are exclusively recommended or mandated. By contrast, in many other jurisdictions the use of standard forms is not widespread and there is much greater use of bespoke construction contracts.[694]

Standard forms can create confidence through familiarity, and their potential strengths include:

- Accessibility for users if the language is clear and straightforward and if they are supported by guidance[695]
- Flexibility if consistent forms cover multiple options in respect of warranties and pricing[696]
- Completeness if there are corresponding consultant appointments, subcontracts and post-completion operation, repair and maintenance contracts.[697]

Standard form construction contracts are not always seen as supporting increased efficiency. Concern has been expressed that they can be too focused on creating excuses for failure in performance, including for example late project completion, cost overruns and defects in the completed project.[698] Other possible weaknesses in standard forms include:

- Complexity if they do not use language that is clear and straightforward
- Inflexibility if they require adoption of different forms for different project sizes, warranties or pricing options[699]

---

694 As explained, for example, in Chapters 22 and 24.
695 Latham, M. (1994) recommended 'easily comprehensible language', Section 5.18.4.
696 Latham, M. (1994) recommended 'a wholly interrelated package of documents…which is suitable for all types of project and any procurement route', Section 5.18.3.
697 Latham, M. (1994) recommended a contract which 'clearly defines the roles and duties of all involved', Section 5.18.3.
698 Bennett, J. (2000), 174.
699 For example, it is arguable that the wide range of alternative JCT (2016) contract forms is unnecessary.

- Confusion if they are not integrated with corresponding consultant appointments, subcontracts and post-completion term contracts governing operation, maintenance and repair[700]
- Disappointment if they are amended in ways that undermine their intended effect.[701]

The style and length of standard form contracts varies significantly, and one of the attractions of the NEC contracts, for example, is how they use brevity wherever possible. By contrast, the FIDIC 2017 General Conditions run to 123 pages, compared to 68 pages in FIDIC 1999. Whether the additions are justifiable requires full analysis of the two FIDIC editions, but commentators have pointed out that:

- 'Adding numerous specific provisions to cover behaviour in more distinct eventualities means that there are more boundaries that actual circumstances may lie near and more questions about which provisions apply'
- 'For this reason, adding many detailed provisions to a contract can make disputes even more likely'.[702]

9.3 How Can Standard Form Contracts Support Collaborative Procurement?

Standard form construction contracts can support collaborative procurement. However, they will not provide a good starting point for collaborative relationships if they are seen as unfair by the parties on whom they are imposed, if they treat different team members inconsistently or if they do not describe collaborative processes. In reviewing the collaborative features of standard form construction contracts, it is helpful to consider the following relevant Latham recommendations which are discussed below by reference to each of the above standard forms:

- 'A specific duty for all parties to deal fairly with each other, and with their subcontractors, specialists and suppliers, in an atmosphere of mutual cooperation'[703]
- 'Clearly defined work stages, including milestones or other forms of activity schedule'[704]
- 'Integration of the work of designers and specialists'[705]
- A 'specific and formal partnering agreement' that is 'not limited to a particular project'[706]
- Partnering arrangements that 'include mutually agreed and measurable targets for productivity improvements'[707]

---

700 For example, the absence of a JCT (2016) private sector consultant appointment and the absence of a FIDIC term contract are serious omissions.
701 'The experience of contractors and quantity surveyors is that…it is very rare for an unamended form of contract to be used', Procuring for Value 2018), 33.
702 Milgrom & Roberts (1992), 131.
703 Latham, M. (1994), Section 5.18.1.
704 Latham, M. (1994), Section 5.17, 4.b.
705 Latham, M. (1994), Section 4.3.
706 Latham, M. (1994), Section 6.43.
707 Latham, M. (1994), Section 6.47.

- 'Shared financial motivation' and 'incentives for exceptional performance'[708]
- 'Taking all possible steps to avoid conflict on site'.[709]

The UK Office of Government Commerce commissioned an independent report from Arup Project Management to compare the respective merits of NEC3, PPC2000 and JCT CE in satisfying the principles set out in its Achieving Excellence in Construction initiative.[710] The report was published in 2008 and concluded that all three forms of contract would enable parties, using them correctly, to meet the Achieving Excellence in Construction standards.[711]

In 2012 the UK Procurement/Lean Client Task Force report recommended that only collaborative forms of contract should be used for Trial Projects, and identified JCT Constructing Excellence, NEC3 Option C, and PPC2000 as suitable project contract forms for early contractor involvement (ECI), collaborative working and BIM.[712] In each case, they recommended that these contracts should have an 'absolute minimum of amendments, with no changes to risk allocation or payment terms except where they are improved' and that 'effort should be taken to avoid the use of liquidated damages, retentions, parent company guarantees and performance bonds'.[713]

The Trial Projects were recommended to use different standard contract forms in order to 'examine how they were applied and the real experiences of the teams'.[714] Trial Project case studies reported that:

- The Environment Agency 'used the NEC Contract on this project, utilising a master template they have set up as part of their existing framework. This contract has been selected due to its focus on partnering, as part of the Environment Agency's drive for integration and collaborative working'[715]
- On Cookham Wood the Ministry of Justice 'adopted the PPC2000 standard form of contract (with minimum amendments) to define its processes for Early Contractor Involvement (ECI) under Two Stage Open Book'[716]
- Supply Chain Management Group (SCMG) confirmed that 'PPC2000 has been used with minimum amendments to the published form and with no liquidated damages, retentions, general liability caps or performance bonds'.[717]

The Construction Leadership Council reported in 2018 that:

- 'In the last 10 years, the industry has improved the way projects and programmes of work are procured and delivered, particularly in the public and regulated sector, with adoption of partnering contracts such as NEC3 and PPC2000'

---

708 Latham, M. (1994), Sections 5.18.2 and 5.18.12.
709 Latham, M. (1994) Section 5.18.11.
710 Arup (2008).
711 Arup (2008), 1 by reference to, for example, OGC (2007), AE (1999).
712 Procurement/Lean Client Task Group (2012), 21, 22.
713 Procurement/Lean Client Task Group (2012), 22.
714 Procurement/Lean Client Task Group (2012), 21.
715 Rye Harbour Trial Project case study.
716 Cookham Wood Trial Project case study.
717 SCMG Trial Project case study.

- 'This is continuing to develop with the issue of NEC4 and the TAC-1 Term Alliance Contract and FAC-1 Framework Alliance Contract'.[718]

9.4 How Does FIDIC 2017 Support Collaborative Procurement?

FIDIC provides contracts that are well-known and designed to be usable in any jurisdiction, including its 'Red Book' Conditions of Contract for Construction.[719] The long-waited second edition of the FIDIC Red Book was published in December 2017.[720] The FIDIC contract forms include the 'White Book' 2017 Client/Consultant Model Services Agreemen[721] but do not provide for the early conditional appointment of a main contractor or specialist subcontractors during the preconstruction phase of a project.

FIDIC 2017 can be assessed as follows by reference to the Latham criteria:

- A specific duty for all parties to deal fairly with each other, and with their subcontractors, specialists and suppliers, in an atmosphere of mutual cooperation:
  There is no general obligation of fair dealing in FIDIC 2017, although the FIDIC White Book governing the appointment of consultants, includes an obligation that 'In all dealings under the Agreement the Client and the Consultant shall act in good faith and in a spirit of mutual trust'.[722] There are also specific obligations in FIDIC 2017:
  – For the client to provide assistance to the contractor in obtaining permits and approvals[723]
  – For the client and contractor to cooperate with each other and with third parties.[724]
- Clearly defined work stages, including milestones or other forms of activity schedule:
  The FIDIC 2017 programming provisions are considered further in Section 17.4 and provide for:
  – An initial programme to be approved by the FIDIC project manager (known as the 'Engineer') after the contract is entered into, although this programme governs only the obligations of the main contractor and any nominated subcontractors and makes no connections to corresponding deadlines for other team members[725]
  – The possibility of milestones for provision of 'a part of the Plant and/or a part of the Works stated in the Contract Data (if any)', but with no recognition of corresponding milestones for the client and consultants[726]

---

718 Procuring for Value (2018), 33.
719 There is extensive published guidance on the FIDIC Red Book, including for example Glover and Hughes (2018).
720 FIDIC (2017), the first edition of which was published in 1999. In December 2017, FIDIC also published the second edition of its 'Silver Book' Conditions of Contract for EPC/Turnkey Projects and its 'Yellow Book' Conditions of Contract for Plant & Design-Build.
721 FIDIC White Book (2017).
722 FIDIC White Book (2017) clause 1.16.1.
723 FIDIC (2017) clause 2.2.
724 FIDIC (2017) clauses 2.3 and 4.6.
725 FIDIC (2017) clause 8.3.
726 FIDIC (2017) Special Provisions clause 4.22.

- The main contractor's right to give notice to the Engineer 'whenever the Works are likely to be delayed or disrupted if any necessary drawing or instruction is not issued to the Contractor within a particular time, which shall be reasonable', but there is no provision for agreement of the dates or periods of time against which any alleged lateness will be measured.[727]

- Integration of the work of designers and specialists:
  There are no processes in FIDIC 2017 for early appointment of the main contractor or specialist subcontractors to undertake any design contributions or reviews during a period prior to start on site, and therefore FIDIC 2017 creates no contractual system for the main contractor and/or specialist subcontractors to integrate their design work with the work of design consultants prior to start on site[728]

- A specific and formal partnering agreement that is not limited to a particular project:
  There is no FIDIC framework alliance contract that connects multiple projects, although it is possible to create a multi-project framework alliance by using FIDIC 2017 in conjunction with FAC-1

- Partnering arrangements that include mutually agreed and measurable targets for productivity improvements:
  There are no provisions in FIDIC 2017 for mutually agreed and measurable targets. For the measurement of improved productivity over multiple projects it is possible to use FIDIC 2017 in conjunction with FAC-1[729]

- Shared financial motivation and incentives for exceptional performance:
  FIDIC 2017 invites value engineering proposals from the main contractor which:
  - Are prepared at the main contractor's cost after start on site
  - Are subject to approval by the Engineer at its sole discretion
  - Are costed in a variation implementing an approved proposal which 'shall include consideration by the Engineer of the sharing (if any) of any benefit, costs and/or delay between the Parties stated in the Particular Conditions'.[730]

- Taking all possible steps to avoid conflict on site:
  FIDIC 2017 provides for early warning[731] but does not provide for joint risk management. It includes detailed provisions governing the role of a 'Dispute Avoidance/Adjudication Board'[732] and includes an obligation to attempt amicable settlement before referring a dispute to arbitration.[733]

9.5 How Does ICC 2014 Support Collaborative Procurement?

The Infrastructure Conditions of Contract ('ICC') do not include a consultant appointment, and they do not provide for the early conditional appointment of a main contractor or any specialist subcontractors during the preconstruction phase of a project.

---

727 FIDIC (2017) clause 1.9.
728 A bank headquarters project case study in Section 19.10 refers to the bespoke adaptation of FIDIC for early contractor involvement.
729 It has been reported that FIDIC is being used with FAC-1 in Kazakhstan on a mining project, and in Bulgaria on a German embassy project, both referred to in Section 10.9.
730 FIDIC (2017) clause 13.2.
731 FIDIC (2017) clause 8.4 as considered in Section 19.6.
732 FIDIC (2017) clauses 21.1 to 21.4.
733 FIDIC (2017) clause 21.5. See also Chapter 19.

The ICC 2011 contracts introduced an optional Partnering Addendum, which is designed to be added to each two-party ICC contract and which provides for:

- A 'Core Group' whose members reach decisions 'by unanimous vote of all members in attendance'[734]
- A 'consolidated programme showing the proposed timing of the contributions of the Partners as set out in the Partners' Bi-Party Contracts'[735]
- 'Partner Risk Managing Arrangements' to be set out in a schedule[736]
- 'Partner KPI/Incentive Arrangements' to be set out in a schedule.[737]

ICC 2014 does not refer to the 2011 Partnering Addendum but the ICC 2014 introduction states that it includes 'provisions requiring a collaborative approach'.[738] These can be assessed as follows by reference to the Latham criteria:

- A specific duty for all parties to deal fairly with each other, and with their subcontractors, specialists and suppliers, in an atmosphere of mutual cooperation:
  ICC 2014 introduces an obligation on the part of the contractor, the client and the engineer (on the client's behalf) to 'collaborate in a spirit of trust and mutual support in the interests of the timely, economic and successful completion of the Works'[739]
- Clearly defined work stages, including milestones or other forms of activity schedule:
  ICC 2014 provides for a programme to be created and agreed after the contract is entered into,[740] and Guidance Notes state that this programme 'should not be included in the Contract Documents'.[741] In the absence of an ICC consultant appointment, there is no provision for integration of agreed main contractor deadlines with any deadlines for related consultant services
- Integration of the work of designers and specialists:
  There are no provisions in ICC 2014 for early appointment of the main contractor or subcontractors to undertake any design contributions or reviews during a period prior to start on site, and therefore ICC 2014 does not create any system for the main contractor or subcontractors to integrate their design work with the work of design consultants prior to start on site. ICC 2014 provides that 'the Contractor may during the course of the Works submit proposals to improve the sustainability of the Works or their delivery', and for these to be instructed as changes.[742] However, the parties may be reluctant to propose and agree these improvements due to the increasing risk and cost of their effect on other work that is already in progress
- A specific and formal partnering agreement that is not limited to a particular project:

---

734 ICC Partnering Addendum (2011) P2(3). This provision follows very closely the approach used in PPC2000 clauses 3.5 and 3.6.
735 ICC Partnering Addendum (2011) P4(7). Reference to 'proposed' timing leaves it unclear as to whether the consolidated programme creates any binding commitments.
736 ICC Partnering Addendum (2011) P5. A guidance note suggests that agreed risk managing arrangements may be subject to advice from professional indemnity insurers.
737 ICC Partnering Addendum (2011) P6. A guidance note suggests that possible incentives include a 'Share of under/over-expenditure' and a 'Bonus for meeting targets'.
738 ICC Guidance Notes (2014), Introduction, 1.
739 ICC (2014) clause 6.1 as considered in Section 7.5.
740 ICC (2014) clause 9.1.
741 ICC Guidance Notes (2014), 6.
742 ICC (2014) clause 22.3.

There is no ICC framework alliance contract that connects multiple projects, although it is possible to create a multi-project framework alliance by using ICC 2014 in conjunction with FAC-1

- Partnering arrangements that include mutually agreed and measurable targets for productivity improvements:
  ICC 2014 does not provide for mutually agreed and measurable targets. For the measurement of improved productivity over multiple projects it is possible to use ICC 2014 in conjunction with FAC-1
- Shared financial motivation and incentives for exceptional performance:
  ICC 2014 does not provide for shared financial motivation or incentives for exceptional performance, although the ICC 2011 contracts include a Target Cost Version which provides for shared savings and shared cost overruns[743]
- Taking all possible steps to avoid conflict on site:
  ICC 2014 provides for early warning and joint risk management through 'a meeting of appropriately authorised persons to consider actions or measures in response to the matter so notified'.[744] It also provides the option for the parties to agree that a dispute shall be referred to conciliation or mediation.[745]

9.6 How Does JCT 2016 Support Collaborative Procurement?

The JCT contracts are themselves the product of a collaborative endeavour that goes back to 1931, when the JCT (joint contracts tribunal) was formed so that architects and contractors could create a mutually acceptable contract form. Since then, JCT contracts have been the medium for testing many legal issues in court, although it is unfair to conclude that these disputes arose from deficiencies in the contract forms themselves.

JCT 2016 comprises a complementary set of main contracts and subcontracts, plus a corresponding consultant appointment added in 2008 which is stated to be for public sector clients only.[746] The JCT 2016 contracts do not provide for the conditional appointment of a main contractor or specialist subcontractors during the preconstruction phase of a project, with the limited exception of the design submission procedure contemplated by the JCT MPCC (2016) Major Project Construction Contract.[747] Instead, the JCT PCSA 2016 and JCT PCSA(SP) 2016 comprise separate forms of pre-construction services agreement for a main contractor and for a specialist subcontractor.[748]

The 2016 JCT building contracts, preconstruction services agreements and public sector consultant appointment can be assessed as follows by reference to the Latham criteria:

- A specific duty for all parties to deal fairly with each other, and with their subcontractors, specialists and suppliers, in an atmosphere of mutual cooperation:

---

743 ICC Target Cost Version (2011) clause 60(5).
744 ICC (2014) clause 6.1(a).
745 ICC (2014) clause 19.3. See also Chapter 19.
746 This was the first JCT consultant appointment and it is hard to see why it is still stated to be suitable only for public sector clients.
747 JCT MPCC (2016) clause 12.
748 JCT PCSA (2016) and JCT PCSA (SP) (2016).

There is an obligation of this type in JCT 2016 building contracts such as JCT SBC/Q 2016 which is expressed as an option in Schedule 8 (Supplemental Provisions)[749] but there are no corresponding provisions in the JCT 2016 pre-construction services agreements. The JCT consultant appointment has a duty to 'liaise and cooperate fully with the other members of the Project Team'.[750]

- Clearly defined work stages, including milestones or other forms of activity schedule: The JCT 2016 programming provisions are considered further in Section 17.4. They do not create a consistent picture in respect of client, main contractor, and consultant rights and obligations and comprise:
  - A binding 'Programme' in the JCT 2016 pre-construction services agreements[751]
  - A 'master programme' in JCT SBC/Q 2016, agreed after the contract is entered into, and stated not to be binding[752]
  - The option of a binding 'Information Release Schedule' in JCT SBC/Q 2016, setting out the dates for release of design details to the main contractor.[753] The Information Release Schedule is not stated to be not integrated with a separate 'Design Submission Procedure'[754] which provides in general terms for design submissions by the main contractor 'in sufficient time to allow any comments of the Architect/Contract Administrator to be incorporated'[755]
  - A binding 'Programme and/or Information Release Schedule' under the JCT 2016 consultant appointment.[756]
- Integration of the work of designers and specialists: The JCT 2016 pre-construction services agreements enable early appointment of the main contractor and/or specialist subcontractors during the preconstruction phase of a project so that they can integrate their design work with the work of design consultants prior to start on site. However, JCT PCSA (2016) also provides that the contractor 'shall duly prepare and submit his Second Stage Tender in accordance with the Second Stage Tender Requirements', which suggests that the main contractor and its subcontractors must compete with others for the right to undertake the construction phase of a project and, therefore, could lose the right to construct the project into which their work has been integrated.[757]
- A specific and formal partnering agreement that is not limited to a particular project: The JCT 2016 Framework Agreement contains a range of collaborative provisions[758] and governs multiple projects, but it does not have the features of a framework alliance contract.[759] Instead, it is possible to create a multi-project framework alliance using JCT 2016 contracts in conjunction with FAC-1.[760]

---

749 JCT SBC/Q (2016) Schedule 8 Clause 1 'Collaborative Working', as considered in Section 7.5.
750 JCT CA (2016) clauses 2.1 and 2.3.
751 JCT PCSA (2016) clause 2.1.
752 JCT SBC/Q (2016) clause 2.9.
753 JCT SBC/Q (2016) clause 2.11.
754 JCT SBC/Q (2016) clause 2.9.5 and Schedule 1.
755 JCT SBC/Q (2016) Schedule 1 clause 1.
756 JCT CA (2016), clause 2.3.
757 JCT PCSA (2016) clause 2.7.
758 As considered in Section 4.3.
759 As considered in Section 4.4.
760 Examples of JCT contracts being used with FAC-1 are included in Sections 10.9 and 10.10.

- Partnering arrangements that include mutually agreed and measurable targets for productivity improvements:
  There are provisions for performance measurement indicators to be agreed between the client and main contractor under in JCT SBC/Q 2016 Schedule 8[761] but there are no equivalent provisions in the JCT 2016 pre-construction services agreements or JCT 2016 consultant appointment. For measurement of improved productivity over multiple projects it is possible to use JCT contracts in conjunction with FAC-1.
- Shared financial motivation and incentives for exceptional performance:
  There are no provisions for incentives in the JCT 2016 contracts[762] other than by deduction of liquidated damages for delay.[763]
- Taking all possible steps to avoid conflict on site:
  None of the JCT 2016 forms provide for early warning or for joint risk management. JCT SBC/Q 2016 provides for 'serious consideration' to be given to a request for mediation[764] and there is an equivalent provision in JCT 2016 consultant appointment but not in the JCT 2016 pre-construction services agreements.[765]

The JCT 2016 Constructing Excellence Contract[766] is considered separately in Section 9.9 as a standard form project alliance contract.

9.7   How Does NEC4 Support Collaborative Procurement?

The NEC4 contract forms comprise a complementary set of consultant appointments, contracts, subcontracts and term contracts.[767] The NEC contracts were strongly supported by Latham and certain provisions were amended in order to reflect the recommendations of his 1994 report. The NEC approach to their contracts:

- 'Sought to introduce a detailed management system which brings disputes to the surface at an early stage and which avoids the need for the parties to revert to the traditional battle at final account stage'
- 'Places the management of the contract at the forefront and seeks to avoid the disputes which created such a lack of confidence and trust in the construction industry in the last half of the twentieth century'.[768]

The language used in the NEC contracts, and the claims made as to their impact, have attracted mixed views.[769] The use of the present tense 'shows that the underlying

---

761 JCT SBC/Q (2016) Schedule 8 clause 5.
762 Except in JCT CE (2016), as considered in Section 9.9.
763 JCT SBC/Q (2016) clause 2.32.
764 JCT SBC/Q (2016) clause 9.1.
765 JCT CA (2016) clause 12.1. See also Chapter 19.
766 JCT CE (2016).
767 NEC have created extensive guidance to support the use of their contracts. The NEC3 contract forms are also analysed in Thomas, D. (2012) and Eggleston, B. (2006).
768 Thomas, D. (2012), (ix).
769 For example, Uff, J. (2018), 186, 'Apart from the early warning procedures in NEC, it is difficult to identify the provisions which may constitute good management or which can be said to motivate the parties towards improving the management of the project… Certainly the hype with which NEC is promoted does nothing to establish its credibility'.

purpose of the form is not to seek to mirror traditional language but to take a different approach to the way in which employers and contractors carry out projects'.[770] NEC3 contracts have been used successfully to deliver many complex projects and programmes of work.[771]

The NEC4 contracts can be assessed as follows by reference to the Latham criteria:

- A specific duty for all parties to deal fairly with each other, and with their subcontractors, specialists and suppliers, in an atmosphere of mutual cooperation:
  This obligation is included in all NEC4 contract forms.[772]
- Clearly defined work stages, including milestones or other forms of activity schedule:
  The 'Accepted Programme' is a central feature of all NEC4 contracts,[773] as are the 'Key Dates'.[774] Both are binding documents and integrate deadlines for the actions and responses of consultants, contractors and subcontractors, and are considered further in Section 17.4.
- Integration of the work of designers and specialists:
  NEC4 published its early contractor involvement Option X22 in July 2017 and this is also considered in Sections 3.5 and 8.3. Option X22 does not describe processes for preconstruction phase design development, for subcontractor appointments or for joint risk management by reference to a preconstruction phase accepted programme or key dates. These omissions mean that there is no way to ensure that early engagement of the main contractor also brings in the specialists whose design input can complement consultant designs and who can participate in thorough reviews of cost, time, and risk options.

  Option X22 summarises preconstruction phase processes as requirements for the main contractor to submit 'design proposals for Stage Two to the Project Manager for acceptance' plus 'the Contractor's forecast of the effect of the design proposal on the Project Cost and the Accepted Programme'.[775] It leaves a wide discretion for the project manager to:

  - Not accept a Stage One cost because 'it includes work which is not necessary for Stage One'[776]
  - Not accept a Stage Two submission because 'it will cause the Client to incur unnecessary costs to Others'[777]
  - Not accept a Stage Two submission because 'the Project Manager is not satisfied that the Prices or any changes to the Prices have been properly assessed'.[778]
  These rights of rejection create the possibility of a stand-off in finalising agreed costs, and may put team members at odds in a way that is unlikely to sustain a collaborative approach.

- A specific and formal partnering agreement that is not limited to a particular project:

---

770 Thomas, D. (2012), (ix).
771 For example, NEC3 was used as the contract for the 2012 London Olympics.
772 NEC4 (2017) clause 10.2, as considered in Section 7.5.
773 NEC4 (2017) clauses 31 and 32.
774 NEC4 (2017) clause 30.3.
775 NEC4 (2017) Option X22.3.
776 NEC4 (2017) Option X22.2(2).
777 NEC4 (2017) X22.3(3).
778 NEC4 (2017) X22.3(3).

The NEC4 Framework Contract does not have the features of a framework alliance contract[779] but it is possible to create a multi-project framework alliance by using NEC4 contracts in conjunction with FAC-1.[780]

- Partnering arrangements that include mutually agreed and measurable targets for productivity improvements:
  There are provisions for mutually agreed performance measurement targets in the NEC4 contracts under Option X12 and Option 20, and it is also possible to measure improved productivity over multiple projects using NEC4 contracts in conjunction with FAC-1

- Shared financial motivation and incentives for exceptional performance:
  There are provisions for incentives in the NEC4 contracts under Option X6 (Bonus for Early Completion), under Option X12 as referred to below and under the shared pain/gain provisions in NEC4 Option C and Option E[781]

- Taking all possible steps to avoid conflict on site:
  The NEC4 contracts provide for early warning and joint risk management.[782] They also provide dispute avoidance roles for 'Senior Representatives' and a 'Dispute Avoidance Board'[783] and, under the NEC4 Alliance Contract, for an 'Alliance Board' and an independent expert[784]

The NEC contracts added a Partnering Option X12 in 2003, renamed in 2017 as 'Option X12 (Multi-party collaboration)', which:

- Stops short of a completely integrated set of contract relationships as it does not create a multi-party contract, but it provides that 'a Partner may ask another Partner to provide information that it needs to carry out work in its Own Contract'.[785] It has been noted that 'it is unclear how this is to be enforced between team members who are not in a contractual relationship'[786]

- Provides for agreed incentives, including performance indicators and associated payments[787]

- Refers to the operation of a cross-contract core group,[788] but does not state how its meetings are organised, what constitutes a quorum or how it reaches decisions.

NEC4 Option X12 includes potential clashes with NEC4 core clauses:

- Option X12 provides for a partner to give 'early warning to the other Partners when it becomes aware of any matter that could affect the achievement of another Partner's objectives stated in the Schedule of Partners'.[789] However, there is no indication of

---

779 As considered in Section 4.3.
780 Examples of NEC4 being used with FAC-1 are included in Sections 10.9 and 10.10.
781 As considered in Section 16.8.
782 NEC4 (2017) clause 15.
783 NEC4 (2017) Options W1 and W3. See also Chapter 19.
784 NEC4 Alliance Contract (2018) clauses 21 and 96.
785 NEC4 (2017) Option X12.3(2).
786 Baker, E. (2007), 348.
787 NEC4 (2017) Option X12.4.
788 NEC4 (2017) Option X12.2.
789 NEC4 (2017) Option X12.3(3).

what happens next, and there are no links to the separate early warning provisions that operate as between the parties to each NEC4 contract[790]

- Option X12 provides that 'the Core Group may give an instruction to the Partners to change the Partnering Information', namely the information which specifies how the partners will work together, on the basis that 'each such change to the Partnering Information is a compensation event which may lead to reduced Prices'.[791] This allows the Option X12 core group to make decisions that could have financial effects on a range of NEC4 contracts, but with no rules governing how they put this into practice
- Option X12 provides that 'the Core Group prepares and maintains a timetable showing the proposed timing of the contributions of the Partners', and states that 'the Contractor changes its programme if it is necessary to do so in order to comply with the revised timetable'.[792] This implies that initially the Option X12 timetable will match the 'Accepted Programme' in each NEC4 contract but also empowers the Option X12 core group to require amendment of any Accepted Programme, each change being a 'compensation event which may lead to reduced Prices'.[793] These core group powers appear to clash with other NEC4 programme provisions, for example the authority of the NEC4 project manager to review and accept updates to the Accepted Programme.[794]

The NEC4 Alliance Contract is considered separately in Section 9.9 as a standard form project alliance contract.

9.8 How Does PPC2000 Support Collaborative Procurement?

The PPC2000 Project Partnering Contract was published by the Association of Consultant Architects and reflects detailed recommendations of the Construction Industry Council Partnering Task Force.[795] The two-stage, multi-party structure of PPC2000 integrates the appointment of all consultants with the appointment of the main contractor and its subcontracted supply chain members.[796] There are corresponding long- and short-form subcontracts[797] and a term partnering contract[798] that has now been developed into a term alliance contract.[799] Latham commented:

- 'I was consulted on the first draft of PPC2000 prior to its launch at the Building Centre 10 years ago and have described it previously as the full monty of partnering and modern best practice'[800]
- 'The successes of PPC2000, and more recently TPC2005, rest on their clear commitment to integration of the partnering team around a single contractual hub and,

---

790 NEC4 (2017) clause 15.
791 NEC4 (2017) Option X12.3(6).
792 NEC4 (2017) Option X12.3(7).
793 NEC4 (2017) Option X12.3(7).
794 NEC4 (2017) clause 32.
795 CIC (2002).
796 Guidance on the structure of PPC2000 is set out in Mosey, D. (2001) and Mosey, D. (2003/2).
797 SPC2000 (2008) and SPC2000 Short Form (2010).
798 TPC2005 (2013).
799 TAC-1 Term Alliance Contract (2016).
800 Association of Consultant Architects (2010), 1.

equally importantly, their clear description of the procurement, communication and project management processes upon which successful partnering depends'.[801]

PPC2000 can be assessed as follows by reference to the Latham criteria:

- A specific duty for all parties to deal fairly with each other, and with their subcontractors, specialists and suppliers, in an atmosphere of mutual cooperation:
  This is a mutual obligation in all PPC2000 contracts[802]
- Clearly defined work stages, including milestones or other forms of activity schedule:
  The PPC2000 preconstruction phase Partnering Timetable and construction phase Project Timetable are considered further in Section 17.4. They are binding between all team members under the multi-party structure, and integrate deadlines for the agreed actions and responses of the consultants, main contractor, and those subcontractors who also sign PPC2000,[803] with corresponding timetables under the subcontract forms
- Integration of the work of designers and specialists:
  PPC2000 provides for early appointment of the main contractor and specialist supply chain members that allows them to integrate their design work with the work of design consultants prior to start on site.[804] The two-stage structure of PPC2000 is considered further in Sections 3.5, 8.3 and 9.9.
  PPC2000 omits a provisional consultant appointment to govern the preparation of the project brief, budget and price framework that will be used for the selection of other team members. In practice PPC2000 users award provisional consultant appointments under an informal exchange of letters that makes clear:
  - The limited scope of the early pre-PPC2000 services and fees
  - That the early services are governed by the PPC2000 terms
  - The commitment of the client and the consultant to sign PPC2000 as soon as the other team members have been selected
- A specific and formal partnering agreement that is not limited to a particular project:
  In order to create a multi-project alliance it is possible to use PPC2000 in conjunction with the FAC-1 Framework Alliance Contract[805]
- Partnering arrangements that include mutually agreed and measurable targets for productivity improvements:
  PPC2000 includes performance measurement and targets for all team members.[806] For measurement of improved productivity over multiple projects it is possible to use PPC2000 in conjunction with FAC-1
- Shared financial motivation and incentives for exceptional performance:
  PPC2000 provides for a range of incentives, including shared savings, shared added value, and pain/gain[807]
- Taking all possible steps to avoid conflict on site:

---

801 Association of Consultant Architects (2010), 1.
802 For example, PPC2000 (2013) clause 1.3, as considered in Section 7.5.
803 PPC2000 (2013) clause 6.
804 PPC2000 (2013) clauses 8.3(iv) and 10.
805 Examples of PPC2000 (2013) being used with FAC-1 are included in Sections 10.9 and 10.10.
806 PPC2000 (2013) clauses 4.2, 13.3 and 23 and Appendix 8.
807 PPC2000 (2013) clauses 13.1, 13.2 and 13.3.

PPC2000 provides for early warning[808] and joint risk management.[809] It also provides for dispute avoidance and resolution by a core group and by conciliation.[810]

PPC2000 is also considered in Section 9.9 as a standard form project alliance contract.

9.9 Which Standard Form Contracts Support a Project Alliance?

It is recognised that 'although the philosophy of alliancing is non-adversarial, the alliance is a commercial transaction and the alliance legal agreements must be appropriate to that commercial transaction'.[811] The published standard forms of alliance contract are:

- ConsensusDocs 300[812]
- The FAC-1 Framework Alliance Contract and the TAC-1 Term Alliance Contract
- The PPC2000 Project Partnering Contract which, with the related PPC International contract, were the first standard forms expressly designed for project alliances
- The JCT CE 2016 Constructing Excellence Project Team Agreement which includes some project alliance features
- The NEC4 Alliance Contract for complex projects, published in a consultative version in July 2017 and in its final form in June 2018.

FAC-1 and TAC-1 are considered in Chapters 10 and 11, and ConsensusDocs300 is referred to in other chapters. The project alliance features of PPC2000, JCT CE and the NEC4 Alliance Contract are considered below.

PPC2000 has been described as 'one of the more successful tools for delivering improvement in procurement which…combines two-stage procurement structure (an early, conditional appointment, followed by an unconditional one) in conjunction with a single contractual hub that everybody signs'.[813] An independent review for the UK Government reported that:

- 'The PPC2000 documentation represents a complete procurement and delivery system that is distinct from other forms of contract available'
- 'The impetus of the PPC form is for early contractor involvement'
- 'This should result in the Client procuring his Constructor at a point in the process where his specialist construction and management skills can have a great impact on the project'.[814]

As a two-stage multi-party contract, PPC2000 provides the enterprise contract machinery that supports a project alliance, and it has been recognised by US proponents of project alliances using integrated project delivery ('IPD') as 'a partnered contract that in many ways resembles an IPD agreement'.[815] Its preconstruction and

---

808 PPC2000 (2013) clause 3.7.
809 PPC2000 (2013) clause 18.1.
810 PPC2000 (2013) clauses 27.3 and 27.4. See also Chapter 19.
811 Department of Treasury and Finance, Victoria (2009), 155.
812 ConsensusDocs 300 (2016).
813 All-Party Group for Excellence in the Built Environment (2012), 23.
814 Arup (2008), 40 and 41.
815 Fischer et al. (2017), 387.

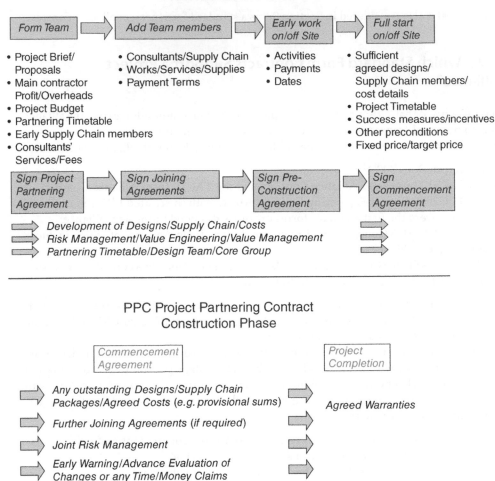

Diagram 7 PPC2000 Preconstruction phase and construction phase processes.

construction phase processes are illustrated in Diagram 7. Its collaborative governance system using a 'Core Group' is considered in Sections 12.7 and 19.7.

PPC2000 fulfils the following requirements for creating an alliance as set out in Section 3.3:

- How is the alliance created, between which members and can additional members be added?
  PPC2000 is signed by the client, by the main contractor, by all consultants providing services to the client, and by subcontracted supply chain members fulfilling key

roles. Together they comprise a project alliance, with the facility for other alliance members to join the multi-party contract at a later stage if agreed by existing alliance members.[816] The subcontracted supply chain members signing a PPC2000 contract participate in direct engagement with the client and consultants but they can also sign individual subcontracts with the main contractor,[817] governing for example payment and performance management.

- Why is the alliance created and what are the measures and targets for its success?
  PPC2000 sets out agreed objectives of all alliance members, with performance indicators linked to the success of the project[818]
- How is each stage of the agreed scope of works, services and supplies authorised, in what stages and to which alliance members?
  PPC2000 establishes the authority for alliance members to proceed from stage to stage under the project brief, consultant services schedules and specialist contracts by reference to the Partnering Timetable and Project Timetable, under a conditional contract governing both the preconstruction phase and construction phase of a project. It provides that, following satisfaction of listed preconditions and agreement of a fixed price or target cost, the alliance members are authorised to proceed with the construction phase of the project.[819]
- What will alliance members do together or individually in order to improve economic and social value, by means of what contributions and by what deadlines?
  PPC2000 provides for the joint activities of all alliance members in addition to their individual roles and responsibilities, for example:
  - Joint design development with main contractor and subcontractor input[820]
  - Supply Chain Collaboration[821]
  - Joint risk management[822]
  - Agreement of the construction phase Project Timetable.[823]
- How will alliance members be rewarded for their work?
  PPC2000 incentivises performance through payments related to results, for example through shared savings, shared added value, and pain/gain.[824]

The JCT 2016 Constructing Excellence contracts have some project alliance features. They are supported by guidance with a clear focus on collaborative working, and they comprise a set of identical two-party forms that can be used for client/contractor, client/consultant, and contractor/subcontractor relationships. These bilateral contracts can be supplemented by an optional multi-party 'Project Team Agreement', and each requires the parties to work together 'in a cooperative and collaborative manner in good faith and in the spirit of trust and respect'.[825]

JCT CE does not create all the machinery necessary to act as an enterprise contract, and some of its processes require significant clarification. For example:

---

816 PPC2000 (2013) clause 10.2 and pro forma in Appendix 2.
817 SPC2000 (2008) and SPC2000 Short Form (2010).
818 PPC2000 (2013) clauses 4.1, 4.2 and 23.
819 PPC2000 (2013) clauses 14 and 15.1 and pro forma Commencement Agreement in Appendix 3 Part 2.
820 PPC2000 (2013) clause 8.
821 PPC2000 (2013) clauses 10 and 12.
822 PPC2000 (2013) clauses 12.9 and 18.1 and pro forma Risk Register in Appendix 7.
823 PPC2000 (2013) clause 6.2.
824 PPC2000 (2013) clauses 13.1, 13.2 and 13.3.
825 JCT CE (2016) clause 2.1 and JCT Project Team Agreement (2016) clause 2.1.

- The JCT CE Project Team Agreement states that its rights and obligations override inconsistent rights and obligations in any other JCT CE contract,[826] but also states that no duty of care is created under the Project Team Agreement in contract or tort, with no liability arising under the Project Team Agreement for any act or omission except under its risk and reward provisions[827]
- The JCT CE Project Team Agreement refers to the 'Project Team' which has only advisory status,[828] and gives them no prescribed decision-making processes
- The JCT CE Project Team Agreement creates scope for confusion by also providing for the creation of an additional non-binding 'Project Protocol'[829]
- JCT CE guidance recommends early involvement of contractors and key specialists under 'a two-stage appointment with separate forms of agreement for a preconstruction stage for services (stage 1) and for the construction stage (stage 2)'.[830] However, the JCT CE contracts themselves do not create connections between the preconstruction and construction phases, nor do they set out any preconstruction phase processes for design development, for Supply Chain Collaboration or for development of a construction phase programme. JCT CE needs to be developed in order to govern these processes and to state the preconditions governing award of the construction phase JCT CE contracts
- The JCT CE 'Risk Allocation Schedule' suggests that the parties will have worked together during a pre-contract period to create a risk register[831] and that they will embody the outputs of their work in the Risk Allocation Schedule,[832] including agreement of time/cost effects of a 'Relief Event'.[833] However, the JCT CE contracts do not set out any preconstruction phase processes for joint risk management.

The NEC4 Alliance Contract was published in July 2018 'for use on major projects or programmes of work where longer-term collaborative ways of working are to be created'.[834] This is the first NEC multi-party contract form and includes an obligation to 'act in a spirit of mutual trust and co-operation' agreed directly among all alliance members,[835] plus the commitment of all alliance members 'to collaborate with each other to achieve the Alliance Objectives and partner objectives of every Partner'.[836]

The NEC4 Alliance Contract provides for collective responsibility of alliance members 'to support the delivery of the contract on a best for project basis'.[837] It provides for the role of an 'Alliance Manager'[838] and for governance by an 'Alliance Board', which is responsible for managing the Alliance and may issue binding instructions by unanimous

---

826 JCT CE Project Team Agreement (2016) clause 1.5.
827 JCT CE Project Team Agreement (2016) clause 2.9.
828 JCT CE Project Team Agreement (2016) clause 2.6, for example as to the impact of a relief event under JCT CE (2016) clause 5.10.
829 JCT CE Project Team Agreement (2016) clause 2.8.
830 JCT CE Guide (2016), 6.
831 JCT CE (2016) clause 5.1.
832 JCT CE (2016) clause 5.3.
833 JCT CE Project Team Agreement (2016) clauses 5.7 to 5.16.
834 NEC4 Alliance Contract (2018) title page.
835 NEC4 Alliance Contract (2018) clause 10.2.
836 NEC4 Alliance Contract (2018) clause 20.1.
837 NEC4 Alliance Contract (2018) clause 20.1.
838 NEC4 Alliance Contract (2018) clause 22. The Alliance Manager role is considered in Section 12.4.

agreement achieved at a meeting of all its members or by their written confirmation.[839] The role and terms of reference of the Alliance Board are considered further in Sections 12.4, 12.7 and 19.7, including the risk of Alliance Board members blocking decisions by non-attendance at a meeting.

The NEC4 Alliance Contract fulfils the following requirements for creating an alliance as set out in Section 3.3:

- How is the alliance created, between which members and can additional members be added?

  The NEC4 Alliance Contract is signed by the client, by the main contractor, by all consultants providing services to the alliance, and by certain specialists, with the facility for other parties to join the multi-party contract at a later stage if agreed by the Alliance Board.[840] The specialists signing an NEC4 Alliance Contract participate in direct engagement with the client and consultants but there is no corresponding form of specialist subcontract with the main contractor governing, for example, payment and performance management.

- Why is the alliance created and what are the measures and targets for its success?

  The NEC4 Alliance Contract sets out agreed objectives and performance indicators linked to the success of the project.[841] A central document is the 'Performance Table'[842] which measures the success of the alliance. It 'includes the targets for performance stated for the Alliance Objectives, sets out the adjustment to payment if a measured performance is higher or lower than the target and sets out how the adjustments are shared between the Partners'.[843]

- How is each stage of the agreed scope of works, services and supplies authorised, in what stages and to which alliance members?

  We have seen how improved value is created primarily during the preconstruction phase of an alliance but the NEC4 Alliance Contract treats early contractor involvement only as an option.[844] It is not reasonable or realistic to expect, in the absence of early contractor involvement, that an alliance can be formed immediately before start on site. Without early contractor involvement there is no time for the alliance members to benefit from joint working on designs, costs and risk management, for a collaborative culture to be established and for the Alliance Board to agree important documents such as the Implementation Plan and the Performance Table.

  Where there is early contractor involvement under Option X22, the Alliance Board makes the crucial decision of when to start on site.[845] This is in contrast to Option X22 in the other NEC4 forms where the decision to proceed to Stage 2 is made only by the project manager.[846] The Alliance Board may decide not to commence Stage Two, in which case 'the Client may appoint another person or organisation to complete

---

839 NEC4 Alliance Contract (2018) clause 21.3.
840 NEC4 Alliance Contract (2018) clause 21.8.
841 NEC4 Alliance Contract (2018) clauses 20 and 53.
842 NEC4 Alliance Contract (2018) clause 53.
843 NEC4 Alliance Contract (2018) clause 11.2(25).
844 NEC4 Alliance Contract (2018) Option X22.
845 NEC4 Alliance Contract (2018) Option X22.2.
846 NEC4 (2017) Option X22.5.

Stage Two and can use any material, information or design which the Alliance has provided'.[847] This gives the client as an Alliance Board member a way to veto Stage Two, to dismantle the alliance and to use all its Stage One work to complete the project using another team, whether or not that other team is an alliance.

- What will alliance members do together or individually in order to improve economic and social value, by means of what contributions and by what deadlines?

A central document in the NEC4 Alliance Contract is the 'Implementation Plan' which describes:

 – the management structure of the Alliance
 – the roles and responsibilities of the members of the Alliance
 – delegation by the Alliance Board
 – the use of common systems and processes and
 – 'other information which the Alliance Board requires to be included'.[848]

The Implementation Plan lies at the heart of the NEC4 Alliance Contract and is integrated with a programme governing the work of all alliance members.[849] It is surprising that the Implementation Plan can be set out in the Contract Data without a contractual system of prior consultation among alliance members or can be issued as 'an instruction given in accordance with the contract'.[850]

The NEC4 Alliance Contract includes specialists as alliance members, but it does not provide for collaborative working with any supply chain members outside the alliance and guidance suggests that they do not have much to offer.[851] This approach limits the scope for the alliance to work with subcontractors and suppliers who are not specialist alliance members. It also allows non-collaborative practices to be adopted in dealing with those subcontractors and suppliers.

- How will alliance members be rewarded for their work?

The NEC4 Alliance Contract incentivises performance through payments related to results, for example through alliance members sharing agreed cost savings.[852]

Both PPC2000 and the NEC 4 Alliance Contract provide the option for a project to be subdivided into sections or schemes,[853] and PPC2000 also refers to the agreement among alliance members of 'joint initiatives and strategic alliancing'.[854] However, neither form has the features of a framework alliance contract or term alliance contract as considered in Chapters 10 and 11.

---

847 NEC4 Alliance Contract (2018) Option X22.9.
848 NEC4 Alliance Contract (2018) clause 11.2(20).
849 NEC4 Alliance Contract (2018) clauses 32 and 33.
850 NEC4 Alliance Contract (2018) clause 11.2(20).
851 Subcontractors are approved by the Alliance Manager under clause 26.1 and there is no provision for them to engage with other alliance members. In addition, the definition of subcontractors means that some do not even need Alliance Manager approval, apparently because 'The main purpose is to exclude persons or organisations who are ordinary suppliers of Plant and Materials and Equipment and instead to allow the parties to focus on those persons and organisations providing something particular on the project,' NEC4 *managing an alliance contract* (2018), 7 by reference to NEC4 Alliance Contract (2018) clause 11.2(31).
852 NEC4 Alliance Contract (2018) clause 53 and the Performance Table.
853 PPC2000 (2013) clause 6.3 and NEC4 Alliance Contract (2018) Option X26.
854 PPC2000 (2013) clause 24.

9.10 How Are Collaborative Project Contracts Used in Practice?

The use of standard form construction contracts in the UK was subject to national surveys in 2012, 2015 and 2018: [855]

- Traditional JCT contracts remained dominant in the 2018 survey, being the forms used most often by 62% of respondents, an increase from a low point of 39% in 2015 and from 60% in 2012. JCT contracts were used sometimes by 57% of respondents in 2015 and by 72% in 2012
- JCT Constructing Excellence was used most often by only 1% of survey respondents in 2018, an increase from 0% in 2015 and 0% in 2012. JCT Constructing Excellence was used sometimes by 3% of respondents in 2015 and by 2% in 2012
- NEC contract forms were used most often by 14% of the 2018 survey respondents, a decrease from 30% in 2015 but an increase from 16% in 2012. NEC contracts were used sometimes by 53% of respondents in 2015 and by 29% in 2012
- There is no reported evidence as to use of NEC Option X12 (Multi-party Collaboration)
- PPC2000 was the form most often used by 3% of 2018 survey respondents, an increase from 1% in 2015 and 2% in 2012. PPC2000 was used sometimes by 5% of respondents in 2015 and 6% in 2012.

The 2018 survey reported that partnering and alliancing are shown to be most frequently used by only 3% of respondents, in contrast to 46% of respondents mostly using traditional procurement and 41% mostly using design and build.[856] The language of the report is unhelpful on this point as the terms 'traditional' and 'design and build' describe only the allocation of design and construction responsibilities, and this same allocation of responsibilities has to be made when embarking on partnering and alliancing.

63% of respondents reported that they used a contract that 'included an ethos of mutual trust and cooperation'[857] but it is hard to reconcile this with the procurement model adopted by those who did not use a partnering or alliancing contract. Possibly, for the others, the collaborative ethos was only symbolic, and the survey rightly questioned 'whether an ethos is sufficiently robust to maintain collaboration'.[858] We might also conclude that the 38% of 2018 respondents who considered disputes are still increasing, and the 49% who thought they are staying the same,[859] would agree that just adding a collaborative gloss is not sufficient.

The 2018 survey recognised that there is a risk of collaboration 'falling apart at the first hurdle if that collaboration is not described in contracts', and recognised that collaborative contracts answer the question of 'who is responsible for what and when, and with whom do they collaborate?'[860] That leads us to reflect on why in 2018 bespoke consultant appointments were reported to be used by 37% of the industry, compared to bespoke

---

855 NBS (2012), NBS (2015) and NBS (2018).
856 NBS (2018), 10.
857 NBS (2018), 14.
858 NBS (2018), 14.
859 NBS (2018), 24.
860 NBS (2018), 13.

construction contracts being used by only 5%.[861] The drafters of bespoke contracts are often tempted to be one-sided rather than collaborative: otherwise why would they not use the matching standard forms that are readily available?[862]

The project case studies that appear in this book describe how project alliance contracts have been used, and how other contract forms have been adapted, in order to support and integrate the teams, their relationships and their agreed activities. Users of NEC3 referred to in case studies include Anglian Water, Connect Plus, Eden Project, Environment Agency, Highways England, Property Services Cluster and Transport for London. Users of PPC2000 referred to in case studies include City of London Corporation, Department for Work and Pensions, Glasgow Housing Association, Jarvis Hotels, Land Securities Trillium, London Borough of Hackney, Ministry of Justice, Maudsley Charity and Poole Hospital NHS Trust.

The construction contracts adopted in other jurisdictions are considered in Sections 20.3 (Australia), 21.3 (Brazil), 22.3 (Bulgaria), 23.1 (Germany), 24.3 (Italy) and 25.3 (USA). For example, the published ConsensusDocs 300 project alliance contract form is used in the USA,[863] although there is evidence of bespoke forms such as the IPD Agreement produced by Hanson Bridgett also having significant impact.[864] The standard forms used in Australia were subject to a review in 2009,[865] but the project alliances in Australia have been implemented entirely using bespoke contract forms.[866]

Relatively few construction contracts are designed for use in multiple jurisdictions and many include references to specific laws of the jurisdiction where they are published. However, most contract terms do not depend on references to national laws, and most national laws will apply whether or not they are paraphrased in contract clauses. That said, there are likely to be implied terms and minimum requirements imposed under any legal system that affect the drafting of a construction contract used in the relevant jurisdiction.

For example, the international version of PPC2000 has been examined by the Deutscher Baugerichtstag as a possible basis for collaborative working in Germany[867] and has been translated into German.[868] The translation process gave rise to questions that required careful consideration by reference to the specific requirements of German law. This in turn led to a number of amendments in the German version, for example as to the appropriate duty of care, the available types of insurance and the definition of insolvency. However, the work undertaken on the translation and adaptation of PPC2000 for use in Germany has shown how a contract form developed in a common law jurisdiction can be adapted to reflect the different requirements of a civil law system.[869]

---

861 NBS (2018), 17 and 18.

862 JCT do not meet the need for a full set of matching contracts for so long as they do not offer a consultant appointment for private sector clients.

863 Fischer et al. (2017), 383.

864 As explained in Chapter 25 and in Fischer et al. (2017), 383.

865 Bell, M. (2009).

866 As explained in Chapter 20.

867 Deutscher Baugerichtstag (2018), 67.

868 The German version of PPC2000 is available at https://shop.bundesanzeiger-verlag.de/bau-und-architektenrecht-hoai/ppc-international-e-book.

869 Mosey et al. (2018). See also Chapter 23.

10

How Does the FAC-1 Framework Alliance Contract Operate?

10.1 Overview

The FAC-1 Framework Alliance Contract was developed through consultation with clients and construction industry representatives and through trialling by early users. It is designed to connect multiple projects, to integrate framework alliance members and to set out processes for improving value. It combines the features of:

- A framework contract 'the purpose of which is to establish the terms governing contracts to be awarded during a given period' but which is not itself a construction or engineering contract[870]
- An alliance contract recording 'long term partnering on a project [or programme of works] in which a financial incentive scheme links the rewards of each of the alliance members to specific and agreed overall outcomes'.[871]

The 2012 Effectiveness of Frameworks Report found that 'the general lack of standard-form framework arrangements makes it difficult for clients to procure frameworks on a consistent basis'.[872] In addition, the UK Infrastructure Client Group recognised that 'a horizontal agreement between the respective partners [captures] the principles within the commercial model, particularly those that jointly incentivise performance and create collaboration'.[873] Australian alliance research noted that 'there is a need for government to establish a standard form contract that is robust, tested and clearly understood by all parties. This would improve legal certainty and transaction efficiency for government and NOPs (Non-Owner Participants)'.[874]

When the KCL Centre of Construction Law was researching new alliance contract forms, our Dean, Professor David D. Caron, suggested that we might consider a collaborative alternative to FIDIC. We drafted and consulted on a new project alliance form, which received very positive feedback across 14 jurisdictions. However, we concluded that our priority should be creation of a framework alliance form to help plan and

---

870 Public Contracts Regulations 2015, Reg. 33(2).
871 European Construction Institute as quoted in Infrastructure Client Group (2014).
872 Effectiveness of Frameworks (2012), 54. For example, as described in Chapter 4, the JCT 2016 and NEC4 standard form framework contracts do not include the features of a framework alliance: neither form integrates multi-party relationships, and neither form describes systems for integrating the work of framework members or systems for pursuing agreed objectives that could improve value under related project contracts.
873 Infrastructure Client Group (2015).
874 Department of Treasury and Finance, Victoria (2009), 154.

Collaborative Construction Procurement and Improved Value, First Edition. David Mosey.
© 2019 John Wiley & Sons Ltd. Published 2019 by John Wiley & Sons Ltd.

integrate the works, services and supplies governed by any number or type of project contract forms, embracing the forms produced by FIDIC and other publishers.

The FAC-1 drafting team used processes and provisions drawn from successful framework alliances that were successfully established by Trial Project clients including the Ministry of Justice,[875] Hackney Homes with Homes for Haringey[876] and Surrey County Council.[877] A consultation draft was sent to 120 organisations in 14 jurisdictions. Consultation responses included a range of proposals for amendment which were adopted in the published version. These are summarised in Appendix F.

Feedback from consultation group members was discussed at a consultative conference attended by over 80 delegates in October 2015 and at a second consultative conference in February 2016 which attracted over 200 delegates. Further recommended amendments emerging from both conferences comprised the final stage of consultation leading to publication of FAC-1 in June 2016. The published version of FAC-1 was endorsed by the UK Construction Industry Council and by Constructing Excellence.

This chapter considers the key features of FAC-1, how it awards work and seeks improved value and how it supports collaborative procurement. It examines how FAC-1 operates as a framework contract and an alliance contract, and also the extent to which FAC-1 has the characteristics of an enterprise contract as described in Sections 8.5 and 8.7. This chapter also considers the flexibility of FAC-1 to accommodate different client requirements and its adaptation for use in common law and civil law jurisdictions. Case studies at Section 10.10 illustrate how FAC-1 is being used in practice.

10.2   What Are the Key Features of FAC-1?

FAC-1 is a new type of contract that acts as an umbrella to enable, connect, and enhance multiple project appointments. Its structure is illustrated in Diagram 8. FAC-1 describes enterprise planning activities that comprise successive actions, reactions and interactions through which the alliance members create enough data to award, compare and integrate project contracts, and to improve on the assumptions made when the framework alliance contract was entered into. FAC-1 can be used in many different ways, for example:

- To integrate the projects or programmes of works, services or supplies required by one or more clients and undertaken by one or more consultants, contractors, subcontractors, suppliers, manufacturers and operators
- To support and integrate the award of project contracts for a programme of related or unrelated projects
- To support and integrate the project contracts awarded in relation to a complex project
- To support and integrate the project contracts awarded in relation to a project using Building Information Modelling (BIM)
- To support and integrate the work of a central purchasing body that enables the award of project contracts by multiple clients

---

875 Effectiveness of Frameworks (2012), 103–4.
876 SCMG Trial Project case study.
877 Project Horizon Trial Project case study.

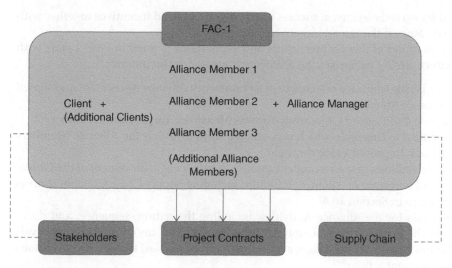

Diagram 8 Structure of FAC-1.

- To support and integrate the work of a joint venture, a consortium or a public private partnership.

FAC-1 was developed so as to provide a contractual medium for:

- Integrating the roles of alliance members through joint objectives and performance measurement linked to agreed incentives
- Awarding work on any number or range of projects
- Planning the early engagement of alliance members and supply chain members in advance of projects commencing on site
- Agreeing the means to achieve improved value with the benefit of long-term relationships
- Capturing learning and improvement from one project to another
- Creating systems that work in any jurisdiction and in conjunction with any form of project contract.

A report on the first use of FAC-1 noted that 'a housing association in the East Midlands is the first to use a new framework contract that connects competitors and encourages them to work together. Futures Housing Group has adopted the Framework Alliance Contract (FAC-1), which not only makes it easier for small businesses to bid for deals, it sets shared objectives for those on the framework. Alliance members are expected to work together to improve value, share information on suppliers, review and compare prices and tender or renegotiate subcontracts'.[878]

FAC-1 sets out a multi-party structure connecting alliance members that include one or more clients, an in-house or external alliance manager and any combination of other alliance members with a facility to add additional alliance members by agreement at any time.[879] It records the reasons why the framework alliance is being created, stating its

---

878 Supply Management (2016). A case study appears in Section 10.10.
879 FAC-1 Framework Alliance Contract (2016) clause 1.11.

scope and its agreed objectives, success measures, targets and incentives together with agreed exit routes if the agreed targets are not fulfilled.[880]

In the remainder of this chapter words and expressions defined in FAC-1 start with capital letters. FAC-1 comprises the following Framework Documents:

- Contract Terms which are referred to in a Framework Alliance Agreement and signed by all Alliance Members
- In Schedule 1 the agreed Objectives, Success Measures, Targets and Incentives[881]
- In Schedule 2 a Timetable which sets out the ways in which the Alliance Members agree to seek Improved Value through:
  - agreed deadlines, gateways, and milestones in respect of achievement of the Objectives and in respect of Alliance Activities, including Supply Chain Collaboration as referred to in Section 10.4
  - timescales for the Alliance Activities, including the nature, sequence, and duration of the agreed actions of each Alliance Member and any consents or approvals (whether required from Alliance Members or third parties) that are preconditions to subsequent actions[882]
- In Schedule 3 a Risk Register, as referred to in Section 10.6[883]
- In Schedule 4 a Direct Award Procedure and/or Competitive Award Procedure, as referred to in Section 10.3
- In Schedule 5 a set of Template Project Documents, as referred to in Section 10.3
- In Schedule 6 any Legal Requirements and Special Terms that give rise to additions and amendments to the Contract Terms, as referred to in Section 10.7
- A Framework Brief provided by the Client to all other Alliance Members as the basis for their selection and which sets out:
  - all technical, management and commercial requirements in relation to the Framework Programme
  - the required approach to design, Supply Chain engagement, costing, Risk Management and programming
  - all other relevant procedures and expected outcomes
  - where appropriate the required approach to BIM
- Framework Prices and Framework Proposals submitted by each Alliance Member in response to the Framework Brief, which are binding only between the Client, the Alliance Manager and the other individual Alliance Member who submits them, and which are confidential between that limited group of Alliance Members[884]
- If the Alliance Manager is an independent consultant, an Alliance Manager Services Schedule and Alliance Manager Payment Terms.

Guidance notes on completion of the FAC-1 Framework Alliance Agreement and Framework Documents are set out in Appendix G. The responsibility of each Alliance Member is limited to the Framework Documents that it prepares or contributes to, except to the extent of its stated reliance on information provided by other Alliance

---

880 FAC-1 Framework Alliance Contract (2016) clauses 2 and 14.2 and Schedule 1.
881 As illustrated in the Futures Housing Group, Football Foundation, LHC and Crown Commercial Service case studies in Section 10.10.
882 As considered further in Chapter 17.
883 And as considered further in Chapter 18.
884 FAC-1 Framework Alliance Contract (2016) clauses 1.3.3 and 13.3.2.

Members.[885] The Framework Documents are complementary and FAC-1 sets out an Early Warning procedure and a default order of priority in the event of any error, omission or discrepancy.[886] It also provides that Project Contracts take precedence over FAC-1 if a discrepancy cannot be resolved by agreement.[887]

10.3   How Does FAC-1 Award Work?

It is fundamental to the FAC-1 alliance relationships that Alliance Members understand not only the prospective scope and nature of the work covered by the Framework Programme but also how that work will be awarded. FAC-1 provides two options for the criteria and related procedures that lead to the award of Project Contracts, namely the Direct Award Procedure and the Competitive Award Procedure, as illustrated in Diagram 9.

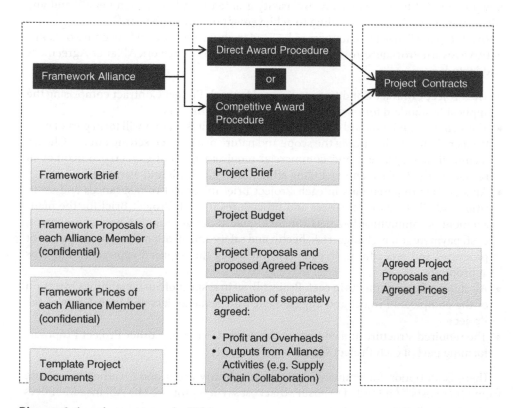

Diagram 9 Award processes under FAC-1.

---

885 FAC-1 Framework Alliance Contract (2016) clause 1.4.
886 FAC-1 Framework Alliance Contract (2016) clause 1.5.
887 FAC-1 Framework Alliance Contract (2016) clause 1.5.3.

Each procedure should:

- State the agreed procedure and timescales for the issue of each Project Brief, for the submission of proposed Agreed Prices and other Project Proposals relating to that Project and for all steps leading to award of Project Contracts, including the method, rules and criteria for evaluation of proposed Agreed Prices and other Project Proposals and the required format for proposed Agreed Prices and other Project Proposals
- Explain how the Project Contracts and other Template Project Documents will be applied to each Project
- Explain the procurement model for each Project, including the sources and timing of all contributions to design, Supply Chain engagement, costing, programming and Risk Management and incorporating the required approach to BIM as appropriate
- Explain all procedures relating to the conditional and/or unconditional award of Project Contracts, including any intended Orders for Pre-Contract Activities.[888]

FAC-1 states any minimum value or type of Project Contracts that will be awarded to Alliance Members so that they have a clear picture as to the level of certainty in the flow of work.[889] It also states any exclusivity granted to Alliance Members[890] and any adjustment of exclusivity according to achievement of agreed Targets.[891]

The Template Project Documents to be used in any Direct Award Procedure or Competitive Award Procedure are listed in Schedule 5 to the Framework Alliance Agreement and include:

- The Project Contract Conditions to be used for each Project Contract comprising the applicable standard forms of contract and any amendments
- The structure and standard components of the Project Brief that will form part of each Project Contract, describing the scope and nature of a Project, setting out the Client's technical, management, and commercial requirements and expected outcomes in respect of the Project, and including all required standards and warranties
- All standard requirements in each Project Brief in respect of insurances and securities and all standard processes and procedures in each Project Brief for the management of communication, performance, quality, design, Supply Chain engagement, cost, payment, time, change, risk, health and safety and all other Project management processes and procedures, in each case including the required approach to BIM as appropriate
- All standard requirements in each Project Brief in respect of Sustainability, Operation of the completed Project and engagement with Stakeholders and with Users of the Project
- The required structure and content of the Agreed Prices and other Project Proposals forming part of each Project Contract.

There is an option for Framework Prices to identify Profit and Overheads separate from other costs,[892] intended to ensure that cost savings proposed by Alliance Members do not erode their own margins.

---

888 Guidance notes in FAC-1 Framework Alliance Contract (2016) Schedule 4.
889 FAC-1 Framework Alliance Contract (2016) clause 5.6.
890 FAC-1 Framework Alliance Contract (2016) clause 5.7.
891 FAC-1 Framework Alliance Contract (2016) Schedule 1 Part 2.
892 FAC-1 Framework Alliance Contract (2016) clause 4.3.

10.4   How Does FAC-1 Support Supply Chain Collaboration?

FAC-1 provides for agreed Alliance Activities[893] linked to agreed deadlines under an agreed Timetable.[894] Alliance Activities can include Supply Chain Collaboration. FAC-1 clause 6.3 describes the process of Supply Chain Collaboration whereby 'if stated in the Framework Alliance Agreement or the Framework Brief or otherwise agreed by Alliance Members, the Alliance Activities shall include Supply Chain Collaboration in order to achieve Improved Value through more consistent, longer term, larger scale Supply Chain Contracts and through other improved Supply Chain commitments and working practices'.[895]

The process is set out in detail as follows:

- 'Agreeing through the Core Group, if not already set out in the Framework Brief, the basis for sharing information between Alliance Members in relation to their current and proposed Supply Chain Contracts and, if not already set out in the Timetable, the timescales for each stage of Supply Chain Collaboration'
- 'Reviewing and comparing the value offered by each Alliance Member's current and proposed Supply Chain'
- 'Reviewing the potential for more consistent, longer term, larger scale Supply Chain Contracts and for other improved Supply Chain commitments and working practices'
- 'Jointly re-negotiating Supply Chain Contracts or undertaking joint Supply Chain tender processes, in each case through procedures to be approved by the Core Group, to be led by one or more agreed Alliance Members and to be organised, monitored and supported by the Alliance Manager'
- 'Subject to approval by the Client and any Additional Clients of the Improved Value resulting from Supply Chain Collaboration, agreeing and entering into more consistent, longer term, larger scale Supply Chain Contracts and other improved Supply Chain commitments and working practices'.[896]

FAC-1 provides commercial protections for Alliance Members in respect of their exchanges of information by means of:

- Provision for Framework Prices and other Framework Proposals to be confidential between the Client, the Alliance Manager and the other Alliance Member who provides them,[897] and for amended confidentiality provisions if agreed[898]
- Provisions governing mutual Intellectual Property Rights.[899]

These commercial protections ensure that there is clear understanding as to which data is the exclusive property of one Alliance Member, for example because it represents the results of research and development intended to create a competitive advantage. Where a client using FAC-1 wishes to ensure that data is not retained confidentially by one Alliance Member, and instead is shared for the purposes of seeking Improved Value across the Framework Programme, it will be necessary to make this clear in the

---

893 FAC-1 Framework Alliance Contract (2016) clause 6.
894 FAC-1 Framework Alliance Contract (2016) Schedule 2.
895 FAC-1 Framework Alliance Contract (2016) clause 6.3.
896 FAC-1 Framework Alliance Contract (2016) clause 6.3.
897 FAC-1 Framework Alliance Contract (2016) clause 13.3.2.
898 FAC-1 Framework Alliance Contract (2016) clause 13.3 and any amended confidentiality provisions stated in the Framework Alliance Agreement.
899 FAC-1 Framework Alliance Contract (2016) clause 11.

Framework Brief and in the Framework Alliance Agreement by reference to clauses 6 and 13 of the Contract Terms:

- As regards specific requested Framework Proposals
- As regards the outputs from Supply Chain Collaboration and other Alliance Activities
- As regards lessons learned on individual Projects.

For example, the Football Foundation agreed to use FAC-1 as a means:

- 'To integrate specialists and consultants rather than depend on main contractors
- To create flexibility through Joining Agreements for Additional Clients who are primarily applicants for funding from the Football Foundation
- To develop common supply chains, integrated processes and to participate in value engineering exercises, all designed to improve the way in which changing rooms and pitches are delivered'.[900]

10.5 How Does FAC-1 Measure and Reward Performance?

FAC-1 provides flexible foundations that a client and its advisers can use when developing their procurement strategy. Its multi-party structure provides a shared system of open performance measurement and rewards agreed by all Alliance Members. FAC-1 states the agreed overall Objectives of the Alliance and the Framework Programme, and the individual Objectives of the Alliance Members.[901] These in turn form the basis for Success Measures that determine the success of the Alliance and the Framework Programme and that measure the performance of the Alliance Members.[902] FAC-1 states the agreed Targets for each Success Measure, which should include the method of recording relevant data, the Alliance Member responsible for measuring against that data and the system for reporting to the other Alliance Members.[903]

For example, the Objectives of the Crown Commercial Service FAC-1 consultant framework alliance were to:

- 'Share and monitor best practice intelligence
- Share and monitor learning between Projects and programmes of work
- Establish, agree and monitor consistent and more efficient working practices
- Agree and monitor techniques for better team integration
- Agree and monitor improved procurement and delivery systems on Projects and programmes of work
- Share and monitor other improvement initiatives created with contractors and other Supply Chain members'.[904]

On a smaller scale, Epping Forest District Council created an FAC-1 framework alliance to deliver a £22 million house building programme of new homes, integrating a multi-disciplinary team that comprised the design consultants, a cost consultant and a group of four main contractors. The agreed FAC-1 Objectives were:

---

900 Case study summarised in Section 10.10.
901 FAC-1 Framework Alliance Contract (2016) clause 2.1 and Schedule 1 Part 1.
902 By reference to FAC-1 Framework Alliance Contract (2016) clause 2.3 and Schedule 1 Part 2.
903 Guidance note in FAC-1 Framework Alliance Contract (2016) Schedule 1 Part 2.
904 Case study summarised in Section 10.10.

- 'To deliver high levels of User satisfaction that improve and enhance the lives of those living in their new homes
- To deliver homes that are sustainable for the Client and Users
- To demonstrate value for money through both capital investment and whole life costs
- To learn from shared experiences on the Alliance and to adapt, develop and improve the quality of new homes over the life of the Framework Alliance Contract'.[905]

FAC-1 provides for agreed Incentives,[906] which can include for example:

- Additional payments including shares of savings achieved through Supply Chain Collaboration and other Alliance Activities[907]
- Adjustment of any exclusivity in the award of Project Contracts[908]
- Extension of the scope of the Framework Programme and/or the duration of the Alliance.[909]

Each Alliance Member needs to understand how the agreed Objectives and Success Measures may affect the future award of Projects, the continuation of its role as an Alliance Member or the continuation of the Alliance as a whole. FAC-1 provides clarity as to which Targets are so important that a failure to meet them will require urgent action and may ultimately determine whether an Alliance Member's appointment may be terminated.[910] It also includes an Early Warning system[911] enabling notification of the reasons behind any issues or obstacles that are encountered.

10.6 How Does FAC-1 Support Collaborative Risk Management?

FAC-1 provides for consensus-based governance by a Core Group of individuals representing Alliance Members in support of their agreed mutual commitments.[912] The FAC-1 Core Group makes its decisions by the unanimous agreement of a quorum of all members present, so that there is no risk of Core Group members being overruled by a majority and so that individual Core Group members cannot frustrate collaborative working by refusing to attend a meeting.[913] The FAC-1 Core Group is the forum where Alliance Members can consider and agree proposals for Improved Value, and where they can raise issues with each other in order to resolve problems before they become disputes.[914]

---

905 www.allianceforms.co.uk. (News and Users) as submitted by Neil Thody of Cameron Consulting.
906 To be set out in FAC-1 Framework Alliance Contract (2016) Schedule 1 Part 3 by reference to clause 2.4.
907 Which can be payable under a Project Contract or under FAC-1 Framework Alliance Contract (2016) clause 8.
908 Under FAC-1 Framework Alliance Contract (2016) clause 5.7 and Schedule 1 Part 2.
909 Linked to FAC-1 Framework Alliance Contract (2016) clause 14.1.
910 FAC-1 Framework Alliance Contract (2016) Schedule 1 and clause 14.2.
911 FAC-1 Framework Alliance Contract (2016) clause 1.8.
912 FAC-1 Framework Alliance Contract (2016) clauses 1.6 and 1.7, as considered in Section 12.7.
913 FAC-1 Framework Alliance Contract (2016) clause 1.7.3. This system has been used successfully under the PPC2000 (2013) and TPC2005 (2013) contract forms, has been influential on the improved value achieved using these contract forms and has enabled their successful avoidance of disputes, as evidenced by only two court cases in 18 years.
914 FAC-1 Term Alliance Contract (2016) clause 15.1, as considered in Section 19.7.

FAC-1 provides for an in-house or external Alliance Manager who integrates the work of the Alliance Members and who:

- Implements the Direct Award Procedure and Competitive Award Procedure
- Monitors and supports achievement of the agreed Objectives, Success Measures and Targets
- Monitors and supports achievement of deadlines, gateways and milestones in the Timetable and prepares updates for Core Group approval
- Organises Core Group meetings, including in response to Early Warning
- Organises, supports and monitors Supply Chain Collaboration and other Alliance Activities
- Manages agreed payments and Incentives
- Monitors and supports Risk Management in accordance with the Risk Register and prepares updates for Core Group approval at intervals stated in the Timetable.[915]

Risk Management is defined in FAC-1 as 'a structured approach to ensure that risks are identified at the earliest opportunity, that their potential impacts are allowed for and that by agreed actions such risks and/or their impacts are eliminated, reduced, insured, shared or apportioned'.[916] The Alliance Members agree to 'undertake Risk Management together and individually in accordance with the Framework Documents in order to analyse and manage those risks using the most effective methods'.[917]

The FAC-1 Risk Register[918] records the commitments of Alliance Members to undertake agreed Risk Management actions, including:

- The nature of each risk, its likelihood, and its impact on the Framework Programme and/or achievement of the agreed Objectives and/or agreed Alliance Activities, including any anticipated financial impact
- The Alliance Members responsible for each Risk Management action
- The agreed Risk Management actions, including actions to reduce the likelihood of each risk and to reduce its financial and other impact
- The agreed periods or deadlines for completing those actions.[919]

This system for joint management of risks provides new opportunities to analyse, understand and reduce risks and to mitigate their potential effects, as considered in Chapter 18. The scope to agree joint and individual Risk Management actions can enable Alliance Members to reduce risks, to share risks as agreed and to price more accurately in the Direct Award Procedure and Competitive Award Procedure any risks that cannot be reduced or shared.[920]

Where problems arise that might otherwise escalate into disputes, the additional information shared among Alliance Members through joint Risk Management provides a stable basis for non-adversarial dispute resolution by the Core Group.[921]

---

915 FAC-1 Framework Alliance Contract (2016) clause 3.1.
916 FAC-1 Framework Alliance Contract (2016) Appendix 1.
917 FAC-1 Framework Alliance Contract (2016) clause 9.2.
918 FAC-1 Framework Alliance Contract (2016) clause 9.3 and Schedule 3.
919 Guidance note in FAC-1 Framework Alliance Contract (2016) Schedule 3.
920 Joint risk management is considered in Chapter 18.
921 FAC-1 Framework Alliance Contract (2016) clause 15.1 as considered in Section 19.7.

10.7   How Can FAC-1 Reflect Differing Requirements?

Different clients and their legal advisers need flexibility in adapting a standard form framework alliance contract that enables them to develop and include details and variations that reflect their specific requirements. FAC-1 is designed as a flexible form that can accommodate a wide range of specific requirements.

FAC-1 provides scope for the bespoke drafting needed to reflect the details of the Alliance relationships and the underlying Project Contracts. These details can be set out in:

- The Framework Alliance Agreement
- The Framework Brief and other Framework Documents
- The agreed Objectives, Success Measures, Targets and Incentives
- The Timetable of agreed Alliance Activities
- The agreed award procedures for Project Contracts and the Template Project Documents
- Specific amendments to the Contract Terms by way of Legal Requirements or Special Terms.

The FAC-1 Direct Award Procedure and Competitive Award Procedure are intended to be compatible with any combination of roles, responsibilities and warranties among consultants, contractors and other Alliance Members and Supply Chain members engaged on each Project. They can set out options in respect of:

- Integrating the contributors to Project Contracts that govern a traditional or design and build or construction management model, with any level or combination of design, supply, construction and operation contributions from design consultants, contractors, subcontractors, suppliers, manufacturers and operators
- Integrating activities under related Project Contracts where these contribute to a single complex Project or combination of Projects
- Integrating the capital and operational phases of each Project.

A framework alliance under FAC-1 can be combined with any allocation of design, supply, construction and operation responsibilities. However, it is fundamental to an effective framework alliance that appointments and relationships are structured in a way that enables early integrated contributions from all Alliance Members and from other Supply Chain members.

When awarding Project Contracts FAC-1 provides flexibility so that Clients and other Alliance Members can use different Project Contract Conditions which are suitable to the location, requirements, structure, costing, type and size of each Project. FAC-1 is designed for use with any one or more types of Project Contracts, and cross-reference to the applicable Project Contract Conditions should be included in the description of the Template Project Documents.[922] For example, FAC-1 has been used in combination with FIDIC, ICC, JCT, NEC and PPC2000 and a range of consultant appointments.[923]

---

922 FAC-1 Framework Alliance Contract (2016) clause 5.3 and Schedule 5.
923 www.allianceforms.co.uk. (News and Users).

FAC-1 clauses on payment[924] and adjudication[925] are drafted to conform to the requirements of the UK Housing Grants, Construction and Regeneration Act 1996 as amended, but without specifically referring to that legislation. When completing FAC-1 Schedule 6 Part 1, Clients and other Alliance Members need to decide whether it is necessary for specific legislative provisions to be restated in the Legal Requirements,[926] for example in order to clarify particular legal responsibilities and procedures such as:

- Roles of Alliance Members in relation to health and safety under the UK Construction (Design and Management) Regulations 2015
- Payment arrangements in respect of UK Value Added Tax.

FAC-1 provides for the Contract Terms to be supplemented or amended by Special Terms that reflect the particular needs of the Client or other Alliance Members or are required by reason of the nature of the Framework Programme.[927] If required, FAC-1 Special Terms can be used to align certain Contract Terms with the selected Project Contract Conditions, for example the Schedule 1 definitions of Insolvency Events or Intellectual Property Rights or the clause 15 and Appendix 4 options in respect of dispute resolution.

FAC-1 provides for a duty of reasonable skill and care linked to each Alliance Member's agreed role, expertise and responsibilities,[928] with options for agreed amendments, for example:

- To reflect the wording of the duty of care owed under the selected Project Contract Conditions
- To incorporate a 'no blame' clause or other agreed limits on liability[929]
- To agree a fitness for purpose obligation.[930]

The FAC-1 duty of care can be extended to parties other than Alliance Members as agreed, for example by a third-party collateral warranty or by creation of other third-party rights.[931] The duty of care under Project Contracts is stated in the relevant Project Contract Conditions as set out or referred to in the Template Project Documents.[932] Where third-party rights are to be created in respect of Project Contracts, these are stated in the Project Contract Conditions or other Template Project Documents.

Each jurisdiction develops its own procurement models and its own standard form construction and engineering contracts, recognising the requirements of national law as well as practices prevailing in the national commercial environment. However, the principles and practices of alliances can be applied in any jurisdiction and should not be constrained by differences between, for example, a common law system and a civil law

---

924 FAC-1 Framework Alliance Contract (2016) clause 8.
925 FAC-1 Framework Alliance Contract (2016) clause 15.3 and Appendix 4 Part 2.
926 Stated in the FAC-1 Framework Alliance Contract (2016) Framework Alliance Agreement by reference to clauses 13.4 and 15.5.
927 FAC-1 Framework Alliance Contract (2016) clause 13.5 and Schedule 6 Part 2.
928 FAC-1 Framework Alliance Contract (2016) clause 10.1.
929 The legal implications of which are considered in Sections 7.7 and 18.7.
930 The legal implications of which are considered in Section 5.3.
931 FAC-1 Framework Alliance Contract (2016) clause 10.6.
932 FAC-1 Framework Alliance Contract (2016) Schedule 5.

system. FAC-1 was consulted on in 14 jurisdictions and is designed to be suitable for use in any country under any legal system.

Adaptation of an alliance model for use in different jurisdictions may be subject to legal issues specific to that jurisdiction, and examples are considered in Sections 20.5 (Australia), 21.5 (Brazil), 22.5 (Bulgaria), 23.5 (Germany), 24.5 (Italy) and 25.5 (USA). Adaptation of an alliance model will be more straightforward if the framework alliance contract can operate in conjunction with national Project Contract Conditions. FAC-1 is designed for use with the Project Contract Conditions of any jurisdiction and also:

- Provides for the choice of the applicable law[933]
- Does not include terms that restate specific provisions of English law
- Provides for the Contract Terms to be supplemented or amended to state particular Legal Requirements in order to comply with the laws of the jurisdiction stated in the Framework Alliance Agreement.[934]

10.8 How Does FAC-1 Reflect UK Government and Industry Recommendations?

FAC-1 is being used in conjunction with a range of UK Government and construction industry procurement recommendations, including:

- Project procurement using Two Stage Open Book, Cost Led Procurement and Integrated Project Insurance
- Adoption of BIM
- Use of Project Bank Accounts.[935]

The 2012 Effectiveness of Frameworks Report defined an effective framework in the following terms, all of which are reflected in FAC-1:

- 'Has a demonstrable business need' – FAC-1 Schedule 1 Part 1 sets agreed Objectives and Schedule 1 Part 2 sets agreed Success Measures and Targets, all linked to an agreed understanding of Improved Value
- 'Has effective governance processes, active stakeholder engagement and client leadership' – FAC-1 clauses 1.6 and 1.7 establish terms of reference for the Core Group and clause 1.10 establishes engagement with Stakeholders
- 'Actively supports its clients throughout the project lifecycle, ensuring that clients and the supply chain receive a legacy of improvement' – The FAC-1 definition of Improved Value includes 'improved quality, improved Operation, improved staff and other resources, improved health and safety and other working procedures, improved Sustainability…and other benefits to Alliance Members, Users and Stakeholders'
- 'Is driven by aggregated demand to create volume and generate efficiencies, and provides sufficient work opportunities to cover supplier investment' – FAC-1 provides for multiple Projects and Additional Clients so as to enable demand to be aggregated, and clauses 5.6 and 5.7 state any agreed minimum amounts of work and any exclusivity in the award of Project Contracts

---

933 FAC-1 Framework Alliance Contract (2016) clauses 13.4 and 15.5.
934 FAC-1 Framework Alliance Contract (2016) clause 13.4 and Schedule 6 Part 1.
935 As considered in Section 18.9.

- 'Maintains competitive tension in terms of value, quality and performance during its life' – FAC-1 Schedule 4 sets out the agreed Direct Award Procedure and Competitive Award Procedure and Schedule 1 Part 2 provides for adjustment of amounts of work awarded according to achievement of agreed Targets
- 'Is designed and managed to deliver the required outcomes and continuously improve upon them' – FAC-1 clause 2.1 records the commitment of Alliance Members to seek to achieve the agreed Objectives and clause 2.2 records the agreement of Alliance Members to investigate and submit proposals for Improved Value
- 'Can demonstrate greater value for money for the taxpayer' – The FAC-1 definition of Improved Value includes 'improved cost and/or time certainty' and 'cost and/or time savings' among other measures
- 'Pays fairly for the work done and the risks taken' – The FAC-1 payment provisions are set out in clause 8 and are consistent with UK regulatory requirements
- 'Contributes to the development of an effective and efficient construction market' – The FAC-1 definition of Improved Value includes 'improved efficiency, improved profitability and other benefits to Alliance Members'
- 'Harnesses the power of public sector procurement to provide jobs and skills, local employment and enables SMEs to prosper' – The FAC-1 definition of Sustainability includes 'employment and training opportunities' and clause 6.3 Supply Chain Collaboration provides a new system for improved engagement with *small and medium enterprises* (SMEs)
- 'Ensures supply chains are engaged from the earliest stages of a project' – FAC-1 clause 6.3 sets out the system of Supply Chain Collaboration that maximises early engagement of Supply Chain members
- 'Ensures transparency and collaborative values flow down the supply chain to produce supply chains that clients can have confidence in' – FAC-1 clause 6.2 provides for complementary relationships with Supply Chain members, and clause 6.3 describes Supply Chain Collaboration.

The ICG Alliancing Code of Practice[936] sets out the following key principles for an alliance agreement, all of which are reflected in terms and options under FAC-1:[937]

- 'A clear definition of how partner return will be generated and how this is aligned with project outcomes (back to back)': the FAC-1 Schedule 1 Objectives, Success Measures, Targets and Incentives are agreed by all Alliance Members, and clause 6.2 provides for complementary Supply Chain relationships
- 'A clear definition of the risks that will be collectively assumed by all alliance parties (and any that will be assumed by individual participants – the fewer the better)': FAC-1 clause 9.2 records agreement of the Alliance Members 'to undertake Risk Management together and individually', and the Schedule 3 Risk Register is agreed by all Alliance Members
- 'How incentive mechanisms will work', 'How shared return arrangements will work across all alliance parties – shared pain/gain', 'How target costs and incentive thresholds will be set': the FAC-1 Schedule 1 Part 3 Incentives are agreed by all Alliance Members and include provision for shared savings

---

936 Infrastructure Client Group (2015) Section 3.4.
937 These features also appear in the TAC-1 Term Alliance Contract considered in Chapter 11.

- 'A no blame – no dispute approach (wilful default being the only direct route to legal process)': FAC-1 clauses 10.1 and 10.2 provide options to amend the duty of care that can establish a no blame – no dispute approach if required
- 'That all parties will act fairly and on a "Best for Project" basis': FAC-1 clause 1.1 provides for Alliance Members to work 'together and individually in the spirit of trust, fairness and mutual cooperation for the benefit of the Framework Programme'
- 'Full open book costing': FAC-1 clause 4.3 separates Profit and Overheads from other costs
- 'The establishment of collective leadership teams and management groups', 'Collective and unanimous decision-making responsibility (with any exceptions clearly defined)': FAC-1 clauses 1.6 and 1.7 establish these terms of reference for the Core Group.

10.9 How Is FAC-1 Being Used in Practice?

FAC-1 been used by public and private sector clients, for construction and engineering projects, and for frameworks and alliances in a wide range of sectors. It has been adopted on procurements of varying sizes governing a wide range of works, services and supplies.

Some of the first users of FAC-1 were alliance teams in the housing sector. For example:

- Futures Housing Group, led by John Thornhill, was the first user of FAC-1 and put in place a framework alliance that integrated the work of 23 small contractors under a £30 million works programme, saving an average of over 9% against previous prices. Futures Housing also set out in FAC-1 their new initiatives for training and employment, for rapid payment and for discounted access to supplies from the builder's merchant Travis Perkins[938]
- The first UK local government user of FAC-1 was Royal Borough of Greenwich on a £20 million programme of mechanical and electrical works[939]
- One of the smallest FAC-1 programmes was a £7.5 million procurement by Southern Housing Group.[940]

Other early FAC-1 housing users include Cartren Conwy, Gentoo, Homes in Sedgemoor, Liverpool City Council, Places for People and Your Housing Group. FAC-1 has been used for modular social housing programmes, for example by North Devon Homes on £40 million of modular new build projects. It also provided the basis for the UK Construction Leadership Council and the Department for Business Energy and Industrial Strategy initiative to create model form strategic agreements that will accelerate the construction of affordable housing. For this purpose, King's was appointed in 2017 to develop a model form contract based on FAC-1 for long-term strategic relationships that focus on off-site and modular solutions. The resultant model form is designed:

- For use in strategic procurements by clients who wish to evaluate the quality, cost and flexibility of alternative solutions
- To integrate the roles of clients, consultants, contractors and manufacturers and the influence of stakeholders and regulators

---

938 Case study www.allianceforms.co.uk and summarised in Section 10.10.
939 www.allianceforms.co.uk (News and Users).
940 www.allianceforms.co.uk (News and Users).

- To integrate other contracts governing site purchase, funding, design, planning, construction and manufacture.[941]

In the private sector FAC-1 has been adopted on alliances led by:

- The Football Foundation, who integrated the work of five contractors and modular suppliers with two consultant firms under a £150 million national programme for construction of new sports ground changing rooms[942]
- The Football Foundation, who also procured a £60 million national programme of mini-pitches
- Kier Services, who combined FAC-1 with a term alliance in order to create an integrated Supply Chain engaged on a combination of capital works and ongoing operation, repair and maintenance.[943]

FAC-1 has been adopted on a range of multi-client programmes establishing framework alliances among multiple contractors and consultants, including:

- A £150 million housing programme awarded by Optivo to Axis Europe, Keepmoat, Kier Services, Mears and Osborne[944]
- The £2.8 billion Crown Commercial Service national consultant frameworks for project managers and architects, their £1.2 billion modular framework alliance and their £30 billion contractor framework alliance[945]
- The nationwide LHC local government consortium on programmes of schools and community buildings with projected aggregate values of up to £5 billion
- LHC on regional programmes of new build housing with projected aggregate values of up to £1.5 billion and related consultant services with projected aggregate values of up to £150 million.[946]

FAC-1 has been translated for use in Brazil, Bulgaria, Germany and Italy. Collaborative work on these translations has shown that relatively few amendments are necessary for FAC-1 to reflect the requirements of a range of civil law jurisdictions, and this is illustrated by the schedules set out at the end of Chapters 21, 22 and 24. FAC-1 is already being used on projects outside the UK, for example:

- FAC-1 was chosen by Ecological Sequestration Trust and Resilience Brokers to support engagement with and integration of their partner organisations financing and implementing Global Goals in Human Settlements and City-Regions[947]
- The Italian translation of FAC-1 has been adopted in order to integrate contributions to BIM on a new build school project for the Milan Union of Municipalities Adda Martesana, and is also being used for a major privately financed project at the Milan University Campus[948]

---

941 www.allianceforms.co.uk (News and Users).
942 Case study summarised in Section 10.10.
943 Case study summarised in Section 11.10.
944 www.allianceforms.co.uk (News and Users).
945 Case study summarised in Section 10.10.
946 Case study summarised in Section 10.10. LHC have also used FAC-1 to procure contractors and consultants in other regions, who will undertake a series of housing programmes for multiple clients.
947 Roadmap 2030, 110 and 112.
948 With supporting advice and research led by Professor Angelo Ciribini, Professor Giuseppe Di Guida and Professor Sara Valaguzza, and as summarised in Section 10.10 and considered further in Section 14.8 and in Chapter 24.

- The European Bank for Reconstruction and Development obtained a briefing on FAC-1 and decided to review its potential value under pilots with selected clients. They reported that, as part of their $350 million syndicated loan facility to the JSC Shalkiya Zinc mining company in Kazakhstan, FAC-1 will be adopted as the means to integrate multiple FIDIC contracts governing expansion of the Kazakhstan Shalkiya Zinc and Lead mine and construction of a new processing plant on the same site[949]
- The German version of FAC-1 will be used on two projects in 2019, firstly to integrate the Supply Chain members engaged on a mega-project (combined office/residential) in Germany, as an overarching umbrella to create harmonised collaboration between the parties to a multi-party main contract and the subcontractors who are not part of the main contract, and secondly on the reconstruction of the German embassy in Sofia, as a multi-party integrator for the supply chain members of contractors engaged under FIDIC Silver Book contracts.[950]

10.10 FAC-1 Framework Alliance Case Studies

This section sets out case studies of FAC-1 framework alliances that have been or are being established:

- By Futures Housing Group on a housing programme with 23 small and medium-sized contractors and with a national builder's merchant. Their work is also recognised in case studies prepared by the Housing Forum[951] and the Construction Leadership Council[952]
- By the Football Foundation on framework alliances governing their national programme of modular changing rooms and a national programme of mini-pitches
- By LHC as a central purchasing body on behalf of public sector clients, appointing contractors and consultants on a national programme of schools and community buildings and on regional programmes of modular and other new build housing
- By Milan Union of Municipalities Adda Martesana on a new build school project and by University of Milan on a campus Private Finance Initiative (PFI) project
- By Crown Commercial Service in order to integrate the work of designers and project management consultants engaged on a range of construction projects for Government Departments and the wider public sector.

FAC-1 has been closely scrutinised by the UK Government's procurement body, Crown Commercial Service. Following their use of FAC-1 for the successful integration of consultant services Crown Commercial Service also adopted FAC-1 as the standard form that will underpin procurement of:

- Their £1.2 billion modular framework alliance, with a focus on the education and health sectors, as advertised in November 2018

---

949 Conference presentation on 1 March 2018 at King's College London, provided by Evgeny Smirnov of the European Bank for Reconstruction and Development.
950 As noted in Section 23.2.
951 Housing Forum (2018), 23.
952 Procuring for Value (2018), 34.

- Their £30 billion collaborative contractor framework alliances, advertised on 31 January 2019 and estimated to be the largest construction procurement ever undertaken in the UK.

The Crown Commercial Service procurements are designed for use by a wide range of public sector clients, and FAC-1 is being used in conjunction with JCT 2016, NEC4, and PPC2000 in order to establish:

- Encouragement of Supply Chain Collaboration by appointed contractors, either working with groups of users or with individual users undertaking multiple projects
- Encouragement for users to adopt BIM, to use Project Bank Accounts[953] and to integrate capital works with ongoing operation, repair and maintenance
- Options for users to adopt Two Stage Open Book, Cost Led Procurement or Integrated Project Insurance when calling off one or more projects
- Sub-alliances under additional FAC-1 contracts through which one or more users integrate their programmes of work.

The range of FAC-1 alliances is growing, and in 2018 I had discussions with clients and consultants in the UK and other jurisdictions who plan to use FAC-1 for:

- Airport infrastructure and facilities
- Commercial developments
- National and regional highways
- Oil refineries and other power plants
- Urban transport networks.

Futures Housing Group Framework Alliance

Futures Housing Group adopted the Framework Contract Alliance (FAC-1), which not only makes it easier for small businesses to bid for deals, it sets shared objectives for those on the framework. Alliance members are expected to work together to improve value, share information on suppliers, review and compare prices and tender or renegotiate subcontracts – Supply Management (2016).

Futures Housing Group embarked on its framework alliance with predicted outcomes that included:

- Better service delivery through a stronger partnering approach built upon closer ties to the successful bidders
- Efficiency gains from the innovative approach to selecting and managing individual works
- Greater emphasis on employment and training and a greater impact on local economies
- New standard contract terms providing lower risk through closer engagement criteria
- Continued success in spending over 75% of contractor budget within 25 miles of the client's two regional offices, contributing to their 'Lean and Local' ambitions.

The tender included measures attractive to SMEs (small and medium enterprises) but open to all, with cashflow-easing features that included the provision of key materials in certain lots on a free issue basis, such as heating installations, bathroom installations and

---

953 As considered in Section 18.9.

electrical works. It also included proposals to have embedded payment cards with the contactors, enabling them to claim payment immediately on agreement of final account.

The tender was published in the OJEU on 16 December 2015 after consultation with leaseholders, as required by legislation. A Bidders Day was held by the Group where all comers could ask questions about the process and about the tender. Attended by over 120 potential bidders, questions and answers were published and made available to all enquiries. Throughout the tender process, the emphasis was on a collaborative approach via FAC-1. Of the 23 contractors invited to join the framework alliance, only two were non-SME. The average size of the other companies was less than 25 employees.

Various tools were used to introduce the framework alliance, including a speed-dating session, ensuring that each contractor met their client counterpart and all the other contractors in the framework alliance. Every contractor welcomed the new Framework Alliance Contract, its plain language and sequential approach to each stage making it easy to understand.

The client reported: The sense of truly belonging to a collaborative group is already apparent. Since launching on 1 August 2016, the framework has seen the first call off contracts by direct award for pilot programmes of heating installation, kitchen installation and bathroom installation. Costs returned have shown an average saving of over 9% against the previous framework.

Reported to KCL Centre of Construction Law by John Thornhill of Futures Housing Group and recorded in www.allianceforms.co.uk (News and Users).

Football Foundation Modular Changing Rooms and Mini-Pitch Programmes

The Football Foundation, (with involvement from The Football Association and Sport England), advised by Cameron Consulting, used FAC-1 to procure two national works programmes worth £150 million and £60 million.

Their drivers were the Football Association Chairman's Report in 2014 which cited that 'players and coaches repeatedly tell us that the quality of pitches (and their associated facilities) is a barrier to both recruiting and retaining players', and that 'wider social & community benefits from increased physical activity including the fight against preventable diseases, improving cohesion, and personal development for young people'.

The first programme adopted FAC-1, combined with JCT building contracts, for Modular Build Changing Rooms using:

- £150 million investment
- Match funding from Applicants.

An FAC-1 framework alliance was created between the Football Foundation and:

- Gleeds and QMP acting in the role of Framework Managing Consultant
- Western Building Systems, Integra Buildings, Extraspace Solutions, Ashley House and Portakabin as the Modular Suppliers
- Cameron Consulting as the Independent Adviser.

The second FAC-1 procurement was for a Mini-Pitch Programme using:

(Continued)

- £60 million investment
- Integration with the Changing Rooms facility through an interface protocol.

An FAC-1 framework alliance was created between the Football Foundation and:

- Labosport acting in the role of Framework Managing Consultant
- Support in Sport (SIS Pitches) as the 3G Pitch Supplier
- Cameron Consulting as the Independent Adviser.

The clients have used FAC-1:

- To integrate specialists and consultants rather than depend on main contractors
- To create flexibility through Joining Agreements for Additional Clients who are primarily applicants for funding from the Football Foundation
- To develop common supply chains, integrated processes and to participate in value engineering exercises, all designed to improve the way in which changing rooms and pitches are delivered.

Reported to KCL Centre of Construction Law by Neil Thody of Cameron Consulting and recorded in www
.allianceforms.co.uk (News and Users).

LHC Central Purchasing Programmes for Schools, Community Buildings and Housing Programmes

LHC was established in 1966 by 13 London boroughs with objectives that included 'the coordination of industrialised building projects' and 'the establishment of common social and technical standards'. LHC is now a nationwide central purchasing body and has adopted FAC-1 on procurement programmes that include:

- Schools and community buildings with a projected aggregate value of over £5 billion
- New build housing with partners such as the Scottish Procurement Alliance, covering projects with a projected aggregate value of £1.5 billion, including sheltered accommodation, care homes and student accommodation plus associated works including land purchases
- Related project management, cost consultancy, architectural, structural engineering, building services engineering and other consultancy services with a projected aggregate value of £150 million.

LHC used FAC-1 to create large-scale multi-client framework alliances with the appointed companies on each LHC framework and with the LHC clients who sign project registration documents to use each LHC framework. By this means any publicly funded organisation throughout England and Wales or Scotland may become an FAC-1 Additional Client.

The agreed FAC-1 Objectives are:

- For LHC as a central purchasing body to operate the Framework Alliance Contract in a way that is accessible to a wide range of Additional Clients covering a broad Framework Programme

- To deliver the Framework Programme in order to achieve Improved Value for LHC and Additional Clients on the terms agreed with all other Alliance Members
- To generate employment and training opportunities for priority groups
- Vocational training
- To up-skill the existing workforce
- Equality and diversity initiatives
- To make subcontracting opportunities available to SMEs, the third sector and supported businesses
- Supply chain development activity
- To build capacity in community organisations
- Educational support initiatives
- To work with schools, colleges and universities to offer work experience
- To minimise negative environmental impacts, for example impacts associated with vehicle movements and/or associated emissions and impacts on protected areas, buildings or sites.

Reported to KCL Centre of Construction Law by John Skivington of LHC and recorded in www .allianceforms.co.uk (News and Users).

Crown Commercial Service Consultant Alliance

FAC-1 has been used by the UK Crown Commercial Service on their £2.8 billion national frameworks for the multi-disciplinary services comprising their Project Management and Full Design Team Services ('PMFDTS'). The appointed firms entered into FAC-1 contracts with each other in order to agree ways to deliver better value services to their Government clients. Crown Commercial Service entered into an FAC-1 framework alliance with AECOM, AHR Architects, AMEC Foster Wheeler Environmental and Infrastructure, Arcadis, Capita, Faithful & Gould, Gardiner & Theobald, Gleeds, Kier Business Services, Mace, McBains, Mott McDonald, Ridge, Turner & Townsend and WYG.

The agreed FAC-1 Objectives were 'to deliver Improved Value for the Client and for Users of the PMFDTS Framework Agreement and for this purpose to:

- share and monitor best practice intelligence
- share and monitor learning between Projects and programmes of work
- establish, agree and monitor consistent and more efficient working practices
- agree and monitor techniques for better team integration
- agree and monitor improved procurement and delivery systems on Projects and programmes of work
- share and monitor other improvement initiatives created with contractors and other Supply Chain members'.

The agreed FAC-1 Success Measures and Targets were:

- 'Evidence of progress and completion of Alliance Activities by the Core Group, by Special Interest Groups and by Alliance Members, in accordance with the Timetable

(Continued)

- Enhancement of the reputation of the PMFDTS Framework Agreement
- Additional users of the PMFDTS Framework Agreement
- Additional work awarded by users of the PMFDTS Framework Agreement.'

The consultant alliance members formed Special Interest Groups through which they examined the potential for shared best practice and consistent approaches to early contractor involvement and BIM.

Crown Commercial Service have since decided to adopt FAC-1 for the procurement of their £1.2 billion modular framework alliance and for their £30 billion collaborative contractor framework alliances. These framework alliances encourage Supply Chain Collaboration by appointed contractors, either working with groups of users or with individual users undertaking multiple projects, plus the adoption of BIM, the use of Project Bank Accounts and the integration of capital works with ongoing operation, repair and maintenance.

Reported by the author to www.allianceforms.co.uk (News and Users).

Milan Union of Municipalities Adda Martesana Liscate School

The Milan Union of Municipalities Adda Martesana created the first FAC-1 contract in Italy, governing the construction of a high school with an approximate value of €5 million. Tenders were evaluated on proposals in relation to the building envelope, availability of construction materials, technical characteristics, environmental requirements, functionality of components, efficient use of resources, safety, site management and maintenance. FAC-1 was explained to bidders, and the contractor was selected according to criteria for the most economically advantageous tender (MEAT) (80% technical, 10% financial, 10% time).

The FAC-1 alliance activities include data-sharing, BIM model management and maximum supply chain involvement. Objectives include the joint management of unforeseen events so as to minimise delays and additional costs, with agreed FAC-1 measures for monitoring cost and time throughout the project. Contractor profit and overheads are fixed at 10% profit and 13–17% overheads.

The alliance members are the client, the contractor, the design team, the construction manager and the safety coordinator. Subcontractors and suppliers will be invited to join the alliance but not the FAC-1 Core Group. The Alliance Manager is an officer of the client, the BIM consultant and coordinator is Politecnico of Milan, and the FAC-1 Independent Adviser is the Italian Centre of Construction Law and Management (CCLM), who also completed the legal elements of the documents thanks to the work of researchers at the University of Milan.

University of Milan campus PFI project

The second FAC-1 project in Italy is a €335 million project for the design, construction and operation of a campus for University of Milan. This privately funded project will involve Lendlease as the 30-year concessionaire, noting their experience of collaborative projects such as North Wales Prison. Other alliance members will include the client, the publicly-owned company responsible for urban regeneration (Arexpo), the construction manager and the safety coordinator for the realisation of the urban regeneration plan for the surrounding area. Stakeholders will include project funders and public bodies involved in the grant of permits and authorisations.

FAC-1 will govern the complex interactions between the alliance members and stakeholders during the design and construction phases, in order to ensure that the designs respond to the needs of the client, in order to measure time and cost and in order to avoid modifications arising from design errors or insufficient study of the project. The Italian National Anti-Bribery Authority will be monitoring the use of FAC-1 on this project in order to assess the potential for improved transparency and improved value through collaborative construction procurement.

Reported to KCL Centre of Construction Law by Professor Sara Valaguzza and recorded in www .allianceforms.co.uk (News and Users).

11

How Does the TAC-1 Term Alliance Contract Operate?

11.1   Overview

Term alliances provide a collaborative model for a wide range of asset management activities, and for the integration of these activities with capital projects so as to establish a whole life approach to procurement. Term alliances can also provide an efficient system under which to call off capital project works and supplies that are manufactured off-site. The potential of term alliances to deliver improved outcomes has been proven on programmes of works, services and supplies undertaken in sectors such as housing, offices, public buildings, education, highways and leisure.[954]

Term alliance contracts offer significant scope and duration for their members to invest in collaborative relationships, and for them to invest in new ways of working that generate improved value. However, an enhanced duty of good faith is not implied in a term contract simply because of its potentially long duration, and the culture underpinning many term contracts has been, at best, arm's length.[955]

In order to create and sustain a collaborative approach to the operation, repair and maintenance of a built facility, and in order to embed fair and open systems for delivering improved value, team members can benefit from a standard form contract that fulfils the alliance features of:

- Shared objectives, success measures, targets and incentives
- Systems through which alliance members build up and manage shared design, time, cost and operational data
- Activities through which alliance members can improve value
- Clear timeframes and deadlines for alliance members' activities, responses and approvals
- Systems for joint management of risks and the agreed avoidance and resolution of disputes.

This chapter reviews the development the TAC-1 Term Alliance Contract and how it builds on the features of the TPC2005 term partnering contract. It considers the key features of TAC-1, how it awards work, how it supports Supply Chain Collaboration

---

954 For example, as described in Association of Consultant Architects (2010).
955 *Bedfordshire County Council v Fitzpatrick* [1998] 62 Con LR 64 (TCC) where a term contract for highway maintenance was found not to imply a special 'duty of trust and confidence'.

Collaborative Construction Procurement and Improved Value, First Edition. David Mosey.
© 2019 John Wiley & Sons Ltd. Published 2019 by John Wiley & Sons Ltd.

and how it supports collaborative risk management. It examines how TAC-1 operates as a term contract and an alliance contract, and also the extent to which TAC-1 has the characteristics of an enterprise contract as described in Section 8.5. This chapter also looks at the ways in which TAC-1 can meet varying client requirements, how it differs from TPC2005 and how it is being used in practice.

11.2 How Was TAC-1 Developed?

The TAC-1 Term Alliance Contract was developed in parallel with the FAC-1 Framework Alliance Contract. The drafting of FAC-1 drew on many principles of the TPC2005 form, and the research team rapidly concluded that term alliances could benefit from a refreshed version of TPC2005 that is fully aligned with FAC-1.

TAC-1 is designed to set out processes for mobilisation and ordering of tasks, for integrating any number of alliance members and for improving value over the course of a long-term relationship. It combines the features of a term contract and an alliance and describes enterprise contract activities that comprise successive actions, reactions and interactions through which the parties create enough data:

- To enable the award of each task
- To improve on the commercial assumptions made by alliance members when the TAC-1 contract was entered into
- To improve value so as to justify continuation of the alliance.

The TAC-1 drafting team used processes and provisions drawn from successful term alliances that were successfully established by clients that included Royal Borough of Greenwich,[956] Welwyn Hatfield Council,[957] the Sheffield City Council/Kier joint venture[958] and the Surrey County Council highways programme.[959] Drafts of TAC-1 were reviewed at the same time as drafts of FAC-1 by the same consultation group of 120 organisations in 14 jurisdictions and using the same consultation process. TAC-1 was published in 2016, and the published form was endorsed by the UK Construction Industry Council and by Constructing Excellence.

TAC-1 is designed to provide a contractual medium for:

- Integrating the roles of alliance members through joint objectives and performance measurement linked to agreed incentives
- Awarding work through orders in respect of any number or range of tasks
- Planning the early engagement of alliance members in advance of task commencement
- Agreeing the means to achieve improved value with the benefit of long-term relationships
- Capturing learning and improvement from one task to another
- Creating systems that work in any jurisdiction.

---

956 Summarised in Section 5.10.
957 Association of Consultant Architects (2010), 45 and referred to in Section 5.7.
958 Summarised in Section 8.10.
959 Summarised in Section 5.10.

TAC-1 is also designed to make clear:

- Why the term alliance is being created and what measures will determine its success
- Who the alliance members are and how they will work with each other and with the wider supply chain to improve efficiency and add value
- How work will be awarded to alliance members using open processes and standard order forms
- How alliance members will develop an alliance culture that will help them to manage risks and avoid disputes.

The structure of TAC-1 in many ways mirrors the structure of FAC-1 and many clauses are identical in these two contract forms. However, TAC-1 differs from FAC-1 in the following respects:

- FAC-1 creates a framework alliance that operates in conjunction with the award of multiple project contracts for services, works and supplies, whereas TAC-1 creates a self-contained term alliance that operates through the issue of orders for services, works and supplies
- FAC-1 provides for project contracts to be awarded by one or more clients to one or more consultants, contractors and suppliers, using a direct award procedure and/or competitive award procedure, whereas TAC-1 provides for the issue of orders in respect of tasks to be implemented by a single provider
- FAC-1 provides for the agreement of framework prices that, through the direct award or competitive award procedure, are developed into agreed prices for each project, whereas TAC-1 provides for task prices to be established in advance subject to agreed adjustments, subject to clarifications for individual Tasks and subject to agreed savings through Supply Chain Collaboration.

11.3 What Are the Key Features of TAC-1?

In the remainder of this chapter words and expressions defined in TAC-1 start with capital letters. TAC-1 comprises the following Term Documents:

- Contract Terms which are referred to in a Term Alliance Agreement signed by all Alliance Members
- Orders issues to the Provider for the implementation of Tasks
- In Schedule 1 the agreed Objectives, Success Measures, Targets and Incentives
- In Schedule 2 a Timetable, which sets out the ways in which the Alliance Members agree to seek Improved Value through:
  - agreed deadlines, gateways and milestones in respect of achievement of the Objectives and in respect of Alliance Activities, including Supply Chain Collaboration as referred to in Section 11.5
  - timescales for the Alliance Activities, including the nature, sequence and duration of the agreed actions of each Alliance Member and any consents or approvals (whether required from Alliance Members or third parties) that are preconditions to subsequent actions.
- In Schedule 3 a Risk Register, as referred to in Section 11.7
- In Schedule 4 an Order Procedure, as referred to in Section 11.4

- In Schedule 5 a set of Template Order Documents, as referred to in Section 11.4
- In Schedule 6 any Legal Requirements and Special Terms that give rise to additions and amendments to the Contract Terms, as referred to in Section 11.8
- In Schedule 7 the Term Brief which is sent out by the Client to the Provider as the basis for its selection and which is described in more detail below
- In Schedule 8 the Term Prices and Term Proposals which are submitted by the Provider in response to the Term Brief, and which are binding only between the Client and the Provider.[960] There is an option for Term Prices to identify Profit and Overheads separate from other costs,[961] intended to ensure that cost savings proposed by the Provider or another Alliance Member do not erode its own margins
- If the Alliance Manager is an independent consultant, an Alliance Manager Services Schedule and Alliance Manager Payment Terms.

The responsibility of each Alliance Member is limited to the Term Documents that it prepares or contributes to, except to the extent of its stated reliance on information provided by other Alliance Members.[962] The Term Documents are stated to be complementary, and in the event of any error, omission or discrepancy TAC-1 sets out an Early Warning procedure and a default order of priority.[963] It also provides that Orders take precedence over the other Term Documents if a discrepancy cannot be resolved by agreement.[964]

The TAC-1 Term Brief comprises one or more documents contained or described in Schedule 7 and sets out the Client's technical, management and commercial requirements and expected outcomes in respect of the Term Programme and each type of Task, including (to the extent not stated in the Term Alliance Agreement or Order Procedure):

- The scope and nature of the Term Programme and the different types of Task
- The location and nature of the Sites
- Any different parts of the Term Brief applicable to different types of Task
- Contributions by Alliance Members and Supply Chain members to design, Building Information Modelling (BIM), Risk Management and other aspects of each type of Task
- Arrangements for engagement with Stakeholders under clause 1.10
- Requirements for Sustainability and Operation under clause 2.2, including in respect of Improved Value
- Whether Task Briefs will be issued for different types of Task under clause 5.2.1
- Whether Task Proposals are required and whether any other preconditions to implementation of each Order will apply under clause 5.2.3
- How Task Prices will be calculated and how any stated Budget will apply to each Order under clause 5.2.4
- Task Deadlines or how they will be calculated under clause 5.2.5
- Details of Supply Chain Collaboration and other Alliance Activities under clauses 6.1 and 6.3 and by reference to the Timetable and Risk Register

---

960 TAC-1 Term Alliance Contract (2016) clause 1.3.3.
961 TAC-1 Term Alliance Contract (2016) clause 4.3.
962 TAC-1 Term Alliance Contract (2016) clause 1.4.
963 TAC-1 Term Alliance Contract (2016) clause 1.5.
964 TAC-1 Term Alliance Contract (2016) clause 1.5.3.

- All required standards and warranties and the Quality Management System applicable to each Task under clause 7.4
- Responsibilities for assets made available by the Client under clause 7.7
- Constraints and procedures in respect of access to and possession of the Sites and arrangements for cooperation with Users and other parties undertaking activities on the Sites under clause 7.8
- Responsibilities for the security, state and condition of the Sites under clause 7.9
- Responsibilities for health and safety and Site welfare under clause 7.10
- Fluctuations in Term Prices and Task Prices under clauses 4.2 and 8.1
- Any revised payment intervals under clause 8.3
- Supporting information in respect of Provider payment applications under clause 8.3.1
- Insurances under clause 12
- Records to be kept by Alliance Members under clause 13.7
- Handover actions of Alliance Members under clause 14.9.[965]

11.4 How Does TAC-1 Award Work?

TAC-1 is a self-contained contract and does not award work by creating separate consultant appointments, construction contracts or supply contracts. Instead, TAC-1 provides for the issue of Orders for a range of Tasks, using an agreed Order Procedure which:

- States the agreed procedure and timescales for the issue of each Order, for the build-up of additional data in relation to more complex Tasks and for all steps leading to unconditional authorisation of each Task
- Explains how the Template Order Documents will be applied to each Task
- Explains the roles and responsibilities of Alliance Members in respect of each Task, including the sources and timing of all contributions to design, Supply Chain engagement, costing, programming and Risk Management, incorporating the required approach to BIM as appropriate
- Explains all other procedures relating to the conditional and/or unconditional award of Orders for Tasks.[966]

TAC-1 provides a standard Order Form with provision for details to be inserted in respect of the relevant Tasks, Order Procedure preconditions, Task Deadlines and Task Prices.[967] This form is designed for signature on behalf of the Client and the Provider, and is likely to be varied where no signatures are required because an electronic order system is put in place.

The TAC-1 Order Procedure should set out agreed procedures and timescales in relation to each type of Order:

- For the issue of each Order which in relation to simpler Tasks may itself state all required information[968]

---

965 TAC-1 Term Alliance Contract (2016) Guidance note in Schedule 7.
966 TAC-1 Term Alliance Contract (2016) Guidance note in Schedule 4.
967 TAC-1 Term Alliance Contract (2016) clause 5.2 and Appendix 3. Neither JCT MTC (2016) nor NEC4 TSC (2017) includes an order form.
968 Under TAC-1 Term Alliance Contract (2016) clause 5.

- For the creation of a Task Brief required for a more complex Task[969]
- For the submission of any Task Proposals and for satisfying any other preconditions to implementation of an Order, with applicable timescales in each case, which may be required for more complex Tasks in order to supplement the TAC-1 Term Brief and Term Proposals[970]
- For the calculation of Task Prices if not stated in an Order and how any stated Budget applies to an Order, although for simpler Tasks the Task Prices are calculated using amounts agreed in the TAC-1 Term Prices and are stated in the Order itself[971]
- For the calculation of Task Deadlines, although in respect of simpler Tasks the Task Deadlines can be calculated using periods agreed in the TAC-1 Term Brief and Term Proposals and can be stated in the Order itself[972]
- For any other requirements under an Order including the sources and timing of Supply Chain contributions[973]
- For access to and possession of each Site, and for cooperation with Users, Stakeholders and other parties undertaking activities on the Sites.[974]

A two-stage Order Form may be suitable for certain types of Task. For example, 'Southern developed a two-stage TAC-1 ordering process comprising:

- Order 1 for design development and preparation of a "Project Task Proposal"
- Order 2 to undertake the works in accordance with the "Project Task Proposal", agreed price and timetable'.[975]

It is fundamental to the TAC-1 alliance relationships that Alliance Members understand not only the prospective scope and nature of the work covered by the Term Programme but also how that work will be awarded. TAC-1 provides for any minimum value or type of Orders that will be issued to the Provider so that it has a clear picture as to the level of certainty in the flow of work.[976] It also states any exclusivity granted to the Provider[977] and any adjustment of exclusivity according to achievement of agreed Targets.[978]

The Template Order Documents to be used in any Order[979] for a more complex Task include:

- The structure and standard components of the Task Brief that forms part of the Order describing the scope and nature of the Task and setting out the Client's technical, management, and commercial requirements and expected outcomes in respect of that Task, and including all required standards and warranties
- All standard requirements in the Task Brief in respect of insurances and securities and all standard processes and procedures in the Task Brief for the management of

---

969 Under TAC-1 Term Alliance Contract (2016) clause 5.2.1.
970 Under TAC-1 Term Alliance Contract (2016) clause 5.2.3.
971 Under TAC-1 Term Alliance Contract (2016) clause 5.2.4.
972 Under TAC-1 Term Alliance Contract (2016) clause 5.2.5.
973 Under TAC-1 Term Alliance Contract (2016) clause 6.2.
974 Under TAC-1 Term Alliance Contract (2016) clause 7.8.
975 Case study summarised in Section 11.10.
976 TAC-1 Term Alliance Contract (2016) clause 5.6.
977 TAC-1 Term Alliance Contract (2016) clause 5.7.
978 TAC-1 Term Alliance Contract (2016) Schedule 1 Part 2.
979 Based on the list in TAC-1 Term Alliance Contract (2016) Schedule 5.

communication, performance, quality, design, Supply Chain engagement, cost, payment, time, change, risk, health and safety and all other Task management processes and procedures, in each case including the required approach to BIM as appropriate
- All standard requirements in the Task Brief in respect of Sustainability, Operation of the completed Task and engagement with Stakeholders and with Users
- The required structure and content of the Task Prices and Task Proposals to be submitted by the Provider and incorporated in the Order.

11.5 How Does TAC-1 Support Supply Chain Collaboration?

TAC-1 provides for Alliance Activities by which the Alliance Members agree how they will seek Improved Value.[980] These Alliance Activities can include Supply Chain Collaboration and TAC-1 clause 6.3 describes the process of Supply Chain Collaboration whereby 'if stated in the Term Alliance Agreement or the Term Brief or otherwise agreed by Alliance Members, the Alliance Activities shall include Supply Chain Collaboration in order to achieve Improved Value through more consistent, longer term, larger scale Supply Chain Contracts and through other improved Supply Chain commitments and working practices'.

The process of Supply Chain Collaboration under TAC-1 follows the same sequence as in FAC-1, as described in Section 10.4. TAC-1 provides commercial protections for Alliance Members in respect of the information that they exchange through Supply Chain Collaboration, by means of:

- Confidentiality provisions that can be amended[981]
- Provisions governing mutual Intellectual Property Rights.[982]

The benefits of Supply Chain Collaboration under a term alliance are illustrated in Diagram 10, taken from a report provided by the Project Horizon Trial Project team, who stated that:

- 'Each scheme was designed with an expectation of longevity and appropriate whole life costing, but also with an agreed material warranty to back up the investment being made by Surrey and a bid to keep the roads "pot-hole free"'
- 'This added a commitment from the Alliance team to add specific value based on the premium materials being utilised'
- 'This design commitment has resulted in just 28 remedial works to date since the start of the project in 2013, which is approximately 4% of the works carried out over 5 years'.[983]

---

980 TAC-1 Term Alliance Contract (2016) clause 6.
981 TAC-1 Term Alliance Contract (2016) clause 13.3, with scope for amended confidentiality provisions stated in the Term Alliance Agreement.
982 TAC-1 Term Alliance Contract (2016) clause 11.
983 Extracts from May 2018 Trial Project report prepared by Jane Young of Surrey County Council for KCL Centre of Construction Law following the completion of their five-year Trial Project. See also the Project Horizon Trial Project case study in Section 5.10.

	5-year total	5-year total eligible for discount	Percentage of Total
Total number of schemes	642	517	
10-year warranty		393	76%
5- to 8-year warranty		114	22%
2-year warranty		10	2%
Number of schemes requiring remedial work		28	
Percentage of schemes requiring remedial work			5%

Diagram 10 Warranties and defects on Project Horizon.

	Budget	Discount saving	VE saving	Total saving
13/14	£ 31 000 000	£ 2 000 000	£ 1 120 000	£ 3 120 000
14/15	£ 26 000 000	£ 1 400 000	£ 1 900 000	£ 3 300 000
15/16	£ 15 000 000	£ 1 573 545	£ 1 852 662	£ 3 426 207
16/17	£ 9 910 000*	£ 450 930	£ 453 868	£ 904 798
17/18	£ 7 130 000*	£ 112 041	£ 307 128	£ 419 169
TOTALS	**£ 89 040 000**	**£ 5 536 516**	**£ 5 633 658**	**£ 11 170 174**

Diagram 11 Savings on Project Horizon.

The Surrey report noted that:

- 'The table above shows the percentage of warranties agreed for the whole 5-year programme'
- 'The usual contractual warranty on carriageway construction is 2 years for workmanship and materials, therefore for a Local Authority to achieve up to a 10-year warranty was unprecedented'
- 'This further illustrated the benefit of capital investment and long-term programmes to add longer-term value to the client and stakeholders'.[984]

Diagram 11 is taken from the same report and shows the cumulative Supply Chain Collaboration savings agreed over a five-year period by the Project Horizon term alliance.

Surrey reported that:

- 'The above table shows the detail of the annual budgets (gross), the discount savings and Value Engineering savings across the whole 5 years of the project. The figures were agreed at the end of the financial year by all parties – Kier, Surrey, Aggregate Industries and Marshall Surfacing – and the savings were allotted to an "Opportunity Pot" which was then drawn down to help finance the next year's programme'

---

984 Extracts from May 2018 Trial Project report prepared by Surrey County Council for KCL Centre of Construction Law.

- 'Time and cost predictability have been strong positive factors in being able to forward programme and budget monitor effectively. The list of schemes for each district (11 districts in Surrey) were agreed at each of the 11 Local Committees by the Members, mostly prior to Year 1 of construction; however, there were a couple of areas whose Members did not agree their district programmes until 3 months after delivery had started – this was due to the political cycle'
- 'Having a known 5-year programme is undoubtedly an advantage when forward planning and programming. This ensured that the 2 suppliers, Aggregate Industries and Marshall Surfacing, had early sight of each financial year, enabling them to resource, price and order materials in advance. All design and build walkthroughs were undertaken by a "Task Team" which involved a representative from Surrey, Kier and Aggregate Industries or Marshall Surfacing – the design decision was taken as a collective for each scheme'.[985]

11.6 How Does TAC-1 Measure and Reward Performance?

Each client in each sector has particular objectives and requirements, and TAC-1 provides flexible foundations that a client and its advisers can use when developing their term procurement strategy. Its multi-party structure provides a shared system of open performance measurement and rewards agreed by all Alliance Members.

Following the same approach as FAC-1, TAC-1 states the agreed overall Objectives of the Alliance and the Term Programme, and the individual Objectives of the Alliance Members.[986] These form the basis for the Success Measures that determine the success of the Alliance and the Term Programme and that measure the performance of the Alliance Members.[987] TAC-1 states the agreed Targets for each Success Measure, which should include the method of recording relevant data, the Alliance Member responsible for measuring against that data and the system for reporting to the other Alliance Members.[988]

For example, one agreed target of the Kier Highways supply chain alliance 'is that 90% of the projects ordered by Surrey as capital schemes are able to evidence the use of ECI, with a record of ECI benefits maintained by the Alliance Manager to demonstrate material benefits at either feasibility or construction stage'.[989]

TAC-1 provides for agreed Incentives,[990] which can include for example:

- Additional payments including shares of savings achieved through Supply Chain Collaboration and other agreed Alliance Activities[991]
- Adjustment of the scope of the Term Programme and/or of any exclusivity in the award of Orders[992]
- Extension of the duration of the Alliance.[993]

---

985 Extracts from May 2018 report prepared by Surrey County Council for KCL Centre of Construction Law.
986 TAC-1 Term Alliance Contract (2016) clause 2.1 and Schedule 1 Part 1.
987 By reference to TAC-1 Term Alliance Contract (2016) clause 2.3 and Schedule 1 Part 2.
988 Guidance notes in TAC-1 Term Alliance Contract (2016) Schedule 1 Part 2.
989 Case study summarised in Section 11.10.
990 To be set out in TAC-1 Term Alliance Contract (2016) Schedule 1 Part 3 by reference to clause 2.4.
991 Payable under TAC-1 Term Alliance Contract (2016) clause 8.
992 Under TAC-1 Term Alliance Contract (2016) clause 5.7 and Schedule 1 Part 2.
993 Linked to TAC-1 Term Alliance Contract (2016) clause 14.1.

Each Alliance Member needs to understand how the Objectives and Success Measures may affect the continued issue of Orders, the continuation of its role as an Alliance Member or the continuation of the Alliance as a whole. TAC-1 provides clarity as to which Targets are so important that a failure to meet them will require urgent action and may ultimately determine whether an appointment under TAC-1 may be terminated.[994] It also includes an Early Warning system[995] enabling notification of the reasons behind any issues or obstacles that are encountered.

The most powerful incentive under a TAC-1 contract is the prospect of additional work:

- By expansion of the scope of the Term Programme, as illustrated by the increased amount of work offered under the Project Horizon term alliance contract in exchange for the improved value achieved through Supply Chain Collaboration[996]
- By extension of the term, as illustrated by the potential contract extension that motivated Kier to create Improved Value through its new Supply Chain alliance.[997]

11.7   How Does TAC-1 Support Collaborative Risk Management?

TAC-1 provides for consensus-based governance by a Core Group of individuals representing Alliance Members in support of their agreed mutual commitments.[998] The TAC-1 Core Group uses the same collaborative procedures as FAC-1, making its decisions by the unanimous agreement of a quorum of all members present, so that there is no risk of Core Group members being overruled by a majority and so that individual Core Group members cannot frustrate collaborative working by refusing to attend a meeting.[999] The TAC-1 Core Group is the forum where Alliance Members can consider and agree proposals for Improved Value, and where they can raise issues with each other in order to resolve problems before they become disputes.[1000]

TAC-1 provides for an in-house or external Alliance Manager who integrates the work of the Alliance Members and who:

- Issues Orders and implements the Order Procedure
- Monitors and supports achievement of the agreed Objectives, Success Measures and Targets
- Monitors and supports achievement of deadlines, gateways and milestones in the Timetable and prepares updates for Core Group approval
- Organises Core Group meetings, including those called in response to Early Warning
- Organises, supports and monitors Supply Chain Collaboration and other Alliance Activities

---

994 TAC-1 Term Alliance Contract (2016) Schedule 1 Part 2 and clause 14.2.
995 TAC-1 Term Alliance Contract (2016) clause 1.8.
996 Case study in Section 5.10.
997 Case study in Section 11.10.
998 TAC-1 Term Alliance Contract (2016) clauses 1.6 and 1.7, as considered in Section 12.7.
999 TAC-1 Term Alliance Contract (2016) clause 1.7.3. This system has been used for many years under TPC2005.
1000 TAC-1 Term Alliance Contract (2016) clause 15.1, as considered in Section 19.7.

- Manages agreed payments and Incentives
- Monitors and supports Risk Management in accordance with the TAC-1 Risk Register and prepares updates of the Risk Register for Core Group approval at intervals stated in the Timetable.[1001]

The TAC-1 system of collaborative Risk Management provides new opportunities for Alliance Members to mitigate risks and their potential effects, as considered in Chapter 18. Risk Management, the scope for joint and individual actions and the use of a Risk Register are described in TAC-1 in the same terms as those used in FAC-1, as explored in Section 10.6.

Agreed joint and individual Risk Management actions under TAC-1 can avoid the need for Alliance Members to gamble on unknown matters when implementing Orders and can avoid wasteful risk pricing and the dumping of risk down the Supply Chain. These actions also enable Alliance Members to price more accurately in the Order Procedure any risks that cannot be eliminated and/or to share such risks as agreed. Where problems arise that might otherwise escalate into disputes, the additional information shared among Alliance Members through Risk Management provides a stable basis for non-adversarial dispute resolution.

11.8 How Can TAC-1 Reflect Differing Requirements?

Different clients and their legal advisers need flexibility in adapting a standard form term alliance contract so that they can develop and include details and variations that reflect their specific requirements. TAC-1 is a flexible form that, in the same ways as FAC-1, is designed to accommodate a wide range of requirements.

TAC-1 provides scope for the bespoke drafting needed to reflect the details of the Alliance relationships and underlying Orders. These details can be set out in:

- The Term Alliance Agreement
- The Term Brief and other Term Documents
- The agreed Objectives, Success Measures, Targets and Incentives
- The Timetable of agreed Alliance Activities
- The agreed Order Procedure and the Template Order Documents
- Specific amendments to the Contract Terms by way of Legal Requirements or Special Terms.

The TAC-1 Order Procedure is intended to be compatible with any combination of roles, responsibilities and warranties undertaken by the Provider and by other Alliance Members and Supply Chain members engaged on each Task. An Alliance under TAC-1 can create any allocation of design, supply and construction responsibilities. However, it is fundamental to an effective term alliance that appointments and relationships are structured in a way that enables integrated design and other contributions from all Alliance Members and from other Supply Chain members.

TAC-1 provides for a duty of reasonable skill and care linked to each Alliance Member's agreed role, expertise and responsibilities with options for agreed amendments.[1002]

1001 TAC-1 Term Alliance Contract (2016) clause 3.1.
1002 TAC-1 Term Alliance Contract (2016) clause 10.1.

The TAC-1 duty of care can be extended to parties other than Alliance Members by the means stated in the Term Alliance Agreement or as otherwise agreed,[1003] for example by third-party collateral warranties or by creation of other third-party rights.[1004]

TAC-1 provides for its Contract Terms to be supplemented or amended by Legal Requirements,[1005] and by Special Terms that reflect the particular needs of the Client or other Alliance members or are required by reason of the nature of the Term Programme.[1006] TAC-1 is designed for use in any jurisdiction and does not contain provisions taken from English law. The payment and adjudication provisions in TAC-1 clause 8 (Payment) and clause 15.3/Appendix 4 Part 2 (Adjudication) are drafted to be compliant with the UK Housing Grants, Construction and Regeneration Act 1996 as amended, without specifically referencing any provisions of that legislation.

Other specific provisions required for compliance with English law need to be set out in Schedule 6 Part 1 (Legal Requirements). For example, these can include the following provisions taken from TPC2005:

- Cross-reference between TAC-1 clause 3.4.3 and the UK Transfer of Undertakings (Protection of Employment) Regulations 2006 and any amendments (TPC2005 clause 11.2)
- Cross-reference between TAC-1 clause 7.10 and the UK Construction (Design and Management) Regulations 2015 and any amendments (TPC2005 clause 4.1)
- Cross-reference between TAC-1 clauses 8.1/8.6 and UK Value Added Tax (TPC2005 clauses 7.1 and 7.6)
- Cross-reference between TAC-1 clause 8.11 and the UK Late Payment of Commercial Debts (Interest) Act 1998 and any amendments (TPC2005 clause 7.6)
- Cross-reference between TAC-1 clause 8 and the UK Finance Act 2004 and any amendments (TPC2005 clause 7.12)
- Cross-reference between TAC-1 clause 14 and breach of the UK Local Government Act 1972 and any amendments (TPC2005 clause 13.7).

11.9 How Does TAC-1 Differ from TPC2005?

TAC-1 is based on lessons learned from analysis as to how successful term alliances deliver agreed cost savings and other improved value, including the success of TPC2005. It follows closely the provisions of TPC2005 and develops provisions that have helped to deliver Improved Value including:

- Significant cost savings and greater cost and time certainty
- Improved performance and extended warranties
- Improved employment and training opportunities
- Improved engagement of local and regional Supply Chain members
- Improved Client and Stakeholder satisfaction
- New Sustainability initiatives.[1007]

1003 TAC-1 Term Alliance Contract (2016) clause 10.6.
1004 Although creation of third-party rights in respect of specific Orders would require a provision in TAC-1 Term Alliance Contract Appendix 3 Section 7.
1005 TAC-1 Term Alliance Contract (2016) clause 13.4 and Schedule 6 Part 1.
1006 TAC-1 Term Alliance Contract (2016) clause 13.5 and Schedule 6 Part 2.
1007 As referred to in the case studies in Sections 5.10 and 11.10 and in additional case studies published at http://ppc2000.co.uk/case-studies.

Many Clients, Providers and other Alliance Members are familiar with TPC2005 and wish to know what improvements have been introduced in TAC-1 compared to TPC2005 (2013). These include:

- Clearer connections between agreed Objectives, Success Measures, Targets and Incentives in TAC-1 Schedule 1, with supporting guidance notes
- Details of the Order Procedure in TAC-1 Schedule 4 and provision for Template Order Documents in TAC-1 Schedule 5, with supporting guidance notes
- Guidance for completion of the Term Brief in Schedule 7 and for completion of the Term Prices and Term Proposals in Schedule 8
- A more detailed Order Form in TAC-1 Appendix 3, offering greater flexibility as to types of Task and Order for which TAC-1 can be used, with supporting guidance notes
- Updated communications provisions under TAC-1 clause 1.9
- Greater focus and a simpler approach in respect of Alliance Activities under TAC-1 clause 6, particularly Supply Chain Collaboration under TAC-1 clause 6.3, consolidating a range of provisions set out in the TPC2005 clause 2.2 (processes), clause 5.5 (business cases), clause 5.6 (specialist tenders) and clause 5.7 (volume supply agreements)
- Provision for the Provider to cooperate with Users and others undertaking activities on Sites under TAC-1 clause 7.8
- The option of specific grounds for extension of time/additional cost under TAC-1 clause 9.6 (of which details can be added in Term Alliance Agreement)
- New Intellectual Property Rights provisions under TAC-1 clause 11
- Provision for confidentiality under TAC-1 clause 13.3, with the option to amend this provision in the TAC-1 Term Alliance Agreement
- Details under TAC-1 clause 14.2 as to procedures/options following failure to meet agreed Targets
- Provision for handover between old and new Providers under TAC-1 clause 14.9
- Dispute Board and arbitration options under TAC-1 clauses 15.2 and 15.4.

11.10 How Are TPC2005 and TAC-1 Used in Practice?

The range of TPC2005 users includes those described in the case studies in Section 5.10 and in additional case studies[1008] relating to collaborative Term Programmes in sectors such as housing, offices, public buildings, education, highways and leisure. Section 13.6 refers to the use of TPC2005 in conjunction with BIM.

TAC-1 has been adopted by a range of clients. The first user was Southern Housing Group on a £230 million stock reinvestment Term Programme, as summarised in the case study below. Other early users of TAC-1 are notified under 'News and Users' at www.allianceforms.co.uk and include:

- Catalyst Housing Group, who selected TAC-1 for a Term Programme of responsive, gas, planned and cyclical works with a total value of £210 million, starting at £10 million per annum and rising to £40 million per annum

---

1008 For example, as set out at http://ppc2000.co.uk/case-studies.

- Central Bedfordshire, using TAC-1 for a £200 million Term Programme of capital works
- South Liverpool Housing, using TAC-1 for a Term Programme joint venture with Penny Lane Builders governing repairs and maintenance with an approximate value of £103.5 million
- Royal Borough of Greenwich, using TAC-1 for their £300 million, 10 year programme of modular/off-site manufactured new build housing.

The Project Horizon Trial Project case study summarised in Section 5.10 describes a term alliance prototype that demonstrates how TAC-1 is designed to operate. The Alliance created for the Project Horizon Trial Project was recognised in:

- The HM Treasury Procurement Routemap[1009]
- The Supply Chain Collaboration Toolkit published by the Highways Maintenance Efficiency Programme[1010]
- The P13 Blueprint, which recognised how Surrey as 'enterprise leader' and Kier as 'integrator' developed the roll-out of their alliance 'beyond road maintenance to the whole capital maintenance portfolio' and then 'extended their supplier engagement to build a supply chain alliance'.[1011]

The Kier case study summarised below shows how the combination of TPC2005 with FAC-1 in a multi-party supply chain alliance can be adopted and led by a main contractor at any stage in the lifecycle of a term programme.

Southern Housing Group Term Alliance

Southern Housing Group has a £23 million annual spend on its stock reinvestment programme, comprising planned maintenance (£20 million) and cyclical decorations (£3 million). It needed to set up a contract delivery model based on regional approach in four contract areas and to achieve better integration between planned maintenance, cyclical decorations, development and maintenance, plus strategic collaboration to reduce risk and maximise benefits.

Southern used the TAC-1 Term Alliance Contract to:

- Reduce the number of contracts and contractors it works with
- Reduce procurement activity
- Reduce reliance on external consultants and build an 'in-house' delivery team
- Improve efficiency in project delivery
- Maximise benefits for collaborative working
- Develop common supply chains.

1009 Infrastructure and Projects Authority (2014).
1010 HMEP Supply Chain Collaboration Toolkit (2014).
1011 P13 Blueprint (2018). The P13 Blueprint states that Surrey 'utilised learning from others through the Alliancing Code of Practice', albeit that the Code of Practice was published in 2015 and Surrey's procurement strategy was created in 2011.

The agreed TAC-1 alliance member objectives were:

- To work with the Client in a spirit of trust, fairness, mutual cooperation, dedication to agreed common goals and an understanding of each other's expectations and values
- To work with the Client to maximise its investment in its assets in line with the Group Asset Management strategy
- To work with the Client to meet its objectives in respect of the Public Services (Social Value) Act 2012
- To maintain and improve homes to improve lives for the customer
- To work collaboratively with the Client and its customers to excel in customer service
- To work collaboratively to achieve value for money, efficiency and best working practices.

Southern had undertaken successful previous projects of a similar nature under TPC2005, and received a very positive industry response as to the use of TPC2005 in its soft market test undertaken in 2015/16 and also in its 'Meet the Buyer' meetings.

Southern developed a two-stage TAC-1 ordering process comprising:

- Order 1 for design development and preparation of a 'Project Task Proposal'
- Order 2 to undertake the works in accordance with the 'Project Task Proposal', agreed price and timetable.

The TAC-1 Price Framework comprised:

- Basket Rates for Component Renewal
- Fixed prices for 'Property Archetypes' for Cyclical Decorations
- Schedules of rates for all other works
- Percentage bandings for site overheads, central office overheads and profit.

The Southern TAC-1 provided for:

- 'Included' design covered in site overheads
- 'Excluded' design
- Early supplier involvement in design, for example kitchen suppliers.

Reported to KCL Centre of Construction Law by Neil Thody of Cameron Consulting and recorded in www.allianceforms.co.uk (News and Users).

Kier Highways Services Supply Chain Alliance

Surrey County Council (SCC) and Kier were parties to a Core Highways Maintenance Contract, having entered into a six-year term partnering contract running until 2017, with an option to extend by up to four years and with an original scope of immediate response, safety defects, winter service and minor planned works.

(Continued)

The option was exercised to extend the contract to 2021, with a new TPC2005 Partnering Timetable

- Incorporating a revised commercial model
- Including mutual commitments to seek improved value through re-procurement of the Kier supply chain for capital schemes (up to 70% of the annual budget).

FAC-1 was adopted by Kier alongside TPC2005 for its supply chain re-procurement in order to deliver the strategic goals of:

- Increasing collaboration between SCC, Kier and the supply chain
- Achieving the objectives of the Surrey Business Plan
- Demonstrating value for money, targeting a 2.5% saving off the 15/16 capital programme expenditure
- Developing a sustainable supply chain through to 2021.

The re-procurement provided an opportunity to broaden use of alliances so as to:

- Consolidate Project Horizon experience
- Maximise benefits for Surrey
- Strengthen all parties' experience in alliances and collaboration
- Provide the supply chain with visibility and continuity
- Improve supply chain relationships and confidence.

Steps were taken to ensure the Surrey/Kier team were aware of FAC-1 requirements, including engagement with the Kier in-house legal adviser who confirmed the suitability of FAC-1 to co-exist with TPC2005/STPC2005 main contract and subcontract arrangements. Kier informed the supply chain of its alliance intentions, which gained unanimous support, and provided a user guidance for alliance activities.

Alliance members committed to:

- Adopt and participate in early contractor involvement ('ECI')
- Share and/or improve working practices for the benefit of Surrey and the Term Programme
- Attend other framework review meetings as required in order to achieve improved value
- Implement social value proposals.

The agreed target was that 90% of the projects ordered by Surrey as capital schemes will be able to evidence the use of ECI, with a record of ECI benefits maintained by the Alliance Manager to demonstrate material benefits at either feasibility or construction stage.

The immediate benefit of the re-procurement exercise was a saving of circa **8%** against prices under the previous contract model. There was also a level of social value put forward by the supply chain although this is turnover-dependent. The predominant areas were (i) developing local supply chains, (ii) increasing spend with local suppliers, (iii) creating local employment/skills development opportunities, and (iv) encouraging local recruitment.

Reported to KCL Centre of Construction Law by Nigel Owers of Kier Highways and recorded in www .allianceforms.co.uk (News and Users).

12

How Is a Collaborative Culture Created?

12.1 Overview

We often emphasise the importance of a collaborative culture, but it can be very hard to define. It has been described as 'the willingness of a party to be vulnerable to the actions of another party based on positive expectations regarding the other party's motivation and/or behaviour', where:

- 'Perceptions of goodwill entail attributions regarding the intention of another party to behave in a trustworthy manner'
- 'Perceptions of competence entail attributions regarding the other party's ability to behave or perform as expected'.[1012]

However, there is no universal business morality that creates collaborative norms of behaviour through which we can be sure that goodwill and competence will be translated into actions. Therefore, team members need guidance in order to establish a clear and balanced understanding of what a collaborative culture means in practical terms and how they are expected to sustain it.

Collaborative procurement is 'the lubrication for the machinery of successful project delivery, not an end in itself',[1013] and it is important that this lubrication also encourages and sustains a suitable culture.[1014] Commentators have suggested that:

- 'Maintaining the link between the commercial model and the required behaviours is an important responsibility for the leadership group
- The effectiveness of the commercial model is a collective responsibility – not something undertaken separately by the client'.[1015]

Alliance members should agree at an early stage the active steps they will take to create, embed and sustain a culture of collaborative working at all levels of seniority and in all activities affecting the project or programme of work.[1016] If these matters are

1012 Malhotra & Lumineau (2011), 992.
1013 JCT CE Guide (2016), Section 16.
1014 'A particular project or series of projects will not be successful simply because the parties have signed up to work collaboratively if there is no underlying basis of trust', JCT CE Guide (2016), Section 15.
1015 Infrastructure Client Group (2015), Section 4.4.
1016 For example, 'Two Stage Open Book enabled a culture of collaborative working at all levels of the supply chain as well as the creation of integrated project teams with better defined roles for individuals employed by the client, Tier 1 contractor and Tier 2/3 supply chain members', Project Horizon Trial Project case study.

Collaborative Construction Procurement and Improved Value, First Edition. David Mosey.
© 2019 John Wiley & Sons Ltd. Published 2019 by John Wiley & Sons Ltd.

expressed only in vague aspirational terms, or if they are one-sided expectations, then a collaborative team will encounter unintended obstacles such as:

- The cynicism generated where behavioural change is promoted by facilitators in workshops that appear to lead only to more workshops, and that are not connected to specific, agreed collaborative commitments. Any facilitator offering collaborative exercises needs to provide a clear explanation of their purposes and recognisable connections to the team members' commercial objectives. Otherwise the facilitator and the exercises will be seen as wasted time and can invite negative views best summarised as 'never trust a hippy!'[1017]
- The cynicism generated if all team members are not willing to make changes in order to work more efficiently together. To expect other team members to adopt new collaborative behaviour without having to change your own behaviour is not a fair or credible approach. It is equivalent to encouraging sustainability by expecting others to use your old bathwater, rather than by everyone agreeing to use a shower.[1018]

In this chapter we will look at the ways in which a balanced collaborative culture can be developed and sustained through leadership, management and communication. We will consider how an alliance culture can reach all stakeholders, how it can be supported by a 'Core Group' or 'Alliance Board' and how it can be facilitated by training, workshops and appointment of an independent adviser.

12.2   What Is the Role of People in a Collaborative Culture?

Commentators note that 'when considering the project objectives, it is easy but dangerous to forget that no objective can be achieved without people'.[1019] Successful collaborative working requires leadership by appropriate individuals appointed by each alliance member. Alliance members should 'identify the key individuals and their roles in supporting each particular collaborative opportunity', including all individuals whose 'competence and collaborative behaviour have been determined to have a significant impact upon the effectiveness of collaborative working'.[1020]

Achieving continuity in the key individuals engaged on a collaborative project or programme of work is a related challenge that it is important to tackle as the accumulated knowledge, trust and goodwill achieved by any individual is not easy to pass on. It is always tempting for an organisation to put its best people onto winning the next appointment, transferring them away from the appointment they have just won, and for this reason some alliance contracts restrict changes in important personnel without justification or other alliance members' prior consent.

Collaborative relationships are personal and are not assignable, and continuity can be broken when people move to another project or are promoted or leave their employer. A strong collaborative provision in NEC4 Option X22 is the requirement for the contractor not to replace any key person unless instructed by the project manager or unless that

---

1017 Wise words from the lyrics of 'Who killed Bambi?', Edward Tudor-Pole and Vivienne Westwood.
1018 This actually happened to me while staying at the home of an eminent construction lawyer who specialised in water sector projects. He was very surprised that I declined his water-saving suggestion.
1019 Lock, D. (2000), 11.
1020 ISO 44001 Section 8.3.3.

person is unable to continue to act.[1021] This requirement for continuity is valuable and could logically be extended to the key persons working for the NEC4 project manager and for other team members.

An important step in creating and sustaining a collaborative culture is for team members not only to agree the appointment of key individuals but also to make provision for their replacement if they do not perform their agreed roles. It is possible that certain individuals will not be suitable to work in a collaborative culture. For example:

- On integrated project delivery (IPD) projects in the USA 'it was not uncommon for owners to have to remove or replace members of their own team that couldn't adapt to a collaborative process', and it was reported that a team can be 'strengthened by the process of realising they had a team member that didn't fit and then jointly taking action to reconfigure the team'[1022]
- PPC2000 provides that 'if any individual employed by a Partnering Team member or for whom it is responsible disrupts or otherwise adversely affects the Project then, after Consultation with the Core Group, the Client may require exclusion of that individual from the Project and the Site and the relevant Partnering Team member shall engage a suitable replacement and notify the Core Group accordingly'[1023]
- The NEC4 Alliance Contract requires that the alliance as a whole provides 'each key person to do the job stated in the Contract Data, or provides a replacement person who has been accepted by the Alliance Manager'.[1024] However, it is difficult to see how in practice an alliance can collectively fulfil this responsibility because in the absence of a special purpose vehicle each individual is employed by an alliance member rather than by the alliance as a separate entity.

As regards the individuals who sit on a Core Group or Alliance Board, as considered in Section 12.7, PPC2000 provides that it is for each alliance member to 'ensure that any of its employees who are Core Group members shall attend Core Group meetings and fulfil the agreed functions of a Core Group member'.[1025] There is no equivalent wording recognising that alliance members are responsible for their employees as Alliance Board members under the NEC4 Alliance Contract, and it could be argued that this increases their risk of personal liability and raises the question of whether they fall outside the no blame clause considered in Section 7.7.

Personal commitment to a collaborative culture may not be balanced, and the underperformance or non-cooperation of any team member can have a significant negative effect. This has been described as the 'free rider problem',[1026] for example where a team member might calculate that its own failure to contribute or cooperate on any aspect of the project would be a way of keeping its own costs at a minimum. This can undermine the success of joint working, and the skills of leadership, consultation and persuasion all play a part in avoiding this risk.

1021 NEC4 Option X22.4.
1022 Fischer et al. (2017), 24.
1023 PPC2000 (2013) clause 7.5.
1024 NEC4 Alliance Contract (2018) clause 24.1.
1025 PPC2000 (2013) clause 3.4.
1026 Milgrom & Roberts (1992), 161.

Problems such as a free rider can only be addressed if a failure by a team member can be identified objectively and if, by an open consultative process, the other team members can consider what actions are required for the failing team member to avoid being replaced. The rights and procedures required to achieve this can be set out in a contract that makes clear the agreed success measures and targets, the agreed forum for performance reviews and the basis for taking actions that result from identification of a problem.

Identifying, retaining and empowering the right individuals to take key roles on behalf of each team member is only part of the picture. These individuals need to be connected and supported by leadership and by a management system designed to integrate alliance members and to 'demonstrate the relationship's performance', for example in terms of:

- 'Delivery performance and outputs
- Improved risk profile
- Continued alignment of objectives
- Behaviour and trust
- Enhanced collaborative profile/skills
- Additional value created
- Issue management'.[1027]

12.3   Who Leads a Collaborative Culture?

Many of the case studies summarised in this book include the name of a client lead, underlining how collaborative procurement benefits from a person in the client organisation being willing to take an active lead role. An absent or arm's length client leaves a missing piece in the team structure whereas a client's personal interest and support can be very motivational. For some clients, collaboration is part of their own business model, for example the Honda principles of 'advancement, challenge, quality and partnership'.[1028]

That is not to say that collaborative procurement is only open to clients who have an established collaborative approach based on their own procurement experience. Any client is entitled to depend on the expertise of other team members, and the multi-party contractual links created by a collaborative model can avoid the client being left to resolve competing claims arising under separate appointments. However, the potential success of collaborative construction procurement is significantly increased when there is a visible client lead who attends meetings and is known to other team members.[1029] In this way clients can show other team members that they are genuinely committed to collaborative procurement. UK Government reports recognise that:

- 'Clients should take an active role to ensure that key elements of management (such as risk management) are put in place early during pre-project planning'[1030]

1027 ISO 44001, Section 8.8.4.
1028 Bayfield and Roberts (2004), 3.
1029 'The best performing alliances have members of the client organisation as a core part of their team. These team members are seen to play a crucial role as part of the team, they also bring an understanding of stakeholder and customer issues. Having client members integrated into delivery teams sets the tone for integration and helps develop relationships across the alliance', Infrastructure Client Group (2014), 18.
1030 ICE/DETR (2001), 27.

- 'Well-focused and capable public sector construction clients' need to offer 'intelligent central support' and provide 'effective and consistent leadership throughout the course of construction projects'.[1031]

Commentators have noted that:

- 'An alliance will inevitably require cultural change in all participating organisations; however, this is most important for the client organisation as a reversion to traditional behaviours by the client inevitably has the most significant impact on the development of the alliance'[1032]
- 'Direction and impetus must come from clients'[1033]
- 'PPC2000 requires the active involvement of the client with full and open communication between all parties',[1034] and 'implies that the client takes an active part in the problem-solving processes and continues to participate in building the team during the construction phase'.[1035]

In less collaborative models it is suggested that 'keeping the client at arm's length over the selection of the contractor's team' should help clarify the allocation of risks.[1036] However, this arm's length approach does not help the management of risks, and other commentators have noted how it is 'important that as integration is developed a clear client role is maintained to manage the contract and associated governance'.[1037]

Clients may be concerned that their involvement in collaborative procurement will dilute or confuse the roles and responsibilities of other team members. It is therefore important to describe and limit the client's role and its relationships with other team members, for example:

- 'To act as a focal point' to integrate its interests
- In conjunction with other members as appropriate 'to define the scope and objectives of the project...and to agree upon the methods of proceeding'
- 'To create a clear brief for the designers' and to assist in its development
- 'To react swiftly in obtaining any necessary strategic client decisions required during the currency of the design or construction phases'
- 'To monitor the overall progress and performance on the project'.[1038]

Collaborative procurement requires commitment 'from the client (who must not be regarded as being outside the team) through his advisers, to the contractor and the contractor's man on the site'.[1039] Latham suggested that 'partnering is a collaborative effort, involving all the construction team. It must be led by clients and should

---

1031 NAO (2005), Section 3, 9. A 1995 report stated that public sector clients should 'provide the necessary leadership and be clearly accountable for delivering the project requirements in accordance with the approvals given', including establishing the budget, the organisation structure and communication processes, ensuring involvement of users and stakeholders, ensuring that an appropriate brief is developed, establishing a progress and reporting procedure and also approving changes and dealing with problems and disputes through to post-completion evaluation, Efficiency Unit (1995), Sections 75 and 76.
1032 Infrastructure Client Group (2015), Section 2.1.
1033 Egan, J. (1998), Section 14, 13.
1034 Gruneberg & Hughes (2006), 15.
1035 Gruneberg & Hughes (2006), 48.
1036 Rawlinson, S. (2008), 69.
1037 Infrastructure Client Group (2014), 19.
1038 NEDO (1975), Section 3.7, 26.
1039 Banwell Report (1964), 5, Section 2.8.

also include architects, engineers, cost consultants, contractors, subcontractors and manufacturers'.[1040] For example:

- On the Cookham Wood Trial Project 'leadership has been shown by individuals appointed by the client, architect, Tier 1 main contractor and Tier 2 key subcontractors so as to embed new behaviours'[1041]
- On Project Horizon Surrey County Council 'required personal leadership and the recognition of different stakeholder interests in client and contractor organisations who need to be identified, consulted and persuaded to adopt a new and bolder approach'[1042]
- On the Property Services Cluster Trial Project 'each contractor is taking a lead role in specific areas such as programming, supply chain procurement or design. Some of the projects will continue to be fully designed by the relevant client, whilst many of the others will be handed over to contractor-led design teams at mutually agreed points'.[1043]

Whoever takes a lead role, the coordination and motivation of an alliance or any collaborative team as a multi-party group also needs management, because 'when there is a divergence of interests, even moderate-sized groups often find it impossible in practice to reach a unanimously acceptable decision'.[1044]

12.4   Who Manages a Collaborative Culture?

The successful delivery of any construction project requires coordination and motivation of multi-party groups through management roles taken on by several different organisations, for example:

- The project manager managing the coordination of the design consultants, cost consultant and main contractor
- The lead designer managing the other design contributors
- The main contractor managing the subcontracted supply chain members.

The management of collaborative procurement depends on delegated authority being made clear in contracts so that its extent and how it is exercised are understood and accepted by all team members. The management of a collaborative culture should be integrated with the management roles exercised by the project manager, design team leader and main contractor, and each team member exercising a management role should do so in a way that is consistent with creating and sustaining a collaborative culture.

Collaborative procurement should also encourage self-regulation by team members, motivated by their shared objectives, success measures and agreed incentives. For example, the Anglian Water @one alliance reported how:

1040 Association of Consultant Architects (2010), 1.
1041 Cookham Wood Trial Project case study.
1042 Project Horizon Trial Project case study.
1043 Property Services Cluster Trial Project case study.
1044 Milgrom & Roberts (1992), 145.

- 'Teams regularly review their progress and performance using the HPT framework. This provides teams with a set of characteristics that describe high performance and a framework for them to collectively undertake self-assessments
- A development plan is agreed covering leadership, behaviours, how the team interact, clarity on roles and responsibilities, through to development plans to support individual team members
- The overall process continues to learn and develop as more teams undertake self-assessments, allowing the characteristics of high performance to be periodically reviewed and updated. Teams are, therefore, always comparing themselves against a current view of best practice'.[1045]

The manager of an alliance has some functions equivalent to those of any project manager insofar as 'any project manager worthy of the title will want to make certain that whenever possible his or her tactics are preventative rather than curative'.[1046] The importance of collaborative management responsibilities is illustrated by a UK case where a project manager was held liable for adopting early contractor involvement but not using it as a means to build up the agreed design, cost, risk and supply chain data within the client's budget and brief, instead using the preconstruction phase as a period for satisfying the particular requirements of one prospective tenant of the completed project.[1047]

Problems can arise if an alliance manager acts only as an arm's length project manager, issuing instructions without consultation, when he or she should instead adopt a consultative approach. Where an alliance manager has been accustomed to a traditional leadership role, 'increased participation can lead to feelings of vulnerability and exposure on the part of managers who were formerly accustomed to leadership. This is partly because participation in close working relationships demystifies the competence of senior personnel'.[1048] While an alliance manager may have decision-making powers equivalent to those of a project manager, it also has a collaborative role in drawing the alliance members together and in helping them to reconcile their differing interests.

The FAC-1 Framework Alliance Contract and the TAC-1 Term Alliance Contract follow the approach initiated in the 'Client Representative' role under PPC2000.[1049] They each provide for an 'Alliance Manager' who is an alliance member itself, whose role is to integrate the work of the other alliance members and whose responsibilities are quoted in Section 10.6. If the Client Representative under PPC2000, or the Alliance Manager under FAC-1 and TAC-1, is an employee of the client, then the client should sign the multi-party contract both as the client and in the capacity of the Client Representative or Alliance Manager. This ensures that it there always is an alliance member contractually responsible to the other alliance members for the performance of the Client Representative's or Alliance Manager's role and responsibilities.[1050]

1045 Infrastructure Client Group (2014), 14.
1046 Lock, D. (2000), 507.
1047 *Plymouth and South West Co-operative Society v Architecture Structure and Management* [2006] EWHC 5 (TCC).
1048 Barlow et al. (1997), 16. This is illustrated by the defensive attitude of the project engineer in the Bahrain Bank Headquarters case study summarised in Section 19.10.
1049 The term 'Client Representative' appears to imply a one-sided approach but PPC2000 states that the Client Representative shall exercise 'any discretion fairly and constructively', PPC2000 (2013) clause 5.1 (i).
1050 For example, on the Bath and North East Somerset and Bermondsey Academy projects summarised in Sections 12.10 and 18.10, the client took on the alliance manager role whereas, for example, in the Cookham

The NEC4 Alliance Contract provides for the appointment of an 'Alliance Manager'[1051] who also undertakes 'some aspects of the contractor's role'.[1052] Guidance suggests that the Alliance Manager may be 'an employee of a Partner or the Client',[1053] although there is no express requirement for the appointing alliance member to ensure that the Alliance Manager does what he or she is required to do. This individual carries significant responsibilities such as:

- Approving subcontractors proposed by any alliance member[1054]
- Issuing a first programme if one is not included in the Contract Data and preparing all revised programmes[1055]
- Certifying the taking over of the works[1056]
- Producing a quality plan[1057]
- Assessing payments due in consultation with the partners[1058]
- Assessing quotations for compensation events[1059]
- Exercising judgement independently and in accordance with the contract.[1060]

The NEC4 Alliance Manager is also required to implement instructions such as those:

- From the client to change the 'Client's Requirements'[1061]
- From the Alliance Board to resolve 'an ambiguity or inconsistency in or between documents'[1062]
- From the Alliance Board 'to change the Scope'[1063]
- From the Alliance Board to deal with an act of prevention.[1064]

In light of these responsibilities, it is surprising that the NEC4 Alliance Manager is an individual[1065] and also that he or she is not a member of the alliance.[1066] It seems illogical to place a party with such responsibilities outside the alliance and to leave him or her apparently unprotected by all agreed limitations of liability, particularly as the work done

Wood and North Wales Prison projects summarised in Section 14.10 this role was performed by an independent consultant. In all cases the relevant party was also a member of the Core Group and in no cases was an individual exposed to contractual liability.

1051 NEC4 Alliance Contract (2018) clause 22.
1052 Introducing the new NEC4 Alliance Contract (2018).
1053 NEC4 managing an alliance contract (2018), 5 by reference to NEC4 Alliance Contract (2018) clause 11.1(4).
1054 NEC4 Alliance Contract (2018) clause 26.1.
1055 NEC4 Alliance Contract (2018) clauses 32.1 and 33.
1056 NEC4 Alliance Contract (2018) clause 36.3.
1057 NEC4 Alliance Contract (2018) clause 40.2.
1058 NEC4 Alliance Contract (2018) clause 50.
1059 NEC4 Alliance Contract (2018) clause 62.
1060 NEC4 managing an alliance contract (2018) 20 by reference to clause 22.1.
1061 NEC4 Alliance Contract (2018) clause 14.2.
1062 NEC4 Alliance Contract (2018) clause 16.1.
1063 NEC4 Alliance Contract (2018) clause 16.2.
1064 NEC4 Alliance Contract (2018) clause 18.1.
1065 NEC4 Alliance Contract (2018) clause 11.2(3) refers to 'the person appointed by the Alliance Board' and NEC4 managing an alliance contract (2018) 20, by reference to NEC4 Alliance Contract (2018) clause 22.1, refers to 'a key individual'.
1066 For example, NEC4 Alliance Contract (2018) clauses 10.1 and 10.2 distinguish between 'The Alliance and the Alliance Manager...' and clause 15.1 distinguishes between 'A member of the Alliance and the Alliance Manager...'.

by the Alliance Manager relieves the alliance members from significant aspects of their collective responsibility. Although the Alliance Manager is answerable to the Alliance Board, the NEC4 view is that 'it would not seem appropriate or practical for the Alliance Manager to be part of the Alliance Board'.[1067]

ConsensusDocs 300 provides that 'the Core Group may appoint such intermediate project managers of the IPD Team that it deems necessary for effectively managing the Project'.[1068] The responsibility of the Core Group for the acts and omissions of the managers they appoint is limited under ConsensusDocs 300, in that:

- 'The Core Group shall not have any duties of supervision over or control of any person employed or retained on the Project'
- 'Each IPD Team member is alone responsible for supervising its own employees'.[1069]

There is a tradition of trilateral governance linked to 'the interests of the principals in sustaining the relation [which] are especially great for highly idiosyncratic transactions', for example 'the use of an architect as a relatively independent expert'.[1070] Similarly, the management of an alliance can include the delegation of authority that is entrusted to one or more individuals acting independently as alliance manager. However, this delegated authority should not be relied upon to the exclusion of other collaborative decision-making, bearing in mind for example that:

- 'Economic analysis typically fails to deal adequately with the propensity of humans to seek power as a goal in its own right'[1071]
- Reliance on third-party determination of performance by 'an expert relatively independent of the parties' provides 'no guarantee of smooth performance; witness the fairly large amount of litigation under the AIA [construction] contracts with respect to delays, payments and completion of the work'.[1072]

Therefore, an alliance should also be supported by a collective governance system that gives authority to all team members, acting through a forum such as a Core Group or Alliance Board as considered in Section 12.7.

12.5 How Can a Collaborative Culture Include all Stakeholders?

A collaborative culture should extend to all members of the team and also to stakeholders outside the team who are affected by a project or programme of work.[1073] Stakeholders who can be involved more closely through collaborative procurement may include users and occupiers of the relevant projects. For example:

1067 NEC4 managing an alliance contract (2018), 20 by reference to NEC4 Alliance Contract (2018) clause 21.
1068 ConsensusDocs 300 (2016) Article 3.5 (Management of the IPD Team).
1069 ConsensusDocs 300 (2016) Article 3.7 (Supervision).
1070 Williamson O.E. (1979), 249, 250.
1071 Macneil, I.R. (1981), 1055.
1072 Macneil, I.R. (1978), 866.
1073 PPC2000 (2013) provides for the rights of 'Interested Parties' in clause 3.9, and FAC-1 Framework Alliance Contract (2016) and TAC-1 Term Alliance Contract (2016) provide for the interests of 'Stakeholders' in clause 1.10.

- On the Hampstead Heath Ponds project alliance, 'BAM attended early stakeholder meetings, crucial in explaining to the many interested parties how the impact of the works would be minimised. Introducing key individuals such as the plant drivers at an early stage was an approach carried forward to the construction phase so as to ensure the continual building of relationships'[1074]
- On the Bath and North East Somerset Council residential care homes project, 'residents, their relatives and unit managers were actively engaged in the design process through focus groups'[1075]
- On a three-year £600 000 contract for the maintenance of gas appliances in 1250 dwellings and a number of commercial properties, Havelok Homes and their service provider PH Jones ensured that 'the client's tenants' liaison group was recognised as an "Interested Party" [in the contract] and [they] provided invaluable assistance and liaison'[1076]
- On a London Borough of Harrow schools programme 'the multi-party integration of full design services at day one' reduced the preconstruction programme by an estimated 9 months, facilitated planning applications and secured agreement to the scope of schemes with schools recognised in [the contract] as 'Interested Parties', and 'relations between team members and the schools' head teachers have remained positive in what could have been stressful circumstances'[1077]
- The Project Horizon Trial Project term alliance reported 'improved stakeholder satisfaction through better communications with Surrey's members and with residents as regards access and delegated authority, for example allowing Kier to alter diversions', and 'analysis of the first eight months of work on site showed over 50 complimentary letters received from residents and Council members'.[1078] It emerged that this was the first time that Surrey Highways had received any complimentary letters at all.

Industry regulators can be recognised stakeholders, for example on the Macclesfield Station project alliance where:

- 'West Coast Trains/Virgin Trains were particularly conscious of the wide range of stakeholders who could influence the timing of preconstruction activities and construction activities'
- 'The preconstruction phase "Partnering Timetable" was used to ensure that deadlines for all planning and design stages were agreed and committed to by team members, and that the dates for all on-site activities were visible to and approved by Network Rail, the Strategic Rail Authority and the station manager'
- The [team] secured 'maximum involvement of third party stakeholders as "Interested Parties" in approving preconstruction outputs and in agreeing an appropriate approach to the construction phase'.[1079]

Stakeholder consultation can contribute improved value, for example on the Rye Harbour Trial Project, where the team 'liaised extensively with Natural England to

---

1074 Case study summarised in Section 3.10.
1075 Association of Consultant Architects (2010), 10, summarised in Section 12.10.
1076 Association of Consultant Architects (2010), 23.
1077 Association of Consultant Architects (2010), 22.
1078 Project Horizon Trial Project case study.
1079 Association of Consultant Architects (2010), 31, summarised in Section 17.10.

develop solutions that would mitigate some of the impacts of necessary works. In many cases this also had benefits for the project in terms of reducing timescales and saving costs, as outlined below:

- Some precious intertidal plants from salt marsh were relocated and transplanted, avoiding loss of vegetation
- Natural England accepted habitat they had created at Rye Harbour Farm as mitigation for the mudflats lost when they had to drive approximately 1000 metres of piling
- Close working with marine ecology teams meant they prevented any mudflat washed in the Rother being seen as wasted and damaging the environment. This saved a lot of money in waste disposal
- Extensive negotiations with Natural England enabled them to continue working through the bird breeding season. This was achieved through the utilisation of a vibro-piling innovation, which reduced the impact of noise on the site so that the birds were not disturbed. This was a massive innovation, which avoided huge de-mobilisation and remobilisation costs of £117 000. The programme was also reduced by a number of weeks. This solution was developed by Jackson Civil Engineering and Team Van Oord (Tier 2)
- To mobilise plant around the site, a raised track was constructed along the line of an existing stone track. This helped to reduce both short and long-term impacts to the site'.[1080]

12.6 How Can a Collaborative Culture Improve Communication?

Relationships between individuals and the organisations for which they work need the support of clear contractual communications systems so as to function efficiently. For this purpose, a collaborative team should 'establish, maintain and actively manage an effective communication process, including the messages for key stakeholders (including all collaborative parties), the vision, the objectives behind the collaboration and how concerns will be managed'.[1081]

ISO 44001 suggests that in any organisation should 'determine the need for internal and external communications relevant to the collaborative business relationship management system, including: on what it will communicate; when to communicate; with whom to communicate and how to communicate'.[1082]

People communicate constantly during a construction or engineering project, and it is not possible to prescribe all the ways that this takes place. Communication can also be adversarial if it is aggressive or if it is used as part of a strategy to overwhelm and confuse other parties. For example, communication on a construction project can go wrong when it descends into a process of mutual bombardment with hostile instructions, complaints, clarifications and requests for information that are all really preparation for claims and counter-claims. This approach to communication can best be described as 'rather provoking', a term that my mother used when recalling how

---

1080 Rye Harbour Trial Project case study.
1081 ISO 44001, Section 8.6.2.6.
1082 ISO 44001 Section 7.4.

she returned from her lunch break in 1943 to find that a bomb had destroyed her office.[1083] While team members may have no choice but to tolerate being bombarded with hostile communications, they will then of course respond in kind, and these escalating exchanges do nothing to cultivate or sustain a collaborative culture.

Collaborative working can sometimes involve a disproportionate increase in the amount of time spent communicating, particularly if it creates excessive numbers of points of contact.[1084] Problems arise if team members believe they are spending time at meetings for which they have not made any financial allowance when costing the project.[1085] Therefore, it is suggested that:

- 'An explicit communication strategy is developed, and the necessary channels between the project participants are established', and 'rules for the use of these channels must also be put in place'[1086]
- 'The links in a communication system should be "planned and monitored"'.[1087]

For example, under the Connect Plus Trial Project framework, 'the project teams have a clear process for exchanging information on a collaborative basis at an early stage, with participants in early contractor involvement meetings working together to agree solutions that promote the best method of delivering the project. Often such discussions are led by the tier 1 contractor (with tier 2/3 support), so as to utilise experience from recent similar projects and to offer clear and well considered methods for the efficient delivery of the works'.[1088]

In communications between team members, 'for every instruction which is sent out (on a project), a resulting feedback signal must be generated. Otherwise there will be no way of knowing when corrective actions are needed'.[1089] It is recognised that a system of 'feedback is absolutely crucial for construction to achieve improvements in its performance. It operates at every level and all teams should use systematic feedback to control their performance'.[1090]

Communication can be assisted by structured meetings of the team members' representatives who are authorised and required to address issues when they arise, for example through the forums considered in Section 12.7. Clear procedures and terms of reference for these meetings, linked to mechanisms for the incremental agreement of new information, increase the chances of preserving the relationships between team members while also respecting their individual commercial interests. However, the style

---

1083 During World War II my mother Vera (1919–2012) worked for the Air Ministry in Bush House on the Aldwych, London, in a building that, coincidentally, is now occupied by King's College London. My uncle Peter Mosey, who went on to be a senior construction manager for Bovis, recalled how he learned the importance of collaborative working with subcontractors while engaged on bomb damage repairs to unstable buildings in London during the late 1940s.

1084 Barlow et al. (1997), note 4, 55.

1085 Concerns have been expressed that collaborative working can lead to an excessive number of points of contact involving more senior staff than would normally be appropriate, for example where specialists are required to attend meetings that are not relevant to their trades because a consultant thinks it necessary for collaboration that all meetings should involve everyone engaged on the project, Barlow et al. (1997), 55.

1086 Smith N.J. (2002), 247.

1087 Smith N.J. (2002), 12.

1088 Connect Plus Trial Project case study.

1089 Lock, D. (2000), 482.

1090 Bennett, J. (2000), 187.

adopted in the management of meetings can unintentionally undermine the team members' confidence in a collaborative culture, for example:

- If meetings start with explanations of delay or inefficiency and end with promises as to how matters will be dealt with differently – only to lead to a further round of excuses at the next meeting[1091]
- If individuals 'read and react to signals differently, particularly verbal signals in a meeting or conversation, even in circumstances where the declared purpose of the parties is to achieve "a timely and compatible combined response"'.[1092]

One way to create an environment that improves communication is the co-location of team members. In exploring experience of alliance projects in the USA it was reported that 'to achieve both an integrated team and integrated processes, a team must work in the same physical space at least for some significant time or for important decisions, using specific collaborative practices'.[1093] This was successfully adopted on Trial Projects such as Project Horizon, North Wales Prison and Rye Harbour. For example, on the Rye Harbour Trial Project 'a virtual office was set up so that everyone had access to the most relevant and up-to-date information available. Some of the supply chain partners also co-located (piling contractor, designer), ensuring that decisions could be made much more quickly'.[1094]

12.7 What Is the Role of a Core Group or Alliance Board?

The governance of a typical company is managed by a combination of its board of directors and the delegation of authority to officers with specific functions and a clear hierarchy. By contrast, a construction project is managed almost entirely by delegating authority to individuals representing individual team members such as the lead designer, project manager, main contractor and others, but with a less clear and frequently changing hierarchy. There is no project equivalent of a board of directors to step in and provide collective guidance, support and decisions. This is a significant omission because:

- 'A collaborative arrangement does not reduce the need for effective and timely governance. In fact, the self-assurance processes likely to be used in an alliance model will only be effective if they are supported by timely and visible governance'[1095]
- 'Losing sight of the need for contract management and governance has been a clear lesson from some of the early integrated alliances'[1096]
- 'A standard governance arrangement would result in improved understanding of roles and authorities and more effective and efficient project delivery'.[1097]

Collaborative procurement can be supported by a group that is committed to joint review of issues emerging through the contract machinery, ranging from the agreement

---

1091 Lock, D. (2000), 510.
1092 Williamson, O.E. (1993), 48.
1093 Fischer et al. (2017), 43.
1094 Rye Harbour Trial Project case study. A virtual office is a practical arrangement distinct from the concept of a virtual organisation considered in Section 7.6.
1095 Infrastructure Client Group (2014), 26.
1096 Infrastructure Client Group (2014), 19.
1097 Department of Treasury and Finance, Victoria (2009), 154.

of proposals for improved value to the resolution of differences between team members. A system for joint issue resolution:

- 'Defines a decision-making hierarchy
- Identifies and resolves issues at the earliest practicable opportunity
- Assigns importance, priority and/or timeframe, and responsibility for resolution at the optimum level
- Tracks the status of the issue: e.g. open, investigating, escalated, resolved
- Aligns with any agreement and/or contracting approach and integrates with lessons learned'.[1098]

For example, Glasgow Housing Association led a £1 billion programme of stock refurbishment and new build for over 40 000 properties which required the coordination of 63 housing associations with 24 constructors and 27 framework consultants. Its alliance contracts provided for:

- Use of a contractual Core Group as 'an essential means for joint problem solving and strategic decision making'
- A supply chain structure which 'allowed GHA to create supplier framework agreements with key components suppliers, so they had representation on Core Groups and were full members of the partnering team'.[1099]

Other examples include;

- The Greenwich Council term alliance where 'the Core Group and Partnering Team structures promoted communication which ensured the right people were dealing with issues at appropriate levels'[1100]
- The Whitefriars framework alliance who used:
  - 'Core Group exchange of information and shared best practice, leading to use of the most economical common kitchen supplier'
  - 'Regular Core Group consultation to identify opportunities for improved efficiency leading to more rapid turnaround on site'.[1101]

There is no Core Group or equivalent forum under the FIDIC, ICC or JCT contract forms. A 'Core Group' is provided for in the PPC2000, FAC-1 and TAC-1 contract forms, and an equivalent group is provided for in ConsensusDocs 300, in NEC4 Option X12 and through the NEC4 Alliance Contract 'Alliance Board'. For example, in addition to non-adversarial dispute resolution as considered in Section 19.7, the FAC-1 and TAC-1 Core Group undertake:

- Review of proposals for Supply Chain Collaboration and other joint activities intended to achieve Improved Value[1102] including agreement of the basis for alliance members to share information in order to enable Supply Chain Collaboration[1103]
- Approval of updates to the agreed Timetable[1104]

1098 ISO 44001 Section 8.6.8.
1099 Association of Consultant Architects (2010), 18.
1100 Association of Consultant Architects (2010), 19, summarised in Section 5.10.
1101 Association of Consultant Architects (2010), 46, summarised in Section 4.10.
1102 FAC-1 Framework Alliance Contract (2016) clause 2.2 and TAC-1 Term Alliance Contract (2016) clause 2.2.
1103 FAC-1 Framework Alliance Contract (2016) clause 6.3.1. and TAC-1 Term Alliance Contract (2016) clause 6.3.1.
1104 FAC-1 Framework Alliance Contract (2016) clause 2.6 and TAC-1 Term Alliance Contract (2016) clause 2.6.

- Agreement of procedures for alliance members to jointly renegotiate or tender Supply Chain Contracts in order to enable Supply Chain Collaboration[1105]
- Approval of updates to the Risk Register.[1106]

The NEC4 Alliance Contract Alliance Board has more wide-ranging powers, and:

- 'Sets the strategy for the achievement of the Alliance Objectives and all partner objectives
- Agrees the allocation of work within the Alliance
- Appoints and instructs the Alliance Manager
- Resolves disputes between members of the Alliance'.[1107]

Contract provisions should state who are the Core Group or Alliance Board members, on what grounds they can be replaced, who is entitled to call a meeting and what decision-making powers they have as a group. For example:

- Core Group terms of reference stating who calls a meeting and when it is called are all set out in PPC2000,[1108] FAC-1,[1109] TAC-1,[1110] ConsensusDocs 300[1111] and in the NEC4 Alliance Contract[1112] but not in NEC4 Option X12
- Similarly, the Core Group quorum and the requirement for decisions to be made by a unanimous vote are set out in PPC2000, FAC-1 and TAC-1, and in the NEC4 Alliance Contract but not in NEC4 Option X12.

A Core Group or Alliance Board should meet regularly, particularly during the pre-construction phase, in order to build effective relationships and ensure the maximum opportunities to agree improved value.[1113] Successful collaboration is the result of consensus achieved through persuasion rather than majority voting or the unilateral exercise of power, and Core Group or Alliance Board members should be encouraged to meet even when they do not want to. It is frustrating if there is no opportunity to put important facts, options or decisions in front of a Core Group or Alliance Board member because they fail to attend a meeting. In order to address this risk FAC-1 and TAC-1 adopted an approach devised by the CIC Partnering Task Force[1114] and used successfully under PPC2000 and TPC2005, namely providing for Core Group decisions to be made unanimously by consensus of those members present at a meeting. By this means:

- Core Group members are encouraged to attend meetings and to participate actively in Core Group decision-making

---

1105 FAC-1 Framework Alliance Contract (2016) clause 6.3.4 and TAC-1 Term Alliance Contract (2016) clause 6.3.4.

1106 FAC-1 Framework Alliance Contract (2016) clause 9.4. and TAC-1 Term Alliance Contract (2016) clause 9.4.

1107 NEC4 Alliance Contract (2018) clause 21.5.

1108 PPC2000 (2013) clauses 3.5 and 3.6.

1109 FAC-1 Framework Alliance Contract (2016) clause 1.7.

1110 TAC-1 Term Alliance Contract (2016) clause 1.7.

1111 ConsensusDocs 300 (2016) Articles 3.12.1 and 3.12.2.

1112 NEC4 Alliance Contract (2018) clause 21.

1113 For example, ConsensusDocs 300 (2016) Article 3.12.4 provides that 'during the Preconstruction Phase the Core Group shall meet regularly (at least every other week) and shall schedule regular meetings for the IPD Team to facilitate collaboration regarding all Project elements'.

1114 CIC (2002).

- The requirement for unanimity means that every Core Group member in attendance must reach agreement and that no attendees can be outvoted by others
- A quorum comprising those members in attendance means that Core Group decisions cannot be blocked simply by a member's non-attendance.[1115]

However, the NEC4 Alliance Contract provides that the Alliance Board can only make a decision by agreement of all members of the Alliance Board, either at a meeting 'if all members are present' or 'by a communication from each member of the Alliance Board confirming their agreement'.[1116] This approach runs the risk of important decisions being blocked or delayed because a member of the Alliance Board simply does not attend or communicate in the way required.

The operation of Core Groups, both at a strategic level and at a project level, was a system used successfully on the Ministry of Justice framework alliance where:

- 'The Alliance has a solid governance structure through a Strategic Core Group comprising representatives from the MoJ and the Alliance suppliers'
- 'Information on the delivery pipeline and updates on the MoJ ways of working, challenges, initiatives, etc. are discussed as part of Strategic Core Group meetings'
- 'A [project] Core Group comprising representatives from the MoJ and the Alliance suppliers deals with any issues that may arise on projects as part of a defined structured hierarchy for project governance applicable to each project'.[1117]

Core Group and Alliance Board members can only work with the information that is provided to them, whether it is set out in or developed pursuant to the contract, and they will find it difficult to maintain consensus if called upon to reach subjective decisions or to make uninformed judgments as to the team members' roles and responsibilities. However, if the right individuals are chosen, and if the contract terms governing their terms of reference are clear, then the collaborative support, decision-making and problem-solving work of a Core Group or Alliance Board can save the client and other team members a great deal of time and money.

12.8 What Is the Value of Training and Workshops?

Latham recognised the need for 'serious training, deep culture change led from the top and continuous reinforcement'.[1118] He emphasised that 'clients and contractors cannot go to bed on Friday night as an adversarial client or contractor and wake up on Monday morning as a partnering convert'.[1119] Latham perceived that progress depends on the need to challenge a 'cynics bestiary' of those 'who do not believe in partnering', comprising six fundamental types: 'the stick-in-the-mud', 'the jobsworth', 'the one who just doesn't get it', 'the die-hard sceptic', 'the control freak' and 'the young people who don't believe in partnering because they have been fed a poisoned account'.[1120] Training in

---

1115 FAC-1 Framework Alliance Contract (2016) clause 1.7 and TAC-1 Term Alliance Contract (2016) clause 1.7.
1116 NEC4 Alliance Contract (2018) clause 21.3.
1117 Effectiveness of Frameworks (2012), 104.
1118 Latham, M. (2004).
1119 Association of Consultant Architects (2010), 1.
1120 Latham, M. (2004).

collaborative procurement systems can help to inform team members as to how these systems operate in practice and can help to overcome bias, preconceptions and illogical objections.[1121]

The UK Construction Leadership Council reported that 'the experience of contractors and quantity surveyors is that new forms of contract are often poorly understood within the supply chain'.[1122] Representatives of all team members need to develop knowledge, experience and evidence in order to implement collaborative procurement processes, including:

- An informed commitment to support the agreed alliance processes
- A full understanding of how and why alliance processes work
- The capability to make prompt decisions in accordance with the agreed governance system.

ISO 44001 describes how:

- 'Organizations will need to determine the necessary competence of people doing work that, under its control, affects the management system's performance, its ability to fulfil its obligations and ensure they receive the appropriate training
- In addition, organizations need to ensure that all people doing work under the organization's control are aware of the collaborative relationships policy, how their work may impact this and implications of not conforming with the collaborative business relationship management system'.[1123]

The role of the contract as a practical guide through agreed collaborative activities is a new concept for the construction industry. For those who perceive the contract primarily as a means to pursue or defend claims, the establishment of an enterprise contract as a routemap for enterprise planning is likely to be counterintuitive. This highlights the need for training that is practical and project-focused in order to illustrate the benefits of collaborative relationships and new contractual processes. Coaching and facilitation in relation to behavioural change can be combined with practical training and guidance on the agreed contractual systems of collaborative procurement.[1124] This is necessary to avoid a twin-track approach whereby collaborative working is reserved for the workshops while traditional behaviours govern implementation of the day to day project processes.[1125]

Concerns were expressed that, although early contractor involvement had been central to the Highways Agency (HA) procurement of major projects for several years, 'there has been very little training provided resulting in a lack of commitment from HA staff at all levels. This has resulted in HA lacking the ability to set sensible budgets, challenge Target Prices and manage the process effectively'.[1126] By contrast, Connect

1121 Smith, R.J. (1995), 69.
1122 Procuring for Value (2018), 33.
1123 ISO 44001 Implementation Guide (2017), 6.
1124 For example, Lindsey Henshaw, who leads a course in Contract Management as part of the Major Projects Programme MSc at the Saïd Business School, University of Oxford, said, 'I use the PPC2000 contract as the focus for the course assignment. The students are required to familiarise themselves with the contract and analyse its partnering features from a particular perspective. They are required to use their skills of analysis in an unfamiliar context and they find the challenge a rewarding one'. Email to the author, 2018.
1125 'The SCMG relationships and structures are sustained by training and support to embed a collaborative culture', SCMG Trial Project case study.
1126 Nichols (2007), 330.

Plus engaged a collaborative change consultant 'in order to identify what was required to adopt an integrated and sustainable collaborative approach', and 'a number of framework facilitators were then selected from key influential roles (such as project managers and commercial managers)…from across the supply chain to ensure that there are sufficient appropriate individuals with the attitude and competency to deliver a collaborative working culture'.[1127]

An inclusive approach to collaborative procurement can be enhanced by involving all stakeholders in facilitated workshops, for example on the Havelok Homes gas maintenance term alliance where:

- 'Workshops facilitated by the contractual Partnering Adviser clarified each party's role and flagged up some vital pre-commencement tasks to be added to the agreed TPC Partnering Timetable'
- 'These encouraged a climate of good faith that eliminated mistrust and promoted a "can-do" attitude. The parties agreed a range of KPIs with targets and monthly reporting as to safety, customer satisfaction, audit progress, quality of paperwork, timeliness of completing jobs, keeping appointments, doing the work right first time and getting paid on time'.[1128]

Workshops can fulfil the following functions:

- 'Team building…to form the partners into a team
- Team maintaining…as the project progresses
- Team completing…to assess the project after completion
- Team repairing…if needed, to ease conflict and gain focus'.[1129]

It is important that workshop facilitators are trained themselves, and in establishing the Connect Plus collaborative model, the client arranged 'the training and personal development of people to work in accordance with a collaborative skill-set' so as to:

- 'Build and maintain a "community" culture supporting collaborative supply chain relationships led from the top'
- 'Create a group of trained and accredited facilitators from throughout the supply chain that are responsible for promoting and maintaining a collaborative culture'
- Adopt an 'in-depth approach to people selection, skills training and psychological development in these key roles [which] has proven critical to the outcomes achieved on the ground'.[1130]

12.9   What Is the Role of an Independent Adviser?

It is tempting to assume that if an alliance needs independent advice, then it sounds like it is in trouble. However, the CIC Partnering Task Force, when considering how to disseminate partnering knowledge and best practice, concluded that it is unrealistic to assume that project teams can adopt new approaches to procurement, project management and

---

1127   Connect Plus Trial Project case study.
1128   Association of Consultant Architects (2010), 23.
1129   CIC (2005), 3.
1130   Connect Plus Trial Project case study.

partnering without the benefit of advice.[1131] Their view was that, while the ideal number of advisers to support new project processes is zero, the next best number is one, namely an independent adviser who can be seen 'as an "ombudsman" for the team'.[1132]

The CIC proposed the appointment of an independent adviser accountable to all team members who could 'prepare (on an even-handed basis) the documents that record the team's commitments, procedures and expectations'.[1133] This idea was carried through into the PPC2000 and TPC2005 contracts under which an independent 'Partnering Adviser' is tasked with the 'provision of fair and constructive advice as to the partnering process, the development of the partnering relationships and the operation of the Partnering Contract'.[1134]

The Arup report for OGC recognised that 'the process for the development of the Partnering Documents is supported by the Partnering Adviser who can assist in resolving differences and how the PPC method is to be adopted'.[1135] For example, on the Havelok Homes £600 000 term alliance, the team reported that:

- 'Workshops facilitated by the TPC Partnering Adviser clarified each party's role and flagged up some vital pre-commencement tasks to be added to the agreed Partnering Timetable'
- 'These encouraged a climate of good faith that eliminated mistrust and promoted a can-do attitude'.[1136]

FAC-1 and TAC-1 adopt a similar approach and provide the option for alliance members to appoint an 'Independent Adviser to provide impartial and constructive advice and support' in relation to 'the implementation of the Framework Alliance Contract and the avoidance or resolution of any dispute'.[1137] The value of any independent adviser will depend on his or her acceptability to all team members as a source of balanced advice and support, and this in turn will be linked to the experience that the adviser can bring from other projects in order to illustrate how other teams have dealt with similar issues.

The CIC suggested that an independent adviser could be 'from any construction profession, or even the legal profession, and much will depend on the skills and expertise of the individual rather than the organisation'.[1138] Lawyers can be credible independent advisers if they can provide an authoritative bridge between any one-sided or overly cautious advice offered by other advisers who are accountable only to individual team members. There are many instances of a prominent role played by lawyers as advisers on collaborative projects, such as Howard Ashcraft in the USA and Marko Misko in Australia, both in creating collaborative procurement models and in helping to deliver collaborative projects.[1139]

An independent adviser should always offer a positive and constructive approach in his or her advice to team members. When I was in-house legal adviser to His Excellency

---

1131 CIC (2002).
1132 CIC (2005), 5.
1133 CIC (2002), 16, Note 13.
1134 For example, in PPC2000 (2013) clause 5.6(iv).
1135 Arup (2008), 44, 45.
1136 Association of Consultant Architects (2010), 23.
1137 FAC-1 Framework Alliance Contract (2016)/TAC-1 Term Alliance Contract (2016) clause 3.3 and Appendix 1.
1138 CIC (2002), 16.
1139 As explained in Chapters 20 and 25.

Yousuf Ahmed Al-Shirawi, the Bahrain Minister of Development and Industry,[1140] I once told him that an innovative joint venture structure he proposed was not possible because it did not conform to Bahrain law. A few hours later his driver delivered a handwritten letter to my office, politely reminding me that 'a legal adviser, to be effective, must have the acumen to interpret his client's expectations in a way that makes them possible'.

A constructive and neutral advisory role is not easy, and an independent adviser may encounter the following obstacles:

- Clients and other team members who are reluctant to spend money on independent advice
- Advisers to individual team members who say that no independent advice is needed, because this might call into question the reasonableness of their own advice
- Project managers who say that no independent advice is needed because they see this as duplicating their own work and calling into question their objectivity
- Team members who assume that independent advice is required only for the resolution of disputes.

As regards the last point, an independent adviser is expected to assist 'in the solving of problems and the avoidance or resolution of disputes',[1141] but it would be a missed opportunity if an independent adviser is treated only as a resource for non-adversarial dispute resolution. Where an adviser is only a dispute resolver, he or she 'will be conservative and will not move far from the status quo…[even in those] dispute resolution processes governed by the norm of continuing the relation'.[1142]

The range of work undertaken by an independent adviser will depend in part on the experience of team members and their familiarity with the collaborative procurement model. For example, the Dudley College team implementing the integrated project insurance (IPI) procurement model obtained advice from 'an Independent Facilitator who works with the client from early in the process to establish the project along IPI lines, assists in procuring Alliance members, and generally facilitates the process of collaborative working through to project completion'.[1143] It was reported that this role was crucial to 'establishing and maintaining a collaborative working environment, guiding the Alliance through the IPI Model and in helping Members take ownership and responsibility for the project'.[1144]

12.10 Collaborative Culture Case Studies

A collaborative culture is led, managed and sustained by individuals rather than organisations. It cannot be imposed or taken for granted, but it can be supported by selection processes that create an integrated team and by contract provisions that establish the

---

1140 H.E. Yousuf Ahmed Al-Shirawi (1923–2004) was Bahrain's first Minister of Development and Industry: his guiding principle was to imagine what did not exist and ask: 'Why not?' He led diversification of the Bahrain economy into a range of aluminium industries, based on the idea that a smelter could use electricity generated from the gas reserves in Bahrain's declining oil fields.
1141 For example under PPC2000 (2013) clause 5.6(vi) and as considered in Section 19.8.
1142 Macneil, I.R. (1978), 896.
1143 University of Reading (2018), 14.
1144 University of Reading (2018), 179.

basis for involvement of all stakeholders, for improved communication and for consensual governance. All the case studies in this book involved the creation and support of a collaborative culture.

The first case study below describes an inclusive collaborative culture established by the Bath and North East Somerset Council project alliance where the client took a strong lead and ensured that residents, their relatives and unit managers were actively engaged in the design process through focus groups. The team used the contract to underpin the collaborative culture, for example in establishing close integration between the roles of the architect, the main contractor and its subcontracted supply chain members, and in developing close engagement with end users and with other stakeholders.

The second case study summarises the Connect Plus M25 Trial Project, which created a sustainable business culture using:

- Facilitation and training offered by a collaborative change consultant
- 360° collaborative performance review processes
- Continuous improvement dialogues, including a collaborative cost review
- A balanced scorecard process, allowing the whole supply chain to monitor and score performance by reference to critical success factors.

The Connect Plus framework was cited in the P13 Blueprint as a case study of contractual arrangements underpinning the role of an integrator.[1145]

Bath and North East Somerset Council Care Homes

A £12 million programme of four residential homes and related facilities for the elderly was undertaken by a project alliance that included Bath and North East Somerset Council, Leadbitter Construction, PRP Architects and Halcrow Engineers. Bath and North East Somerset Council selected PPC2000 in order to create an integrated multi-party team comprising its lead designer, main contractor and housing association partner Somer Housing Group, so that they could address all cost and risk issues jointly in advance under the preconstruction phase Partnering Timetable.

The residential care homes programme comprised a mixture of new build and refurbishment projects with demanding deadlines and decant arrangements, as well as an expectation of high-quality design.

Residents, their relatives and unit managers were actively engaged in the design process through focus groups. The alliance achieved:

- Establishment of a clearly defined design/supply/construction process with a single integrated Partnering Timetable and Project Timetable governing the preconstruction and construction phase of each project
- By this means, close integration between the roles of the architect as lead designer and that of the main contractor and its supply chain members
- Close engagement with end users and with other stakeholders as team members or recognised in the contract as 'Interested Parties'

(Continued)

1145 P13 Blueprint, 19.

- Regular measurement of performance and feedback to team members using key performance indicators
- Close cost control under Open-book pricing throughout the programme.

Leadbitter Construction reported that they were 'involved in the whole process from inception to completion, assisting the client and design team to find the most cost-effective solution throughout each stage of the design'.

Association of Consultant Architects (2010), 10.

Connect Plus M25

Connect Plus created integrated teams under a collaborative framework to deliver a £350 million highways asset management programme using Two Stage Open Book within a 30-year concession awarded by Highways England. The Trial Project case study examined parts of this framework programme, comprising M25 Concrete Paving Works, M25 Gade Valley Maurer Joint Replacement and M4 Strengthening Programme. The teams comprised Connect Plus (M25) and Highways England (clients), Jackson Civil Engineering, Aggregate Industries, Lafarge Tarmac, Balfour Beatty, (Geoffrey) Osborne, Skanska UK (main contractors), Atkins Consultancy Services, Flint & Neill, Parsons Brinckerhoff, Connect Plus Services (lead designers and contract managers) and Temporal Consulting (collaborative change consultant).

Connect Plus created and implemented an innovative Sustainable Business Culture Model ('the Model') through which it delivered a highways asset management programme in an efficient and collaborative manner, agreeing cost savings of **8%** together with design improvements that are expected to reduce whole life costs and improve long-term reliability. The culture promoted continuous improvement dialogues, including a collaborative cost review, ensuring that good practice was repeated and that lessons were learned and applied to subsequent projects.

The Model was developed to enable:

- A focus not just on what is delivered but also on how it is delivered
- Alignment of the supply chain to a common vision, values and behaviours
- In depth face-to-face engagement
- Mature relationship management and capability
- Robust and timely feedback
- Evidence-based performance monitoring on a range of issues, including relationships.

Connect Plus established a balanced scorecard process to allow the whole supply chain to input into monitoring performance across the Connect Plus community. The supply chain was regularly invited to score performance against agreed objectives by reference to critical success factors. Scores were given on a 1–12 scale, allowing the whole M25 community to give feedback on the programme, understand areas of strength and improve areas of weakness. Crucially, the balanced scorecard process also generated greater understanding of why things were working well or not through greater understanding of the participants' perspectives and subsequent relationships.

The balanced scorecard approach enabled Connect Plus to understand and measure progress towards its declared objectives of:

- Creating and maintaining a group of directors and facilitators empowered with the skills and behaviours to support and lead the cultural change and role model collaboration
- Delivering a whole life approach
- Minimising the impact of maintenance works
- Maintaining project facilities
- Enhancing knowledge of project facilities
- Respect for the environment
- Reduced risk.

Connect Plus Trial Project case study.

13

How Can BIM Support Collaborative Procurement?

13.1 Overview

Digital transactions and technological advances enable the rapid creation and sharing of data. They can help a construction project team to connect a range of established production processes with innovations, prototypes and the unique features of each project and site. Each project relies on the coordination of a diverse network of people, products, services and works, and requires the integration of a huge number of interconnected delivery and payment processes. Digital technology can improve this coordination and integration, and in these ways is closely linked to collaborative construction procurement.

The KCL Centre of Construction Law formed a research group which analysed the links between Building Information Modelling (BIM), procurement and contracts over a 24-month period to July 2016.[1146] Mark Bew[1147] stated that the resultant KCL BIM report placed 'a sharp focus on the performance of traditional working methods that should have been addressed many years ago'.[1148] Mark had once told me, on a conference call, 'I am not sure whether you have got the picture yet'. I was concerned that he thought I had not grasped the basics of digital technology, but in fact he had just emailed me a diagram and was checking whether it had arrived.

The progress of BIM is described in different ways in various jurisdictions, and in the UK:

- 'Level 1 comprises both 2D and 3D work in an agreed format with shared standards and information, but with work by each team member created and maintained separately'
- 'Level 2 comprises a move towards more collaborative working by combining the information prepared by each team member into a single Common Data Environment, accessible for sharing and exchange of data. Each team member's 3D models are set up through a common file format for analysis, checking, coordination and integration via the creation of a Federated Model and are capable of export to a common file format such as IFC (the 'Industry Foundation Class' international standard for sharing BIM data across different software platforms) and COBie (the 'Construction

---

1146 Details of the research group, and of the projects teams and specialists they interviewed, are set out in Appendix D. Their findings are considered in Mosey et al. (2016).
1147 Chair of the UK Government BIM Working Group and author of Digital Built Britain (2015).
1148 KCL Centre of Construction Law (2016).

Collaborative Construction Procurement and Improved Value, First Edition. David Mosey.
© 2019 John Wiley & Sons Ltd. Published 2019 by John Wiley & Sons Ltd.

Operations Building Information Exchange' format for recording project operation and maintenance data)'
- 'Level 3 signifies full collaboration by the project team members and anticipates the use of a single BIM model held by all project team members to access, use and modify at any time within a centrally held Common Data Environment'.[1149]

BIM is gaining ground, with 21% of respondents to a UK survey in 2018 saying it is fully integrated in their contracts (up from 14% in 2015) and a further 40% saying it is 'referenced in their contracts'.[1150] However, these statistics invite us to ask whether the 40% are just nodding in the direction of BIM technology rather than working out how to make it part of an integrated approach to procurement. BIM will benefit from procurement models that can ensure it 'is collaboratively created and developed, and evolves through the design, build (and maintain) lifecycle'.[1151] The evolution, adoption and impact of BIM in a range of other jurisdictions is considered in Sections 20.2 (Australia), 21.2 (Brazil), 22.2 (Bulgaria), 23.4 (Germany), 24.2 (Italy) and 25.2 (USA).

This chapter will consider how BIM and related digital technologies can influence collaborative business models by creating more effective connections, more reliable data and better communications. It will assess how BIM can support the multiple interactions involved in any alliance contract that has the features of an enterprise contract as explored in Chapter 8, including:

- How BIM can capture the data that is developed, shared, commented on and agreed among the members of a design team
- How BIM can capture the build-up of agreed costs and timescales agreed among alliance members
- How BIM can create a clearer audit trail of changes and their knock-on effects
- How BIM data enables the operation of completed projects and the capturing of lessons learned from one project to another.

This chapter will also look at the potential for digital transactions implemented by 'smart contracts' to expedite the sale, purchase and delivery of materials and equipment, and will consider whether or not contracts that exclude human intervention can have a broader role when placed in the context of collaborative decisions and the exercise of personal judgement.

13.2   What Is the Impact of Digital Technology?

BIM and other digital technologies are helping to create 'an environment where technology and working with technology is second nature in construction'.[1152] At a practical level 'efficiencies are beginning to be seen as the process is a much better way of working,

---

1149 KCL Centre of Construction Law (2016), 57, 58.
1150 NBS (2018), 16.
1151 NBS (2018), 16.
1152 Digital Built Britain (2015), 13.

for example significantly reducing the number of contractor requests for information or reducing waste or re-work on site'.[1153]

However, the contracting practices of the construction industry do not always keep up with communication technology. For example, despite the ubiquity of smart phones, it was only in December 2017 that FIDIC dropped the option for its Engineer to issue verbal instructions, with all the unnecessary risks of misunderstandings that this option involved.[1154]

Construction project processes require systems for information exchange, generating and communicating data between many individuals and organisations which establish:

- How ideas, services, works and supplies are developed, costed, programmed and competed for
- How products are assembled and how they are delivered, coordinated and integrated in the built asset
- How the completed asset is used and maintained and how it performs.

Digital technology can enable the creation and management of design, cost and time data that supports asset creation and asset maintenance.[1155] Measurable benefits created through BIM level 2 have been demonstrated in exemplar projects,[1156] and have added value both to asset creation and to asset operation, repair and maintenance. Research has shown how the potential impact of BIM is affected by the choice of a procurement and delivery model.[1157]

BIM sits alongside a range of other digital technology. For example, the internet of things ('IoT') describes direct exchange of data between household and commercial devices, intended to create connectivity that can enable a better understanding of buildings and their inhabitants. This phenomenon has an increasing impact on construction projects and particularly on the creation of more efficient systems for their operation, repair and maintenance. In 2014 the UK Government Office for Science reported that buildings connected via IoT 'create significant opportunities for cost minimisation through energy optimisation and predictive maintenance'.[1158]

Artificial intelligence ('AI') can be applied on construction projects, for example using drones that monitor and capture data from multiple angles and elevations to determine progress on site. Where AI can detect early signs of defective work or other problems, and particularly where AI can assess their implications, it is possible to improve the correction of errors or omissions and to mitigate the effects of delay and disruption.

Other emerging initiatives include 'smart contracts', as considered in Sections 13.8 and 13.9. These are programmes that can implement transactions automatically, using a self-contained and secure database known as a 'distributed ledger' with the ability to share data across a network of multiple geographical locations and users and through

---

1153 Ralph Montague, Managing Partner, ArcDox Architecture, Dublin, KCL Centre of Construction Law (2016).

1154 As previously permitted under FIDIC (1999) clause 3.3. Verbal instructions remain an option under JCT SBC/Q (2016) clause 3.12.1.

1155 As considered in Section 13.6.

1156 As illustrated in the Trial Project case studies summarised in Section 14.10 and in the additional case studies reviewed in KCL Centre of Construction Law (2016) and listed in Appendix D.

1157 KCL Centre of Construction Law (2016).

1158 Government Office for Science (2014).

which any transaction or other change automatically updates all distributed copies of the ledger.[1159]

13.3   What Is the Impact of BIM?

BIM increases the scope and speed of data exchanges, which highlights procurement and contractual questions as to who provides what data, when it is best provided and how it is used and relied upon. These questions are brought into sharp focus because BIM enables, and arguably depends upon, more integration and collaboration among project team members. However, emerging concerns as to the reliability of BIM computer software programs and as to the licensing of BIM models[1160] appear to have encouraged a defensive contractual approach to legal liability. This apparent caution needs closer scrutiny so as to establish a balance that gives BIM users the confidence to work efficiently while offering appropriate and insurable legal protections.

BIM raises demanding questions that challenge traditional project procurement processes such as:

- The absence of direct connections between team members that appears to foster a 'divide and rule' mentality and the dependence on a project manager as an intermediary
- The lack of clarity in the timing and integration of consultant design deliverables
- Fragmented responses to early warning of a problem
- The slow progress of payments down the supply chain
- The use of project information systems primarily as a source of evidence to support later claims rather than as a forecasting tool and as a rapid response system to avoid, manage and resolve disputes.

Unhelpfully, the features of BIM are described using terminology that does not match familiar construction contract wording. This has created confusion as to how BIM documents fit with other procurement documents. For example:

- The BIM 'Employer's Information Requirements' set out:
  - The client's BIM requirements and specifications, which form part of the client's brief to any consultant or contractor at tender stage, and which should also form part of their respective contracts[1161]

1159 In 'a distributed ledger' assets can be 'financial, legal, physical or electronic' and can be stored 'cryptographically' with controlled access, Government Office for Science (2016), 14.

1160 'There is a perceived lack of confidence in technology… Whilst attempts have been made to resolve issues of interoperability as between different parametric modelling software packages, a lack of confidence still remains as to whether the original file is presented as intended outside its native file format' Gibbs et al. (2015), 167–179.

1161 'In summary the EIRs establish a consistent digital format for providing data that can be reviewed by the client in order to assess and approve the project's progress and viability at each work stage'. 'In effect the pre-contract EIRs constitute a consistently structured Request for Proposals relating to BIM', Shepherd, D. (2015), 19 and 21.

- BIM standards,[1162] which form part of the client's brief issued to any consultant or contractor at tender stage, and which should also form part of their respective contracts
- The 'Common Data Environment'[1163] created through BIM, which includes agreed software and formats, with cross-reference to a 'BIM Execution Plan'[1164,1165]
- 'Project Procedures' which describe processes for matters such as coordination, inconsistency, project standards, methods and procedures, level of definition, asset information model, software, amendments and specified information[1166]
- The 'BIM Execution Plan':
  - Formulates each consultant's and contractor's intended processes to fulfil and achieve the Employer's Information Requirements
  - Can also set out the timing of each party's BIM contributions
  - Should form part of the commitments offered by each team member, firstly in their respective tender submissions and secondly in their contracts.[1167]

It has been suggested that adopting BIM 'progressively reduces the cost, time and uncertainty of design, construction and operation of buildings by making previously laborious and ambiguous processes quicker and more accurate'.[1168] To achieve these benefits depends on team members cutting through the hype, translating the jargon and aligning BIM with properly integrated procurement models.

13.4 How Can BIM Enable Collaborative Procurement?

It is an error to assume that any digital medium automatically improves communication, or that digital technology is a substitute for collaborative construction procurement. To try to implement one without the other is a serious missed opportunity, and commentators have suggested that:

- 'As BIM and data management technology drive new approaches to the design and construction process, the need to replace traditional competitive procurement and tendering processes with more collaborative structures and arrangements becomes ever more acute'[1169]

1162 Such as Uniclass 2015, the classification of building information for all stages of the project lifecycle. The Employer's Information Requirements are referred to as the 'Project Information Requirements' and the 'Exchange Information Requirements' in BS EN ISO 19650 (2019).
1163 Defined in KCL Centre of Construction Law (2016), Appendix D as 'a central source of information for the project that is used to collect, manage and disseminate graphical models and nongraphical data for the whole project team'.
1164 As considered in Section 14.3.
1165 Pre-Contract Award BIM Execution Plan (2013) and Post- Contract Award BIM Execution Plan (2013).
1166 Appendix 2 of the CIC BIM Protocol (2018), as described in Section 13.4, lists these contents of the Project Procedures and refers to them as separate from the Employer's Information Requirements, yet the two documents are closely related.
1167 For example, in a set of JCT Design and Build (2016) 'Contractors' Proposals' or the PPC2000 (2013) 'Project Proposals'.
1168 Saxon, R. (2013), 3.9.
1169 Sir John Armitt, Pinsent Masons (2016), 3.

- 'What partnering needed to succeed was BIM and this risk-managing collaboration concept will probably return to favour in supply chain relationships'[1170]
- 'The industry's route map to collaboration and high efficiency new delivery models can only be underpinned by BIM and the importance of its adoption cannot be over-estimated'[1171]
- 'Establishing a "single source of truth" on projects for monitoring projects early, potentially supported by collaborative technology, helps to minimize misalignments and enable corrective action'.[1172]

The Cookham Wood Trial Project adopted a collaborative approach to procurement, for example by providing for agreed individuals comprising the Core Group to meet and review design development proposals and to resolve questions and problems arising from BIM clash detection.[1173] King's BIM research interviewees gave other examples of collaborative working through BIM, such as the use of models to assist joint working by design consultants and in helping to explain design proposals to a client.[1174] However, these collaborative exercises form only a small part of the picture. In order for BIM to enable collaborative procurement, 'information needs to persist through the whole lifecycle of development and needs to be made available or shared to a wide range of communities including the asset owner, financiers, developers, architects, residents and local council'.[1175]

If BIM is intended to support an integrated and transparent approach, it should be used with procurement and delivery models that obtain early BIM model contributions from the main contractor and from subcontractors, suppliers, manufacturers and operators without causing delay or fragmenting the warranties relied on by the client. This is no longer an optional extra in a world where:

- Incorrect advice on procurement models can create liability for advisers[1176]
- Miscalculations by bidders resulting from software errors in procurement can give rise to significant disputes.[1177]

King's BIM research interviewees reported that BIM provides a major advance for collaborative working through joint risk management.[1178] However, they also expressed concerns as to the risk of poor interoperability between different types of BIM software, and offered differing views as to who should manage the BIM 'Common Data Environment' and the 'BIM Level 2 Federated Model'.[1179]

1170 Saxon, R. (2013), 5.27.
1171 Farmer, M. (2016), 36.
1172 McKinsey Global Institute (2017), 8.
1173 Under PPC2000 (2013) clauses 3.3 (Core Group and members), 3.4 (Responsibility for Core Group members), and 3.5 (Core Group meetings).
1174 KCL Centre of Construction Law (2016), Appendix A Part 2.
1175 BSI (2015).
1176 For example, consultant liability for lack of fully informed procurement recommendations in *Plymouth and South West Co-operative Society v Architecture Structure and Management* [2006] EWHC 5 (TCC).
1177 For example, the US case of *M.A. Mortenson Company Inc. v Timberline Software Corporation et al.* (1999) 93 Wash. App 819 where a software error resulted in a bid being too low.
1178 For example, Tom Oulton, BIM Manager, Turner & Townsend stated: 'I have seen time–cost–quality benefits as well as risk reduction in all the projects I have worked on', KCL Centre of Construction Law (2016). Also, the Pinsent Masons (2014) BIM survey reported that 57% of respondents considered a major benefit of BIM to be risk mitigation.
1179 KCL Centre of Construction Law (2016), Appendix A Part 3 Questions 1.3, 1.4, and 2.1.

Effective collaboration through BIM can be developed by adopting a procurement model:

- That brings all BIM contributors onto the team at the optimum time
- That uses BIM to build reliable shared data and mutual confidence
- That considers the operational impact of BIM on those who will operate, repair and maintain the completed capital project
- That creates a set of contracts which integrate all the team members' roles.

In April 2018 the CIC published the Second Edition of their Building Information Modelling Protocol,[1180] designed to supplement construction contracts and consultant appointments and to establish consistent provisions for adoption of BIM level 2. The First Edition in 2013[1181] was intended to encourage the industry towards adopting BIM through provisions that in some cases appeared to favour designers and contractors at the expense of the client, for example in relation to liability, time management and data integrity as considered in Sections 14.2, 14.3 and 14.5. The 2018 CIC BIM Protocol acknowledged recommendations on these issues in the King's BIM research report and created a more balanced medium through which BIM can connect a series of two-party contracts. However, the 2018 CIC BIM Protocol makes no reference to the connections between successful BIM and collaborative procurement, for example through the early appointment of an integrated team.

13.5 How Can BIM Support Early Contractor Involvement?

The 2011 UK Government Construction Strategy recommended BIM in conjunction with early contractor involvement and collaborative working.[1182] The 2016 UK Government Construction Strategy confirmed that 'BIM is a way of working that facilitates early contractor involvement, underpinned by the digital technologies which unlock more efficient methods of designing, creating and maintaining our assets'.[1183] The majority of projects reviewed in the King's BIM research adopted a procurement model that involved early contractor involvement.[1184] Interviewee comments linked successful BIM to a procurement model under which 'early contractor involvement and bringing tier 2 & tier 3 in early to advise on design has brought in efficiencies'.[1185] This evidence and the Trial Project case studies suggest that BIM needs planned and integrated contributions from the main contractor and specialist trades working alongside consultants, and that it is difficult, expensive and wasteful to try to obtain the benefits of BIM through a traditional single-stage procurement system.

Two Stage Open Book has been adopted on projects that combined BIM with early contractor involvement and collaborative working, two of which were among those examined in the King's BIM research. Each team included the main contractor and

---

1180 CIC BIM Protocol (2018).
1181 CIC BIM Protocol (2013).
1182 Government Construction Strategy (2011).
1183 Government Construction Strategy (2016–2020), Section 22.
1184 KCL Centre of Construction Law (2016), 30 and 31.
1185 Paul Davis, Information Modelling & Management Capacity Programme (IMMCP) Delivery Team, Transport for London quoted in KCL Centre of Construction Law (2016).

certain subcontractors appointed early under a conditional contract and a binding timetable which set out how they agreed to work together in contributing designs, costs and other data to the development and integration of BIM models.[1186]

Those King's BIM research interviewees who work with private sector developers described the benefits of using BIM in conjunction with the procurement of individual trade packages through construction management.[1187] They emphasised the need for different treatment of BIM according to the complexity and capabilities of different trades and explained the ways that this can be accommodated in a construction management procurement model.[1188]

Digital Built Britain (DBB) described incremental progression in the development of BIM where, at Level 3A, 'collaborative models of working facilitated by data will permit greater engagement with lower tier suppliers'.[1189] The future of BIM outlined in BIM 2050 includes the prediction that 'design consultants and principal contractors will be appointed simultaneously, early in the lifecycle, to enable concurrent working at outline business case stage'.[1190]

In order to achieve these objectives, BIM can be more closely connected to integrated procurement models such as project alliances, framework alliances and term alliances. For example, on the Cookham Wood project alliance:

- 'The implementation of BIM has created improved value in the pre-commencement and construction phases of the project'
- 'Virtual and actual prototypes have been produced to engineer out potential defects and clashes'
- 'It is also envisaged that the data that BIM will capture will positively inform the future facilities management of the project'.[1191]

13.6 How Can BIM Support Whole Life Asset Management?

BIM, IoT and AI can promote the data capture required to deliver a better asset both at the capital expenditure stage of a project and in its operational lifecycle. The potential benefits of digital technology can extend long beyond completion of a capital project if it is used to create accessible data that supports interactive operation, repair and maintenance systems, informing the development of an 'AIM' or 'Asset Information Model'[1192] in place of traditional operation and maintenance manuals. The operational lifecycle of an asset is likely to be far more costly than the capital expenditure stage, and AI combined with BIM can help to define, measure and manage this cost so as to 'create a direct data chain between design, construction, commissioning and operation of assets

---

1186 As illustrated in the case studies summarised in Section 14.10.
1187 KCL Centre of Construction Law (2016), 30 and 31.
1188 Construction management is a procurement model under which there is no main contractor. Instead the different trades are appointed direct by the client alongside a consultant construction manager, as considered in the case studies summarised in Section 7.10.
1189 Digital Built Britain (2015), 23.
1190 Built Environment (2014), 23.
1191 Cookham Wood Trial Project case study, the impact of which is considered in Mosey, D. (2014).
1192 An AIM or Asset Information Model is the data and information contained within a model used for managing, maintaining and operating an asset.

to enhance social outcomes and, through data feedback mechanism, provide a basis for continued improvement in asset design and performance'.[1193]

The closer integration of the design, construction and operation phases in the life of a built asset formed part of the UK Government's approach to BIM Level 3. Digital Built Britain included proposals at BIM Level 3A for the 'development of BIM and asset data enabled FM and AM Contracts – including the FM and AM roles in using and maintaining BIM models'.[1194] However, the development of contract models that help to fulfil the operational potential of BIM data also depends on the adoption of procurement models that invite the market to submit whole life asset management proposals.

The King's BIM research interviewees recognised the importance of as-built BIM models for the purposes of long-term asset management, particularly for clients who retain ownership of their completed projects.[1195] They explained how BIM is increasingly being used to define, measure and manage this cost, including work by Ministry of Justice (MoJ) using BIM to survey and data-log over six million assets for condition, life to replacement, priority ranking and impact.[1196] It was reported that this has helped to reduce unplanned maintenance by over 20% and to identify critical asset maintenance programmes.[1197]

Asset management innovations achieved through the use of BIM on the Cookham Wood project alliance included:

- 'Development by Arup and EMCOR of a model for service ducts and cell risers [that] could be serviced more quickly and reliably by repair and maintenance engineers without reference to a computer. It is expected that this will result in whole life benefits'
- Data available to the FM database that 'will allow more efficient repairs and maintenance that in turn should save Ministry of Justice money'.[1198]

An initiative to use BIM as a means to integrate the capital and operational phases of a project was the UK Government Soft Landings' ('GSL') programme.[1199] The GSL objectives included 'supporting collaboration in the supply chain including Designers, Suppliers, Constructors and Asset Managers throughout the whole asset lifecycle…[for which purpose] the contractor shall allow for the participation of appropriate subcontractors in design reviews, and record and act on identified access, commissioning and potential maintenance risks, where appropriate'.[1200] Trials of GSL included projects at Liverpool Prison (Ministry of Justice) and Shonks Mill Flood Storage Area (Environment Agency) which used BIM in conjunction with GSL to reduce operational costs.[1201]

Digital Built Britain recommended a 'new protocol to address certainty associated with asset performance – including validated data and digital briefing, building

---

1193 Department for Culture, Media & Sport (2017).
1194 Digital Built Britain (2015), 24.
1195 KCL Centre of Construction Law (2016), Appendix A Part 3 Question 2.9.
1196 KCL Centre of Construction Law (2016), 37.
1197 KCL Centre of Construction Law (2016), 37.
1198 Cookham Wood Trial Project case study.
1199 GSL (2014).
1200 Introduction and Section D2, GSL Employer's Requirements, Design and Construction Team Specification http://www.bimtaskgroup.org/gsl-policy-2.
1201 For example, http://www.bimtaskgroup.org/wp-content/uploads/2013/11/Liverpool-Prison-Case-Study-Final-18-11-13-web.pdf.

on the foundations of Government Soft Landings'.[1202] Successful use of BIM for asset management has been linked to the interfaces and systems established in the procurement model and contract terms, and some BIM projects have used a multi-party term alliance contract. For example, the City of London Corporation used TPC2005 in conjunction with BIM to procure the replacement of mechanical and electrical systems at the UK Central Criminal Court, where BIM was adopted to retrofit digital designs and data that will support the future repair, maintenance and facilities management of the building.[1203]

13.7 How Do Collaborative Teams Use BIM in Practice?

Collaborative procurement through early contractor involvement under a project alliance structure provides new opportunities to develop detailed knowledge of project components and to ensure that BIM models are accurately costed.[1204] Under the Property Services Cluster framework 'the BIM approach that was piloted on West of Waterlooville is being rolled out by Hampshire County Council on subsequent schemes including on Aldershot and Park Prewett '....' Using the early appointed supply chain to bring certainty and knowledge to the viability of design/delivery solutions'.[1205]

The Cookham Wood case study was independently audited and reported 20% agreed savings, namely a cost of £2332 per square metre against a baseline benchmark of £2910 per square metre. For example, the main contractor and its specialist subcontractor submitted a precast volumetric cell proposal in response to the client brief which was developed through BIM by the wider design team and which led to a time saving of six weeks and a saving in overheads of £85 000.[1206]

Other innovations created through the use of BIM with early contractor involvement under a collaborative contract included:

- Use of solid precast floor slabs in place of pre-stressed floor slabs, resulting in a time saving of 12 days
- Creation of more resilient lighting in the education block through a bespoke solution proposed by EMCOR which also created a significant cost saving.[1207]

Under a multi-party project alliance contract, each Cookham Wood alliance member agreed directly with the others to be responsible for errors, omissions and discrepancies in the BIM models it prepared or contributed to, except to the extent of its stated reliance on any contribution or information provided by any one or more other alliance members.[1208] As the sequence and nature of each contribution were spelled out, each alliance member was contractually entitled to rely on all earlier designs and was made aware of who will rely on the completeness, accuracy and timeliness of its contributions.

1202 Digital Built Britain (2015), 24.
1203 KCL Centre of Construction Law (2016), 36.
1204 Sometimes known as '5D BIM': KCL Centre of Construction Law (2016), Glossary, Appendix D.
1205 Property Services Cluster Trial Project case study.
1206 Cookham Wood Trial Project case study.
1207 Cookham Wood Trial Project case study.
1208 PPC2000 (2013) clause 2.4 (Responsibility for Partnering Documents).

The North Wales Prison project alliance applied the same procurement and contract approach,[1209] and used lessons learned on Cookham Wood in agreeing an extended preconstruction appointment of 38 weeks with alliance members that included the main contractor and the mechanical and electrical subcontractor. This project achieved agreed cost savings of 26%, again using BIM in conjunction with conditional preconstruction phase appointments of the main contractor and subcontractors that enabled the team to focus 'jointly on identifying cost savings through value engineering without eroding the quality or function of the Project being delivered'. These included:

- 'A more efficient way to achieve BREEAM (Building Research Establishment Environmental Assessment Method) excellence which resulted in a saving of more than £1 million
- A reduced footprint for the Entry Building/Energy Centre, using lessons learned from the Oakwood prison project and consultation with operational colleagues
- Re-location of the multi-faith building to provide a more operationally efficient layout of the site and a reduction in construction costs'
- 'Applying Value Engineering to the mechanical and electrical solution which resulted in an alternative lighting solution previously used in Scotland … which brought savings [and] which the MoJ can now use throughout the estate'.[1210]

The Dudley College Trial Project adopted BIM and reported:

- The use of 'a single BIM Execution Plan… representing the whole of the supply chain process'
- That 'the collaborative design approach using BIM has reduced the amount of documentation that typically needs to be produced… in particular the need for tender drawings and specifications'.[1211]

13.8 What Is the Impact of Smart Contracts?

While BIM provides a medium to improve the quality of data shared among people and the efficiency of how it is shared, smart contracts raise the question of whether human inefficiency should result in people being removed from certain transactions. A smart contract[1212] has been referred to as 'the ultimate automation of trust',[1213] the objectives of which 'are to satisfy common contractual conditions (such as payment terms), minimise expectations and minimise the need for trusted intermediaries'.[1214] A smart contract is a self-implementing and digitally formatted process which is embedded as a code in hardware and software language through which payments can be automatically released when due without any further decision being made by a person.[1215]

1209 North Wales Prison Trial Project case study.
1210 North Wales Prison Trial Project case study.
1211 Dudley College Trial Project case study.
1212 This is a term popularised by Nick Szabo, a cryptographer, who defined it as 'a computerised transaction protocol that executes the terms of a contract', Bitcoin Magazine (2016).
1213 BBVA (2015).
1214 Bitcoin Magazine (2016).
1215 BBVA (2015), 'Smart contracts are written as computer programmes rather than in legal language on a printed document'.

Self-implementing technology through a distributed ledger enables a process where the programme commands sequential events to occur and it is not possible to stop the initiated process. The transactions under a smart contract are irrevocable because performance is assessed automatically with cross-verification preventing the system from being abused.[1216] However, smart contracts may create challenges for regulators because they decentralise control by connecting the parties directly, handling payment without a need for any intermediary whose conduct could be subject to regulatory compliance.

The introduction of distributed ledger technology into the construction industry could streamline the identification and authentication processes of numerous supply chain transactions. Data in relation to processes, products and parties can include financial transactions plus copyright records and building component authentication. It may also be possible to use distributed ledger technology as a database tool for recording transactions so as to verify the sequence of events that need to take place prior to settlement of monies due. In tracking the movement of funds, it is likely that distributed ledger technology will also have a role in identifying money-laundering, the use of unsustainable sources of materials and the false labelling of goods.

However, smart contracts and distributed ledger technology are essentially systems for complying with agreed mutual obligations as to delivery and payment. In order to establish the self-executing transactions of a smart contract, the parties need to express their underlying contractual commitments according to conventional contractual rules. A conventional contract will also govern the obligations of the software programmer, a party who (subject to the small print) carries significant responsibility for smart contracts functioning as intended. For example, 'the fact that acceptance was automatically generated by a computer software cannot in any manner exonerate the defendant from responsibility [when] the defendant programmed the software'.[1217]

13.9 What Are the Limits of Smart Contracts?

A construction project relies not only on multiple transactions but also on complex contributions and interactions among many individuals and organisations. The information and management processes within the BIM information delivery cycle are reliant on project management, risk management and change management, all of which involve multiple decisions, key milestones and complex information.[1218]

Smart contracts can only self-implement once the programmed software recognises that identified activities have taken place. Without further development, they are not designed to accommodate the series of interconnected decisions and the exercise of discretionary judgement that relate to:

- Innovation through the review and development of designs
- Collaborative decisions such as the approval of new ideas
- Joint risk management decisions and actions
- Change procedures and decisions in relation to other optional matters
- Problem-solving and dispute resolution.

1216 Mason, J. (2016).
1217 *Software Solutions Partners v H.M. Customs and Excise* [2007] EWHC 971.
1218 As considered in BS EN ISO 19650 (2019).

It is possible for certain of the above processes to be automated with the benefit of significantly more preparatory information. For example, it may be possible automatically to restrict the parameters that a manager uses when exercising a discretion rather than leaving this entirely to his or her personal judgement. However, to embed this in smart contracts would require information to be defined at the outset of the project in respect of the pass and fail criteria for each of the relevant parameters and for these criteria to be combined with an evaluation model.

It may also be possible to introduce additional AI to the software programming associated with smart contracts, such as the capability to sort data, find relevant patterns, make predictions and develop machine learning so as to incorporate the variations that accommodate agreed changes at each payment point. AI or IoT sensors could track and record transport, delivery and installation of materials, goods and equipment and the verification of work completed and in use. However, AI and IoT do not appear to be capable of dealing with the more complex quality and workmanship issues which arise regularly during the construction process.

It has been noted how 'contracts usually require judgement and the use of discretion which require subtlety and richness in the language which is extremely different to code' and which justify 'maintaining a human involvement in the process' of their implementation.[1219] Rather than seek to automate all project processes, it is arguable that the construction industry would achieve better results by pursuing the development of collaborative procurement through BIM in tandem with the development of smart contracts, IoT and AI, so as to optimise the roles of digital technologies that:

- Inform and support human judgements and interactions where these are required in order to agree the optimum approaches to design, cost, time and risk management, in order to explore value-adding proposals and in order to undertake other collaborative processes
- Inform and support automatic supply chain transactions where all required decisions have been made and where human judgments and interfaces no longer have the potential to increase efficiency.

13.10 BIM Research Results

The following extracts from the King's BIM research report illustrate the range of procurement and contracting practices adopted on 12 leading BIM-enabled projects and by 40 leading practitioners:

How did the contract documents set out the process/procedure for managing a common data environment (CDE) (Diagram 12)?

- 7 interviewees stated that the client manages the CDE during both the design development and construction stages
- 13 interviewees stated that the contractor manages the CDE once appointed
- Only 2 interviewees stated that lead designers manage the CDE
- 15 interviewees stated that it is all dependent on client and/or project.

How has the practice of BIM changed your perception of data sharing?

1219 Mason, J. (2016).

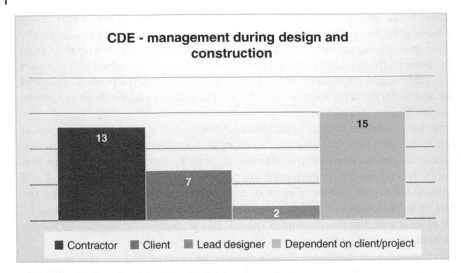

Diagram 12 CDE – management during design and construction.

- Many considered that data sharing has been 'done for years', with comments such as 'At the end of the day it is still all about the teams and how well they work together'
- 3 interviewees suggested it needs to be 'taught' so as to ensure a methodical approach under BIM, with comments such as: 'It is all about collaboration. People need to be helped to collaborate'
- 9 interviewees admitted that information in the model is not being properly shared, with comments such as: 'Some struggle as to obtaining the information which team members say they are not going to provide', and 'Industry is holding back information in the model due to concerns over IP loss'.

In what ways did the project involve BIM models prepared or contributed to by the main contractor and/or trade contractors/subcontractors/suppliers, and/or in what ways did these team members provide comments/feedback on consultant BIM models?

- Most of the project teams interviewed created a procurement model and contract terms specifically designed to involve main contractors and key subcontractors at an early stage in order to obtain their BIM contributions during the tender process and/or during the preconstruction and/or construction phases
- 8 interviewees expressed the opinion that the tier 2 and tier 3 'supply chain is not yet ready to embed the new processes required for BIM'.

What documents described the deadlines for each team member to achieve progress in its BIM model development to the agreed levels of detail and in its comments/feedback on other BM models (Diagram 13)?

- 19 interviewees referred to use of a traditional design development programme, and then a construction programme, but without expressing a clear view as to their enforceability or contractual status

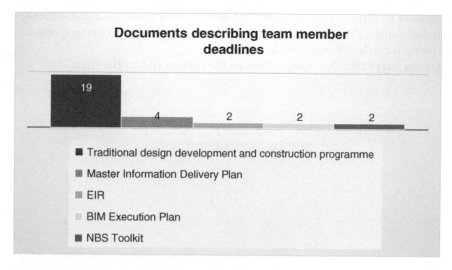

Diagram 13 Documents describing team member deadlines.

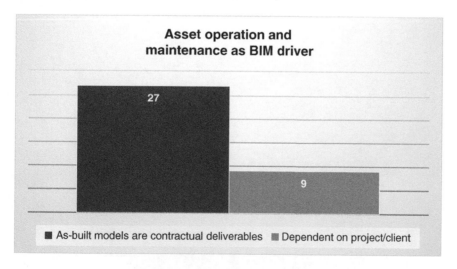

Diagram 14 Asset and operation and maintenance as BIM driver.

- Others mentioned timelines expressed in:
  - the Master Information Delivery Plan (4)
  - the NBS BIM Toolkit (2)
  - the EIRs (2)
  - the BIM Execution Plan (2).

To what extent was management of the asset and its operation and maintenance the driver in the use of BIM (Diagram 14)?

- Client owner interviewees with a long-term interest in their assets acknowledged the importance of 'as-built' BIM models, with comments such as 'Ease of asset replacement is an obvious benefit during operation and maintenance'

- 27 interviewees confirmed that 'as-built' BIM models were contractual deliverables, with comments such as 'There was a specific request (within building contract) requiring the contractor to hand over as-built model'
- 9 interviewees stated that this is dependent on the project and/or the client.

14

How Does BIM Support Collaborative Contracts?

14.1 Overview

We have considered in Chapter 13 how collaborative procurement can help to ensure that the right parties contribute Building Information Modelling (BIM) design, cost, time and operational data at the right level of detail and at the right time. This chapter will look further into the relationship between BIM, alliances and other collaborative contracts.

To search for references to BIM in a contract is not the most effective way of determining whether it supports the adoption of BIM in practice. It is more useful to assess the potential for BIM to operate in tandem with enterprise planning and decision-making processes of the types explored in other chapters. This chapter will explore how BIM can be enabled and supported through specific contractual provisions governing:

- The impact of BIM on agreed deadlines
- Mutual intellectual property rights among team members
- Reliance by team members on BIM data
- Responsibility for managing BIM data
- Links between the BIM data used for design, construction and operation.

In all these respects it is important to consider whether any contract provisions relating to BIM should increase or reduce the rights or duties of team members or have any other effects that may compromise normal commercial expectations. This chapter will explore how BIM is treated in standard form contracts and how it deals with problems in practice.

Commentators also regularly link BIM to collaborative working[1220] and there is a natural fit between BIM and collaborative contracts that provide for early contractor involvement and Supply Chain Collaboration. This chapter will consider how BIM is treated in alliance contracts. However, we will look firstly at the basic legal question of whether BIM affects the duty of care owed by team members.

1220 'The industry's route map to collaboration and high efficiency new delivery models can only be underpinned by BIM and the importance of its adoption cannot be overestimated' Farmer, M. (2016), 36.

Collaborative Construction Procurement and Improved Value, First Edition. David Mosey.
© 2019 John Wiley & Sons Ltd. Published 2019 by John Wiley & Sons Ltd.

14.2   How Does BIM Affect a Duty of Care?

A consultant has a duty to use reasonable skill and care in creating a design or providing other advice,[1221] and in producing a design that is buildable,[1222] unless there is express agreement that the duty of a consultant goes further and makes it responsible for producing a result that is fit for its purpose.[1223] There is no evidence that this duty will be increased or reduced by the adoption of BIM. For example, there is no implication that using BIM would raise a consultant's duty of care to fitness for purpose in respect of what a design or other advice will achieve.

Fitness for purpose is an implied duty only for a design and build contractor,[1224] or for the supplier or manufacturer of a product,[1225] and again this duty should not be increased or reduced by adoption of BIM. It can only be reduced to a duty of reasonable skill and care through clear contract wording.[1226]

The 2013 CIC BIM Protocol reduced the duty of care of team members to 'reasonable endeavours',[1227] apparently as a means to encourage adoption of BIM by design consultants. However, none of the consultants included among King's BIM research interviewees considered there was any need for a reduced duty of care by reason of adopting BIM.[1228] In line with recommendations in the King's BIM research report, the 2018 second edition of the CIC BIM Protocol revised the duty of care to correspond to 'the relevant level of skill and care applicable to its equivalent obligations in the Agreement'.[1229]

In establishing the level of a duty of reasonable skill and care:

- A consultant, in its approach to BIM, as in any other activity, is expected 'not to lag behind other ordinarily assiduous and intelligent members of his profession in knowledge of the new advances, discoveries and developments in his field'.[1230] This means keeping up with the profession, firstly in advising on the benefits and risks of adopting BIM, and secondly in applying BIM to the design process and to related costing, programming and project management services[1231]
- Professional knowledge and practices relating to BIM are rapidly evolving and a consultant is entitled to claim that it applied 'the state of the art' at the time of giving its

---

1221 As confirmed in *Greaves & Co (Contractors) v Baynham Meikle & Partners* [1975] 1 WLR 1095 (CA); UK Supply of Goods and Services Act 1982, Section 13.

1222 As confirmed in *Equitable Debenture Assets Corporation v William Moss Group* [1984] 2 Con LR 1 (QB).

1223 As confirmed in *George Hawkins v Chrysler (UK) and Burne Associates* (1986) 38 BLR 36 (CA).

1224 As confirmed in *Viking Grain Storage v TH White Installations* (1985) 33 BLR 103 (QB); *Tesco v Costain* [2003] EWHC 1487 (TCC); UK Supply of Goods and Services Act 1982, Sections 4(5) and 4(6).

1225 UK Sale of Goods Act 1979, 1994, section 14(3).

1226 Of the type found in standard form construction contract provisions such as JCT Design and Build (2016) clause 2.17.1, NEC4 Option X15 and PPC2000 (2013) clause 22.1.

1227 CIC BIM Protocol (2013) clause 4.1.2.

1228 KCL Centre of Construction Law (2016), Appendix A Part 3 Question 2.6.

1229 CIC BIM Protocol (2018) clause 4.1.

1230 *Eckersley T E & Others v Binnie & Partners* [1988] 18 Con LR 1 (CA).

1231 'Those who do not follow recommended best practice are likely to be exposed by BIM. Claims will therefore still form part of BIM projects, but it is hoped that the process of working in a "Level 2" environment will help reduce the likelihood and severity of these developing into disputes', Gibbs et al. (2015), 179.

advice, although this defence will be judged by reference to the guidance and publications available to the profession as a whole at that time[1232]

- A contractual commitment to comply with statutory obligations relevant to BIM can enhance a duty of care, for example under the UK CDM Regulations 2015 which provide that a designer has a duty when 'preparing or modifying a design' to 'eliminate, so far as is reasonably practicable, foreseeable risks to the health and safety of any person' or otherwise to reduce or control those risks, and 'to take all reasonable steps to provide, with the design, sufficient information about the design, construction or maintenance of the structure to adequately assist the client, other designers and contractors to comply with their duties'.[1233]

As regards a designer's duty to use reasonable skill and care in reviewing and checking its own designs and those of other designers at each stage of BIM model development:

- Increased access to BIM data emerging throughout the life of a project could increase the likelihood that a designer has become aware, or should become aware, of the need to reconsider an earlier design[1234]
- Increased access through BIM to other team members' designs could affect a designer's duty to warn of errors or problems that it notices in another team member's work[1235]
- Contractors may find that the increased accessibility of BIM data increases their duty to check, and even to validate, the designs provided by a consultant[1236]
- A duty to review and work to a design is likely to include an enforceable obligation to use BIM models if made available.[1237]

As regards the quality of BIM software, English case law has distinguished between software supplied in tangible form on a disk (which is capable of being a sale of 'goods') and software delivered purely in electronic form (which is not a sale of 'goods').[1238] For example, the UK Consumer Rights Act 2015 defines 'digital content' as 'data which are produced and supplied in digital form' and recites that contracts for the sale of digital content should not be classified as sales contracts or service contracts. It seems likely that this approach will be applied to other UK legislation relating to the sale of goods, and in particular the Sale of Goods Act 1979. If so, unless and until the gap is filled by legislation, there is no scope for the implication of the statutory implied terms in BIM software licensing contracts as to satisfactory quality or fitness for purpose.

1232 As illustrated in *Wimpey Construction UK v Poole* [1984] 2 Lloyd's Rep 499 (QB).
1233 UK Construction (Design and Management) Regulations 2015, Regulation 9.
1234 For the purposes of the tests in *Samuel Payne v John Setchell* [2002] BLR 489 (TCC).
1235 For the purposes of the tests in *Tesco Stores v The Norman Hitchcox Partnership* (1997) 56 Con LR 42 (QB) and in *J Murphy & Sons v Johnston Precast (formerly Johnston Pipes)* [2008] EWHC 3104 (TCC).
1236 In the case of *Cooperative Insurance Society v Henry Boot Scotland* [2002] EWHC 1270 (TCC) Judge Seymour stated: 'Someone who undertakes…an obligation to complete a design begun by someone else agrees that the result, however much of the design work was done before the process of completion commenced, would have been prepared with reasonable skill and care'.
1237 A US case that was settled before it went to court concerned a life-sciences building where the mechanical and electrical engineer did not inform the main contractor that the building's M&E systems needed to be very tightly installed into the ceiling plenum. The main contractor worked off 2D drawings despite the availability of a BIM model, and ran out of space with only 70% of the sequence complete: Matthews (2011).
1238 The Court of Appeal in *Computer Associates UK v The Software Incubator* [2018] EWCA Civ 518 held that the supply of software in the form of a download is not a 'sale of goods'.

Clarification of each team member's responsibilities requires clear drafting in an integrated set of contracts. For example, the 2018 CIC BIM Protocol introduces a 'Responsibility Matrix' linked to 'defined Project stages'[1239] and creates scope for important clarifications, although this will benefit from guidance regarding completion of Appendix 1 where the 'Responsibility Matrix' details are to be set out.

14.3   How Does BIM Affect Agreed Deadlines and Interfaces?

The efficient development of BIM models requires clarity as to when each level of detail will be provided in each BIM model by each team member at each stage of design, supply and construction. In the King's BIM research nearly half the interviewees reported that they relied only on existing contract provisions to govern the timing of BIM contributions. However, the treatment of timing under some standard form construction contracts leaves gaps or mismatches in the mutual commitments of multiple team members:

- To produce and deliver their BIM models to the agreed level of detail by agreed deadlines at each stage
- To provide comments and approvals by agreed deadlines at each stage
- To specify what matters may prevent agreed deadlines being met.[1240]

When completing the Construction Industry Council (CIC) Responsibility Matrix[1241] team members should go further in forward planning and mutual collaborative commitments than the CIC recommended templates for a 'BIM Execution Plan' ('BEP'), which include only a succession of agreements to agree successive layers of timing rather than any clear deadlines. The Pre-Contract BIM Execution Plan refers to 'Major project milestones' for information delivery, and envisages that 'a more detailed and co-ordinated Master Information and Delivery Plan must be developed and agreed with the stakeholders following contract award and included in the post-contract award BEP'.[1242]

Equivalent wording appears in the Post Contract-Award BIM Execution Plan, plus a template for a 'Task Information Delivery Plan' (or 'TIDP')[1243] with delivery dates for tasks at milestones but without defining tasks.[1244] A 'Master Information Delivery Plan' (or 'MIDP')[1245] is then 'to be developed from the separate TIDPs produced for each task within the project. This more detailed, coordinated MIDP must be developed and agreed with the stakeholders. When completed the MIDP must be published in this document appendix and on the project extranet as a project plan'.[1246]

1239 CIC BIM Protocol (2018) clause 10.26 and Appendix 1.
1240 The impact of programmes and timetables on collaborative construction procurement is explored in Chapter 17, as are the relevant provisions of standard form contracts.
1241 CIC BIM Protocol (2018) Appendix 1.
1242 Pre-Contract Award BIM Execution Plan (2013), Section 5.
1243 Post-Contract Award BIM Execution Plan (2013), Sections 3.2 and 4.4.
1244 Links to the 'Agreed matrix of responsibilities across the supply chain' are set out at Section 4.3 of the Post-Contract Award BIM Execution Plan (2013) in respect of each stage in the Plan of Work but without further explanation.
1245 KCL Centre of Construction Law (2016), Glossary at Appendix D.
1246 Post-Contract Award BIM Execution Plan (2013).

This approach postpones agreement of the deadlines and interfaces on which successful BIM and successful alliances depend. Only two King's BIM interviewees used the BIM Execution Plans as a means to describe agreed deadlines for BIM contributions, comments and feedback, while only four interviewees used the MIDP for this purpose.[1247] Closer attention needs to be given to agreement of clear and binding timescales so that mutual reliance on the timing of all parties' contributions will enable BIM to help achieve greater efficiency and to support collaborative procurement.[1248]

The 2013 CIC BIM Protocol allowed a team member an extension of time in delivery of BIM models and related data, to the agreed level of detail and in accordance with the Information Requirements, for any reason due to 'events outside its reasonable control'.[1249] As the protocol was stated to take precedence over all other aspects of each project contract,[1250] the effect of this provision was to sweep aside all specified grounds for extension of time and the detailed procedures for agreeing those extensions of time, as set out in detail in standard form construction contracts. The King's BIM research highlighted this anomaly and as a result the 2018 CIC BIM Protocol second edition expressly linked extensions of time to 'any events or circumstances which entitle the Project Team Member to an extension of time or additional time under the Agreement'.[1251]

An illustration of how accurate project planning can support BIM was provided on the Cookham Wood Trial Project, where the alliance members agreed a preconstruction phase BIM Execution Plan integrating the timing of their model contributions through a single set of agreed periods and dates. During the preconstruction phase the main contractor developed a construction phase BIM Execution Plan, again governed by a single set of integrated timelines, for review and approval by the wider project alliance team.[1252]

14.4   How Does BIM Affect Intellectual Property Rights?

In any team structure, 'fear of opportunism may deter parties from relying on one another as much as they should for efficiency'.[1253] Perceived threats to intellectual property rights have been seen as a serious obstacle to the adoption of BIM in that:

- 'There is a concern in some innovative Tier 2s that their IPR can be stolen more easily through BIM as it is easy to copy files that show how innovative approaches work' and that 'at present some contributors are reluctant to participate without a better way to prevent their original ideas being captured, before or after contract, for later use by others with access to BIM'[1254]
- 'Ultimately, innovation-led modernisation continues to be inhibited at all levels by the lack of industry-wide strategic leadership with a more integrated client and industry

---

1247 KCL Centre of Construction Law (2016), Appendix A Part 3 Question 2.4.
1248 KCL Centre of Construction Law (2016), Section 5.1 Treatment of deadlines and interfaces.
1249 CIC BIM Protocol (2013) clause 4.1.2.
1250 CIC BIM Protocol (2013) clause 2.1.
1251 CIC BIM Protocol (2018) clause 4.1.2.
1252 Cookham Wood Trial Project case study.
1253 Milgrom & Roberts (1992), 128.
1254 Saxon, R. (2013), Sections 5.21 and 6.10.

agenda. It is also critically undermined by a fundamental unwillingness to collaborate if this involves divulging competitive advantage or intellectual property. There are deep-seated perceptions in the supply chain of short-term threats to market share and dilution of returns'.[1255]

King's BIM interviewees took differing views as to who owns and should own the intellectual property rights in BIM models, some accepting that these rights are owned by the client and others stating that they refuse to hand them over.[1256] Concerns as to loss of intellectual property rights were reported by a number of interviewees as a reason for failure to share BIM data,[1257] although these may be primarily commercial concerns rather than evidence of a new legal issue. Intellectual property rights should not need additional legal protections by virtue of attaching to BIM models given that, for example, existing UK statutory copyright protection already covers graphic and non-graphic design work plus 'computer programs' and 'preparatory design material for a computer program'.[1258]

Commentators have noted that intellectual property rights should not present any major roadblocks to BIM adoption if there is a clear understanding as to:

- Each team member's ownership or permission in respect of all contributions to models
- Grant of limited, non-exclusive licences to reproduce, distribute, display or otherwise use those contributions
- Equivalent clarity in respect of contractor and subcontractor contributions
- The use of models for facilities management during the operation and maintenance phase.[1259]

A distinction has been drawn between intellectual property rights at BIM Level 2, where contributions to BIM models can be traced to their authors, and the position at BIM Level 3, where contributions may become indistinguishable. Commentators have raised concerns as to the intellectual property implications of BIM Level 3 if contributions cannot be separated and if a contributor cannot prevent or even see amendments made to its work by another contributor. The difference is more a question of insurable liability than intellectual property as it is possible to recognise joint authorship if that is how a BIM Level 3 model is to be owned.[1260]

For the purposes of BIM Level 2, team members need to rely on each other's ownership of intellectual property in their respective BIM contributions, and those rights should be licensed in a way that supports the completion, use and operation of the project. Most standard form construction contracts create intellectual property rights only between the client and contractor and separately between the client and each consultant, but not directly between the main contractor and consultants.[1261] PPC2000 and

---

1255 Farmer, M. (2016), 36.
1256 KCL Centre of Construction Law (2016), Appendix A Part 3 Question 2.7.
1257 KCL Centre of Construction Law (2016), Appendix A Part 3 Questions 2.1 and 2.7.
1258 UK Copyright Designs and Patents Act 1988, section 3(1).
1259 RICS (2015), Section 4.1, 56.
1260 As noted in Cooper, W. (2013).
1261 For example, in FIDIC (2017) clause 1.10, ICC (2014), clause 4.11, JCT SBC/Q (2016) clause 2.41 and NEC4 (2018) clause 22.1.

the NEC4 Alliance Contract both create direct mutual rights and licences between the client, the main contractor, each consultant and each other party who becomes a party to the project alliance contract.[1262] For example, the multi-party contract structure of PPC2000 enables the alliance members to agree that their mutual licences cover 'any purpose relating to the completion of the Project and (only in regard to the Client) the Operation of the Project'.[1263]

As regards liability arising from the licensing of intellectual property rights, a multi-party contract enables direct licences between alliance members and can limit those licences and related liability as appropriate. For example, PPC2000 excludes liability of an alliance member 'for the use of any design or document that it prepares for any purpose other than that for which it was agreed to be prepared as stated in, or reasonably inferred from, the Partnering Documents'.[1264]

The CIC BIM Protocol is designed to be added in identical terms to all consultant appointments and construction contracts and to those subcontracts involving a contribution to design.[1265] In this way it can create a matrix on consistent mutual intellectual property licences.[1266] The 2018 edition recognises that the provisions governing mutual intellectual property rights will not apply if these matters are already covered in the contracts that it supplements.[1267] The 2018 CIC BIM Protocol also removes the previously contentious reference to suspension or revocation of intellectual property licences for non-payment.[1268]

14.5 How Does BIM Affect Reliance on Data?

When faced with human imperfections, it is tempting to consider whether a computer could do a better job. We can easily become confused by the intoxicating effect of new technology that is constantly available to the construction industry. However, in the context of creating accurate data for the purposes of safe and efficient construction, we should remember that computers are not accountable for their errors, and it is humans who make and programme a computer and who interpret that data.

Reliance on data is fundamental to the ability of project team members to fulfil their contractual obligations. Computer-aided design has been widely used for many years, as have online document management and project management systems, without attracting defensive terms in construction contracts. However, the wider reliance on software for the successful implementation of BIM has given rise to significant liability concerns. King's BIM research interviewees expressed differing views as to whether the

---

1262 PPC2000 (2013) clause 9 and NEC4 Alliance Contract (2018) clause 23.
1263 PPC2000 (2013) clause 9.2 (Licence to copy and use).
1264 PPC2000 (2013) clause 9.3 (Liability for use of designs and documents). The NEC4 Alliance Contract (2018) clauses 23.4 and 23.6 set out similar provisions.
1265 CIC BIM Protocol (2018) clauses 3.1.1 and 4.2.
1266 CIC BIM Protocol (2018) clause 6.
1267 CIC BIM Protocol (2018) clause 6.1.1.
1268 CIC BIM Protocol (2013) clause 6.4.

client or the design team should select BIM software, while the majority noted particular concern that the interoperability of BIM models is not dealt with satisfactorily.[1269]

Standard form construction contracts and consultant appointments do not typically limit did exclude liability for the accuracy of two-dimensional drawings or computer-aided three-dimensional designs. Hence, contractual silence on this issue is the starting point in the FIDIC 2017, ICC 2014, JCT 2016, NEC4 and PPC2000 standard forms. By contrast, the 2013 CIC BIM Protocol excluded any warranty as to the integrity of electronic data transmission, and also excluded any liability for corruption or alteration occurring after transmission.[1270] These exclusions placed all the excluded risks with the client, and did not require project team members to pass any liability at all to BIM software providers.[1271] The 2018 second edition of the CIC BIM protocol reflected the concerns raised in the King's research and reduced the first exclusion so that it applies only in relation to any software used by a team member being 'compatible with any software or software format used by or on behalf of' any other team member.[1272]

The importance of this issue is illustrated where loss or damage results from a software error but where the typical exclusions of liability used by software providers prevent any remedy.[1273] For example, the standard Autodesk Terms of Use limit the duration of any warranty to a 90-day period (or the licence term if shorter) and limit Autodesk's liability (to the extent permitted by law) so that the user's sole and exclusive remedy for Autodesk's breach of this warranty will be for Autodesk, at its option:

- 'To attempt reasonably to remedy the breach or'
- 'To refund amounts received for the affected subscription and terminate such subscription'.[1274]

The Autodesk terms are also subject to extensive disclaimers, including:

- The exclusion of any warranty that 'the Offerings or Output, or the access thereto or use thereof, will be available, uninterrupted, error-free, secure, accurate, reliable or complete',[1275]
- The exclusion of liability for any 'incidental, special, indirect, consequential or punitive damages; loss of profits or revenue; business interruption or loss of use'.[1276]

Exclusions of this type significantly reduce the ability of clients to rely on BIM software, and questions as to the reliability of BIM software lead to questions as to its benefits. Contract documents governing the provision of BIM-related software should provide for balanced risk allocation and a reasonable level of user protection.[1277]

1269 KCL Centre of Construction Law (2016), Appendix A Part 3 Questions 1.4 and 2.8.
1270 CIC BIM Protocol (2018) clause 5.
1271 'Residual liability for the integrity of the electronic data rests with the employer' Lewis, S. (2014).
1272 CIC BIM Protocol (2013) clause 5.1.2.
1273 For example, the US case of *M.A. Mortenson Company Inc. v Timberline Software Corporation et al.* (1999) 93 Wash. App 819 where a software error resulted in a bid being too low but the software provider's wide limitation clause excluded liability for 'any damages of any type' and was upheld by the court.
1274 Autodesk Terms of Use (2018) clause 18.1.
1275 Autodesk Terms of Use (2018) clause 18.2.
1276 Autodesk Terms of Use (2018) clause 18.3.
1277 'The promoter will want to ensure as far as possible that the licence terms under which the BIM software is supplied make allowance for a reasonable degree of liability on the part of the software provider

14.6 Who Manages BIM Data?

Arguably a decision as to who should take a lead role in managing the BIM Common Data Environment (CDE) should be determined alongside and according to the choice of procurement model. Surprisingly, the views of King's BIM interviewees varied widely as to whether this responsibility should rest with the lead designer, the main contractor or the client.[1278] King's BIM research interviewees recognised the potential for improved data management through BIM[1279] and adopted varying approaches in using specialist BIM expertise, including the appointment of a BIM 'Information Manager' or recruitment of their own specialist employees, and there was general consensus among interviewees that more guidance is required as to the BIM Information Manager role.[1280]

The ability of team members to rely on the specialist BIM expertise of a BIM Information Manager will depend on:

- Clarifying exactly what services the BIM Information Manager provides[1281]
- Other team members acting reasonably in their reliance on the BIM Information Manager's expertise.[1282]

The CIC 'Outline Scope of Services for Information Management' included requirements for the BIM Information Manager to:

- 'Initiate, agree and implement the Project Information Plan and Asset Information Plan'
- 'Enable integration of information within the Project Team and co-ordination of information by the Design Lead'
- 'Provide the services to host the Common Data Environment'.[1283]

The UK BIM Task Group recommended that 'the Information Manager has a key role in setting up and managing the Common Data Environment ("CDE"). The CDE is a key tool for effective collaboration, quality control and avoidance of waste'. However, the CIC Protocol 2013 guidance stated that 'the Information Manager has no design related duties. Clash detection and model coordination activities associated with a "BIM coordinator" remain the responsibility of the design lead'.[1284]

Whichever team member's appointment includes the BIM Information Manager role, there was little evidence from King's research interviewees that this role was seen as central to any aspect of BIM. Many reported greater reliance on the use of BIM consultants to support and inform the use of BIM, but generally these consultants were not

---

for losses incurred by participants and losses to the project overall attributable to failings in the software', ICE proceedings (2014/2), 115.

1278 KCL Centre of Construction Law (2016), Appendix A Part 3 Question 1.3.

1279 Sonia Zahiroddiny, BIM Strategy Manager at HS2 suggested that 'The key with BIM is realising value from our digital assets through clear and consistent ways of defining data requirements, consistency in data management and improved data procurement, which gives us greater provenance and assurances in the quality of data we receive' KCL Centre of Construction Law (2016), 26.

1280 KCL Centre of Construction Law (2016), 28 and Appendix A Part 3 Question 2.5.

1281 Noting, for example, the Outline Scope of Services for Information Management (2013) and BS EN ISO 19650 (2019).

1282 The expectation that team members will act reasonably in their reliance on an expert was considered in *Cooperative Group v John Allen Associates* [2010] EWHC 2300 (TCC).

1283 Outline Scope of Services for Information Management (2013).

1284 CIC BIM Protocol (2013) Guidance Note 4, vi.

appointed to take responsibility for model coordination and clash detection in the way that the BIM Task Group envisaged.[1285] Rather than being appointed as team members with a schedule of project deliverables, some BIM consultants were brought in only to facilitate the performance and interfaces of the team members.[1286]

The responses from King's BIM interviewees suggested that there remained a lack of clarity as to who should take on the role of BIM Information Manager and as to how this should interface with the role of the design lead as the party responsible for BIM model coordination. As long as these roles are not fully understood and as long as team members each rely primarily on their own staff, or on BIM consultants who are not themselves members of the team with clear services, there will be limits on the extent to which BIM can support integrated teams.

The BIM data management role on the Dudley College Trial Project was allocated to fullest extent possible who was 'expected to lead on design/BIM management' and who was 'primarily responsible for enabling BIM and motivating the IPT (integrated project team) to utilize it to the fullest extent possible'.[1287]

14.7 How Is BIM Treated in Construction Contracts?

The UK Government Construction Clients Group suggested in March 2011 that 'little change is required in the fundamental building blocks of copyright law, contracts or insurance to facilitate working at Level 2 of BIM maturity'.[1288] That statement was presumably intended to encourage the early adoption of BIM Level 2 but it does not reflect the evolving treatment of BIM in contract forms.

In practical terms the contractual treatment of BIM starts with the scope of a consultant's schedule of services or a contractor's project brief, and the following points should be checked:

- Does the contract contain a clear set of obligations as to how the consultant or contractor will be expected to implement BIM, including for example whether this will start with the project procurement process and also continue into post-completion operation?[1289]
- Are there clear representations as to a consultant's or a contractor's level of BIM experience and expertise?[1290]
- Is it made clear what effect the use of BIM will have on the consultant's or contractor's specific duties in respect of design, costing, programming, project management, construction and asset management?[1291]
- Do all parties understand who will be the BIM Information Manager, what duties this role will comprise and how these duties will interface with those of the design lead and the project manager so as to avoid gaps or duplications?[1292]

1285 Lane, T. (2011).

1286 KCL Centre of Construction Law (2016), 28 and Appendix A Part 3 Question 2.5.

1287 Dudley College Trial Project case study.

1288 Government Construction Client Group (2011), paragraph 5.2. However, in the same document the GCCG noted that: 'Contractual and commercial issues have the potential to act as a source of inertia holding back adoption on projects'.

1289 KCL Centre of Construction Law (2016), Chapters 9 and 11.

1290 KCL Centre of Construction Law (2016), Chapter 2.

1291 KCL Centre of Construction Law (2016), Chapters 2 and 5.

1292 The BIM Information Manager is also considered in KCL Centre of Construction Law (2016), Chapter 8 and the management of information is considered in BS EN ISO 19650 (2019).

Standard form construction contracts, consultant appointments and related guidance have used a light touch in relation to BIM, and all standard forms recognise that contracts may need adaptation to reflect rapidly evolving technology. In order to keep their options open, standard forms such as the JCT 2016 and ICC 2014 contracts cross-refer to details set out in a separate BIM protocol, with ICC 2014 providing that the parties 'shall have the benefit of any rights granted to them in the BIM Protocol and of any limitations or exclusions on liability contained within it'.[1293] Some standard forms include specific BIM provisions, for example the CIOB Time and Cost Management Contract Suite.[1294] Others provide only general guidance, such as the FIDIC 2017 advisory notes which identify risks arising from working in a BIM environment, while recognising that 'BIM is founded on a team approach and successful projects utilising BIM encourage collaboration'.[1295] FIDIC 2017 takes a very cautious approach to BIM and lists a total of 25 sub-clauses that 'should be thoroughly reviewed' if BIM is to be adopted,[1296] but it provides no indication as to the possible amendments or additions that should be considered.

NEC4 introduced detailed provisions on information modelling in Option X10 which include:

- Information Model Requirements to be expressed by the client as part of the scope of the project[1297]
- An Information Execution Plan to be submitted by the main contractor and assessed by the project manager[1298]
- Identification of Information Providers who contribute to the Information Model[1299]
- Additional early warning in respect of any matter adversely affecting the Information Model.[1300]

The NEC4 Alliance Contract provides for alliance members to integrate their contributions to BIM through:

- A shared 'Information Execution Plan' which is to be issued and updated by the Alliance Manager[1301]
- A single 'Information Model' which is 'the electronic integration of Project Information in the form stated in the Information Model Requirements', to be created by the Alliance acting collectively,[1302] subject to clarification of each alliance member's roles and responsibilities in the Implementation Plan
- Agreement by alliance members to warn each other of 'any matter which could adversely affect the creation or use of the Information Model'.[1303]

1293 ICC (2014) clause 20 and JCT SBC/Q (2016) clause 1.1.
1294 The Chartered Institute of Building published its 'Complex Projects Contract' in 2013 and relaunched it in June 2016 as the CIOB Time and Cost Management Contract Suite (2015). Both versions include specific BIM provisions but there are no examples of either form being adopted in practice.
1295 FIDIC (2017) Advisory Notes, 53.
1296 FIDIC (2017) Advisory Notes, 53–55.
1297 NEC4 (2017) Option X10.1(4).
1298 NEC4 (2017) Option X10.4.
1299 NEC4 (2017) Option X10 1(5).
1300 NEC4 (2017) Option X10 3.
1301 NEC4 Alliance Contract (2018) Option X10.1(1).
1302 NEC4 Alliance Contract (2018) Option X10.1(3) and 10.2.
1303 NEC4 Alliance Contract (2018) Option X10.4.

PPC2000 states that there is no need to make any amendments at all for it to be a suitable contract for BIM, but also offers in Appendix 10 a number of clarifying amendments that cross-refer to BIM.[1304] Some users of PPC2000 have confirmed that BIM amendments are unnecessary, and on the Cookham Wood Trial Project the 'Ministry of Justice...concluded that PPC2000 does not require any amendments to provide for BIM Level 2 or any contractual BIM Protocol'.[1305] When PPC2000 Appendix 10 was published, the Ministry of Justice (MoJ) exercised the option not to include it and to continue to use PPC2000 unamended.

There has been growing interest in the role that a multi-party, collaborative contract can play in the successful adoption of BIM:

- The Institution of Civil Engineers noted that 'despite the popularity of the NEC3 and JCT suite of contracts in conventional construction projects, PPC2000 was the contract chosen for the UK government's Level 2 BIM trial projects. The multi-party contract was favoured as it governs the duration of the procurement process and promotes collaboration by bringing in key project participants at the design phase of the project'[1306]
- The Royal Institute of British Architects noted that 'given the move towards Integrated building information modelling (iBIM), new procurement models which consider ways of harnessing the skills of all the parties involved in the design, construction and management of a building will need to be developed alongside new collaborative and multi-party contractual documents'[1307]
- The report BIM Built Environment 2050 provided that 'a focus on relational contracting supports this recommendation using multi-party contracts to discourage legal disputes and costly litigation'.[1308]

Many King's BIM interviewees also described their contract forms as collaborative, particularly those who used NEC3 or PPC2000.[1309] It has also been suggested that:

- 'For the purposes of BIM level 2, JCT Constructing Excellence, PPC2000 and NEC3, in order of unfamiliarity, are seen as the vanguard of the contracts of tomorrow'
- 'Learning from the use of these, together with BIM, is expected to allow a vision of the future [from which] level 3 needs to emerge'.[1310]

14.8   How Does BIM Affect Alliances?

Arguably, only the early, direct contractual relationships that can be created between the members of an alliance support the level of team integration required for BIM to fulfil its potential, enabling collective decision-making under 'multi-party contracts to discourage legal disputes and costly litigation'.[1311] An alliance contract also supports a

1304 PPC2000 (2013) Appendix 10, Building Information Modelling.
1305 Cookham Wood Trial Project case study.
1306 ICE proceedings (2014). See also Tyerman, D. (2013).
1307 RIBA (2012).
1308 Built Environment (2014), 23.
1309 KCL Centre of Construction Law (2016), Appendix A Part 2 and Part 3 Question 1.1.
1310 Saxon, R. (2013), Section 5.28.
1311 Built Environment (2014), 23.

procurement model that brings BIM contributors into strategic relationships, that sets out value-adding digital activities and processes, that uses BIM to build reliable shared data and mutual confidence, and that considers the operational impact of BIM on those who will repair, maintain and operate completed projects.

A framework for the adoption of BIM with collaborative procurement in Australia recommends:

- 'Early engagement of facilities management professionals at the design and planning stage to minimise overall operational lifecycle costs of the asset/facility'
- 'Comprehensively contractually binding BIM Management Plans…completed jointly by a project owner representative, design team and contractor'.[1312]

Alliance contracts that have been proven to support BIM include PPC2000 in the UK and comparable multi-party contracts in Australia[1313] and in the USA, the latter including the option of a BIM addendum which provides that 'each Model Contributor shall be responsible for the Contributions it makes to a Model or the data that is developed as a result of that Contributor's access to a Model'.[1314] However, in the short term it is difficult to expect the construction industry and its clients and advisers to abandon bilateral contracts in favour of a full, multi-party project alliance contract.

An alternative suggested in the King's BIM research report was to create an overarching umbrella that brings the project team members under one collaborative roof while leaving their separate two-party appointments intact. A multi-party alliance contract that creates an umbrella of this type can:

- Set out who works with whom and at what level of responsibility, so that the contributions to BIM under bilateral contracts can be drawn together more effectively
- Create mechanisms that ensure stronger commitment to shared objectives and collective self-regulation, as well as improved transparency and efficiency, through the ability to share BIM data on mutually agreed terms.

FAC-1 includes the option for BIM to integrate agreed approaches to design, supply chain engagement, costing, joint risk management and programming in relation to the projects comprising a framework programme or in relation to the components comprising a single project.[1315] Equivalent provisions appear in the TAC-1 Term Alliance Contract in relation to implementation of a term programme and each order.[1316] Other provisions in FAC-1 and TAC-1 that enable alliance members to seek improved value through BIM include clauses and guidance in respect of:

- Data transparency and team integration through direct relationships under the multi-party structure and agreed objectives[1317]

1312 Strategic Forum for the Australasian Building and Construction Industry, 6, as referred to in KCL Centre of Construction Law (2016), Appendix E Part 2.

1313 KCL Centre of Construction Law (2016), Appendix E Part 2.

1314 ConsensusDocs BIM Addendum (2016) clause 5.1 See also Section 25.2.

1315 In the FAC-1 Framework Alliance Contract (2016) provisions for Framework Documents and Template Project Documents.

1316 In the TAC-1 Term Alliance Contract (2016) provisions for Term Documents and Template Order Documents.

1317 Through the FAC-1 Framework Alliance Contract (2016) and TAC-1 Term Alliance Contract (2016) multi-party structure, the Schedule 1 Objectives and the clause 13 limits on confidential information.

- Agreed software and clarity as to reliance on data in the communication systems and template documents[1318]
- Mutual reliance on agreed BIM deadlines, gateways and interfaces in the timetable for agreed alliance activities[1319]
- Flexibility to agree any combination of BIM contributions through the multi-party structure[1320]
- Flexibility to bring in BIM contributions from specialist subcontractors, suppliers, manufacturers and operators through Supply Chain Collaboration[1321]
- Direct mutual licences of intellectual property rights[1322]
- Integration of BIM management with Core Group governance to enable clash resolution
- Flexibility to obtain BIM contributions from additional alliance members involved in the occupation, operation, repair, alteration and demolition of completed projects and tasks[1323]
- Potential for BIM to enable learning and improvement from project to project and from task to task.[1324]

Alliance teams are starting to adopt FAC-1 as a contractual integrator for BIM projects.[1325] In December 2016 at a BIM conference at the University of Milan I suggested that FAC-1 could integrate BIM contributions provided under separate appointments. Fellow speaker Professor Sara Valaguzza immediately offered to create an Italian version of FAC-1 for this purpose and the first Italian FAC-1 was signed less than two years later on 24th October 2018. In this case the FAC-1 contract supported a BIM project alliance to design and construct the Liscate School project in Milan.[1326]

Anglian Water combined the creation of a fully integrated alliance with 'a digital transformation strategy which has seen the alliance design and build everything virtually, including rehearsing and optimising construction in virtual rehearsal suites before going to site'.[1327] As a result, Anglian reported the following benefits:

- 'Not only has this shifted delivery from construction to assembly, it has provided health and safety benefits through off site construction of products'
- 'Digital transformation has also led to more effective engagement with users and operators, with greater involvement in the virtual development of solutions improving operability and operator buy-in'

1318 In the FAC-1 Framework Alliance Contract (2016) and TAC-1 Term Alliance Contract (2016) clause 1.9.3 provision for communication systems, the FAC-1 Schedule 5 provision for Template Project Documents and the TAC-1 provision for Template Order Documents.
1319 In the FAC-1 Framework Alliance Contract (2016) and TAC-1 Term Alliance Contract (2016) Schedule 2 Timetable and the clause 6 Alliance Activities.
1320 Through the Framework Alliance Contract (2016) and TAC-1 Term Alliance Contract (2016) multi-party structure and under the clause 1.11 and the Appendix 2 Joining Agreements.
1321 Through FAC-1 Framework Alliance Contract (2016) and TAC-1 Term Alliance Contract (2016) clause 6.3 Supply Chain Collaboration.
1322 FAC-1 Framework Alliance Contract (2016) and TAC-1 Term Alliance Contract (2016) clause 11.
1323 Under the FAC-1 Framework Alliance Contract (2016) and TAC-1 Term Alliance Contract (2016) clause 1.11 and Appendix 2 Joining Agreements, and the recognition of Operation as a feature of Improved Value.
1324 Under the FAC-1 Framework Alliance Contract (2016) and TAC-1 Term Alliance Contract (2016) Schedule 1 Success Measures and Targets.
1325 BIM+, March 2018.
1326 As summarised in Section 10.10.
1327 Anglian Water case study summarised in Section 4.10.

- 'The progress of digital rehearsal demonstrates the value in delivering through integrated teams, where all the influential parts of the wider supply chain are involved in optioneering and solution development'.[1328]

14.9 How Can BIM Deal with Problems?

We have seen that adoption of BIM should not affect the duty of care of team members for the design, costing and implementation of projects. We have also seen that there is a need for careful professional advice on legal issues such as:

- The status of the contract documents that support BIM and the interfaces between those documents
- The need to safeguard intellectual property rights and reliance on BIM data
- Clarification as to who manages BIM data and the rights of other team members to rely on their expertise.

There is a general obligation for information technology consultants 'to use due care and skill to ensure that the system recommended would carry out the functions for which it was required efficiently and would have an acceptable minimum of operational faults or bugs'.[1329] However, problems may arise from exclusions and limitations of liability in the standard terms and conditions used by software providers,[1330] as a result of which:

- Any equivalent exclusions and limitations in a consultant appointment or construction contract would significantly reduce a client's customary contractual rights
- The mismatch between market norms in a software contract on the one hand, and a consultant appointment or construction contract on the other hand, is an anomaly that can leave consultants and contractors exposed if they do not have recourse to a software provider for loss and damage caused by a software fault.

As regards the management of BIM data, there can also be problems where the custodian of this data abuses its position. For example, on a Mid-Atlantic Power Project in the Falkland Islands there was a payment dispute, and the BIM coordinator revoked another team member's access codes to the CDE.[1331] That team member obtained an injunction overriding this revocation, but the case underlined the power of whoever is in control of access to BIM data if as gatekeeper it can cut off that access and bring the project to a standstill.

14.10 BIM Alliance Case Studies

The following case studies summarise two Trial Projects illustrating the effectiveness of creating a project alliance that combines collaborative working, early contractor involvement and BIM as related systems designed to deliver improved value. These case studies

---

1328 Anglian Water case study summarised in Section 4.10.
1329 *Stephenson Blake (Holdings) v Streets Heaver* [2001] Lloyd's Rep PN44 (QB).
1330 As considered in Section 14.5.
1331 *Trant Engineering v Mott MacDonald* [2017] EWHC 2061 (TCC).

also illustrate how collaborative improvement through a framework alliance, where it overarches and integrates the work of teams engaged on a series of project alliances, can enhance the potential of BIM to deliver improved value.

The Ministry of Justice procures all its projects using BIM, creating multi-party project alliances through PPC2000 linked to strategic consultant and contractor appointments under overarching framework alliances, as summarised in Section 4.10. The following case studies summarise the work undertaken by Ministry of Justice alliance teams on BIM Trial Projects at Cookham Wood[1332] and North Wales Prison. A case study taken from the Cookham Wood Trial Project also appears in the 2018 UK Construction Leadership Council report 'Procuring for Value'.[1333]

Cookham Wood Young Offenders Institution

The Ministry of Justice (MoJ) new build Young Offenders Institution at Cookham Wood in Kent had a project value of £20 million (including construction cost, fees and escorts). The project alliance team comprised MoJ with main contractor/lead designer Interserve, key suppliers SSC (Pre-Cast Volumetric Cell Provider) and EMCOR (Mechanical and Electrical Specialist), plus consultants Faithful + Gould (Client Representative), HLN (Client architect/technical assessor) Fob Design, Tier Consult, Arup, MJ Patch and ICL.

MoJ was keen to establish the ways in which BIM, Government Soft Landings and introduction of a Project Bank Account could support more efficient team working and could benefit its construction and operational needs. This approach was enhanced through the use of BIM as developed for the purpose of team selection as well as ongoing design, construction and asset management.

MoJ considered that PPC2000 actively promotes collaborative working at all stages, and also concluded that PPC2000 did not require any amendments to provide for BIM Level 2 or any contractual BIM Protocol.

Early contractor involvement commenced when Interserve worked with its preferred subcontractors (including SSC and EMCOR) to develop innovative proposals at the point of selection by MoJ from a shortlist of framework contractors. This collaboration continued among Interserve and its subcontractors who worked with MoJ and its consultants as an integrated team to undertake agreed preconstruction phase surveys, design development, risk management, cost reviews, enabling works, and other preconstruction activities that delivered further innovations.

Examples of innovation achieved through BIM included:

- Improved design coordination and change management at an early stage prior to construction, including liaison with the Governor of HMYOI Cookham Wood, who praised the benefits of a 'walk through of the buildings highlighting views into and out of areas that normally I couldn't do until completion'
- Development by Arup and EMCOR of a model for service ducts and cell risers as MoJ had to be satisfied that these could be serviced by repair and maintenance engineers

---

1332 The outcomes from the Cookham Wood Trial Project are also considered in Mosey, D. (2014).
1333 Procuring for Value (2018), 335.

without reference to a computer. It is expected that this will result in whole life benefits.

Taking into account value indicators of similar projects (as analysed by the client and by cost consultant Sweett Group), the project showed agreed cost savings of **20%**. These savings were achieved through:

- Joint working by Interserve with subcontractors in developing innovative proposals at the point of selection
- Further joint working by Interserve and its subcontractors with the client and its consultants throughout the contractual preconstruction appointment
- Additional information provided at the point of selection and throughout the preconstruction phase by the use of.

Cookham Wood Trial Project case study.

North Wales Prison (HMP Berwyn)

HMP Berwyn (formerly known as North Wales Prison) is one of the largest prisons in Europe and was designed and constructed using a Ministry of Justice (MoJ) framework alliance and project alliance in conjunction with BIM level 2. The alliance members included MoJ as client, Lendlease (main contractor), AECOM (Client Representative), Sweett Group (Cost Consultant and CDM Coordinator), WYG (Technical Assessor), Capita Symonds (Architect), TPC Consulting (Civil and Structural Engineers), Hoare Lea (Mechanical and Electrical Engineers) and Crown House (Mechanical and Electrical Specialist).

The alliance undertook a 38-week period of programmed early contractor contributions to design, risk management and finalisation of agreed costs. MoJ applied lessons learned from the Cookham Wood Trial Project which enabled the North Wales Prison team to obtain additional benefits from the use of BIM Level 2 and greater contributions from subcontractors and suppliers, including a specific focus on local/regional SMEs.

The project budget was £212 million, and the alliance agreed innovations and improvements in efficiency ahead of start on site that enabled them to commit to a fixed price of £157 million, representing **26%** of agreed cost savings. The team focused on identifying these cost savings through joint design and risk reviews without eroding the quality or function of the project, and these included:

- A more efficient way to achieve Building Research Establishment Environmental Assessment Method (BREEAM) excellence which resulted in a saving of more than £1 million
- A reduced footprint for the Entry Building/Energy Centre, using lessons learned from the Oakwood prison project and consultation with operational colleagues
- Asbestos mitigation on site which avoided costs of approximately £4 million
- Re-location of the multi-faith building to provide a more operationally efficient layout of the site and a reduction in construction costs

(Continued)

- Challenging the original costed design to incorporate an open 'swale' in place of an attenuation tank, thereby also creating a new environment for wildlife while reducing construction costs
- Applying value engineering to the mechanical and electrical solution which resulted in an alternative lighting solution previously used in Scotland. This solution complied with MoJ standards, brought savings and could be used by MoJ throughout its estate.

The team faced significant early challenges caused by ground conditions on the extensive project site, including asbestos, soft spots and remnants of the concrete foundations of an old munitions factory. Joint working in co-located offices enabled the team to agree a strong response to these issues in finding ways to minimise any delays and cost overruns.

The team used the 'BIM Cave' in Salford to enable a walk through of the virtual project. This offered major operational insights that benefited the appointed Governor and Senior Responsible Officer of North Wales Prison. Early contractor involvement under Two Stage Open Book combined with BIM provided design improvements and enabled effective value management at all stages of the project.

North Wales Prison Trial Project case study.

15

How Can Collaborative Procurement Improve Economic and Social Value?

15.1   Overview

A primary purpose of collaborative construction procurement is to deliver improved value, and therefore team members should agree what they define as value, what improvements they expect, how these improvements will be delivered and how they will be measured.[1334] A collaborative team should commit to their agreed objectives, success measures and targets so that 'properly understood and applied collaborative working can produce a successful and superior project in terms of value and with all those in the supply chain sharing the financial benefits achieved by removing waste from the preconstruction and construction process'.[1335]

ISO44001 'addresses the creation of value concept' and explains that organisations should 'identify external and internal issues and the needs and expectations of…stakeholders and how value is delivered to them'.[1336] The ways of delivering value should also be set out in agreed processes, contributions and deadlines that are supported by contractual commitments.[1337]

The beneficiaries of improved value can include:

- The client who commissions a project
- The occupiers and users of a completed project
- The consultants, contractors, subcontractors, suppliers, manufacturers and operators comprising the team who deliver a project
- Those employees of each team member who are engaged on a project
- The community that is affected by a project.

In order to be sustainable, any collaborative procurement model needs to provide demonstrable benefits for all team members, taking into account that 'as the demand for construction and infrastructure services increases, procurers and suppliers are looking at delivery structures which will provide not only sustainable, long term value to the procurers but also more consistent, better margins for contractors, supply chain members

---

1334 For example, on the Greenwich Council term alliance, 'contract extensions and assessments of achievement were based on performance according to Greenwich's [contract] KPIs against targets agreed in advance with all stakeholders' – Association of Consultant Architects (2010), 19.
1335 JCT CE Guide (2016), Section 14.
1336 ISO 44001 Implementation Guide (2017).
1337 Latham M. suggested that 'PPC2000 and TPC2005 offer exactly the techniques that government and the private sector need to secure "more for less" through a combination of cost savings and added value', Association of Consultant Architects (2010), 1.

Collaborative Construction Procurement and Improved Value, First Edition. David Mosey.
© 2019 John Wiley & Sons Ltd. Published 2019 by John Wiley & Sons Ltd.

and professional teams'.[1338] Sustainable collaborative procurement models also need to balance cost savings with other improvements in economic value and with improvements in social value, using evaluation and performance measurement systems that are clear and transparent.

The UK Social Value Act 2012[1339] required all public sector bodies to include considerations of economic, social and environmental well-being when commissioning public services contracts, but it did not explain the practical means by which this can be achieved. For example, the legislation did not include any integration with public procurement legislation.[1340]

Over the period from 2013 to 2018 evidence was collected through the detailed analysis of Trial Projects,[1341] by reference to stated cost benchmarks,[1342] which showed how the teams working on these projects agreed and delivered significant improvements in economic and social value. Evidence collected by King's from 10 of the Trial Projects, and from over 40 other case studies as listed in Appendix C and quoted throughout this book, illustrates the types of economic and social value that can be achieved through collaborative procurement, including:

- Cost certainty and cost savings
- Improved quality
- Improved supply chain relationships
- Local and regional opportunities
- Improved training and employment outputs
- Improved safety
- Reduced impact on the environment.

This chapter considers each of these categories of improved value and the means by which collaborative procurement has enabled each team to develop and agree relevant proposals for approval. Improvements in time certainty and achievement of time savings are considered in Chapter 17.

15.2   How Can Collaborative Procurement Benefit All Team Members?

An integrated team can improve value through the ways that its members go about their work, for example through steps that improve the ordering and organisation of this work and that facilitate the work of others. However, significant changes to procurement and contracting systems will only be adopted if they offer commercial benefits to all team members. It is important to make clear to all team members the benefits that they can achieve, for example by ensuring that the procurement process provides 'maximum opportunities to learn in detail what matters most to the tier 1 contractor and to each tier 2/3 subcontractor and supplier in how they go about their work, and what steps

---

1338 Sir John Armitt, Pinsent Masons (2016), 3.
1339 The Public Services (Social Value) Act 2012.
1340 Currently the Public Contracts Regulations 2015, as considered in Chapter 6.
1341 As outlined in Appendix B.
1342 Using cost data, provided by HM Treasury, for construction costs benchmarked in 2010/11.

can be taken to improve the ordering and organisation of this work so as to maximise the opportunities for savings and other improved value'.[1343]

Early contractor review and validation of designs and their costs significantly reduce the design risks for consultants, with safety and buildability of designs signed off by all team members. Also, the UK Civil Engineering Contractors Association reported that 'the widespread adoption of ECI by clients would drive down costs for both clients and contractors'.[1344] Other proven benefits of collaborative procurement for contractors include:

- 'Early appointment…that creates a stable basis for preconstruction phase activities leading up to authority for the project to commence on site'
- 'Open Book costing combined with prior agreement of…fees/profit/overheads to ensure that agreed cost savings do not erode margins'
- 'Joint working during the preconstruction phase that enables the … contractor to influence robust programming and early risk management activities, so that the project proceeds to the construction phase on an agreed basis supported by maximum information'
- 'Creation of an environment in which … contractors can demonstrate savings and other improved added value in order to obtain additional work, contract extensions and other agreed incentives such as shared savings'.[1345]

15.3 How Can Collaborative Procurement Create Cost Certainty and Cost Savings?

For a client to obtain cost savings on a project may suggest that someone else will lose out. A cost saving may also imply a reduction or compromise in quality or some other aspect of the project, for example where a consultant or contractor or subcontracted supply chain member seeks ways to deliver cost savings without reducing their expected profit. However, collaborative procurement provides new ways to achieve cost savings that are agreed in advance by all team members and that are not to the financial detriment of any of them. In addition, it can ensure that agreed cost savings do not reduce or compromise quality or any other aspect of the project and shows how cost savings can be combined with other improved value such as extended warranties, community benefits and sustainability initiatives.

It has been reported that consultants working together with contractors earlier than usual during the design phase have 'reduced the costs associated with rework due to poor understanding of the client's requirements',[1346] and that 'collaboration and integrated team working reduces costs and improves performance'.[1347] Trial Project case studies have provided robust evidence of significant cost savings that are attributable to:

- Accelerated mobilisation so as to increase productivity
- Revised designs, for example a more efficient site layout

1343 Two Stage Open Book and Supply Chain Collaboration Guidance (2014), 14.
1344 CECA (2016), 19.
1345 Two Stage Open Book and Supply Chain Collaboration Guidance (2014), 14.
1346 NEDC (1991), 78.
1347 NAO (2005), 68.

- A revised programme with a more economical sequence
- A revised approach to risks, for example following additional site investigations
- Revised working methods to improve efficient interfaces between alliance members.

Trial Project case studies record the following agreed cost savings:

- North Wales Prison 26%
- Archbishop Beck 26%
- Cookham Wood 20%
- SCMG 16%
- Project Horizon 12%
- Anchor Property 9%
- Connect Plus 8%
- Property Services Cluster 7%
- Rye Harbour 6%.[1348]

The Dudley College Trial Project reported a 6% saving but also a small overrun on the agreed target outturn cost, which was picked up through the alliance members' pain share arrangement, but which did not give rise to a claim under the cost overrun cover forming part of the Integrated Project Insurance (IPI) policy.[1349] To put this in context, 'all the participants recorded the project as a very positive experience in collaborative working and one that they would be keen to undertake again'.[1350]

Trial Project cost savings have been benchmarked against other comparable projects, for example where:

- Using Cost Led Procurement, 'the Rye Harbour team along with their cost consultant, Arcadis, then developed an outturn cost for the project. These costs were benchmarked against the Environment Agency's internal project cost tool (made up of previous projects that have been delivered) and they also benchmarked them against other external projects to ensure a reasonable cost'[1351]
- Using Two Stage Open Book on the Cookham Wood project alliance and 'taking into account value indicators of similar projects, cost savings achieved (as analysed by the client and by cost consultant Sweett Group) show cost savings of 20% from the rate of £2910 per square metre anticipated for a comparable project and the rate of £2332 per square metre achieved in relation to Cookham Wood by the time of establishing the agreed maximum price'[1352]
- 'SCMG targeted achievement of substantial cost savings benchmarked against comparable costs incurred by Hackney Homes working with a range of four contractors on its previous Phase II programme of comparable works and using 2009/2010 rates adjusted to reflect inflation. The cost savings achieved include 16.5% by Hackney on its 2010 framework/alliance procurement, plus further savings averaging 14% achieved by Homes for Haringey and Hackney Homes through the application of the SCMG processes from 2010 to 2013 …. In addition, the SCMG Annual Review completed

---

1348 Each referred to in the relevant Trial Project case study.
1349 University of Reading (2018), 37, 38.
1350 University of Reading (2018), 1.
1351 Rye Harbour Trial Project case study.
1352 Cookham Wood Trial Project case study.

March 2014 demonstrated an average price rise in SCMG rates of 1% which is substantially below the tender price inflation forecasts provided by a range of five independent consultants'.[1353]

Trial Project results have shown how savings can be enhanced by learning from project to project and by Supply Chain Collaboration. For example, on the Archbishop Beck Trial Project 'taking into account value indicators for similar projects, cost savings achieved results of 26% from a rate of £1950 per square metre anticipated for a comparable project, to a rate of £1438 per square metre achieved in relation to Archbishop Beck by the time of establishing an agreed price. The means by which these savings were achieved include:

- 'Lessons learned from the Notre Dame School project
- Joint working by Willmott Dixon with Tier 2/3 Subcontractors and Suppliers in developing innovative proposals at the point of selecting Tier 2/3 Subcontractors and Suppliers
- Further joint working by Willmott Dixon with its Tier 2/3 Subcontractors and Suppliers and with Liverpool City Council throughout the contractual preconstruction phase'.[1354]

Certain Trial Projects showed additional savings arising from the process used for the selection of main contractors, and these cost savings may be attributable in part to the state of the market place at the time when bids are invited.[1355] However, they may also be attributable in part to improved competitive processes by which the alliance members are selected, including the confidence created by clear objectives, success measures, targets and incentives. For example, the Ministry of Justice (MoJ) framework alliance was reported to have achieved 'reduced operating costs estimated at £10 million, reduced burden on industry tendering of around £30 million and procurement risk mitigation of about £2 million'.[1356]

Bidders for a framework alliance can save significant costs as a result of the reduced time spent by their bid managers and estimators. They do not have to create prices for all work, service and supply packages and they can, for example, adopt strategic supply chain arrangements already in place as a result of earlier Supply Chain Collaboration. The SCMG Trial Project reported that analysis by a main contractor engaged on the alliance showed how 'bid costs on a procurement supported by Two Stage Open Book under SCMG were far lower than the bid costs they incurred under a comparable single-stage procurement, specifically £719 per £1 million of turnover (under SCMG) as against £4808 per £1 million of turnover (under the comparable traditional bid)'.[1357]

The potential for agreed cost savings is enhanced by long-term mutual commitments, for example where:

- 'British Gas formed alliances with lead partners and members of their onward supply chain in order to deliver their brownfield investment programme. The long-term

---

1353 SCMG Trial Project case study.
1354 Archbishop Beck Trial Project case study.
1355 For example, prior to commencing Supply Chain Collaboration, 'Surrey achieved savings of 16% at the point of selecting Kier in 2011, against prices previously paid for comparable works'. Project Horizon Trial Project case study.
1356 Effectiveness of Frameworks (2012), Case Study, 103–4.
1357 SCMG Trial Project case study.

commitment from each partner to the alliance has seen a turnaround in performance and payback on their investments. This was not immediate, with a long-term view being essential to unlock the considerable performance benefits'

- 'Anglian Water's @one Alliance has delivered complex programmes of projects in the Asset Management Plan 4 (AMP4) and AMP5 regulatory periods, consistently delivering significant efficiency savings. The AMP5 programme of approximately £1bn is on track to deliver ahead of its target saving'
- 'Network Rail projects, including the Reading Capacity programme have been challenged to deliver early and to make significant savings on target cost. An alliance with lead partners and wider stakeholder groups delivered Reading more than a year early and beat the £50m saving target'
- 'The Highways Agency brought their delivery partners together to increase the visibility of their performance and enhance the joint development of solutions through their behaviour focused strategic change programme. Across all its alliances the Highways Agency has saved £1bn'.[1358]

The early users of Framework Alliance Contract (FAC-1) reported significant cost savings, for example 9.3% recorded by Futures Housing Group.[1359] Cost savings have also been achieved through the FAC-1 supply chain alliance created and led by Kier Highways where 'the immediate benefit of the re-procurement exercise is a saving of circa 8% against prices under the previous contract model'.[1360]

15.4   How Can Collaborative Procurement Improve Quality?

Collaborative procurement establishes new lines of communication between the client, consultants, contractors, subcontractors, suppliers, manufacturers and operators. These in turn create opportunities for improved integration, information and innovation which can validate design proposals, ensure budgetary compliance, develop improved design solutions and encourage extended warranties.[1361] For example, the Archbishop Beck Trial Project reported that 'additional benefits achieved through Two Stage Open Book include:

- Appointment of D. Morgan Plc from the city region to provide groundwork and offer innovations through reduced drainage runs and size of attenuation units, also saving cost and time by installing precast ground beams sourced separately by Willmott Dixon. D. Morgan Plc has subsequently opened a Liverpool office
- Assumption of a glulam approach enabled time savings in the Sheppard Robson design process, and early work with B.&K. Ltd as preferred subcontractor enabled the team to design out elements of work that had previously caused issues on site on the Notre Dame School project, improving both the sequencing and health and safety through lessons learned

---

1358 Infrastructure Client Group (2014), 2, 3.
1359 Case study summarised in Section 10.10.
1360 Case study summarised in Section 11.10.
1361 The SCMG housing framework alliance obtained 'Availability of extended warranties above industry standards, managed by suppliers/installers, such as windows warranted for 30 years'. SCMG Trial Project case study.

- Refinement of glulam, steel and crossbeam solution by engaging B.&K. Ltd to undertake design, fabrication and erection with consequent time savings on progressive erection of the superstructure
- Parallel discussions run by Willmott Dixon with the steel-frame fabricators and suppliers (Metsec and Cara Brickwork Ltd), also facilitating direct discussions between them, so as to ensure the optimum solution ahead of formal appointments. Cara Brickwork Ltd have subsequently opened a Liverpool office
- New design solutions offered by Mouchel including a flat soffit to the slabs so that services could travel along the underside of the slabs without obstruction
- Early engagement with a Liverpool-based distributor of bricks so as to ensure compliance with the agreed specification without the risk of increased costs
- Close liaison with finishing contractors and suppliers such as Combined Catering Kitchens, Crown Stage Lighting and Sangwin FFE so as to facilitate end user engagement and achieve greater design certainty and avoidance of escalating costs.'[1362]

Supply Chain Collaboration can improve design quality while at the same time reducing consultant risk as a result of the validation of consultant design solutions that can be achieved through joint reviews with specialist subcontractors and manufacturers. For example:

- On the Rogate House project alliance, 'the collaborative approach to design also allowed agreement of aesthetic improvements such as external metal balconies which were designed and installed in collaboration with the balcony supplier for a cost less than that incurred at Alma House'[1363]
- The Cookham Wood Trial Project reported that 'to address the need for lighting to be indestructible in the Education block, EMCOR provided a bespoke alternative lighting proposal. This solution reduced the amount of trunking and ensured that lighting was located in the centre of the rooms, achieving a significant cost saving and a better lighting solution'[1364]
- The ConnectPlus M25 Trial Project reported how 'improvements to the design were also identified that reduced whole life costs and improved the long-term reliability of the works. Such improvements included rationalisation of bolt sizes, improvements to the noise damping systems and additional packing to base plates'[1365]
- It was reported on a PPC2000 project alliance for design and construction of 'Living Coasts' at Paignton Zoo, Devon, built by Dean & Dyball with design input from waterproofing specialist Tilbury Douglas, how they built 'an estuary on the roof of a restaurant' and 'a cliff on which auks will nest'. The £7 million attraction is home to penguins, puffins, seals and otters, with animal tanks comprising 'a very complicated concrete structure' for which 'the design was under development as the contractor was agreeing a maximum price for the project under the PPC partnering contract'[1366]
- The Core alliance project won an 'Excellence in Collaborative Engineered Design' award, with a key contribution from Swiss contractor Haring who supplied unique glulam beams for the Fibonacci spiral design, and implemented value engineering

---

1362 Archbishop Beck Trial Project case study.
1363 Association of Consultant Architects (2010), 35.
1364 Cookham Wood Trial Project case study.
1365 Connect Plus Trial Project case study.
1366 Pearson, A. (2003), 57.

with the McAlpine joint venture and architect Nicholas Grimshaw. The award judges observed that 'collaborative teamwork has been established over a long period of time at the Eden Project. Special mention should be made of how well Haring integrated into this environment'.[1367] The project also won an award for 'best structural innovation', and it was noted that 'the 100ft span of the glulam timber roof was cut by computer in the form of the equiangular spiral found throughout nature, from sunflower seeds to galaxies. This was combined with the latest environmental technologies to make one of the most complex and sustainable structures ever built in Britain'.[1368]

Under a framework alliance, it can also be agreed that qualitative project improvements will be adopted on future projects. For example, on the Ministry of Justice framework alliance, 'over the course of successive projects, framework contractors build up designs and solutions that they know will meet MoJ specifications, and are able to transfer these from one project type to another'.[1369]

15.5 How Can Collaborative Procurement Improve Supply Chain Relationships?

Supply Chain Collaboration, particularly under a framework alliance or term alliance, enables subcontracted supply chain members to establish how they can earn longer-term, larger-scale supply chain contracts and how they can agree improved commitments and working practices. New opportunities and joint processes can improve efficient working by supply chain members and enable them to offer reduced costs, improved warranties and other improved value.

The Trial Projects have demonstrated how Supply Chain Collaboration can create opportunities for all supply chain members to influence design, cost, risk and programming through joint work with each other and with clients, consultants and main contractors during the preconstruction phase of a project or programme of work. For example:

- The Futures Housing Group FAC-1 framework alliance reported that 'the new contract combined with the latest procurement rules has enabled SMEs to tender for parts of bigger overall frameworks with Futures, creating healthier competition in the market and giving the organisation more resources to draw on. The approach is designed to be easy to understand and to help improve legal compliance as well as better value for money'[1370]
- The Property Services Cluster framework reported that 'during the pre-construction period the design was challenged by the project team during design team meetings where the planners, estimators and entire project team identified alternatives that would potentially save time and/or money. The supply chain also offered better value options including significant savings'[1371]

1367 Building Magazine, 22 March 2007, 27 and Mosey, D. (2009), Case Study, 257–9.
1368 Building Magazine, 20 October 2006.
1369 Cookham Wood Trial Project case study.
1370 Supply Management (2016).
1371 Property Service Cluster Trial Project case study.

- Hackney Homes and Homes for Haringey reported that on their housing framework alliance 'new lines of client contact [were] established with tier 2 and tier 3 supply chain members at an early stage in the preconstruction process so that they can make maximum contributions to design, resident consultation, surveying and programming and can work in conjunction with the client and tier 1 contractors'.[1372]

The Trial Projects have also demonstrated how Supply Chain Collaboration strengthens and improves commercial relationships between main contractors and their subcontracted supply chain members where a framework alliance or term alliance supports a pipeline of work, through:

- The improved likelihood of a successful subcontract bid where the main contractor has already been conditionally appointed, so that there is a significant chance of success (say one in 3), compared to a process where bids are made to a main contractor that is itself still bidding to the client (say as one of 6), with a much reduced chance of success (say one in 18)
- A better understanding of the project/programme of work through direct dialogue with the client(s) and main contractor(s) and an opportunity to achieve competitive advantage by demonstrating proposals for improved design/risk management/programming, and additional benefits such as extended warranties and employment/skills commitments
- The opportunity for subcontracted supply chain members to win larger amounts of work for longer periods than was originally anticipated
- The opportunity in preconstruction phase discussions for subcontracted supply chain members to influence directly the approach taken by the client(s) and/or main contractor(s) to any aspect of the project or programme, so as to improve efficiency and reduce risk in delivery of each subcontracted service/work/supply package.[1373]

The Trial Projects revealed that obstacles to Supply Chain Collaboration may arise where:

- The supply chain management teams of main contractors resist what they see as interference in their private commercial arrangements
- Main contractors hesitate to undertake joint reviews with clients and suppliers because they suspect that clients are not actually committed to a collaborative approach
- Subcontractors and suppliers hesitate to engage in a review system that involves clients because they fear that main contractors are not actually committed to a collaborative approach.

Supply Chain Collaboration enables clients and integrated teams to create a clear contractual path through these obstacles that complies with public procurement regulations[1374] and that does not compromise or undermine any team member's reasonable commercial interests.

1372 SCMG Trial Project case study.
1373 Two Stage Open Book and Supply Chain Collaboration Guidance (2014), 15.
1374 As explained in Section 6.8.

15.6 How Can Collaborative Procurement Create Local and Regional Opportunities?

Local and regional businesses can be selected direct by a client using its own selection processes, for example by sub-dividing a project into specialist packages through collaborative construction management.[1375] It is possible to create a tender that includes 'measures attractive to SMEs but open to all', with features that can benefit small businesses such as 'cashflow easing features that included the provision of key materials in certain lots on a free issue basis, such as heating installations, bathroom installations and electrical works' and 'proposals to have embedded payment cards with the contactors, enabling them to claim payment immediately on agreement of final account'.[1376] As a result of this approach on the Futures Housing Group FAC-1 programme:

- 'Of the 23 contractors invited to join the Framework, only two are non-SME'
- 'The average size of the other companies is less than 25 employees'.[1377]

Where local businesses are not appointed directly by the client, new local opportunities for subcontractors, suppliers, manufacturers and operators can be created using Supply Chain Collaboration systems to ensure that selection by main contractors takes account of the particular benefits that local and regional businesses may offer in terms of cost, quality, sustainability and other relevant factors. This offers a major breakthrough for central and local government and for other public sector clients who wish to support the local and regional economy without infringing public procurement regulations. Trial Project case studies demonstrate how these systems have worked and how social benefits can be obtained:

- Hackney Homes and Homes for Haringey reported that the SCMG housing framework alliance 'specifically targeted the development of opportunities for local tier 2/3 subcontractors and suppliers, building up a pipeline of work… across 30 different disciplines'. They described how:
  - 'The use of local resources and labour… helps break the barriers sometimes felt between local residents and the client or constructor'
  - 'The SCMG systems have demonstrated a breakthrough in enabling public sector clients to deal directly with key subcontractors and suppliers so as to ensure they build up fully integrated working relationships'
  - 'A multi-client, multi-contractor team has engaged with a wide range of SME subcontractors and suppliers under a standardised system'.[1378]
- The Archbishop Beck Trial Project reported:
  - 'Achievement of 60% local spend with Liverpool businesses by Willmott Dixon (compared to 50% target on Notre Dame)'
  - 'Direct liaison, through the City-wide Liverpool City Council scheme, with local supply chain members so as to establish clear commitments in relation to apprentices'.[1379]

1375 As illustrated in the case studies in Section 7.10.
1376 Futures Housing Group case study.
1377 Futures Housing Group case study.
1378 SCMG Trial Project case study.
1379 Archbishop Beck Trial Project case study.

- On the Ministry of Justice framework alliance:
  - 'There are more than 200 suppliers registered in the supply-chains of the Alliance suppliers'
  - 'The MoJ has the ability to influence the supply-chain to incorporate local suppliers and SMEs'.[1380]
- The North Wales Prison Trial Project reported £38.2 million spent on local businesses within a 50-mile radius of the site (against a target of £30 million) and £82.7 million spent on small medium enterprises (against a target of £50 million),[1381] including:
  - 'Steelwork (Hayley Steel, plus steel erection awarded to a local erection contractor)
  - Ground-works (Jones Brothers from Ruthin and Jennings from Colwyn Bay appointed as Specialist subcontractors, plus concrete orders placed with Wrexham and Chester suppliers)
  - Precast vehicles/logistics support and concrete supply (PCS Precast)
  - MEP (Three local subcontractors appointed by Crown House)
  - Roofing (Weatherwise of Deeside)'.[1382]

15.7 How Can Collaborative Procurement Support Employment and Training?

Employment and training opportunities are examples of the social value that can be delivered as part of a construction project. The UK Government has recognised the importance of making 'effective use of public procurement to encourage skills development in construction supply chains'.[1383] Collaborative construction procurement can create and deliver commitments to improve employment and skills opportunities throughout the supply chain[1384] and can use detailed and proportionate benchmarks.[1385] For example:

- On the Whitefriars framework alliance 'establishment of a steady volume of work enabled both constructors to operate using a stable workforce and to increase their efficiency on site…The client, with both constructors and in partnership with Mowlem, established the Whitefriars Housing Plus Agency which secured training opportunities for 38 people in the first year and a total of over 200 during the programme as a whole'.[1386]
- The Maidstone term alliance reported how 'Working in partnership with Mid-Kent College, VINCI Facilities developed a training scheme that equipped a core team with

---

1380 Effectiveness of Frameworks (2012), 103.
1381 North Wales Prison case study and in-house Ministry of Justice publication created jointly by the Welsh Government, Ministry of Justice, Wrexham County Borough Council, Lendlease and AECOM.
1382 North Wales Prison Trial Project case study.
1383 BEIS (2018) where the Construction Sector Deal proposes to 'Continue to use public sector procurement to drive investment in the skills needed to support modernisation and industry-led innovation'.
1384 Glasgow Housing Association reported that contractual 'support of employment and skills outputs led to exceptional number of apprenticeships, recognised in the Scottish Government's report on community benefits', Association of Consultant Architects (2010), 18.
1385 Such as those set out in the Construction Industry Training Board *Client-Based Approach*.
1386 Association of Consultant Architects (2010), 46.

the necessary skills to finish all kitchen and bathroom works on any given property and:
- eliminated the need for co-ordination of multiple trades
- minimised disruption to residents
- reduced completion times from 3 weeks to 10 days'.

They noted that 'a similar partnership with North Kent Construction Skills allowed VINCI Facilities to extend its training outside the business by offering extensive work experience placements, a number of which have now developed into apprenticeships'[1387]

- Hackney Homes and Homes for Haringey used the SCMG housing framework alliance to create 'additional employment and skills opportunities for individuals, for example 46 new apprenticeships over the first 18 months of the Hackney programme', plus 'establishment of the Building Lives Training Academy where apprentices who have got NVQ Level 1 are engaged by constructors/specialists according to demand of ongoing work so as to achieve NVQ Level 2 after 15/18 months'[1388]
- The Project Horizon highways term alliance recorded 'improved apprentice commitments by Kier, Marshall Surfacing and Aggregate Industries, supplementing employment and skills commitments already agreed by Surrey and Kier under an employment and skills plan in accordance with the CITB Client-Based Approach'[1389]
- The Southern Construction Framework ('SCF') reported that, by using Two Stage Open Book, their employment and skills strategy has seen '62 new general and trade apprentices appointed across the SCF region'[1390]
- The North Wales Prison Trial Project reported 100 apprenticeships over the life of the project (as per their target), plus 2150 work experience days (against a 1820 days target), with 66% of the workforce employed from within a 50 mile radius of the site or within a one hour journey (against a target of 50%).[1391] This project was 'granted CITB Skills Academy status from September 2015, being the first Skills Academy in North Wales and the first prison project to have this status'.[1392]

15.8 How Can Collaborative Procurement Support Improved Safety?

Concerns as to the safety of residential properties were highlighted starkly by the Grenfell Tower disaster in 2017 and led to far-reaching recommendations in an independent review by Dame Judith Hackitt.[1393] Fragmented, lowest price procurement practices revealed how, even in the context of health and safety:

- 'The primary motivation is to do things as quickly and cheaply as possible…using the ambiguity of regulations and guidance to game the system'

1387 Association of Consultant Architects (2010), 32.
1388 SCMG Trial Project case study.
1389 Project Horizon Trial Project case study.
1390 Southern Construction Framework Report (2016).
1391 North Wales Prison case study and in-house Ministry of Justice publication created jointly by the Welsh Government, Ministry of Justice, Wrexham County Borough Council, Lendlease and Aecom.
1392 North Wales Prison Trial Project case study.
1393 Building a Safer Future (2018).

- This is 'a cultural issue across the sector, which can be described as a race to the bottom'[1394]
- As a consequence, 'inadequate specification, focus on low cost or adversarial contracting, can make it difficult (and, most likely, more expensive) to produce a safe building'.[1395]

The safe use and occupation of completed projects should be of paramount importance. It is the role of Government to create legal minimum standards of safety and to impose penalties for non-compliance.[1396] However, there is no justification for the continued use of procurement and contracting models that perpetuate the practices criticised in Dame Judith's report. Instead, in order for a client and its team to improve on minimum standards, they should:

- 'Establish procurement processes that allow sufficient time, resources and prioritisation to deliver the core objectives'
- 'Identify how core building safety requirements will be met in the pre-construction phase'
- 'Ensure that information management systems are properly updated and change control mechanisms are utilised'.[1397]

The recommendations in 'Building a Safer Future' include 'tackling poor procurement practices…to drive the right behaviours to make sure that high-safety, low-risk options are prioritised and full life cycle cost is considered when a building is procured',[1398] for which purpose:

- 'The invitation to tender and the bid process must prioritise building safety, and balance the upfront capital cost against quality and effectiveness'[1399]
- Improved resident involvement should enable them to 'play a key part in local and strategic decisions'[1400]
- Building Information Modelling (BIM) should be used to 'ensure accuracy and quality of design and construction'.[1401]

'Building a Safer Future' underlined the role of contractual commitments in driving a change in culture and urged procurers to think carefully about their contracts when establishing 'how to deliver a building with long-term integrity, and the people, products and processes required to do that'.[1402] The concerns and recommendations in 'Building a Safer Future' can all be addressed through the commitments and relationships established in a collaborative procurement model. Evidence of how collaborative procurement of housing can be delivered safely and efficiently, in ways that involve residents

---

1394 Building a Safer Future (2018), 5.
1395 Building a Safer Future (2018), 109.
1396 As considered in Bell, M. (2018).
1397 Building a Safer Future (2018), 34.
1398 Building a Safer Future (2018), 13.
1399 Building a Safer Future (2018), 109.
1400 Building a Safer Future (2018), 67.
1401 Building a Safer Future (2018), 103 'Having BIM enabled data sets during occupation means that dutyholders will have a suitable evidence base through which to deliver their responsibilities and maintain safety and integrity throughout the lifecycle of a building'.
1402 Building a Safer Future (2018), 108.

and that interrogate the quality of each work and supply package, has been provided by the Rogate House case study at a project level[1403] and the SCMG Trial Project at a strategic level.

An example of how collaborative procurement can contribute to health and safety on site was the Portslade School Improvements project alliance, 'a difficult scheme to build above classrooms where children continued with lessons' where:

- 'Both schools needed extra classrooms, yet neither had land for more buildings', and 'existing classrooms were flat roofed, so new first floor extensions could be erected overhead without disturbing the ground floor structure and its inhabitants'
- 'Head teachers became part-time members of the partnering team' and the agreed solution was 'a steel frame with precast concrete plank floor…erected during the summer holiday at both schools creating a safe working platform ready for the autumn term'
- 'That left two full terms over the winter to get the new floors constructed and the whole buildings clad and ready for use by the Easter holiday 2003'.[1404]

15.9   How Can Collaborative Procurement Support Environmental Benefits?

UK company directors have a statutory duty to have regard to 'the impact of the company's operations on the community and the environment',[1405] and there is a pressing need for approaches to procurement that fulfil this duty. For example, the search for ways to maximise energy efficiency and reduce waste can benefit from ideas developed not only by design consultants but also by contractors, subcontractors, manufacturers and operators. In order to evaluate and utilise proposals for sustainability initiatives, it is essential to engage with supply chain members during the planning and preconstruction phase of the project. By inviting these proposals at a time when they can be analysed, costed, agreed and adopted, an alliance can deliver significant reductions in environmental impact.

Relevant contributions could include:

- Proposals as to the most buildable and least wasteful interpretation of consultant designs[1406]
- Proposals in respect of reduced waste and increased recycling[1407]
- Proposals as to efficient use of energy on site, including modern methods of construction such as off-site fabrication[1408]
- Proposals as to efficient use of energy by reduced maintenance and repair in the operation of the built facility.[1409]

1403 Summarised in Section 3.10.
1404 Association of Consultant Architects (2010), 37.
1405 Companies Act 2006, Section 172.1(d).
1406 As illustrated in the Project Horizon, North Wales Prison and Archbishop Beck Trial Project case studies.
1407 As illustrated in the Project Horizon and National Change Agent case studies.
1408 As illustrated in the North Wales Prison and Cookham Wood Trial Project case studies.
1409 As illustrated in the Cookham Wood and SCMG Trial Project case studies.

The alliance team engaged on the £30 million new build Hackney Academy 'focused on improved sustainability and achieved a BREEAM Schools (2006) assessment rating of very good (2 points short of excellent). The project has a ground source heat pump and photovoltaic panels and also makes maximum use of natural daylight and ventilation'. The team reported 'a strong partnering ethos through client leadership...by the City of London Corporation, supporting the team in meeting significant time and cost challenges', 'maximum engagement with stakeholders, including staff and pupils, at all stages of the project', and 'collective team contributions to sustainability improvements, with maximum use of natural light and energy saving techniques'.[1410]

A range of environmental benefits achieved on the Rye Harbour Trial Project are described in Section 12.5. On the North Wales Prison Trial Project:

- The team challenged 'the original costed design to incorporate an open swale in place of an attenuation tank, thereby also creating a new environment for wildlife while reducing construction costs'
- 'In response to the emerging risks created by discovery of asbestos on site which is a MoJ risk, the team worked with Wrexham County Borough Council and the Health and Safety Executive to permit asbestos to be mitigated on site. This resulted in the MoJ avoiding costs of up to £4 million for moving the materials off-site'
- 'A local landscape contractor James Wright provided early services to assist in meeting Wrexham's planning conditions in relation to waste disposal, working with Waste and Resources Action Programme (WRAP) to introduce locally generated green waste solutions that enhanced soil on site and considerably reduced vehicle movements. As a result, James Wright were awarded a significant package contract for landscaping'.[1411]

The Anglian Water @one alliance 'effectively used alliances and collaboration to enable innovation' when it 'was challenged to deliver stretching efficiency and carbon targets – the 50% reduction in embodied carbon being particularly challenging...The @one Alliance experience on carbon demonstrated that collaborative and integrated teams have pooled their combined expertise and their broader partner capability to deliver innovative solutions and have been driven to meet what at first sight looked an unlikely target.'[1412]

The benefits of collaborative procurement are underlined by the conflicting pressures of lowest price and improved value through innovation that are clearly evident in the arena of sustainability. If single-stage bidders are expected to put forward speculative proposals, they may hold back or compromise good ideas in order to reduce their bid prices. In a single-stage bid, there is also the increased risk that clients may reject more

---

1410 Association of Consultant Architects (2010), 20: The project won the Excellence in Building Schools for the Future Award 2009 for 'Innovation and Student Engagement'. The pupils described it as 'the best school ever', and it also won the Constructing Excellence 2010 London and South East Award for 'Integration and Collaborative Working'.
1411 North Wales Prison Trial Project case study, 3, 4.
1412 Infrastructure Client Group (2014), 13: 'As an example, the expansion of the Bedford Water Recycling centre would traditionally have been delivered with extra capacity being built on new sections of the site. However, these would have failed the carbon and cost targets. Following collaborations across the home organisations and the supply chain and close joint development of the solution with the operational team, two new processes were retrofitted to existing civil assets. This led to reduced costs and carbon and a solution the operations team were able to manage and maintain from day one'.

sustainable proposals as unaffordable or unbuildable without the opportunity to investigate them in detail.

By contrast, joint working through early contractor involvement using Supply Chain Collaboration enables the cost and quality benefits of sustainability proposals to be thoroughly developed and assessed by all alliance members, providing clients with 'the means to evaluate the cost of environmental issues…and to balance this against their demonstrable benefits'.[1413] The new lines of communication and the additional time created for joint working on the Trial Projects have led to team members offering more sustainable solutions that are practical and affordable within the client's budget. For example:

- The SCMG multi-contractor housing framework alliance reported 'subcontractor/supplier innovations in proposed new materials and development of specifications, such as future-proofing green roofs at no additional cost and upgrading windows from Grade C to Grade A at no additional cost'[1414]
- The SCMG alliance also reported that supply chain members offered 'improved repairs and maintenance through, for example, self-cleaning glass on high rise blocks'[1415]
- The Horizon highways term alliance reported joint working that delivered 'innovation through collaborative working, for example to increase recycling and reduce landfill'.[1416]

On the A30 Bodmin/Indian Queens new build road project:

- The early appointment of the main contractor enabled it to schedule construction activities so that material was used to the maximum extent on site with very little going to landfill
- In addition, the main contractor was able to propose a number of environmental measures including reunifying the two halves of Goss Moor (a National Nature Reserve) previously divided by the old A30, by diverting the route of the new dual carriageway to the north and then degrading the old A30 back to its sub-base
- While establishing informal agreements with landowners in advance of the required public inquiry, the main contractor arranged for reptile fences to be put up early, so that the relocation of snakes and lizards could proceed in an orderly manner. This was commenced in early spring, as relocation is more difficult during the summer months, and was completed so as to avoid up to six months' slippage in the construction phase
- Finally, the early appointment of the main contractor enabled it to establish its supply chain arrangements early in the design process, so that designs could be developed to suit locally available materials and locally available skills.[1417]

Environmental initiatives can be combined with cost savings and with support for small businesses, particularly where the scale of the programme enables consistent procurement practices and where collaborative systems facilitate exchange of ideas. For

1413 Housing Forum (2010), 10.
1414 SCMG Trial Project case study.
1415 SCMG Trial Project case study.
1416 Project Horizon Trial Project case study.
1417 Mosey, D. (2009), Case Study, 244–6.

example, on a five-year, multi-client programme, the regional National Change Agent consortia achieved:

- 'Efficiency savings totalling £226 million from cumulative expenditure of £1.6 billion'
- 'Over 500 apprentices successfully completing NVQ training to levels 2 and 3 and helped into full employment, with 80% retention'
- 'Establishment of numerous SME businesses and social enterprises'
- 'A joint initiative with WRAP (Waste and Resources Action Programme) to halve waste to landfill'.[1418]

15.10   Benefits of Two Stage Open Book and Supply Chain Collaboration

The improved economic and social value achieved by Trial Projects teams using Two Stage Open Book and Supply Chain Collaboration is summarised[1419] as follows:

- **Cost Savings** – Of up to **20%**, transparently agreed without eroding margins
- **Cost Competition and Control** – Robust, competitive processes to select consultants, tier 1 contractor(s) and tier 2/3 subcontractors and suppliers, for the whole project/programme of work and for each work/supply package, with ongoing cost controls and regular project budget reconciliations
- **Improved Design** – Systematic evaluation and early incorporation of innovations and other design proposals from tier 1 contractor(s) and tier 2/3 subcontractors and suppliers and a natural fit with the objectives, relationships and processes that underpin Building Information Modelling
- **Risk Management** – Preconstruction phase contributions by all integrated team members to identify and reduce/eliminate risks
- **Time Management** – Agreement of clear, binding deadlines at all stages, and new opportunities for tier 1 contractors and tier 2/3 subcontractors and suppliers to propose time savings
- **Extended Warranties** – Early evaluation of extended warranties offered by prospective tier 2/3 subcontractors and suppliers and a natural fit with Government Soft Landings
- **Sustainable Solutions** – Early evaluation of more sustainable materials and working methods proposed by prospective tier 2/3 subcontractors and suppliers
- **Stakeholder Consultation** – Increased opportunities to consult end-users and other third parties, and to take their views into account
- **Appointment of SMEs and Local/Regional Businesses** – A unique opportunity for joint assessment by the client(s) and tier 1 contractor(s) of particular benefits offered by small and medium enterprises and local/regional businesses in relation to particular work/supply packages
- **Employment and Skills Commitments** – The ability to measure and improve the agreed employment and skills commitments of tier 1 contractors and tier 2/3 subcontractors and suppliers.

1418 Association of Consultant Architects (2010), 34.
1419 Two Stage Open Book and Supply Chain Collaboration Guidance (2014), 5.

16

How Is Collaborative Procurement Costed and Incentivised?

16.1 Overview

Creating a fair and transparent basis for costing and incentivisation is central to collaborative construction procurement. Team members can build up agreed costs through the selection processes considered in Chapter 6 and through preconstruction phase cost analysis that enables them to agree all or any of the following:

- A fixed price or target price
- Rates for units of work
- Categories of actual cost incurred by team members.

Team members can commit to prices wholly or partly before or after commencement of construction, and these prices can be subject to adjustments that take account of cost inflation, that take account of agreed provisional sum items and agreed changes, and that take account of claims for unforeseeable events. There are three main options for costing any project:[1420]

- Lump sum fixed prices, where supporting data is set out in bills of quantities and/or schedules of rates but is not used as a basis for adjusting the price of the works in the absence of agreed changes[1421]
- Measure and value, where prices are adjusted by remeasurement according to actual quantities using bills of quantities or schedules of rates which refer only to estimated quantities of any part of the works[1422]
- Cost reimbursement where prices are calculated according to the costs actually incurred by the contractor plus a fee.[1423]

This chapter will explore how collaborative construction procurement can use any cost model according to the features and circumstances of the project or programme of work,[1424] and why the separate agreement of profit, fees, overheads and other costs

1420 German cost models offer an interesting comparison and are summarised in Section 23.1.
1421 For example, JCT SBC/Q (2016) With Quantities, JCT SBC/AQ (2016) With Approximate Quantities and JCT SBC/XQ (2016) Without Quantities; NEC4 (2017) Option A: Priced contract with activity schedule, NEC4 (2017) Option B: Priced contract with bill of quantities; PPC2000 (2013) two-stage build-up of 'Agreed Maximum Price'.
1422 Also known as a 'remeasurement contract'. For example, the ICC Measurement Version (2011).
1423 Also known as 'cost-plus contracts'. For example, under NEC4 (2017) Option E: Cost reimbursable contract.
1424 'Two Stage Open Book…can be implemented in conjunction with any cost model, including fixed prices, target costs, schedules of rates and bills of quantities. The choice of cost model should be governed by

Collaborative Construction Procurement and Improved Value, First Edition. David Mosey.
© 2019 John Wiley & Sons Ltd. Published 2019 by John Wiley & Sons Ltd.

should be seen as a fundamental principle if team members intend to seek cost savings and improved arrangements with subcontracted supply chain members. Although some alliance models are based on cost reimbursement or target costs linked to shared risk and reward incentives, this chapter will also consider how other alliance models can incentivise the agreement of fixed costs following conclusion of agreed preconstruction phase activities.

It is not only the chosen means of expressing the agreed costs that defines collaborative working but also the agreed means of establishing those costs. Whether or not a team uses the build-up of agreed costs to establish a fixed price before start on site, there will be a series of complex interactions between the client, the project cost consultant, the other consultants, the main contractor, subcontractors, suppliers, manufacturers and operators that lead to the costing of each package of works, services and supplies. For these interactions to build up an accurate picture of project costs requires processes that use the machinery of an enterprise contract as considered in Chapter 8.

Whether a team uses collaborative procurement to agree a fixed price or target costs or cost reimbursement, it will need to decide:

- The extent to which elements of a project can or should be sufficiently designed at the point of contractor selection to be costed as part of contractor evaluation, and/or so as to enable the provisional selection of preferred supply chain members at the point of contractor selection
- The extent, if any, to which certain work, service and supply packages can or should be suitable for a contractor to develop a business case for self-delivery
- The breakdown of fees, profit and overheads at the point of contractor selection so as to provide evidence of value attributable to those fees, profit and overheads, and so as to enable, for example, finalisation of overheads to reflect new information emerging from design development and from the selection of subcontracted supply chain members
- The basis for sub-dividing work, service and supply packages and the extent to which it is valuable to agree the fees, profit and overheads payable to subcontracted supply chain members
- The extent to which subcontracted supply chain members can provisionally be approved in line with contractor recommendations, and the activities required to create a business case that finalises accurate and competitive costs for their respective work, service and supply packages during the preconstruction phase
- The timing for obtaining prices for all other work, service and supply packages from prospective supply chain members, and the basis for establishing accurate and competitive costs
- The extent, if any, to which the cost of certain work, service and supply packages should be treated as provisional sums, and should be finalised by a contractor obtaining prices from prospective supply chain members after commencement of the construction phase.[1425]

The chapter will consider options for costing and incentivisation by reference to budgets, fixed prices, targets and actual costs. It will consider the costing and incentivisation

the features and circumstances of each project and programme of work', Two Stage Open Book and Supply Chain Collaboration Guidance (2014), 35, 36.

1425 Two Stage Open Book and Supply Chain Collaboration Guidance (2014), 37.

of individual projects, framework alliances and term alliances and the importance of using Supply Chain Collaboration processes in order to agree cost savings.

16.2 What Is the Impact of Open Book Costing?

The agreement of accurate costs is dependent on the transparent sharing of available data. For example, the client should not withhold information as to obstacles that a contractor may encounter on site, and contractors should not withhold information as to the basis for calculating their risk contingencies in respect of possible obstacles on site.[1426] Hidden discounts and rebates agreed by a contractor with its subcontracted supply chain members are another obstacle to accurate costing. Any discount or rebate should only be permissible if agreed by the client in advance in order 'to ensure that the Client has a clear picture of all costs payable to Specialists and an equally clear picture of the Profit payable to the Constructor'.[1427]

Successful collaborative procurement depends on the chosen cost model identifying separately the fees, profit and overheads payable to a lead consultant or main contractor, so as to establish a transparent basis for the review and agreement of underlying costs.[1428] It may also be possible to extend this open-book model to certain subcontracted supply chain members where they in turn engage sub-sub-contractors for parts of their work. To the extent that costs are distinguished from fees, profit and overheads, collaborative procurement can then ensure that savings do not erode the agreed margins and other income of team members. This also obtains greater control for the client over otherwise invisible benefits that some contractors might expect to obtain through practices such as supplier rebates and prompt payment discounts on amounts that are contractually due to their subcontractors and suppliers.[1429]

The NEC4 cost options include fixed, target and reimbursable costs, and include open-book costing through a 'Schedule of Cost Components' to be used when compiling target costs.[1430] NEC4 includes guidance as to how its cost options are applied, although it provides for early contractor involvement only by reference to two of its six cost options, namely the Option C Target contract with activity schedule[1431] and the Option E cost reimbursable contract.[1432]

PPC2000 defines 'Open-book' as 'involving the declaration of all price components including Profit, Central Office Overheads, Site Overheads and the cost of materials, goods, equipment, work and services, with all and any relevant books of account, correspondence, agreements, orders, invoices, receipts and other relevant documents available for inspection'.[1433]

1426 These issues are explored further in Chapter 18.
1427 PPC2000 Pricing Guide (2008), 2.
1428 Pricing is 'dependent on the chosen cost model identifying separately the fees/ profit/ overheads payable to a Tier 1 Contractor, so as to establish a transparent Open Book basis for the review and agreement of underlying costs', Two Stage Open Book and Supply Chain Collaboration Guidance (2014), 36.
1429 For example, PPC2000 (2013) clause 12.8 prohibits discounts or other benefits payable by any subcontractors or suppliers to the main contractor unless approved in advance by the client.
1430 For example, NEC4 (2017) Option C: Target contract with activity schedule, Schedule of Cost Components, 58–61.
1431 NEC4 (2017) Option C Option X22.
1432 NEC4 (2017) Option E Option X22.
1433 PPC2000 (2013) Appendix 1.

The PPC2000 Introduction to Pricing describes three different ways that team members may agree to apply open-book costing:

- 'During the establishment of the AMP [agreed maximum price]: the early conditional appointment of the Constructor ahead of full design means that, as remaining designs are released, the pricing of those designs can be undertaken on an Open-book basis (under clauses 10.3 and 10.5), whether through proposals put forward by the Constructor as Business Cases or by way of supplementary second-tier tender exercises run by the Constructor with potential Specialist subcontractors, suppliers and sub-consultants'
- 'In respect of provisional sums forming part of the AMP: if for any reason any part of the AMP remains unfixed at the point that a Commencement Agreement is signed authorising start on Site, a further Open-book approach can be applied at any time during the construction phase (under clause 17 and the Project Timetable) in respect of procedures to agree fixed costs for remaining elements of the Project'
- 'On a fully Open-book basis throughout the Project: a third option is to agree (in the Price Framework) that all or part of the AMP will be paid to reimburse actual expenditure by the Constructor whether or not limited or linked to any shared excess or shortfall against a target cost, and in this case the Open-book approach will be used to calculate all payments due'.[1434]

The open-book costing of collaborative procurement should be combined with a selection process that does not tempt bidders to undercut each other on fees, profit and overheads or to prioritise cost savings above other measures of value. Instead, the evaluation criteria used for selecting team members should invite them to demonstrate how their proposed fees, profit and overheads will be deployed in ways that will generate improved value and reduced risks.[1435]

Many alliances have used open-book costing as a new way to build up and agree a fixed price, while others have used it as the basis for a target cost or cost reimbursement. No specific collaborative costing model is compulsory, and for example, on a range of integrated project delivery (IPD) projects, Howard Ashcraft explained that his clients have adopted approximately 35 different commercial models according to their project needs, priorities and preferences.[1436]

16.3 How Can Collaborative Procurement Achieve a Fixed Price?

To build up agreed costs with a common understanding of the factors affecting those costs, and then to establish a point in time when they are translated into a fixed price, offers a major step forward in collaborative procurement and provides all team members with a common understanding of costs that is a world away from the uninformed gambling inherent in lowest price arm's length bids. If established through the collaborative development of cost data and supported by suitable incentives, the agreement of 'a guaranteed maximum price, working to agreed margins with

---

1434 PPC Pricing Guide (2008), 3.
1435 As part of the selection criteria considered in Chapter 6.
1436 KCL conference 'Improving Value Through Alliances, BIM and Collaborative Contracts', 1 March 2018.

full open-book accounting procedures in place' is a model that 'builds trust, helps to overcome the adversarial approach to construction and leads to rapid conflict resolution'.[1437] An informed fixed price can also create 'a high incentive to complete the job as efficiently as possible with high productivity'.[1438]

As illustrated in Chapter 6, a system that establishes a substantiated fixed price for a collaborative project provides a viable alternative to the guesswork by clients, cost consultants and bidders that can undermine single-stage procurement. Instead it uses early appointment of contractors 'to work as part of the team in developing the details of the project and establishing its cost'.[1439] A collaborative team can build up fixed prices that comprise:

- The fees of consultants and the profit and overheads of contractors, preferably agreed as lump sums and subject to increases according to the achievement of agreed targets
- Those cost components that are capable of being priced at the point of main contractor selection, for example where package requirements can be priced accurately by bidders and by their subcontracted supply chain members because their designs and risk assessments have already been tested for their value on other projects
- Those cost components that are to be agreed after main contractor selection, using agreed processes that maximise the contributions of the selected main contractor and its subcontracted supply chain members to establishing the value of proposed designs and risk assessments
- Those cost components that are subject to later equivalent processes during the construction phase.

For example, the Cookham Wood project alliance achieved agreed cost savings of 20% through:

- 'Joint working by Interserve with Tier 2 Subcontractors in developing innovative proposals at the point of selection'
- 'Further joint working by Interserve and its Tier 2 subcontractors with the client and its Consultants throughout the contractual preconstruction appointment'
- 'Additional information provided at the point of selection and throughout the preconstruction phase through the use of BIM'.[1440]

It may be suggested that agreement of a fixed price could discourage continued collaborative working during the construction phase because the risks of working within that price are transferred from the team to the main contractor and its supply chain, leading to defensive steps such as additional risk contingencies. It is recognised, for example, that a main contractor may seek to add last-minute contingencies when finalising a fixed price and may be tempted to maximise the costs of proposed changes during the

---

1437 NAO (2005), 61, Case Example 7 (NHS Estates and Procure 21 – facilitative support for inexperienced clients) offered the example of the Milton Keynes Treatment Centre, where for three months the main contractor worked on a fee basis developing options for the hospital to consider, and then offered a guaranteed maximum price for delivering the approved design, as a result of which a £15 million budget was reduced to a cost of £12 million without comprising user requirements or causing delays.
1438 'In a range of on-site disciplines including steel erection, concrete pouring, piping and wiring, projects using lump-sum contracts rather than cost reimbursable ones had 35 to 88 percent higher productivity', McKinsey Global Institute (2017), 48.
1439 Banwell Report (1964), 10, Section 3.13.
1440 Cookham Wood Trial Project case study.

construction phase. The alliance manager should adopt a robust approach in reminding team members that these tactics are not consistent with their contractual commitments and are not necessary in light of the accurate cost data that they are working with. Provided that alliances are managed fairly and openly using the data and processes agreed, then based on the case studies researched for this book there is no evidence that agreement of an informed fixed price undermines or affects continuing collaborative commitments and behaviours.

For example, on the North Wales Prison project alliance 'Lendlease encountered a major shift in the subcontract market whereby increased workloads and rising prices have led to a scarcity of package contractors willing to make the long-term commitments required by the scale of the project'. In response the team supported Lendlease in:

- 'Recognising the fact of increasing package costs during the preconstruction phase, and utilising Value Engineering measures…in order to reach an Agreed Maximum Price which was fair and robust taking into account Client changes and which will provide value for money'
- 'Approving sub-division of works packages so as to secure the required volume of sub-contract works and supplies, for example: (a) for precast, engaging 8 factories in Germany, Ireland and Belgium as well as England (b) splitting roofing and brickwork each into 2 packages (c) splitting ground-works into 4 packages'.[1441]

If the preconstruction phase costing processes are not set out in a preconstruction phase contract and supported by a timetable, there is the risk that team members will not allocate the time needed for costing their proposals and 'that interests cannot be fully aligned while the client is inclined to take an overly optimistic view of costs and related risks and the contractor is anxious to cover its financial position'.[1442] For example, NEC4 Option X22 does not make clear how costs are built up during the conditional preconstruction phase period and only requires submission of the contractor's forecast of the effect of design proposals on project costs.[1443] Also, NEC4 Option X22 does not enable the agreement of fixed prices because it does not appear in the NEC4 fixed price options.[1444]

16.4　How Do Target Costs and Cost Reimbursement Operate?

Even with the benefit of early contractor involvement, it is not always possible for a team to agree a reliable fixed price, for example where:

- The extent of the materials and works required to complete the project cannot be calculated with complete precision
- Extensive works need to be undertaken underground in conditions that cannot be fully investigated in advance of start on site.[1445]

1441　North Wales Prison Trial Project case study.
1442　Nichols (2007), 32, 33.
1443　NEC4 (2017) Option X22.3(2).
1444　NEC4 Option X22 appears only in Option C (Target cost with activity schedule) and Option E (Cost reimbursable contract).
1445　For example, under the ICC (2011) Measurement Version.

As a consequence, some teams use a measure and value cost model that builds up estimated prices by combining fees, profit and overheads with projected costs based on agreed rates or categories of cost. The projected figures are subject to remeasurement which can be combined with incentives designed to align the commercial interests of team members if and to the extent that fixed costs cannot be calculated accurately in advance.[1446] Measure and value contracts recognise the need to remeasure the extent of the works as they progress and have been typical of many civil engineering projects such as roads, railways, drainage and earth-moving.[1447]

Measure and value contracts can be incentivised by sharing the savings or overruns against target costs. This incentivised approach is used, for example, in the cost model under NEC4 Option C which provides for the main contractor to pay or be paid an agreed share of the difference between the total of:

- The 'Prices', namely the agreed 'lump sum prices for each of the activities on the Activity Schedule' and
- The 'Price for Work Done to Date', namely the 'Defined Cost' comprising the total 'cost of the components in the Schedule of Cost Components less Disallowed Cost'.[1448]

The calculation of what is payable pursuant to a target cost model requires regular analysis of the contractor's books in order to establish the extent of 'Disallowed Cost', including any amounts which are 'not justified by the contractor's accounts and records'.[1449]

Cost reimbursement may be appropriate where the nature as well as the extent of the required work is not clear at the inception of the contract and therefore cannot be priced accurately.[1450] Bills of quantities may be used to show indicative rates or unit prices and indicative volumes of materials and labour, but the sums shown do not represent amounts finally due for performance of the works.[1451] Agreement to reimburse costs incurred under a collaborative contract should be combined with shared pain/gain or other incentives that motivate the contractor to minimise its expenditure.[1452]

Reimbursable costs are limited to actual costs properly incurred, for example as 'Defined Cost' after deduction of 'Disallowed Cost' under the NEC4 Option E (Cost reimbursable contract). Contractors should support their payment applications with evidence of actual costs including invoices, in order to demonstrate that the activities undertaken were appropriate and that the costs were properly and reasonably

---

1446 For example the ICC Measurement Version (2011) can create a system of incentives for the contractor to minimise its expenditure against the agreed rates if the parties also use the ICC Partnering Addendum (2011) P6.
1447 It could be argued that a fixed price lump sum approach is more suited to building projects, particularly those that involve less risk of adjusted quantities arising from risk in the ground.
1448 NEC4 (2017) Option C clauses 11.2(24), 11.2(31), 11.2(32) and 54.
1449 NEC4 (2017) Option C clause 11.2(26).
1450 As illustrated in *Frank W Clifford v Garth* [1956] 1 WLR 570 (CA). 'Cost reimbursable contracts are more likely to be used on large projects with many stakeholders where the time frames – and even the exact form of the final output – may not be fully known when the contract is signed' McKinsey Global Institute (2017), 49.
1451 As considered in *TA Bickerton v North West Metropolitan Regional Hospital Board* [1970] 1 WLR 607 (HL).
1452 As considered in Sections 16.8 and 16.9.

incurred.[1453] The accuracy of cost reimbursement is linked to the amount of detailed cost analysis that precedes approval of payments.

Savings through cost reimbursement can be motivated by including those savings in the criteria that determine the award of further work. This approach was successful in delivering significant savings for Land Securities Trillium and the UK Department of Work and Pensions on their Job Centre Plus programme, where a large number of similar projects were awarded successively to 14 contractors under a multi-party framework alliance, subject to measurement of their performance on previous projects according to criteria that included minimising the costs expended.[1454] However, the use of a cost reimbursement approach on some Australian alliance projects was reported to have led to outturn costs 45–55% higher than the business case estimate, with an average upward adjustment of 5–10% between the initial agreed outturn cost and the final outturn cost.[1455]

Cost reimbursement in the context of collaborative procurement is often linked to a target cost, sometimes developed in stages. An initial project budget can be subject to validation by the team and agreed as an expected cost or early target cost, following which it can be refined to form an agreed target cost or estimated maximum price (EMP) in advance of start on site and then adjusted as appropriate to reflect actual costs incurred through to project completion. For example, the US ConsensusDocs approach to integrated project delivery (IPD) sets out a three-stage approach comprising:

- A 'Validation' phase which continues until the team confirms 'that it is able to commit to design and construct the Owner's Program for an Expected Cost that does not exceed the current Allowable Cost and in a way that meets the other current criteria of the Project Business Case'[1456]
- A preconstruction phase during which a Target Value Design is used as the basis for target value pricing as a 'continuing refinement of the Expected Cost established in the Validation Study and the Target Cost'[1457]
- A construction phase when the Expected Cost can be agreed as an 'Estimated Maximum Price based on the Project requirements [and proposed by the Core Group] when it determines that the Design Documents are sufficiently advanced for that purpose'.[1458]

16.5 How Can Early Contractor Involvement Control Costs?

Collaborative procurement uses a range of techniques to control costs, all of which should be linked wherever possible to a client budget that establishes 'agreed cost limits on subsequent design, risk and procurement activities'.[1459] Cost consultants no longer

---

1453 As considered in Via *Sanantonio Pty Ltd v Walker Corporation* [2009] NSWSC 679.
1454 Job Centre Plus case study summarised in Section 4.10.
1455 Department of Treasury and Finance (2009), ix and xiv.
1456 ConsensusDocs 300 (2016) Article 5.3. The 'Owner's Program' is 'an initial description of the Project and Owner's objectives, including the Allowable Cost, space requirements and relationships, time requirements, flexibility and expandability requirements, special equipment and systems, and Worksite requirements', ConsensusDocs 300 (2016) Article 21.38.
1457 ConsensusDocs 300 (2016) Articles 6.3.2 and 6.6.2.
1458 ConsensusDocs 300 (2016) Article 11.1.
1459 PPC2000 Pricing Guide (2008), 3.

develop only a single set of bills of quantities or schedules of rates for contractor bidders to price. Instead they develop, analyse and manage:

- An appropriate budget based on appropriate cost benchmarks for similar projects
- The systems for bidding and agreeing alliance members' fees, profit, overheads and provisional risk contingencies
- The systems for bidding and agreeing self-delivered work packages and supply chain costs
- The systems for finalising transparent, accurate cost information at each stage in the selection of consultants, of contractors and of other supply chain members
- The systems through which to search for cost savings and the method of assessing the impact of cost savings on other costs.

It can be a precondition 'to commencement of work on Site…that the [alliance] members have achieved an [agreed maximum price] AMP within the Budget subject to any agreed adjustments'.[1460] Guidance states that 'early conditional appointment of the Tier 1 Contractor needs to reflect its ability and commitment to deliver a project/programme of work within the Client's stated financial limits', and that 'it is important for a Client to establish an appropriate Project Budget as a cost ceiling for each project, taking into account the current marketplace and the impact of inflation'.[1461] In creating a budget, benchmarks can 'enable a Client to identify, by reference to its business case and project brief, what a project should cost and whether it can reasonably expect to achieve procurement and delivery of that project within its Project Budget cost ceiling'.[1462]

The RIBA Plan of Work 2013 provided for designers to design according to a 'Project Budget'[1463] and created flexible work stages that recognise how consultants can work with other team members in order to achieve cost savings and other improved value, including through value engineering where appropriate.[1464] Clients may be advised not to declare a budget in the hope that by not sharing this information they are more likely to have the benefit of lower bid prices. On the one hand it is tempting to think that a bid submitted lower than an undeclared budget will be a bonus for the client. On the other hand, a bid that is lower than the budget may not have been thoroughly costed by the bidder and its supply chain and may create cause for concern.

Designing to an agreed project budget is essential to maintain discipline among team members when developing designs, although recent case law suggests that this is not always how a design consultant approaches its brief.[1465] In order to create and maintain effective cost controls throughout a collaborative project, clients and other team members need regular reconciliation of designs and other proposals with the agreed budget. The budget remains the cost ceiling as the team create a more detailed cost plan that establishes agreed costs and prices, allowing the opportunity for redesign or other adjustments if necessary to avoid cost overruns on specific works, services and supply packages.

1460 PPC2000 Pricing Guide (2008), 2.
1461 Two Stage Open Book and Supply Chain Collaboration Guidance (2014), 38.
1462 Two Stage Open Book and Supply Chain Collaboration Guidance (2014), 33.
1463 RIBA (2013) Stage 1.
1464 RIBA (2013) Stages 2, 3, 4.
1465 In *Riva Properties v Foster + Partners* [2017] EWHC 2574 (TCC), the architect designed a project substantially in excess of the client's stated budget but alleged that it was not their responsibility to work within the client's budget or even to make the client aware of a likely cost overrun.

As part of a collaborative procurement system, a client should state its budget so as to be able to require prospective consultants and contractors to state:

- Their expected fees, profit and overheads by reference to that budget
- Evidence of their ability to deliver the project within that budget
- Their proposals for achieving improved value within that budget
- Their proposals for seeking cost savings that could establish an agreed price lower than that budget.[1466]

Concerns may be expressed that a pre-selected main contractor may be less commercially rigorous in its supply chain tender procedures, and that inflated supply chain prices could lead to the total price exceeding the client's budget. These concerns can be addressed by:

- Close monitoring of the contractor's subcontract tendering procedures and documentation by the client and consultants, to ensure that these do not impose excessive demands that could inflate subcontract prices
- A timetable for review of supply chain prices, so as to ensure they add up to a total price within the project budget as a precondition for the construction phase to proceed
- Transparency at each stage whereby the alliance manager receives all documentation prepared and issued by the main contractor, all tender returns and proposals submitted by prospective supply chain members and all related correspondence, and whereby the alliance manager is invited to attend all meetings with prospective subcontractors and suppliers.[1467]

As described in Sections 3.5 and 9.9, PPC2000 sets out a detailed enterprise planning system in respect of each works package, describing business cases and supply chain tenders that are linked to the build-up of costs by reference to agreed deadlines.[1468] A briefer description of early contractor involvement under NEC4 Option X22 provides that:

- 'The Contractor provides detailed forecasts of the total Defined Cost of the work to be done in Stage One for acceptance by the Project Manager'
- 'The Contractor prepares forecasts of the Project Cost in consultation with the Project Manager'.[1469]

We considered in Section 9.7 the wide discretion for the NEC4 Project Manager to reject cost proposals. This discretion, combined with the absence of detailed enterprise planning activities, reduces the potential of NEC4 Option X22 to neutralise the divisive effects of reliance on cost negotiations. For example, there is a right under NEC4 Option X22 for the client to appoint another contractor to complete the Stage Two works if 'the Project Manager and the Contractor have not agreed the total of the Prices for Stage Two'.[1470] This suggests that the project manager can use the threat of replacement of the

---

1466 Two Stage Open Book and Supply Chain Collaboration Guidance (2014), 33.
1467 As provided for in PPC2000 (2013) clause 10.7.
1468 PPC2000 (2013) clauses 10 and 12.
1469 NEC4 (2017) Option X22.2(1) and X22.2(5).
1470 NEC4 (2017) Option X22.5(3).

main contractor as a way to negotiate lower costs, instead of using systems that build up costs through enterprise contract processes.

Without a clear system for developing detailed project costs during the preconstruction phase, the period of early engagement may not be used to reconcile client and contractor cost expectations.[1471] The NEC4 Option C Schedule of Cost Components, and the Schedule of Cost Components that forms part of the NEC4 Alliance Contract, both focus in detail on the costs of self-delivery by the main contractor[1472] but describe the costs of each subcontractor only as a single item.[1473] These omissions suggest that NEC4 cost analysis is only undertaken at main contractor level unless other arrangements are agreed in more detail. For subcontract prices to be established by a process that involves only the main contractor, without a system for the collaborative involvement of the supply chain and other team members, undermines the openness required for successful collaborative working because other team members have no data as to how supply chain costs have been arrived at.

The omission of subcontracted supply chain members from collaborative working and from detailed cost analysis leaves the main contractor free to use non-collaborative practices in its relationships with those supply chain members, such as insisting on cost reductions while excluding them from incentives agreed with the client and other alliance members. For example, a target cost and pain/gain share approach agreed between the client and the main contractor may not be matched by equivalent incentives agreed between the main contractor and any members of its supply chain.[1474]

16.6 How Is a Framework Alliance Costed?

The costing of a framework alliance reflects the likelihood that no specific projects will be awarded at the point of selection of the framework consultants and contractors. A framework alliance should contain systems through which the alliance members create prices that comprise:

- The fees of framework consultants and/or the profit and overheads of framework contractors, preferably agreed as lump sums subject to increase according to the achievement of agreed targets. These fees, profit and overheads may also be subject to variations that reflect the budget for each project, and each budget will be stated only when the client commences its selection procedure for that project under the framework alliance
- Particular overheads that are necessary for the mobilisation and implementation of a framework alliance. These may include, for example, the resources required to participate in Supply Chain Collaboration and other agreed activities that are separate from the delivery of specific projects

---

1471 This was a risk identified in Nichols (2007), 32 and 33.
1472 For example, NEC4 (2017) Option C Schedule of Cost Components clauses 1–3 and 5–8.
1473 For example, NEC4 (2017) Option C Schedule of Cost Components clause 4 (Subcontractors) states these components simply as 'Payments to Subcontractors for work which is subcontracted without taking into account any amounts paid to or retained from the Subcontractor by the Contractor, which would result in the client paying or retaining the amount twice'.
1474 *Alstom Signalling v Jarvis Facilities* [2004] EWHC 1232 (TCC).

- Cost components that are established on sample projects at the point of consultant and contractor selection, and that can be used as a benchmark for agreement of costs for projects when awarded[1475]
- Other cost components that are capable of being priced by bidders and by their preferred supply chain members at the point of selection, for example because the relevant designs and risk assessments are standard requirements for all framework alliance projects and can be priced by consultant or contractor bidders in consultation with subcontractors or suppliers
- Those cost components that are to be agreed during or after consultant or main contractor selection for each project, using agreed processes that maximise the contributions of the main contractor and other supply chain members to improving the value of proposed designs and risk assessments
- Adjustments to reflect cost savings that have been established on previous projects
- Adjustments to reflect cost savings that have been achieved through Supply Chain Collaboration and other alliance activities separate from the delivery of specific projects.[1476]

The procurement of a public sector framework or framework alliance can breach public procurement regulations if it does not involve sufficient comparative cost information at the point of selection or if it does not include sufficient comparative cost information as part of the selection process for specific projects.[1477] Therefore, clients should include in their selection process initial rates and/or sample projects in order to test bidders' costs and other proposals in greater detail, an approach that was adopted for example in the procurement of the Ministry of Justice framework alliances.[1478]

On the Erimus Housing framework alliance:

- 'Initially each contract was an annual schedule of rates'
- 'By year two, all parties agreed to an "open-book" approach and by convergence of best practice and supply chains moved the partners towards adopting common rates'
- 'A pain-gain formula shared overruns and savings 50/50 up to 10% over or under the budget'
- 'The constructors retained or paid any more than the 10% variances and outturns were the basis for the next budget'.[1479]

Cost Led Procurement was developed as a procurement and delivery model for use with frameworks.[1480] Clients can combine features of Cost Led Procurement and Two Stage Open Book in their selection process, for example so that:

1475 As considered in Sections 6.6 and 6.8.

1476 These pricing systems were used in the costing the framework alliances summarised in Section 4.10, including sample projects that were costed for the Ministry of Justice framework alliance and net rates that were costed for the SCMG framework alliance.

1477 *McLaughlin and Harvey v Department of Finance and Personnel* [2008] NIQB 91 and *Henry Bros and Scott* and *Ewing v Department of Education for Northern Ireland* [2008] NIQB 105. In the latter case, the only cost information requested by the client when selecting the members of its framework was their direct fee percentages, and this was considered non-compliant with the requirements of transparency, the equal treatment of tenderers and the development of effective competition in the public sector.

1478 Case study summarised in Section 4.10.

1479 Case study summarised in Section 16.10.

1480 As summarised in Section 2.8.

- They use Cost Led Procurement systems to obtain fixed prices or target costs for elements of a project where they can make available sufficient design and risk data to enable fixed price quotes and where bidders and their preferred subcontractors and suppliers are fully familiar with those elements, for example using data available in a Building Information Modelling (BIM) library or from bidders who have undertaken equivalent work on other projects
- They use Two Stage Open Book systems to test the costs of other packages by means of supply chain tenders led by the main contractor after its selection and conditional appointment to undertake a project.[1481]

16.7 How Is a Term Alliance Costed?

The costing of a term alliance should create sufficient cost certainty to enable orders for tasks to be called off at any time during the agreed term. Under a term alliance, unlike a framework alliance, the rates or costs for works, services and supplies need to be agreed in sufficient detail so that they can be called off with only limited adjustments to reflect specific task designs or risk assessments.

A term alliance can use any combination of the following approaches to costing and can provide for the flexibility to move from one approach to another:

- A schedule of rates is 'stated for each Task or part of a Task which creates cost certainty at the point of issuing Orders but, without other pricing mechanisms, may inhibit systems to achieve efficiency savings'[1482]
- Fixed prices or target costs for all or part of the term programme where 'implementing any Tasks comprising all or part of the Term Programme within one or more fixed prices depends on the accuracy of information in the Term Brief (describing the required works/services/supplies) which will influence the Service Provider's assessment of potential risk'[1483]
- Cost reimbursement where 'for the Service Provider to recover all its costs as openly declared to the Client is likely to depend on Client confidence in the means of cost control such as joint reviews and incentivisation of savings'.[1484]

For example, the Southern Housing term alliance used TAC-1 to create a price framework which comprised a mixture of:

- 'Basket Rates for Component Renewal'
- 'Fixed prices for "Property Archetypes" for Cyclical Decorations'
- 'Schedules of rates for all other works'
- 'Percentage bandings for site overheads, central office overheads and profit'.[1485]

1481 This combined approach was used by Ministry of Justice under their framework alliance summarised in Section 4.10. See also Diagram 3 in Section 2.8.
1482 TPC2005 Introduction to Pricing (2012), 2.1 and 3.6 (Rates and frameworks).
1483 TPC2005 Introduction to Pricing (2012), 2.1, 3.7 (Fixed price/target cost), 3.10 (Price per property), and 3.11 (Price per task and hybrids).
1484 TPC2005 Introduction to Pricing (2012), 2.1 and 3.9 (Open-book).
1485 Case study summarised in Section 11.10.

A term alliance can create prices that comprise:

- The fees of a consultant provider and/or the profit and overheads of a contractor provider, preferably agreed as lump sums subject to increase according to the achievement of agreed targets. These fees, profit and overheads may also be subject to variations that reflect the value of each task, which can be determined only when each order is issued[1486]
- Particular overheads that are necessary for the mobilisation and implementation of the term alliance. These may include, for example, the resources required to participate in Supply Chain Collaboration and other activities that are separate from the delivery of specific tasks. They may also include early mobilisation costs such as expenditure on vehicle preparation, IT configuration, staff training, establishment of call centres and other resources that need to be ready for implementing orders[1487]
- Any cost components priced as a schedule of rates by the provider and by its supply chain at the point of selection. A term alliance may simplify the costing of orders by combining rates into basket rates or by averaging costs into composite rates or by agreeing target rates that are reconciled with actual rates at regular intervals[1488]
- A simplified cost model that establishes, for example, a fixed price or price per property or a price per task irrespective of their actual costs and with or without later reconciliation of these prices with actual costs incurred.[1489] A term alliance may include fixed prices or prices per property in relation to certain cyclical or responsive works, services or supplies and, in these circumstances, it will be important to provide bidders with a well-explained history of the relevant amounts and types of work in previous years
- Any cost components that are to be agreed after the issue of a task order, using agreed processes that maximise the contributions of the provider and its supply chain. A term alliance can provide for the issue of complex orders with their own project alliance cost processes but these processes may have the effect of converting a term alliance into a framework alliance[1490]
- Adjustments that reflect cost savings achieved through Supply Chain Collaboration and other alliance activities separate from the delivery of specific tasks.

For example, on the Maidstone Decent Homes term alliance:

- 'The team was tasked with reducing cost and time. Moving away from the traditional schedule of rates and costs per property, a mechanism of three fixed costs was developed that covered all types of property, facilitating greater transparency and planning, and significantly reducing overheads'
- The team moved 'from schedule of rates to average order values under TPC preconstruction phase costing process'.[1491]

1486 The TPC2005 Introduction to Pricing (2012), 2.4 recommends net rates being agreed with separate fees/profit/ overheads so as to allow for ongoing review of the costs of tier 2/3 subcontractors and suppliers.
1487 TPC2005 Introduction to Pricing (2012), 2.4 (Profit and overheads) and 2.7 (Mobilisation costs).
1488 TPC2005 Introduction to Pricing (2012), 3.2 (Schedule of rates (published)), 3.3 (Schedule of rates (bespoke)), 3.4 (Schedules of rates, profit, and overheads), and 3.5 (Basket or composite rates).
1489 TPC2005 Introduction to Pricing (2012), 3.7 (Fixed price/lump sum), 3.8 (Target costs), 3.10 (Price per property), and 3.11 (Price per task and hybrids).
1490 TPC2005 Introduction to Pricing (2012), 3.6 (Rates and frameworks) and as considered in Section 5.5.
1491 Association of Consultant Architects (2010), 32, 33.

16.8 How Do Shared Pain/Gain Incentives Operate?

The establishment of shared pain/gain incentives should distinguish between:

- The sharing of all or some cost savings, sometimes combined with the sharing of all or some cost overruns, designed to create shared motivation to minimise costs[1492]
- The sharing of all cost savings and overruns arising on a project other than those due to error, omission or delay, sometimes aggregated together with the cost savings and overruns arising on other projects undertaken by the same team, designed to discourage each individual team member from non-collaborative costing such as hidden contingencies[1493]
- The reimbursement or sharing of all cost savings and overruns including those arising from error, omission or delay, under a no blame clause that is designed to encourage collaborative working by removing any fear of claims, as considered in Sections 7.7 and 18.7.

Shared pain/gain incentives are set out in NEC4 Option C where:

- 'If the Price for Work Done to Date is less than the total of the Prices, the Contractor is paid its share of the saving
- If the Price for Work Done to Date is greater than the total of the Prices, the Contractor pays its share of the excess'.[1494]

PPC2000 does not include an equivalent formula but it does extend the principle of shared pain/gain to be an option for the whole team, stating that 'the Partnering Team members shall implement any shared savings, shared value and pain/gain Incentives described in the Project Partnering Agreement and otherwise recommended by the Core Group and approved by the Client'.[1495] Shared savings are a feature of many project alliances using PPC2000, for example the St George's Hospital Keyworker Accommodation scheme where the contract:

- 'Enabled preconstruction work to be carried out at the same time as a final Agreed Maximum Price (AMP) was being agreed in which all risks had been quantified'
- Created 'the incentive to be proactive in managing risk and expenditure so as to earn rewards available through the shared savings mechanism, openly reviewing buying gains obtained through subcontractor and statutory authority orders'.[1496]

Shared savings and/or shared pain and gain can align the commercial interests of the team members if they share in the overall cost performance of the project by reference to its agreed budget or target cost, irrespective of the cause of any cost saving or cost overrun. For example, under the US IPD approach:

- 'Although formulations vary, all or part of the participants' profit is placed at risk and profit may be augmented if project performance is met or exceeded'

1492 As illustrated in the Erimus Housing case study summarised in Section 16.10.
1493 As illustrated in the Anglian Water case study in Section 4.10.
1494 NEC4 (2017) Option C Target contract with activity schedule clause 54.2, as also considered in Section 16.4.
1495 PPC2000 (2013) clause 13.2.
1496 Association of Consultant Architects (2010), 38.

- 'Individual profit is not a function of the amount of work performed, or of individual productivity, but is proportionate to overall project success'.[1497]

This approach is illustrated in ConsensusDocs 300, which provides for a collective 'Risk Pool' of profits put at risk by Risk Pool Members if the Project is delivered for more than the 'EMP (Estimated Maximum Price)' or otherwise fails to achieve established criteria, as more particularly addressed in the 'Risk Pool Plan'.[1498] There may, however, be problems if team members are expected to participate in the Risk Pool Plan by later agreement after they have entered into contract. ConsensusDocs 300 anticipates this by providing that a trade contractor or consultant may 'refuse participation in the Risk Pool Plan' and then 'will instead be compensated on a cost-plus-fee basis subject to a guaranteed maximum price as set forth in the subcontract, or else replaced, at the discretion of the Core Group'.[1499]

On the Anchor Property Trial Project:

- 'Gain share incentives were…established early and it was decided that Anchor would keep the first 10% and the rest would be shared between the client, the contractor and the consultant'
- 'The share basis was client 40%, contractor 40% and consultant 20%. Anchor had decided on a very open and trusting approach with its supply chain and consequently there was to be no pain share'
- 'Accordingly, the savings made were around 9% when projects that failed to improve against targets were taken into account'
- 'Savings were agreed with supply chain members for paint and kitchens and bathrooms, and formal agreements signed'
- 'The real saving was created by a common understanding of cost which was not there before as well as the simple activity of establishing a target cost on a plant, material, labour, subcontract and preliminaries basis'.[1500]

Pain/gain shares provide an incentive to minimise costs, but they should be combined with systems by which team members firstly build up a full set of agreed and reliable cost data. The agreement of pain/gain shares does not itself establish the systems for identifying and testing potential cost savings and other improved value, nor does it ensure that the team will include, or will work collaboratively with, subcontracted supply chain members. If not combined with these systems, the agreement of pain/gain shares leaves open, for example, the risk that the main contractor will achieve its share of savings simply by putting pressure on subcontractors and suppliers to reduce their costs. This would inevitably undermine the commitment of those subcontractors and suppliers to a collaborative approach and leave them more likely to achieve the required savings by compromising quality or safety or some other aspect of the project.

Some commentators suggest that essential features of a modern alliance include a 'target cost or target price with shared pain/gain incentivisation'.[1501] However, the English courts have warned against excessive reliance on a target cost approach as the

---

1497 Fischer et al. (2017), 372.
1498 ConsensusDocs 300 (2016) Article 10.2 (Establishment of the Risk Pool).
1499 ConsensusDocs 300 (2016) Article 10.3 (Establishment of the Risk Pool).
1500 Anchor Property Trial Project case study.
1501 OGC (2007), Procurement Guide, 05, 5.

main foundation for collaborative working if this leads to incorrect assumptions as to what other alliance members are doing. Where a target cost approach was combined with a lack of effective risk management that had disastrous consequences, the judge commented 'I do not find that it was negligent to adopt a target cost form of contract. But I do find, without hesitation, that such a contract having been adopted, the implications of the methane risk for the safety of both the temporary and permanent works were lost sight of'.[1502]

Different models describe the build-up of the figure against which shared savings or shared cost overruns will be calculated:

- The NEC4 Alliance Contract provides for these shares to be calculated by reference to the difference between the 'Budget' and the 'Alliance Cost', the former being part of the 'Contract Data'[1503] and the latter being 'the sum of the Price for Work Done to Date and the Client's Costs'[1504]
- NEC4 Option C provides for these shares to be calculated by reference to 'Defined Costs' incurred above or below 'lump sum Prices' in an 'Activity Schedule.'[1505]

In any case the base point for calculating shared cost savings or overruns should reflect sufficient preparatory work by team members for it to be reasonably accurate. In order to provide a valuable incentive, shared savings should be earned through collaborative activities rather than arrived at automatically as a windfall against an arbitrary figure.

The prospect of shared savings may tempt team members to set the base point for their calculation as high as possible. The JCT CE Guide states that 'normally a client will retain the option not to proceed to the construction stage so as to provide some commercial pressure on the contractor not to pitch his assessment of the target cost for the construction period too high'.[1506] Shared pain/gain clauses need to be carefully drafted in order to avoid unintended consequences, noting for example the English case where a contractor interpreted a pain/gain clause as entitling it to recover actual costs incurred over and above a pain share on the first £140 million cost overrun and also to recover actual costs incurred above the contractor's stated liability for the next £50 million cost overrun.[1507]

Some alliances, such as the Anglian @one alliance, have taken pain/gain shares to a strategic level by agreeing that team members will share in profit and losses according to the overall performance of each annual programme of work. Anglian Water reported how:

- 'The first evolution of the model included partners generating a return from both fee recovery and then gain share from additional out-performance'

1502 *Eckersley v Binnie & Partners* [1988] 18 Con LR 1 (CA).
1503 NEC4 *preparing an alliance contract* (2018), 11.
1504 NEC4 Alliance Contract (2018) clause 11.2(3).
1505 NEC4 (2017) Option C clauses 11.2(31), 11.2(32) and 54, as considered in Section 16.4.
1506 JCT CE Guide (2016), Section 40.
1507 *AMEC v Secretary of State for Defence* [2013] EWHC 110 where the court stated that, despite the problems of analysing a badly drafted clause, it would be extremely strange if a contractor were able to recover from its employer the cost of remedying its own breaches of contract and to be paid to remedy defects it was responsible for in the first place, and that it would be 'commercially surprising' if a contract could go so wrong that a contractor could recover all of its costs 'no matter how unreasonably or improperly incurred'.

- 'The latest evolution is simply based on partners generating a return from collective out-performance'
- 'This has gradually increased the behavioural focus on further levels of out-performance – and removed any incentives around turnover, hours charged, fee recovery and perceived partner self-interest'.[1508]

16.9   What Other Incentives Support Collaborative Procurement?

Target prices and shared pain/gain are not the only available incentives for collaborative procurement. There are other ways to ensure the approach to incentivisation and reward drives effective collaboration and maximum productivity.

If the scope of the team members' agreed services, works and supplies are made clear in collaborative contracts, and if team members are paid a fair profit and an appropriate contribution to their overheads and other costs, this can incentivise them to concentrate their efforts on the best interests of the project or programme of work. It will also reduce the unproductive time spent in devising tactics that prepare the ground for later claims[1509] if team members are motivated by more open, better structured working relationships that demonstrably help them to:

- Avoid losses
- Minimise wasted cost, time and resources
- Enhance their reputation
- Avoid disputes.

Financial incentives can link performance to a range of success measures. For example:

- NEC4 provides for incentives embedded in the shared pain and gain provisions, for other incentives as agreed under Option X12 or Option X20 and also for a bonus on early completion[1510]
- PPC2000 provides for agreement of 'shared saving arrangements and added value incentives' and for consultant and contractor incentives linked 'to achievement of the agreed Date for Completion…or to the achievement of any of the Targets stated in the agreed KPIs'.[1511]

The open-book costing approach can protect profit and overheads by ringfencing them separate from other costs which can then be subject to agreement of savings while they are being are built up into fixed prices. This approach contrasts with the IPD cost model where costs are reimbursed but where profit is placed at risk dependent on achievement of an estimated maximum price.[1512]

Team members need a clear understanding of what work attracts remuneration and what work is undertaken speculatively, and clients should consider the incentive or disincentive determined by the point at which other team members start to be paid for their work. The incentive of a pipeline of work may attract some speculative proposals

---

1508 Case study in Section 4.10.
1509 Bennett and Jayes (1998), 59.
1510 NEC4 (2017) Option C, Option X6, Option X12.4 and Option X20.
1511 PPC Pricing Guide (2008), 3.
1512 As considered in Section 16.8.

for improved value,[1513] but if for example a contractor will only be paid if a project proceeds on site, then commercial logic dictates that its first priority will be ensuring that the project goes ahead rather than the provision of objective advice on how to improve value.

To show that a client is not wasting other team members' time in speculative preconstruction phase work, payment provisions can 'recognise all the contributions being made, and the related risks, responsibilities and rewards, particularly during project development'.[1514] Any agreed preconstruction phase payment entitlements should be clear and not open to different interpretations, as collaborative early contractor involvement can quickly deteriorate into conflict if a team member considers it has been deprived of an agreed payment.[1515]

Incentives for the improved performance of a single project are limited by the scope and value of that project and by concerns as to:

- How far the project cost plan should be developed before it can be the basis for calculating shared savings or overruns
- Whether cost certainty and cash flow may be delayed while incentives are being calculated
- When actions and ideas that generate cost savings should be recognised and rewarded compared to when they should be treated as a team member simply doing its job.

By contrast, the incentives agreed at a strategic level under a framework alliance or term alliance can also be linked to the performance of successive projects or tasks. The most powerful motivators for alliance members to seek improved value are likely to be the prospect of earning additional work through:

- The award of additional projects under a framework alliance contract
- The issue of additional orders under a term alliance contract
- The extension of the duration, or the expansion of the scope, of a framework alliance contract or a term alliance contract.

Incentives relating to the award of additional work can also recognise where performance is impaired by limited capacity, and contractual systems can enable adjustment of the workflow if certain team members become overloaded. A framework alliance contract or term alliance contract should state the success measures and targets that determine the variable award of work, and should also state who evaluates the alliance members' performance and capacity for these purposes.[1516] For example, FAC-1 and TAC-1 provide for incentives which include not only additional payments such as shares

---

1513 As envisaged by the Cost Led Procurement model.

1514 CIRIA (1998), 15.

1515 In *Almacantar (Centre Point) v Sir Robert McAlpine* [2018] EWHC 232 (TCC), a developer engaged a contractor under a two-stage contract to redevelop Centre Point Tower. The Stage One Pre-Construction Services Agreement (PCSA) contained provision that 50% of the fee payable to the contractor for its ECI input would be withheld and released in the first payment under the Stage Two main contract. The parties could not agree the Stage Two contract approach and risk allocation and thus by agreement the PCSA was terminated. McAlpine sought payment of their fee but Almacantar said only 50% of the Stage One fee was payable as the parties had not entered into a Stage Two contract. An adjudicator supported the claim for 100% of the fee but the judge reversed this decision and ordered McAlpine to re-pay 50%.

1516 For example, whether this evaluation is conducted privately by the client or project manager, or openly by a governance body such as a core group.

of savings achieved through agreed Alliance Activities[1517] but also the adjustment of any exclusivity in the award of project contracts or task orders[1518] and the extension of the scope or duration of the agreed appointments.[1519]

16.10 Collaborative Costing Case Studies

The first of the following case studies summarises the Archbishop Beck Trial Project which used a collaborative framework contract in conjunction with an NEC3 Option C project contract and where the client set a strict budget, with main contractor profit and overheads agreed in advance. Agreed collaborative processes were undertaken using systems set out in the framework contract, by which Two Stage Open Book design and risk management enabled cost savings of 26% by the time of establishing an agreed price.

The means by which these savings were achieved included:

- Lessons learned from previous projects
- Joint working with subcontractors and suppliers in developing innovative proposals at the point of their joint selection
- Further joint working with subcontractors and suppliers throughout the preconstruction phase.

The second case study describes how costs and cost savings were agreed by the Erimus Housing, Middlesbrough decent homes alliance using a collaborative framework in conjunction with PPC2000 where appointments were initially made under a schedule of rates, and were then converted by agreement to a fully open-book approach. Savings were identified through preconstruction phase joint working and were motivated by a pain-gain formula.

Archbishop Beck Sports College

The Archbishop Beck Sports College Trial Project had a value £15.9 million (pre-savings) and comprised the construction of a new build school, comprising a three-storey school building, a link building and sports hall with a total approximate gross internal floor area of 12 163 square metres, including associated external works, car park and sub-station. The project team comprised Liverpool City Council (client), Willmott Dixon Construction (main contractor), Sheppard Robson (lead designer) and key suppliers Mouchel (structural, mechanical, and electrical), D Morgan (Groundworks), A&B (M&E specialist works) and Cara & Metsec (steel frame).

Liverpool City Council set a brief for achievement of 20% savings against comparable projects in 2009/2010 so that it could maximise the use of available funding, and this target was achieved. The Council selected Willmott Dixon through the Scape Framework,

---

1517 Which can be payable under a Project Contract or under FAC-1 Framework Alliance Contract (2016) clause 8, and under TAC-1 Term Alliance Contract (2016) clause 8.
1518 Under FAC-1 Framework Alliance Contract (2016) clause 5.7 and Schedule 1 Part 2, and under TAC-1 Term Alliance Contract (2016) clause 5, 7 and Schedule 1 Part 2.
1519 Linked to FAC-1 Framework Alliance Contract (2016) clauses 9.1 and 14.1 and TAC-1 Term Alliance Contract (2016) clauses 9.1 and 14.1.

having previously appointed the same team on design and construction of its Notre Dame School project.

The Council set a strict programme for completion of the Archbishop Beck project, improving on timescales achieved in relation to the Notre Dame School project, and placed particular emphasis on engagement of local businesses and the development of local employment and training opportunities. The project achieved 60% local spend compared to 50% on Notre Dame.

Taking into account value indicators for similar projects, cost savings achieved results of **26%** from a rate of £1950 per square metre anticipated for a comparable project, to a rate of £1438 per square metre achieved by the time of establishing an agreed price.

The means by which these savings were achieved include:

- Lessons learned from the Notre Dame School project
- Joint working by Willmott Dixon with tier 2/3 subcontractors and suppliers in developing innovative proposals at the point of selecting tier 2/3 subcontractors and suppliers
- Further joint working by Willmott Dixon with its tier 2/3 subcontractors and suppliers and with Liverpool City Council throughout the contractual preconstruction phase.

Specific savings included:

- De-bundling certain elements of precast to avoid additional layers of profit and overheads, wastage and to achieve more efficient coordination generated agreed savings of approximately £45 000
- Extensive rounds of value engineering in respect of mechanical and electrical designs, with the preferred specialist subcontractor A&B Engineering and M&E consultant Mouchel contributed to an agreed saving of £258 688
- Removal of suspended ceilings and adding additional acoustic treatments produced a saving of £21 068
- Recycling of waste materials for re-use on site, combined with redesign of external landscaping, resulted in a saving of £55 164
- Installation of under-floor heating in the ground floor slab to allow earlier installation and rationalisation of the heating system produced a saving of £31 234
- Acquisition and re-use of surplus bleacher seating from another school produced a contract saving of £18 564.

Archbishop Beck Trial Project case study.

Erimus Housing Decent Homes Alliance

Erimus Housing procured a £105 million five-year housing improvement framework alliance to upgrade 11 000 houses, working with ROK, Dunelm Property Services and Erimus Building Services. All constructors committed to a single strategic alliancing agreement that allowed benchmarking of performance and promoted exchange of best practice. The programme showed a high degree of cooperation between the strategic partners, for example when one constructor had short-term personnel difficulties and

(Continued)

the other two stepped in to assist. Initially each contract used an annual schedule of rates.

By year two, all parties agreed to an 'open book' costing approach and, by convergence of best practice and supply chains, the partners moved towards adopting common rates. A pain-gain formula shared cost overruns and savings 50/50 up to 10% over or under the budget. The constructors retained or paid any more than the 10% variances, and actual cost outturns were the basis for the next budget.

Residents were involved in consultation and choice through the use of a mobile exhibition unit. The average resident satisfaction score was 9/10.

The Erimus alliance team was commended in the 'Integration and Collaborative Working' category at the Constructing Excellence 2007 awards, and noted that:

- The form and language of PPC made it easy to integrate into an overarching framework agreement and strategic alliancing agreement
- PPC has a built-in provision for 'Open-book' pricing enabling the client and constructors to focus on reducing costs while maintaining ringfenced overheads and profit
- PPC recognised constructor incentives as an effective way to maintain performance.

Ian Gillespie, Investment Manager, Erimus Housing commented: 'I cannot believe the co-operation we are getting. We're consistently under budget. You can read PPC2000 without tremendous legal knowledge. Had we done this with a traditional JCT contract, I know we'd have had many battles. Yet I've worked with PPC since 2000 without a single problem turning into a dispute'.

Association of Consultant Architects (2010), 16.

17

How Does Collaborative Procurement Manage Time and Change?

17.1 Overview

An alliance should agree an 'end to end delivery process…that provides a clear route for the programme or project to progress, including key gateways and milestones'.[1520] No team member acts in isolation and each is reliant at certain points on other team members fulfilling their obligations within specific time limits. All team members also rely on the timely completion of activities by third parties over whom they exercise no control. As a consequence, client advisers should 'make it clear that time spent beforehand in settling the details of the work required and in preparing a timetable of operations, from the availability of the site to the occupation of the completed building, is essential if value for money is to be assured and disputes leading to claims avoided'.[1521]

Good project management requires that 'all significant stages of the project must take place no later than their specified dates',[1522] yet it has been noted that 'insufficient regard is paid to the importance or value of time and its proper use in all aspects of a project, from the client's original decision to build, through the design stages and up to final completion'.[1523]

Failure to agree programmes and timetables in respect of the key activities of all team members at all stages of a project increases the risk of delays and losses. It leaves team members without control over who does what and in what sequence, and ignorant as to the expected timing both of their own contributions and of the contributions of other parties. Collaborative construction procurement should establish commitments to progress a project in agreed stages, using programmes and timetables that integrate design and construction activities and can connect the client, consultant, contractor, subcontractor, supplier, manufacturer and operator inputs.

Alliance members can create a matrix of agreed deadlines which provides for the following enterprise planning activities and events to be identified, planned, monitored, and measured:

- Work authorisations
- Financial authorisations from the client
- Third-party applications and consents

1520 Infrastructure Client Group (2015), Section 3.2.
1521 Banwell Report (1964), 3, Section 2.3.
1522 Lock, D. (2000), 7.
1523 Banwell Report (1964), 3, Section 2.2.

Collaborative Construction Procurement and Improved Value, First Edition. David Mosey.
© 2019 John Wiley & Sons Ltd. Published 2019 by John Wiley & Sons Ltd.

- The start and finish of designs, with separate shorter activities corresponding to design phases
- Release of completed drawings for production or construction
- The start of purchasing activity for each sub-assembly or work package
- Issue of invitations-to-tender or purchase enquiries
- Receipt and analysis of suppliers' or subcontractors' bids
- The issue of a purchase order to each supplier or subcontractor
- Material deliveries
- The starts and completions of manufacturing stages
- The starts and finishes of construction subcontracts, and important intermediate events in such subcontracts
- Handover events of completed work packages.[1524]

The duration and complexity of certain projects may make it difficult to agree a timetable for all activities from the outset. An alliance can sub-divide agreement of a timetable into stages if the planning of a project makes this necessary. For example, on the Tottenham Court Road office development alliance, the client appointed the other alliance members in three stages, with timetables governing their work firstly in seeking planning consent, then in developing designs, costs, and risk management in line with the planning consent when granted, and then in proceeding with construction.[1525]

Agreeing deadlines for the actions, approvals and other interfaces between alliance members is a key feature of an enterprise contract. This chapter will explore how programmes and timetables can support an alliance and how they can contribute to improved value. It will also look at the collaborative management of change. Setting and agreeing the right deadlines for agreed activities is an essential collaborative activity because 'it is very difficult to enforce a plan which is conceived in isolation and it is therefore essential to involve the individuals and organisations responsible for the activities or operations as the plan is developed'.[1526]

17.2   Why Is Collaborative Time Management Important?

Programming is a key function of project management[1527] and of risk management and is not restricted to establishing the timing of the main contractor's performance. Integrated planning also needs to establish the timing of work undertaken by the client and by consultants, subcontractors, manufacturers, suppliers and operators. For example, deadlines that apply to all design contributors should include dates for the submission, review and approval of:

- Each stage of the designs developed before and after submission for planning consent and/or other regulatory approvals
- Each stage of the designs developed before and after prices are submitted by bidding main contractors

---

1524 As described by Lock, D. (2000), 216 who defined this as 'network planning'.
1525 Tottenham Court Road case study, Association of Consultant Architects (2010), 42–43, summarised in Section 6.10.
1526 Smith et al. (2006), 6.
1527 BS 6079 defines project management as the 'planning, monitoring and control of all aspects of a project and the motivation of all those involved to achieve the project objectives on time and to cost, quality and performance'.

- Design details and amendments added before and after committing to the construction of the project
- Design details and amendments added before and after committing to the construction of works packages, including those relating to any approved provisional sum activities
- Design amendments necessary to integrate a design on site.

When undertaking integrated project planning, it has been suggested that 'owners should own the overall masterplan with key milestones that need to be hit' and that 'the contractor is responsible for producing the integrated plan and associated detailed plans and schedules, and for demonstrating how these key milestones will be achieved'.[1528] As regards timetables for consultants, Latham recommended that 'appointment documents for all consultants should contain:

- Details of the duties of the retained consultants
- Specific timescales in which those duties are to be performed
- An arrangement that they should "sign off" their work as properly completed at appropriate stages'.[1529]

In order to integrate the full range of interlocking deadlines, it has been recommended that 'the activities of designers, manufacturers, suppliers, contractors and all other resources must be organised and integrated to meet the objectives set by the client and/or the contractor [so that] sequences of activities will be defined and linked on a timescale to ensure that priorities are identified and that efficient use is made of expensive and/or scarce resources'.[1530] Other commentators have advocated that 'where innovation is needed, milestones are interpreted flexibly on the basis of providing just sufficient information to avoid delaying the project'.[1531]

Clear contract terms setting out the team members' agreed deadlines and interfaces are important because, under English law, there is no implied obligation to proceed with work regularly and diligently or at a particular pace.[1532] Case law has illustrated, for example, how subcontractors are not obliged to comply with main contractor programmes unless this is expressly agreed.[1533]

Even an express obligation to proceed 'regularly and diligently'[1534] has limited impact and is open to varying interpretations for example:

1528 McKinsey Global Institute (2017), 89.
1529 Latham, M. (1994), Section 6.18.4.f, 49.
1530 Smith et al. (2006), 6.
1531 Bennett and Jayes (1998), 80.
1532 In *Leander Construction v Mulalley* [2011] EWHC 3449 (TCC), an 'activity schedule' was not incorporated in a bespoke groundwork subcontract and there was no express requirement to proceed 'regularly and diligently'. In these circumstances a payment withholding notice, based on failure to proceed regularly and diligently, was held to be invalid.
1533 In *Pigott Foundations v Shepherd Construction* [1993] 67 BLR 48 (QB), the JCT 'master programme' was not a main contract document and a subcontractor was not obliged to comply with this main contract programme despite JCT subcontract wording requiring progress 'reasonably in accordance with the progress of the main contract works'. In *Multiplex Constructions v Cleveland Bridge* [2006] EWHC 1341 (TCC), there was a dispute as to whether delay was caused by late contractor design or late briefing. It was held that there was no implied obligation on the subcontractor to enable a main contractor to meet its own main contract programme obligations.
1534 For example, under JCT SBC/Q (2016), clause 2.4 where it is linked to right of termination under clause 8.4.2.

- That 'regularly' means daily attendance with sufficient staff and resources to make substantial progress, and that 'diligently' means applying resources industriously and efficiently[1535]
- That the words imply an obligation to complete work 'efficiently' and to maintain 'orderly progress'[1536]
- That orderly progress can be tested objectively by reference to available work, urgency, purposes and conditions on site.[1537]

Without contractual clarity, the courts assess when design information is actually needed and distinguish this from when a contractor perceives it is needed, although a contractor may be entitled to be issued with drawings promptly on request.[1538] There may also be uncertainty caused by implied obligations:

- That drawings submitted for approval will be considered within a reasonable time[1539]
- That team members will liaise properly with each other[1540]
- That a contractor or subcontractor will finish its work in sufficient time not to delay other later contractors and subcontractors.[1541]

Collaborative procurement can establish clear deadlines and interfaces among all team members. For example, on the Poole Hospital Operating Theatres Programme, collaborative time management included:

- Design input from the main contractor and specialist mechanical and electrical sub-contractor under a preconstruction phase timetable prior to start on site, so as to ensure that a fully designed project was implemented with further design releases not causing delays or misunderstandings
- Organisation of a phased project in a fully operational building under a construction phase timetable jointly prepared and agreed by the client, alliance manager, design consultants, main contractor and mechanical and electrical specialist subcontractor
- Implementation of a complex construction phase timetable, with involvement of end user clinical teams to ensure integration with their clinical programme.[1542]

17.3 Is a Programme the Same as a Timetable?

A programme sets out in detail the timing of each stage of the work of a member of a project team, and creates a critical path showing the sequential impact of activities

---

1535 As considered in *West Faulkner Associates v London Borough of Newham* [1994] 71 BLR 1 (CA) and *JM Hill & Sons v London Borough of Camden* [1980] 18 BLR 31 (CA) (where there were concurrent terminations by the employer for failure to proceed regularly and diligently, and by the contractor for non-payment).

1536 As considered in *Miller v Cannon Hill Estates* [1931] 2 KB 113 and *Hounslow London Borough Council v Twickenham Garden Developments* [1971] 1 Ch 233. In the latter case the employer unsuccessfully sought an injunction to force the contractor to leave site after purported termination for slow progress.

1537 As considered in *Supamarl v Federated Homes* [1981] 9 Con LR 25 and *GLC v Cleveland Bridge and Engineering Company* [1984] 34 BLR 50 (CA).

1538 As considered in *Consarc v Hutch* [2002] All ER (D) 236 (TCC).

1539 As considered in *H Fairweather & Company v London Borough of Wandsworth* (1987) 39 BLR 106 (QB).

1540 As considered in *Vranicki v Architects Registration Board* [2007] EWHC 506 (QB).

1541 As considered in *Duncanson v Scottish County Investment Company* [1915] SC 1106.

1542 Case study summarised in Section 17.10.

on other activities. In practice a programme is considered most often in relation to the work of the main contractor rather than other team members, and in relation to the construction phase of a project rather than its planning and design phase.[1543]

A timetable sets out the deadlines for each stage of a team member's work that affects or is affected by another team member or a third party, during both the preconstruction phase and the construction phase of a project.[1544] To identify the deadlines and interfaces among alliance members, and between alliance members and third parties, a timetable can provide the spine of a successful alliance and can be integrated with the implementation of Building Information Modelling (BIM).

With this in mind we will use the word 'programme' to describe a document detailing the critical path activities of a single alliance member and the word 'timetable' to describe a document setting the agreed deadlines and sequences for the actions of all alliance members:

- Where these actions affect the work of other alliance members
- Where these actions require approvals or contributions or other reactions from other alliance members
- Where these actions require approvals from other parties outside the alliance.

Clients are sometimes advised not to treat a programme as a contract document because this could give contractual status to the main contractor's method statements, as a result of which any variation to these method statements could then give rise to greater time and/or cost entitlements.[1545] This attitude illustrates how the creation and contractual status of a main contractor's programme are often viewed through the defensive lens of how to present evidence when pursuing or resisting claims for additional time and money.

Project managers and contractors may also delay the creation of a programme until just after a construction phase contract has been entered into,[1546] an approach which is reflected in certain of the standard form construction contracts considered in Section 17.4. This delay renders the programme provisions no more than an agreement to agree and creates uncertainties as soon as the project starts on site. The delay is also unnecessary because a main contractor has no additional programming data available just after signing its contract compared to the data available just before signing that contract. These delays and the consequent uncertainties can have adverse effects on team relationships and on the efficient delivery of the project by:

- Creating immediate and unnecessary pressure for the main contractor because it has already committed to a contractual completion date before seeking approval of the programme through which it will meet this commitment

---

1543 For example, the SCL Delay and Disruption Protocol (2017) concentrates only on the programme of the main contractor.

1544 PPC2000 provides for a 'Partnering Timetable' governing deadlines and interfaces during the preconstruction phase of a project and for a 'Project Timetable' governing deadlines and interfaces during the construction phase.

1545 In *Yorkshire Water Authority v Sir Alfred McAlpine & Son (Northern)* (1985) 32 BLR 114 (QB), an approved bar chart and method statement were submitted with the contractor's tender, signed as part of the contract and later assisted the main contractor in claiming it was entitled to follow that method statement or to seek a variation order with appropriate time and cost consequences.

1546 'Time pressures on many projects mean that the set of coordinated method statements, programmes and budgets will still be under development after construction has started' Bennett, J. (2000), 1

- Delaying the ability of the main contractor to use the programme as a tool to help it work in an efficient manner, for example by delaying its ability to create back to back programmes with the members of its supply chain
- Creating the risk of further delay and uncertainty if the main contractor's proposed programme is not accepted by the client or project manager.

These problems are illustrated in the SCL Protocol[1547] which aims to 'assist in managing progress of the works and to reduce the number of disputes relating to delay and disruption'[1548] but which:

- Does not recommend agreement of a programme prior to contract award but only 'as early as possible during the works'.[1549] For certain projects it envisages preparation firstly of an initial programme 'shortly after the contract has been awarded' showing only the first three months of work, then preparation of a proposed programme 'for the entirety of the works'[1550]
- Does not recognise the need for a programme of preconstruction phase activities and hardly mentions the programming of design, alluding only briefly to this as part of 'the key stages of the works' to be covered by a programme together with 'approvals, procurement or manufacturing, installation, construction, commissioning and taking over'[1551]
- Claims to be 'a balanced document, reflecting equally the interests of all parties to the construction process',[1552] but focuses solely on the main contractor's programme and does not, for example, mention any system for programming the activities of consultants.

Caution and delay associated with main contractor programmes can distract attention from the benefits of agreeing a preconstruction phase timetable and a construction phase timetable, in each case as a contractual planning document that expresses deadlines not only between the client and main contractor but also with other team members. The purposes of these integrated timetables are to support collaborative construction procurement by:

- Enabling team members to rely on each other achieving agreed deadlines at each stage of a project
- Connecting the deadlines governing agreed interfaces wherever one team member depends on another completing all or part of its work.

A further function of an integrated timetable supporting collaborative procurement is to record the timelines for all those activities that are preconditions to proceeding with the construction phase, including for example:

- Finalisation of agreed designs, prices and supply chain arrangements
- Establishment of an acceptable understanding regarding project risks

---

1547 SCL Delay and Disruption Protocol (2017).
1548 SCL Delay and Disruption Protocol (2017), Core Principles 1, 5.
1549 SCL Delay and Disruption Protocol (2017), Guidance Part B, Section 1.40, 18 recommends agreement of a programme 'As early as possible during the works' but does not specifically recognise a preconstruction phase, instead envisaging a single programme describing how the main contractor 'plans to carry out the works'.
1550 SCL Delay and Disruption Protocol (2017), Guidance Part B, Section 1.52, 50.
1551 SCL Delay and Disruption Protocol (2017), Guidance Part B, Section 1.40, 18.
1552 SCL Delay and Disruption Protocol (2017), Introduction F.

- Satisfying health and safety requirements
- Agreeing the construction phase timetable
- Fulfilling third-party preconditions such as regulatory consents and project funding.[1553]

In justification of a single-stage procurement process, it is sometimes said that there is no time for early contractor involvement or other collaborative procurement. The case studies repeatedly show how collaborative procurement activities save time rather than create delays, and a well-organised team can establish a timetable governing the 'limited period of time during which to implement preconstruction phase activities without causing a delay in commencement of the construction phase'.[1554] Section 17.5 considers how this challenge can be met, but firstly Section 17.4 considers the relevant provisions of standard form contracts.

17.4 How Are Programmes and Timetables Treated in Construction Contracts?

Standard form construction contracts treat programmes and timetables in very different ways, ranging from a non-binding programme under JCT 2016 to a matrix of connected contract programmes and timetables under NEC4 and PPC2000. For example, the US ConsensusDocs 300 form appears to provide for an integrated timetable in that 'where the work of one IPD Team member is dependent upon the prior performance of another IPD Team member, the IPD Team member whose work follows shall request of, and receive from, the prior performer a commitment as to when the work to be handed off will be finished. Applicable IPD Team members shall agree upon criteria for the hand-off of items of work'.[1555]

FIDIC 2017 provides that:

- The contractor submits an 'initial programme' within 28 days following notice to commence work
- The Engineer (i.e. the project manager) gives notice within 21 days following submission of the initial programme, 'stating the extent to which it does not comply with the Contract…or is otherwise inconsistent with the Contractor's obligations'
- Within 14 days following the Engineer's notice of non-compliance the Contractor submits a 'revised programme' following submission of which the Engineer may give a further notice of non-compliance.[1556]

However, FIDIC 2017 also requires that the contractor is obliged to start work within 42 days from the notice to commence,[1557] at a time when it may not have had any response from the Engineer to its initial programme or may be engaged in a potentially unlimited series of resubmissions. Also, whether an agreed programme has binding legal effect under FIDIC 2017 is not made clear, as the requirement for the contractor to 'proceed in accordance with the Programme' is stated to be 'subject

---

1553 A list of relevant preconditions appears in PPC2000 (2013) clause 14.
1554 Two Stage Open Book and Supply Chain Collaboration Guidance (2014), 31.
1555 ConsensusDocs 300 (2016) Article 2.6 (Pull Planning).
1556 FIDIC (2017) clause 8.3.
1557 FIDIC (2017) clause 8.1.

to the Contractor's other obligations under the Contract'.[1558] The FIDIC programme governs only the obligations of the main contractor and any nominated subcontractors, without connection to corresponding deadlines for the work of consultants and other team members.[1559]

FIDIC 2017 has introduced optional special provisions under which the client can designate milestones to be completed by specific dates for 'a part of the Plant and/or a part of the Works stated in the Contract Data (if any) and described in detail in the Specification as a Milestone'.[1560] Although this optional provision creates additional obligations only for the main contractor, with no express recognition of corresponding milestones for the client and consultants, it represents the first move by FIDIC towards the need for a timetable as well as a programme.

As regards the right of the main contractor to receive timely information, FIDIC 2017 provides that the contractor shall give notice to the Engineer 'whenever the Works are likely to be delayed or disrupted if any necessary drawing or instruction is not issued to the Contractor within a particular time, which shall be reasonable'.[1561] However, it is difficult to see how the contractor will demonstrate the reasonableness of its expectations, and the lateness of the Engineer or any other team member, in the absence of a timetable setting out the agreed dates when drawings and other required instructions will be issued.

ICC 2014 provides for a programme to be created and agreed after the contract is entered into,[1562] and the guidance notes state that this programme 'should not be included in the Contract Documents'.[1563] The ICC Partnering Addendum 2011 refers to a 'consolidated programme showing the proposed timing of the contributions of the Partners',[1564] but the reference to 'proposed' suggests that this too is not binding. Also, there is no standard form ICC consultant appointment and therefore no provision for integration of contractor deadlines with any deadlines for consultant services.

JCT 2016 provides for a non-binding 'master programme' to be agreed after the contract is entered into,[1565] thereby creating an agreement to agree a document that in any event will have no contractual status if and when agreement is reached. The JCT contracts also include the option of a timetable known as an 'Information Release

---

1558 FIDIC (2017) clause 8.3.
1559 FIDIC (2017) clause 8.3 includes a detailed list of the components of the programme. There seems no logic in the requirement at clause 8.3(c) that the programme should state the timing of work undertaken by nominated subcontractors but not include deadlines for the contributions of other team members, beyond references to 'the Review periods for any submissions' (clause 8.3(d)) and 'the sequence and timing of inspections and tests' (clause 8.3(e)).
1560 FIDIC (2017) Special Provision 4.22.
1561 FIDIC (2017) clause 1.9.
1562 ICC (2014) clause 9.1.
1563 ICC Guidance Notes (2014), 6.
1564 ICC Partnering Addendum (2011) P4(7).
1565 JCT SBC/Q (2016) clause 2.9.1.2 requires main contractor to produce a 'master programme', including critical paths and other details specified in the Contract Documents, 'as soon as possible' after execution of the contract, and clause 2.9.1.2 requires updates within 14 days following Architect/Contract Administrator decisions. But under clause 2.9.3 this does not affect the parties' contractual obligations ('no obligation beyond those imposed by the Contract Documents').

Schedule' which sets binding dates for stages in which designs will be released by the client to the main contractor,[1566] although:

- The Information Release Schedule applies only during the construction phase and only to the client and the main contractor
- The Information Release Schedule is not integrated with the contractor's 'Design Submission Procedure'[1567] which provides only in general terms for contractor submissions to be made 'in sufficient time to allow any comments of the Architect/Contract Administrator to be incorporated'.[1568]

However, the Information Release Schedule can be integrated with deadlines for consultant services, and the JCT Consultancy Agreement (Public Sector) includes a requirement for compliance with a 'Programme and/or Information Release Schedule'.[1569] In relation to preconstruction phase activities, the JCT Consultancy Agreement (Public Sector) can be integrated with the JCT Pre-Construction Services Agreement which also includes a requirement for compliance with a 'Programme and/or Information Release Schedule'.[1570] There is also an obligation on subcontractors under the JCT Sub-Contract to undertake their work 'in accordance with the programme details stated in the Sub-Contract Particulars and reasonably in accordance with the main contract works in each relevant Section'.[1571]

In the absence of an Information Release Schedule, JCT 2016 contracts require the Architect/Contract Administrator to provide further drawings or details 'at the time the Contractor reasonably requires them, having regard to the progress of the Works'.[1572] This was amended from the JCT 2011 wording which provided for provision of these further drawings and details 'at the time it is reasonably necessary for the Contractor to receive them, having regard to the progress of the Works'.[1573] Neither provision creates the certainty that team members require, and subjective, conflicting views of what 'reasonably' means are likely to generate misunderstandings and disputes.

NEC4 provides for a timetable of 'Key Dates'[1574] and for an 'Accepted Programme',[1575] both of which are binding and both of which are provided for in the NEC4 construction contracts and in the NEC4 consultant appointment.[1576] NEC4 creates a proactive project management culture of which regular updating of the 'Accepted Programme' is an important and demanding aspect.[1577] 'Key Dates' are defined as 'the date by which

---

1566 JCT SBC/Q (2016) clause 2.11 provides for an 'Information Release Schedule', describing obligation of Architect/Contract Administrator to ensure release of specific information at specific times except where 'prevented by an act or default of the Contractor or any Contractor's Person'.
1567 JCT SBC/Q (2016) clause 2.9.5 and Schedule 1.
1568 JCT SBC/Q (2016) Schedule 1 clause 1.
1569 JCT CA (2016) clauses 2.1 and 2.3.
1570 JCT PCSA (2016) clauses 2.1 and 2.3.
1571 JCT SBC/Sub/C (2016) clause 2.3.
1572 JCT SBC/Q (2016) clause 2.12.2.
1573 JCT SBC/Q (2011) clause 2.12.2.
1574 NEC4 (2017) clause 30.3.
1575 NEC4 (2017) clause 31, although there is an option in clause 31.1 to provide a programme after the contract has been entered into. Note also clause 31.2 as to content and clause 31.3 as to reasons for the Project Manager not to accept the programme.
1576 NEC4 Professional Services Contract (2017).
1577 NEC3 (2013) included an onerous obligation under clause 32.1 for updates to reflect every Compensation Event, but this has been diluted in NEC4 (2018).

work is to meet the Condition stated'[1578] and a remedy for failure to achieve a key date is a deduction from payments otherwise due.[1579] Confusingly, NEC4 Option X12 adds a further 'timetable', to be issued by the Core Group and showing the timing of proposed contributions of each team member,[1580] with an obligation on the main contractor to change its programme (as a compensation event) if necessary to comply with a revised timetable.[1581]

The NEC4 Alliance Contract (2018) provides for a single programme governing and coordinating the work of all alliance members and regularly updated by the Alliance Manager.[1582] However, there are no separate programme or key dates governing early contractor involvement in the NEC4 Alliance Contract, or in NEC4 Option X22, and in this way the NEC4 contracts appear to downplay the importance of collaborative activities during the preconstruction phase of a project.

PPC2000 governs preconstruction phase activities through a 'Partnering Timetable' agreed by all team members at the point of signing the contract,[1583] followed by a 'Project Timetable'[1584] agreed by all team members prior to commencement of the construction phase of the project. These create mutual commitments to agreed deadlines at each stage as between the client, the main contractor, all consultants and any other supply chain members who sign the PPC2000 Project Partnering Agreement and Commencement Agreement or enter into Joining Agreements.[1585]

The Partnering Timetable and Project Timetable are contractually binding on all alliance members, subject to:

- Any agreed preconditions to implementation[1586]
- The agreed contractual change procedure including any agreed arrangements for acceleration or postponement[1587]
- The agreed contractual procedure in respect of events of delay and disruption[1588]
- Statutory rights of suspension for non-payment and agreed rights of suspension or abandonment of the project.[1589]

PPC2000 distinguishes the Project Timetable from 'supporting method statements and procedures' that are to be submitted separately by the main contractor, but which do not have contractual status.[1590]

17.5   What Timetables Support a Collaborative Team?

An integrated timetable setting out agreed deadlines and interfaces between team members sits at the heart of collaborative procurement. Without it, there is the risk that

---

1578 NEC4 (2017) clause 11.2(11).
1579 NEC4 (2017) clause 25.3.
1580 NEC4 (2017) Option X12.3(7).
1581 NEC4 (2017) Option X12.3(7).
1582 NEC4 Alliance Contract (2018), clauses 32 and 33.
1583 PPC2000 (2013) clause 6.1.
1584 PPC2000 (2013) clause 6.5.
1585 PPC2000 (2013) clause 6 and Appendix 6.
1586 PPC2000 (2013) clause 6.1.
1587 PPC2000 (2013) clauses 6.6 and 17.
1588 PPC2000 (2013) clause 18.
1589 PPC2000 (2013) clauses 20.17 and 26.
1590 PPC2000 (2013) clause 6, 2.

contract commitments will be open-ended and will allow delays at any stage. Failure to start and proceed with a particular stage of work on time is a common project risk that can result from delays caused not only by third-party events, but also from problems within the team such as procrastination, shortage of information or lack of resources.[1591] These causes of delay can be identified and averted or limited by means of an integrated multi-party timetable that links directly the deadlines agreed for the mutual commitments of all team members.

An integrated timetable should state each activity that is time-critical, the relevant deadline, the party or parties responsible and the preconditions on which it is dependent.[1592] For example, a preconstruction phase timetable should identify the deadlines and the team members responsible for each of the following:

- Design development and submissions
- Surveys and investigations
- Cost plan development and submissions
- Design contributions and reviews
- Procurement processes for the selection of subcontractors, manufacturers, suppliers and operators
- Joint activities including Supply Chain Collaboration
- Costing of all works, services and supply packages
- Risk management actions linked to a risk register
- Client approvals and comments in response to each submission and proposal by other alliance members
- Applications for and grant of third-party approvals
- Funding, land acquisition, insurances and other preconditions to commencement of work on site
- Satisfaction of health and safety preconditions and other legal and regulatory preconditions to commencement of work on site.

A collaborative team should manage a timetable so that 'when a key target is in danger of being missed, this must be treated as a crisis and clear, effective action taken quickly to get the work back on its planned course. Without this a control system is simply a waste of resources and, once a timetable has been discredited by being ignored, nobody will bother to provide accurate or up-to-date feedback'.[1593] A collaborative timetable should, therefore, be supported by a risk management system that encourages early warning of any delay and its potential consequences.[1594]

Collaborative timetables should set out agreed risk management actions and should recognise where external influences may cause delays outside the control of team members. This can assist team members in identifying opportunities to mitigate the effects of these external influences, and in making 'risk management an integral component of program/project planning and engineering as well as contract administration...by

---

1591 Lock, D. (2000), 7.
1592 Pro forma timetables and guidance notes are set out in PPC2000 (2013) Appendix 6, FAC-1 Framework Alliance Contract (2016) Schedule 2 and TAC-1 Term Alliance Contract (2016) Schedule 2.
1593 Bennett, J. (2000), 186. For example, the PPC2000 (2013) clause 18.3(i) provides that if a client or consultant deadline is missed, then the contractor must give early warning to the client not more than five working days after the expiry of the relevant time limit.
1594 As considered in Chapter 18 and Section 19.6.

a coordinated sequence of activities'.[1595] For example, the Rogate House project alliance reported how 'joint investigation and planning of complex refurbishment works [achieved an] accurate and integrated construction phase [contractual] Project Timetable'.[1596]

Related collaborative techniques sit alongside an integrated timetable and there is growing recognition of the power of 'Lean' processes and tools designed to minimise wasted time.[1597] Lean provides for a master plan of key managerial and contractual milestones, a system of 'pull planning' to agree the timing of activities extracted from the master plan, a six-week 'look ahead' that takes a rolling period of activities from the pull planning, and daily scheduling populated from the six-week look ahead.[1598]

Integrated timetables are management tools that create peer group pressure because all team members are aware of their own deadlines and of the deadlines agreed by other team members. As a result, team members can give early warning to each other as to the consequences of a delay and can attempt to persuade each other to remedy a delay. Compliance with agreed deadlines can also be an agreed success measure, and it can be made clear that failure to comply gives rise to reduced incentive payments or reduced workflow.

However, if agreed deadlines and interfaces are not adhered to and if peer group pressure and incentive systems fail to produce results, then team members may need to take other actions:

- PPC2000 provides a right of termination by the client in the event that the main contractor 'does not commence and continue to fulfil its responsibilities under the Partnering Contract in accordance with the Partnering Timetable and Project Timetable'[1599]
- NEC4 clause 25.3 provides the right to deduct damages for a failure to meet agreed deadlines in the Accepted Programme or Key Dates Schedule, a less draconian approach than the right of termination in PPC2000 and arguably a more effective remedy.

17.6 What Timetables Integrate Collaborative Design?

It is recognised that 'one of the fundamental requirements for effective management of design is an efficient flow of information between the participants in the project. This applies particularly to those that have an input to the design phases of the project, and a list of such participants would include:

- The promoter and appropriate groups within the promoter organisation (such as their design department)
- The users/operators of the project deliverable(s), which may or may not be part of the promoter organisation

---

1595 Smith, R.J. (1995), 66.
1596 Association of Consultant Architects (2010), 34–35, as summarised in Section 3.10.
1597 Lean tools and principles are also noted in Chapter 25.
1598 As noted, for example, in McKinsey Global Institute (2017), 92.
1599 PPC2000 (2013) clause 26.4(i). As regards the consultants who are signatories to PPC2000 (2013), clause 26.3 provides that the client has a right of termination in the event of material breach, which could include failure to adhere to an agreed timetable.

- The project manager
- Team leaders in the project team
- The design manager
- The lead designers
- The design team leaders
- Sub-designers and appropriate team leaders within those groups
- Design approval consultants acting on behalf of the promoter
- Design checking consultants
- Local authorities
- Statutory bodies'.[1600]

It has been suggested that 'effective management of the design process is crucial for the success of the project' and that it should involve:

- The coordination of the consultants, including an interlocking matrix of their appointment documents, which should also have a clear relationship with the construction contract
- A detailed checklist of the design requirements, ensuring that the client fully understands the design proposals
- Particular care over the integration of building services design, and the avoidance of 'fuzzy edges' between consultants and specialist engineering contractors.[1601]

However, an integrated timetable of design activities also needs some flexibility so as to ensure that agreed deadlines do not excessively constrain the creative design process. It has been observed that:

- 'In most engineering design offices and other software groups, highly qualified staff can be found whose creative talents are beyond question. But, while their technical or scientific approach to project tasks might be well motivated and capable of producing excellent results, there is always a danger that these creative souls will not fully appreciate the importance of keeping within time and cost limits'.[1602]
- 'The creative element of the designing process requires a period of synthesis that cannot always be forced, the subconscious mind needs to work on the problem…any designers resent the imposition of a mechanistic management regime, since they feel this constrains their ability to design effectively…Since a fundamental aspect of design is the element of creativity, and the difficulty this brings in terms of an accurate estimate of the time needed to complete a design task, this must be considered in the management regime'.[1603]

Nonetheless, there is a point in the project process when the flexibility required for creative design gives way to the interfaces and deadlines required to complete each agreed level of detail, so that 'everyone in the project team fully understands the process they are working through and have bought into it' and so that 'programme milestones are reliably met'.[1604] For example, the contributions by each team member to each stage

---

1600 Smith, N.J. (2002), 246.
1601 Latham, M. (1994), Section 4.1, 23.
1602 Lock, D. (2000), 479.
1603 Smith, N.J. (2002), 243–244.
1604 Bennett and Jayes (1998), 80.

in the development of BIM models need to be made by agreed dates if BIM is to be adopted successfully.[1605] If interfaces and milestones are clearly understood and agreed, then it becomes much easier to agree that 'where innovation is needed, milestones are interpreted flexibly on the basis of providing just sufficient information to avoid delaying the project'.[1606]

The lead designer should undertake 'coordinating preparation of work stage programmes for the design process',[1607] so that design consultants and other design team members can rely on the timing of each other's contributions and approvals, so that designs will be sufficiently developed to be priced by the main contractor and its supply chain, and so that agreed designs can form the basis for planning consents and other third-party approvals. A multi-party alliance contract can create direct interfaces that fully integrate the timetables in consultant appointments with those in construction contracts.[1608] The following design interfaces have been identified as critical, and an integrated timetable can help a collaborative team to manage them more efficiently:

- Between different design consultants
- Between design consultants and specialist subcontractors and suppliers
- Between the design process and the procurement process
- Between the design process and the construction process
- Between the design process and the operation, repair and maintenance process
- Between the project designs and the designs of other projects.[1609]

17.7 What Timetables Support a Framework Alliance or Term Alliance?

ISO 44001 recommended that 'the collaborative partners shall agree on plans for delivery and performance based on the objectives. The joint management shall:

- Monitor delivery against mutually agreed plans and performance criteria
- Regularly review performance at both executive and operation levels
- Where appropriate, take corrective action to ensure performance is maintained
- Establish, implement and maintain procedures for monitoring and measuring the effectiveness of its collaborative business relationship arrangement
- Establish a process of continual improvement to enhance the effectiveness of its collaborative business relationship management'.[1610]

The integrated timetables that support framework alliance contracts and term alliance contracts can govern agreed activities in respect of:

1605 As considered in Section 14.3.
1606 Bennett and Jayes (1998), 80.
1607 Sinclair, D. (2011), 37.
1608 'The activities of designers, manufacturers, suppliers, contractors, and all other resources must be organised and integrated to meet the objectives set by the client and/or the contractor. In most cases, the programme will form a basis of the plan. Sequences of activities will be defined and linked on a timescale to ensure that priorities are identified and that efficient use is made of expensive and/or scarce resources' Smith et al. (2006), 6.
1609 Smith et al. (2006), 59.
1610 ISO 44001, Section 26.

- The selection of team members for a specific project or task under the agreed award procedures
- Cycles of Core Group meetings
- Engagement with Users and Stakeholders
- Recognition of where activities are dependent on parties or events outside the alliance
- Supply Chain Collaboration processes and other alliance activities
- Performance reviews.

For example, the integrated timetable that supports the FAC-1 framework alliance contract is stated to cover:

- 'Agreed deadlines, gateways and milestones in respect of the Framework Programme and achievement of the Objectives
- The timescales for Alliance Activities, including the nature, sequence, and duration of the agreed actions of each Alliance Member and any consents or approvals (whether required from Alliance Members or third parties) that are pre-conditions to subsequent actions'.[1611]

The SCMG Trial Project framework alliance used an integrated timetable to underpin 'programmed design, risk management, costing, and progressing activities undertaken during the early conditional appointment of an integrated team to seek additional savings and improved value'. As a result, they achieved 'time savings, such as through quicker build-up of prices leading to earlier start on site and reduced client/consultant time/costs'. A subcontractor reported, 'I don't know any other London borough that has managed to carry out so much work in so short a time'.[1612]

An equivalent integrated timetable forms part of the TAC-1 term alliance contract. This enables alliance members to use the time management techniques adopted on the following term alliances:

- The Project Horizon Trial Project term alliance who reported the 'lean programming of individual tasks leading to time savings'[1613]
- The Welwyn Hatfield Council term alliance contract with Mears, where collaborative management of time 'led to better quality for more satisfied customers and created a rigorous system for measuring performance' and 'cut the time that properties were vacant by 66%', with the quicker lettings leading to increased council rent receipts.[1614]

17.8 How Can an Integrated Timetable Improve Value?

It has been asserted that single-stage tendering is quicker than a collaborative two-stage selection process. Not only can a two-stage process save time compared to a single stage tender, but also the extent of claims and disputes arising on traditionally procured projects suggests that a rush to site rarely delivers the required results.

1611 Guidance note, FAC-1 Framework Alliance Contract (2016) Schedule 2 Timetable.
1612 SCMG Trial Project case study.
1613 Project Horizon Trial Project case study.
1614 'The reduced void turnaround time has enabled more families to move in and the tenants really like the viewings which are much friendlier', Peter Sharman, then Deputy Chief Housing Officer, Welwyn Hatfield Council, Association of Consultant Architects (2010), 45.

An integrated timetable for early contractor involvement and Supply Chain Collaboration can run in parallel with the progress of design development so as to avoid any delays at all. In addition, case studies have shown how that the time allocated to preconstruction phase activities under an agreed timetable can give rise to time savings rather than delays. For example:

- On the Rogate House project alliance:
  - 'The entire team agreed the construction phase timetable in advance on the basis of resident consultation to establish an agreed decanting programme'
  - 'Completion of works at Alma House had taken 115 weeks to refurbish 108 flats whereas at Rogate House it took 90 weeks to refurbish 192 flats'
  - 'The Rogate House team had worked at approximately double the speed'.[1615]
- On the Hampstead Heath Ponds project alliance:
  - 'BAM suggested a prefabricated steel frame structure which had been a provisional sum in the tender'
  - 'The prefabricated solution meant that this element of the project was efficiently concluded within six months'
  - BAM also 'reversed the proposed order of the pond works, so that the riskiest pond should be worked on first, in order to create 'float' for later works. This significantly de-risked the project programme'.[1616]
- On the Cookham Wood project:
  - 'Collaboration between Interserve and its specialist SSC resulted in a pre-cast volumetric cell proposal for the construction of the project'
  - 'As a result, the construction programme was reduced from 50 to 44 weeks with a saving of £85 000 in time-related site overheads'.[1617]

Other examples include 'a fit out project in Central London in order to consolidate the University of Westminster's corporate services departments [where] the works involved design of an office fit out, installation of new mechanical and electrical engineering services, new partitions, ceilings and decorations. The use of PPC2000 ensured a fast project turnaround and enabled construction works to commence prior to full completion of the mechanical and electrical services design'.[1618]

Collaborative time management can enable time savings through off-site manufacture, for example on the HMP Bure new build prison where:

- 'Planning drawings were submitted and planning achieved in eight months including the determination period of the local authority'
- 'The first 320 places were delivered in only 40 weeks on site working from when planning was granted. The second phase and remaining 180 places were delivered 6½ weeks ahead of the agreed contract date'
- 'The team used modern methods of construction for all the new build elements, allowing time on site to be reduced to a minimum'
- 'The team used lean planning processes to deliver the agreed Project Timetable'
- 'Joint design changed the traditional build kitchen to a modular solution and reduced costs (this change was instigated, designed, procured and built in 20 weeks)'

1615 Association of Consultant Architects (2010), 35.
1616 Case study summarised in Section 3.10.
1617 Cookham Wood Trial Project case study.
1618 Association of Consultant Architects (2010), 44.

- 'Use of [a contractual] Project Timetable achieved a fully operational prison occupied by its first inmate within 43 weeks from planning'.[1619]

Collaborative time management can address unexpected risks, including the effects of time spent on alliance activities. For example, on the North Wales Prison project:

- 'Collaborative work on Value Engineering to mitigate the threat of cost increases caused an unexpected delay in authorising start on site which in itself threatened to cause a construction delay of up to two months'
- 'The co-located team worked closely together to restructure the construction phase of the programme in a way designed to eliminate this delay while maintaining transparent milestones'
- 'In doing so they established a more logical sequence for phased handover of completed sections of the project'.[1620]

17.9 How Can a Collaborative Team Manage Change?

Team members cannot ignore the possibility of a change in the client's requirements or other new circumstances. A collaborative contract, like any other construction contract, needs procedures for agreeing these changes and their time and cost impacts. Team members need ways of adjusting their original financial arrangements to reflect new information acquired during the course of their relationship.

Provisions for change management are neo-classical features of construction contracts,[1621] with emphasis on the means to agree the cost and time impact of each change.[1622] Wherever possible, it is important to agree these effects before authorising a change although there will be emergencies where action needs to be taken before this new data is available.[1623] The build-up of shared data through alliance activities provides a stable basis for assessing the cost and time impact of a proposed change and for agreeing how to implement a change with minimum adverse effects.

While change is often challenging, some alliance models see it as normal or even desirable. For example, the NEC4 Alliance Contract envisages 'a developing Scope undergoing constantly instructed change causing changes to the Budget, along with collaboration over numerous matters notified through the early warning system'.[1624] Change mechanisms are also the medium through which to agree innovations and other proposals for improved value, and for example under PPC2000 any alliance member can propose a change 'if it is demonstrably in the best interests of the Project'.[1625]

Other collaborative models seek to minimise change, and in the USA 'many IPD projects have a zero-change order goal, and most have far fewer changes than conventional projects'.[1626] In practice this approach discourages rather than prohibits

---

1619 Association of Consultant Architects (2010), 24.
1620 North Wales Prison Trial Project case study.
1621 As considered in Section 8.1.
1622 For example, FIDIC (2017) clause 13, JCT SBC/Q (2016) clause 5, NEC4 clause 60.1(1), PPC2000 (2013) clause 17.
1623 For example, under PPC2000 (2013) clause 17.5.
1624 NEC4 *managing an alliance contract* (2018), 9.
1625 PPC2000 (2013) clause 17.1.
1626 Fischer et al. (2017), 370.

change and relies on the team working together so that, for example, 'design errors and omissions, a fertile ground for contractor claims in traditional projects, are a team risk that needs to be managed by the team together'.[1627] Similarly, in procurement models that support off-site manufacture, team members need to agree designs in advance of fabrication and to minimise later changes that could interrupt efficient and economical production.[1628]

Disputed changes can arise, for example, when a contractor considers that a design consultant has required a change in the client's brief, but the design consultant interprets this not as a change but as the addition of necessary design details. If a contractor is expected to absorb the cost and time effects of additional design details, not signalled in advance and not allowed for in its tender, this can have a negative effect on trust and a knock-on effect on the contractor's supply chain. Avoidance of disputes arising from unexpected changes requires early recognition of the issues and commitment to seeking agreed solutions. For example, on the Jarvis Hotels construction management alliance:

- 'Sub-division for tenancies raised challenges, particularly as to the mechanical and electrical upgrade. Also, the need to keep the hotel operating meant that some parts could not be surveyed in advance of start on site'.
- 'The team used the PPC2000 contractual change mechanisms to minimise cost and time consequences deriving from this continually evolving scope'.
- 'Power cuts, floods and false fire alarms caused disruption. The Core Group was used to resolve concerns and to ensure that work stopped if necessary for customer relations'.[1629]

A client can improve change management if it knows the team members' expected fees, profit and overheads and can then work with the team members to mitigate other aspects of the cost of a proposed change. This knowledge can help to avoid or minimise the scope for disputes because, 'when differences arise against a background of open-book record-keeping and the cooperative exchange of information, the process and disclosure of information can reduce the scope of the difference'.[1630] For example, on the St George's Hospital Keyworker Accommodation project, the team used open-book costing to agree a maximum price, following which 'monthly critical analysis ensured that financial risks could be eliminated or quantified…allowing the client to instruct change instructions which increased the quality of the project further, safe in the knowledge that costs would be confined within the agreed maximum price'.[1631]

When costing a change that commits expenditure of a provisional sum, the shared data available to a project alliance enables its members to follow agreed deadlines and a procedure for establishing a specific amount in place of each provisional sum, namely who will do what in order:

- To achieve completion and approval of remaining designs
- To organise the issue and return of subcontract tenders

1627 Fischer et al. (2017), 371.
1628 'The main issue is that OSM [offsite manufacture] design choices must be made during design development, resulting in limited freedom to make design changes after the construction phase starts' Finnie et al. (2017).
1629 Association of Consultant Architects (2010), 26–27 and case study summarised at Section 7.10.
1630 Comments on PPC2000 in Arup (2008), 50.
1631 Association of Consultant Architects (2010), 38.

- To obtain client approval and authorisation of the quoted fixed amount to allow the relevant work or supply to proceed.[1632]

At a strategic level, an integrated timetable can guide the team members through the process of change. For example, Havelok Homes reported how their timetable 'guided the parties through a process of change during the first years of the Term and underlined their commitment to meet regularly in order to solve problems and explore efficiencies'.[1633] Provisions for agreement of changes in long-term alliance relationships are set out in FAC-1[1634] and TAC-1.[1635]

17.10   Collaborative Time Management Case Studies

The following case studies describe how efficient time management was a key objective for three clients, one a commercial rail operator, the second a hospital trust and the third a Government department responsible for flood defences. In each case the project teams undertook collaborative joint planning and together created a matrix of agreed deadlines and interfaces that enabled them to avoid gaps or misunderstandings in the way they approached their work.

In the first case study West Coast Trains and Virgin Trains led a collaborative team to undertake the refurbishment of Macclesfield Station within a very limited timeframe permitted by the Strategic Rail Authority. When previously using a traditional procurement model, they had experienced delays combined with substantial premiums charged by contractors for meeting time constraints. They were also concerned as to the wide range of stakeholders who could influence the timing of the preconstruction activities and construction activities. On this project they agreed a single preconstruction phase timetable under PPC2000 so as to ensure that all team members understood and committed to the deadlines for each planning and design stage of the project. This approach also ensured that the dates for all on-site activities were visible to and approved by Network Rail, the Strategic Rail Authority and the station manager.

In the second case study Poole Hospital NHS Trust created a project alliance using PPC2000 in order to refurbish six operating theatres and add a seventh within a period of 24 months while maintaining a full operating schedule. It also agreed with the other alliance members arrangements by which the main contractor would stop work immediately in the event of an emergency. Establishing a single preconstruction phase timetable, followed by an integrated construction phase timetable, enabled team members to agree how other work could be carried out on site whenever work on the main theatres had to stop at short notice. The client avoided unacceptable noise levels interfering with clinical activity, and the other team members minimised claims for delay and disruption.

In the third case study the UK Environment Agency used a collaborative framework with NEC3 to govern replacement of a failing structure/steel sheet pile retaining wall. Contractors were asked to bid under Cost Led Procurement, based on the following criteria:

- Demonstrating that they can deliver for the cost

1632 PPC2000 Pricing Guide (2008), 5.
1633 Association of Consultant Architects (2010), 23.
1634 FAC-1 Framework Alliance Contract clause 9.1.
1635 TAC-1 Framework Alliance Contract clauses 9.1 and 9.5.

- Outlining key risks and how they can be mitigated
- Detailing what efficiencies they can deliver based on the cost
- Showing how these efficiencies can be top-sliced from the cost.

As the project was situated in a live harbour, the project team worked extensively with subcontractors and suppliers during design development to utilise their expertise in developing engineering and logistical solutions that would save time. The timeframe that was agreed by the collaborative team enabled the project to go from business case to completion in 14 months, essentially cutting the expected programme time in half.

Macclesfield Station Refurbishment

The Macclesfield Station project alliance comprised the West Coast Trains/Virgin Trains (client), C. Spencer (constructor), Heery International (alliance manager), Atkins (design consultant) and Faithful + Gould (cost consultant). The client had only a very limited time permitted by the Strategic Rail Authority for the refurbishment of Macclesfield Station, namely 40 days of temporary closures over weekends. This led them to establish a PPC2000 pre-construction phase agreement so that, with their constructor, alliance manager and design consultants, they could establish a reliable construction phase timetable that would fit within these tight constraints.

West Coast Trains/Virgin Trains had experience of projects overrunning under construction phase JCT contracts and also of contractors charging substantial premiums for meeting time constraints, and they were particularly conscious of the wide range of stakeholders who could influence the timing of preconstruction activities and construction activities. The integrated preconstruction phase Partnering Timetable was used to ensure that deadlines for all planning and design stages were agreed and committed to by team members, and that the dates for all on-site activities were visible to and approved by Network Rail, the Strategic Rail Authority and the station manager.

The project involved a seven-month preconstruction phase (September 2002 to May 2003) and a five-month construction phase (to September 2003) and was completed on time and within budget. Kevin MacConville of Heery reported that collaborative procurement 'from the initiation stage of the project can and should increase the effectiveness and efficiency of the partnering team including the client decision-making process'.

Association of Consultant Architects (2010), 31–32.

Poole Hospital Operating Theatres

Poole Hospital NHS Trust created a project alliance with Mansell (main contractor), QP Architecture (architect), Whicheloe MacFarlane MPD and Anthony Ward (engineers), Lorne Stewart (M & E specialist subcontractor) and McNaughts (cost consultant) to refurbish six operating theatres and add a seventh within 24 months while enabling its hospital to maintain a full operating schedule. It also needed its constructor to stop work immediately in the event of an emergency, as noise and vibration would not be permitted in the vicinity of the operating theatres.

The process of establishing the preconstruction phase timetable enabled the alliance team members to agree other less urgent works that could be carried out on site when work on the main theatres had to stop at short notice due to unacceptable noise levels interfering with clinical activity. The timetable also allowed for phased possession to enable the client to retain maximum control of the site. Collaborative time management included:

- Constructor and specialist mechanical and electrical subcontractor design input under an integrated timetable governing their work prior to start on site, to ensure that a fully designed project was implemented with further design releases not causing delays or misunderstandings
- Organisation of a phased project in a fully operational building under a construction phase timetable jointly prepared and agreed by the client, alliance manager, design consultants, constructor and mechanical and electrical specialist subcontractor
- Implementation of a complex construction phase timetable, with involvement of end user clinical teams to ensure integration with their clinical programme.

Association of Consultant Architects (2010), 36.

Rye Harbour Flood Defences

The Rye Harbour Trial Project comprised the £9.6 million replacement of a harbour wall by a team comprising Waterways/Environment Agency (client), Jackson Civil Engineering (JCE) (main contractor), Halcrow EC Harris (project manager), Arcadis (cost consultant), Arcelor Mittal (steel sheet pile supplier), Team Van Oord in partnership with Jackson's (civil engineering) and Commercial Marine and Piling (subcontractor).

The project involved the replacement of a failing structure/steel sheet pile retaining wall as part of the Environment Agency's flood defence programme, for which the Environment Agency adopted the Cost Led Procurement route. This enabled them to generate savings of 6% and also to go from business case to project completion in 14 months, essentially cutting the programme time in half. The key driver for selecting the Cost Led Procurement route was the potential to procure and deliver the project within a constrained timeframe.

Early Contractor Involvement (ECI) was established during the development of the bid. Due to the tight timeframe, and some of the complexities of the project (including a number of protection orders on the site), the client, consultants, main contractor and tier 2 and tier 3 supply chain members collaborated extensively to come up with solutions to specific problems and challenges.

The project brief was developed and issued to contractors on the existing Environment Agency framework, and they were asked to bid based on the following criteria:

- Demonstrating that they can deliver for the cost
- Outlining key risks and how they can be mitigated
- Detailing what efficiencies they can deliver based on the cost
- Showing how these efficiencies can be top-sliced from the cost.

Because the client had set out clear costs, this gave the supply chain partners more confidence to put forward innovations within those costs. As the project involved working on

(Continued)

a live harbour, the project team worked extensively with the tier 2 and tier 3 partners during design development to utilise their expertise in developing important environmental solutions, including:

- Some precious intertidal plants from salt marsh were relocated and transplanted, avoiding loss of vegetation
- Natural England accepted the habitat created at Rye Harbour Farm as mitigation for the mudflats lost when they had to drive approximately 1000 m of piling
- Close working with marine ecology teams meant they prevented any mudflat washed in the Rother being seen as wasted and damaging the environment. This saved a lot of money in waste disposal
- Extensive negotiations with Natural England enabled the team to continue working through the bird breeding season. This was achieved through the utilisation of a vibro-piling innovation, which reduced the impact of noise on the site so that the birds were not disturbed. This was a massive innovation, which avoided huge de-mobilisation and remobilisation costs of £117 000. This solution was developed by Jackson Civil Engineering and Team Van Oord (Tier 2), and also led to the programme time being reduced by a number of weeks
- To mobilise plant around the site, a raised track was constructed along the line of an existing stone track. This helped to reduce both short and long-term impacts to the site.

Rye Harbour Trial Project case study.

18

How Can Collaborative Procurement Improve Risk Management?

18.1 Overview

Risk has been defined as the 'effect of uncertainty on objectives',[1636] and as 'the possible adverse consequences of uncertainty'.[1637] Latham commented that:

- 'No construction project is risk free'
- 'Risk can be managed, minimised, shared, transferred or accepted. It cannot be ignored'.[1638]

It has been suggested that 'almost any project will go well if nothing goes wrong, but the test is whether the project can survive adverse conditions'.[1639] Construction and engineering projects carry a wide variety of risks, and can benefit from collaborative assessment of ways to minimise the potential impact of those risks and from agreed risk management actions undertaken by team members working together.

Risks can be managed by a collaborative team if they put the right machinery in place, and 'an effective collaboration is one where the parties share responsibility as far as is practical in supporting the individual risk of the partners'.[1640] However, those 'who focus simply on cultural behaviour changes rather than managing risks may find that rose-tinted glasses obscure their vision. When some of their aspirations start to unravel because the risks that impact time, cost or quality issues have not been addressed in a robust manner, the project and participants suffer'.[1641] Collaborative procurement does not approach risk only in terms of its transfer or allocation. Collaborative risk management involves additional reviews and enquiries through which team members challenge their own and each other's risk assumptions when there is still time to take mitigating actions without causing project delay.

Often, project risk assessments are undertaken by consultants on behalf of the client before a main contractor and its supply chain are selected. For these purposes it is assumed that each bidding contractor will undertake its own risk assessment at its own cost and risk before submitting its tender. To the extent that client and consultant risk assessments are communicated to bidders, these are then left to be interpreted

---

1636 ISO 31000, Section 2.1.
1637 Smith, N.J. (2002), 193.
1638 Latham, M. (1994), Section 3.7, 14.
1639 Fischer et al. (2017), 362.
1640 ISO 44001, Section 8.6.4.
1641 Jones et al. (2003), 189.

Collaborative Construction Procurement and Improved Value, First Edition. David Mosey.
© 2019 John Wiley & Sons Ltd. Published 2019 by John Wiley & Sons Ltd.

by bidders, correctly or incorrectly, when formulating their bid prices for the project. This process does not allow any chance for each bidding contractor or its proposed supply chain to interrogate the client's and consultants' risk assessments. It also does not allow a period of time for the client, its consultants and the selected contractor with its proposed supply chain to work together in trying to reduce risks and the prices attached to those risks.

Until a contractor is selected, systematic risk analysis is compromised by the imbalance of power between the client inviting competitive bids and the bidding contractor, with costs and uncertainty increased among multiple bidders. A client and its consultants have considerable time to assess risks, whereas a bidding contractor has only a period of a few weeks to undertake its separate assessment while at the same time compiling all other aspects of its response to the client's invitation to tender.

The tenderers' risk assessments are likely to be abbreviated and influenced by their wish to submit successful responses to the technical and pricing requirements of the bid. In a single-stage approach, bidding contractors 'will do enough preparatory work to be successful at tender but are unlikely to be able to understand fully all aspects of the project or have sufficient time to identify and consider how to manage the potential risks to the project'.[1642]

The client and its consultants may decide that any bidder proposals describing risk assessments that differ from those in the invitation to tender are in fact techniques to gain a more favourable financial position. Also, each bidding contractor has no right or responsibility to analyse risk jointly with the client and consultants, and this can lead to:

- A cynical assessment by bidders of client and consultant risk assumptions
- The exploitation by the selected contractor of weaknesses in the client's documents
- The exploitation by the selected contractor of any risks which the client has not comprehensively transferred.

For any team to manage risk effectively, it needs to establish the time and processes necessary to undertake joint risk management exercises after contractor selection and prior to start on site, and the ability to make adjustments that reflect the outcome of these exercises. This chapter will examine how risks are priced, how collaborative joint risk management can operate in practice and how this can affect ground risk in particular. It will also look at the effect of managing risk through no blame clauses, through new insurance and through a project bank account ('PBA').

18.2   How Are Risks Priced?

Risk management is not automatically a shared process. Typically it is 'undertaken by both client and contractor organisations, but for different reasons' where:

- 'Clients will usually be concerned with the best use of their capital resources, the likely cost of procuring the facility and their return from their capital investment'
- 'Contractors will be concerned with the decision as to whether to tender for a given project in terms of the returns obtainable, the desired competitiveness of their tender,

---

1642 JCT CE Guide (2016) Section 37.

and the most profitable means of constructing or increasingly designing and building the project'.[1643]

If procurement models, contracts and consultant risk appraisals focus only on the transfer of risk and not on its management, this usually gives rise to a risk premium or risk contingency charged by the party accepting the transferred risk. In addition, 'the fact that both supply chain risk allowances and client contingencies are calculated more by rules of thumb than by a reliable scientific means creates the inherent problem that risks may not be adequately covered by either individual risk allowances or client contingencies'.[1644] For example, it has been noted that 'the true purposes of contingency and risk provisions within the industry have unfortunately been corrupted in many instances from being a pro-active business management tool to one of reactively masking preventable failures and poor planning'.[1645]

The risk premiums or contingencies allowed by main contractors and their supply chain members are likely to be broad estimates, and can prove insufficient to cover the cost of the required remedial actions if and when the risks materialise. However, in those circumstances, it is likely that any contractor, subcontractor or supplier will be unwilling to incur the additional costs necessary to cover the risks as this will erode its profit. As a result, the project will suffer from the claims and counter-claims that arise from the client seeking to impose risk transfer provisions and from the contractor, subcontractors and suppliers resisting costs that make the project unprofitable. In these circumstances, the client and the project are likely to suffer adverse consequences far greater than the cost of the client retaining the risk or agreeing a joint strategy with the contractor and other team members for managing that risk.[1646]

Where unanticipated risks have a later impact on the expected designs or methods of working, the parties are left to argue as to who bears the cost and time consequences and whether the contractor has a claim by reason of an express contract provision in relation to the risk in question or by reason of information provided by the client being incorrect or incomplete. In considering how risks should be managed it is also important to consider the potential risk management liability of consultants, for example in cases where:

- The court observed that 'the first defendant's actual assessment of the risk of methane at the design stage depended on what they actually knew. The competence of their assessment depends also on what they ought to have known'[1647]
- The foundations of a coal-fired power station failed due to the strength and composition of the sub-soil, and the consultant engineers were held to be under a duty to verify the geological conditions on which design assumptions were based.[1648]

There have been attempts to escape the constraints of contract prices by claiming that unforeseen matters change a contract so much that it is frustrated. However, these

---

1643 Smith et al. (2006), 93, 94.
1644 JCT CE Guide (2016) Section 33.
1645 Farmer, M. (2016), 16.
1646 Smith et al. (2006), 62, 63 refers to the premiums charged for the arbitrary transfer of risk.
1647 *Eckersley v Binnie & Partners* [1988] 18 Con LR 1 (CA).
1648 *Mirant Asia-Pacific Construction (Hong Kong) v Ove Arup* [2004] EWHC 1750 (TCC).

claims are not likely to succeed[1649] and the courts usually expect contractors not to rely only on information provided by the client but to 'make provision for a possible worst-case scenario'.[1650]

Contractors can be required to quote any risk contingencies at the point of selection, and a collaborative approach to risk management activities during the preconstruction phase can look for ways to reduce or eliminate these risk contingencies. For example:

- PPC2000 requires that all proposed risk contingencies are notified to the client by other team members prior to their preconstruction phase appointments but are only included in the agreed price if they have been approved after their joint review by all team members and after implementation of agreed risk management activities[1651]
- ConsensusDocs 300 provides that an 'IPD Team Contingency is available to cover Payable Costs from unanticipated events or issues such as design development or refinement, coverage for scope gaps during procurement of subcontractors, correction of defects or omissions in the design or in the installed Work' (where not covered by professional indemnity insurance cover) and also varied market conditions, changes relating to unanticipated site conditions, and resequencing or acceleration to achieve improved outcomes.[1652]

Collaborative risk costing can create:

- A clear understanding of each team member's approach to the provisional costing of risk allowances at the point of its selection, so that collaborative risk management can commence and so that additional risk allowances are not introduced at a later date
- A joint risk management system under which the provisional costing of any matter in a risk register is followed by agreed activities during the preconstruction phase, so that risk allowances can be eliminated or reduced and so that any remaining risk allowances are kept within the overall budget/target costs as a precondition to commencement of the construction phase of the project
- An understanding of where it is appropriate to share the costs of certain risks among team members in agreed proportions so as to discourage risk allowances and so as to encourage collaborative solutions.

18.3 How Can Risks Be Managed Jointly?

ISO 44001 describes joint risk management in the following terms:

- 'The joint risk management team shall establish and record the process to be used for joint risk management within the relationship, recognising the links into each organisation's existing risk management processes'

1649 In *Davis v Fareham* [1956] AC 696, Lord Radcliffe rejected Davis Contractors' claim for additional money and stated 'In my view their safety lies in the insertion of explicit conditions in any fixed price contracts they may undertake; it does not lie in an appeal to the principle of frustration'.
1650 As was required in *Obrascon v HM Attorney General of Gibraltar* [2015] EWCA Civ 712, also referred to in Section 18.6.
1651 PPC2000 (2013) clause 12.9.
1652 ConsensusDocs 300 (2016) Article 5.1.4 (IPD Team Contingency).

- 'The joint risk register shall be reviewed at planned intervals as defined under the governance structure and appropriate actions addressed'
- 'The term "joint" implies that each partner makes a contribution to all of the risk management activities'
- 'An effective collaboration is one where the parties share responsibility as far as is practical in supporting the individual risk of the partners'.[1653]

There are different types of risk[1654] and different ways of approaching risk management. Some risks can be managed by an authoritative approach such as an instruction given by a project manager, but where the answer is less clear other risk management options may be more appropriate such as a collaborative approach or otherwise just 'muddling through'.[1655] Muddling through may be a way to resolve minor issues but a collaborative approach through joint risk management is more prudent where risks could have a serious impact on quality, safety, cost or time.

Joint risk management is a system by which team members have the opportunity to identify risks affecting a project as soon as possible, to agree the status of different types of risk and to agree the strategy to be adopted for dealing with each risk. However, the joint analysis of risk will only benefit the project if actions are undertaken based on the results of that analysis.[1656] Risk management actions can include:

- 'Obtaining additional information
- Performing additional tests/simulations
- Allocating additional resources
- Improving communication and managing organizational interfaces'.[1657]

Joint risk management can create new opportunities for risk mitigating actions, and starts with 'identification of risks that need to be raised with collaborative partners to ensure the most effective approach is adopted'.[1658] These can be set out in a risk register, which 'shall be maintained as documented information and shall be part of the joint risk management process'.[1659]

In order to create joint risk management processes, it is important that 'all parties to a contract have the same appreciation of the identified risks. The contractor and the client should have similar views of the likelihood and potential effects of all risks'.[1660] The UK Office of Government Commerce included in its 'Critical Factors for Success' a system of 'risk and value management that involves the entire project team, actively managed through the project'.[1661]

Joint risk management is enabled partly by early collaborative appointments because 'the earlier the whole project team is appointed the better the risk management process will be. Contractors, consultants and other key suppliers bring knowledge

---

1653 ISO 44001 Section 8.6.4.
1654 For example, Hancock categorised risks as 'tame', 'messes', 'wicked' and 'wicked messes', Hancock, D. (2010), 33.
1655 Hancock, D. (2010), 57–59.
1656 Smith et al. (2006), 34.
1657 Smith et al. (2006), 88.
1658 ISO 44001 Section 8.3.6.3.
1659 ISO 44001 Section 8.3.6.3.
1660 Smith et al. (2006), 94.
1661 OGC (2007), Construction Projects Pocketbook, 1.

and experience of construction, delivery and related financial risks that are helpful in managing risks'.[1662]

A UK report found that Government clients were 'not investing enough effort to address risk when a project is being conceived'[1663] and recommended a project risk management plan which should:

- 'Define acceptable levels of risk in areas of cost, time and quality
- Detail the risk reduction measures to be taken to contain risks within these levels
- Outline cost-effective fall-back plans for implementing if and when specific risks materialise
- Identify the resources to be deployed for managing risk
- Explain the roles and responsibilities of all parties involved in risk management'.[1664]

The Trial Projects have shown how joint risk management can include steps taken by the contractor with other team members to recover construction phase delays. For example, on the Cookham Wood Trial Project, 'due to adverse inclement weather conditions, the project fell behind programme on site, but collaboration between Interserve and SSC has brought construction back on programme'.[1665] Joint risk management also underpinned a collaborative commitment to BIM clash detection and to the notification of pragmatic solutions.

18.4 How Do Construction Contracts Treat Risk Management?

Most standard form construction contracts focus primarily on risk allocation, with provision for contractor entitlements to claim additional time and money for unforeseeable matters that are listed as grounds for loss and expense or compensation events.[1666] Those forms that deal only with the construction phase activities cannot make provision for the earlier involvement of main contractors or specialist supply chain members in collaborative risk management.

The treatment of ground risk illustrates how different standard form contracts treat risk management, and the relevant provisions of FIDIC 2017, ICC 2014, JCT 2016, NEC4 and PPC2000 are considered in Section 18.6. Meanwhile, this Section 18.4 considers the approaches taken in a range of standard forms to the joint management of risk through techniques such as early warning, the use of a risk register and the agreement of actions to avoid or mitigate the effects of risks when they are identified.

The Guide to the JCT CE 2016 contracts gives clear pointers as to how 'a robust system of risk identification and management'[1667] should be undertaken 'ideally at a stage when the proposed design is not complete so that it is possible for the contractor and key specialists to consider ways in which the design can be made easier to build and maintain'.[1668] It states that:

1662 Bennett & Pearce (2006), 249.
1663 Efficiency Unit (1995), Section 154, 56 identified 'two projects which ran into trouble because environmental problems were not identified before construction started'.
1664 Efficiency Unit (1995), 56.
1665 Cookham Wood Trial Project case study.
1666 For example, JCT SBC/Q (2016) clauses 4.20–4.24 and NEC4 (2017) clause 60.
1667 JCT CE Guide (2016) Section 27.
1668 JCT CE Guide (2016) Section 38.

- 'To be effective, the identification and management of risks capable of affecting the project need to involve a collaborative approach, so that risks that cannot be managed by individual organisations can be managed collectively'
- 'The identification of relevant risks should best take place before parties are required to provide prices or estimated target costs for the work they will provide'.[1669]

However, although the JCT CE 2016 contracts provide for a risk register and risk allocation schedule,[1670] the opportunities for early risk management are limited unless a separate JCT CE preconstruction phase contract is put in place, setting put a contractual commitment to undertake risk management activities, a timetable for these activities and links to design development and pricing.

The JCT Framework Agreement 2016 clause 14.1 refers to 'all party pre-contract collaborative risk analysis' with a view to identifying risks, determining their likelihood and 'determining who is best able to manage such risks'. Clauses 14.2 and 14.3 refer to creating and updating a 'risk allocation schedule or matrix', but JCT removed the provision that appeared in the original 2009 edition for the agreement of amendments to the relevant JCT 'Underlying Contract' to reflect a revised balance of risk set out in the agreed risk allocation schedule.

NEC3 introduced the requirement for a risk register, which was amended in NEC4 to an 'Early Warning Register'. This describes risks listed at the point of contract and those which either the project manager or the main contractor notifies as an early warning matter, in each case including 'a description of the matter and the way in which the effects of the matter are to be avoided or reduced'.[1671] Early warnings are to be given by the main contractor or project manager to the other as soon as either becomes aware of any matter which could:

- Increase the total of the Prices
- Delay Completion
- Delay meeting a Key Date
- Impair the performance of the works in use.[1672]

The project manager prepares the first NEC4 Early Warning Register for issue to the main contractor within a week of the start date, and instructs the contractor to attend the first early warning meeting within two weeks from the start date.[1673] NEC4 requires those who attend to cooperate in:

- Making and considering proposals for how the effects of each matter in the Early Warning Register can be avoided or reduced
- Seeking solutions that will bring advantage to all those who will be affected
- Deciding on the actions which will be taken and who, in accordance with the contract, will take them
- Deciding which matters can be removed from the Early Warning Register
- Reviewing actions recorded in the Early Warning Register and deciding if different actions need to be taken and who, in accordance with the contract, will take them.[1674]

1669 JCT CE Guide (2016) Section 34.
1670 JCT CE (2016) Sections 5.1, 5.2, 5.3 and 5.4.
1671 NEC4 (2017) clause 11.2(8).
1672 NEC4 (2017) clause 15.1.
1673 NEC4 (2017) clause 15.2.
1674 NEC4 (2017) clause 15.3.

This NEC4 approach to joint risk management enables a project manager to arrange prompt consultation and agreed actions, and both the project manager and the main contractor can instruct attendance at an early warning meeting by other parties 'if the other agrees'.[1675] The main contractor can instruct its subcontractors to attend but the project manager will have no authority to instruct other consultants to attend unless it obtains the agreement of the 'Service Manager' under the relevant NEC4 professional services contract.[1676]

Surprisingly, there is no express role for the client in NEC4 early warning meetings, even though the outcome of these meetings can be changes to the scope of work.[1677] This omission suggests that the client may be absent from active participation in joint risk management, although the NEC4 guidance mentions the desirability of client attendance. By contrast, the NEC4 Alliance Contract describes an early warning system in which the client is stated to be a participant alongside all other alliance members.[1678]

PPC2000 identifies risk management as a responsibility of alliance members and defines it as 'a structured approach to ensure that risks are identified at the inception of the Project, that their potential impacts are allowed for and that where possible such risks and their impacts are minimised'.[1679] PPC2000 precludes any risk contingency until such time as the relevant risk has been reviewed by all alliance members with proposals for its elimination, reduction, insurance, sharing or apportionment and for removal or reduction of the relevant risk contingency.[1680] It provides for all alliance members 'to analyse and manage risks in the most effective ways including:

(i) identifying risks and their likely costs
(ii) eliminating or reducing risks and their costs
(iii) insuring risks wherever affordable and appropriate
(iv) sharing or apportioning risks according to which one or more Partnering Team members are most able to manage such risk'.[1681]

PPC2000 also provides for early warning notices that lead to a Core Group meeting except where the Core Group members can agree a course of action without a meeting.[1682] It has been noted that the requirement to give early warnings works well in conjunction with the provision for open book record-keeping, with the effect that 'the prospects of a difference becoming a dispute can be mitigated by enabling a focus on the specific issues between the parties and how these tie in to their objectives'.[1683]

For example, on the City of London Corporation Hackney City Academy new build school, the team undertook 'successful risk management when faced with the need to

---

1675 NEC4 (2017) clause 15.2 under which the main contractor and project manager 'may instruct other people to attend an early warning meeting if the other agrees' The wording leaves open the questions of why prior agreement is necessary and what to do in the event of refusal to agree. A subcontractor attends 'if its attendance would assist in deciding the actions to be taken' but this requires the main contractor to instruct attendance under the relevant NEC4 subcontract.
1676 NEC4 Professional Services Contract (2017) clause 15.2.
1677 NEC4 (2017) clause 15.4.
1678 NEC4 Alliance Contract (2018) clause 15.
1679 PPC2000 (2013) Appendix 1.
1680 PPC2000 (2013) clause 12.9.
1681 PPC2000 (2013) clause 18.1.
1682 PPC2000 (2013) clause 3.7.
1683 Arup (2008), 50.

change sites, in adopting a joint strategy that served the needs of the project'.[1684] The client reported that 'the greatest challenge and subsequently the greatest achievement of this project was to completely stop work half way through [RIBA] Stage E, relocate to a new site, start the whole design process all over and still complete 29 months later on the original opening date. It is hard to see how any other form of contract could have focused everyone's energies in such a cohesive, efficient way to deliver such a spectacular building'.[1685]

18.5 How Can Collaborative Procurement Improve Risk Management?

Identification of potential areas of risk and agreement of actions to reduce them are important features of collaborative procurement. Contractual processes can enable alliance members to identify, assess and prioritise risks as early as possible and to establish what actions, if any, can be taken to reduce or eliminate them. For example:

- Main contractors 'should quote any risk contingencies at the point of selection so that joint risk management activities during the Preconstruction Phase can seek ways to reduce or eliminate the need for these risk contingencies'[1686]
- 'The analysis and management of risks relevant to the Project should be by a methodology agreed by the Partnering Team prior to signing the Project Partnering Agreement and reflected in activities described in the Partnering Documents, for example the preparation and agreement of a risk register with an agreed action plan as to how Partnering Team members will deal with the risks identified and any prospective risk contingencies'.[1687]

A collaborative team can use the additional information built up through early exchanges of design, cost, time, and supply chain data to:

- Identify risks and how these risks will be perceived by other parties, for example a consultant putting itself in the place of a contractor and by a contractor putting itself in the place of a subcontractor
- Estimate the likely costs of perceived risks, whether those costs can be accurately identified or will only be guessed at by way of a risk premium, and what additional steps can be taken to identify those costs more accurately
- Establish the steps to be taken to eliminate or reduce risks and their costs, or at least to identify them more accurately
- Provide for insurance of risks wherever affordable and appropriate

---

1684 Association of Consultant Architects (2010), 20 and 21. City of London Corporation created a multi-party project alliance under PPC2000 to deliver the £30 m Hackney City Academy, working with Willmott Dixon (Main Contractor), Northcroft (Cost Consultant) Studio E (Architects), Dewhurst MacFarlane (Structural Engineers) and Max Fordham (Mechanical and Electrical Engineers). The architect, engineers and main contractor were appointed simultaneously under PPC2000 to develop design proposals.
1685 Association of Consultant Architects (2010), 21.
1686 Two Stage Open Book and Supply Chain Collaboration Guidance (2014), 38.
1687 PPC2000 Guide (2003), 39.

- Agree the sharing or apportioning of residual risks according to who is most able to manage those risks and who is most able to afford the cost of risks that cannot be managed
- Recognise that pricing by consultants, contractors, subcontractors and suppliers will take account of how the alliance members approach each of the above actions.

Early joint risk management can generate new shared data that enables the agreed reduction of priced risk contingencies.[1688] It can also create opportunities for the client, contractor and other team members to share the benefits of the resultant savings that are achieved. For example, on the St George's Hospital Keyworker Accommodation project the team agreed for 'preconstruction work to be carried out at the same time as a final Agreed Maximum Price (AMP) was being agreed in which all risks had been quantified'.[1689] This gave the team 'the incentive to be proactive in managing risk and expenditure so as to earn rewards available through the shared savings mechanism, openly reviewing buying gains obtained through subcontractor and statutory authority orders'.

In the joint management of risks, it is important to ensure that team members are not tempted to protect themselves with inflated costs by populating a risk register with unnecessary contingencies. For example, on the Bermondsey Academy project the risk sub-team assumed that their job was to imagine and cost every conceivable possibility, and the Core Group had to work on a more positive and better-informed approach that allowed nearly all the additional contingencies to be removed.[1690] There are other instances where collaborative construction procurement has come under pressure when a main contractor has attempted to add a risk allowance late in the preconstruction phase,[1691] and this can be dealt with by a capable alliance manager:

- Reminding any alliance member who proposes a new risk allowance that preconstruction phase joint risk management is designed to reduce previously declared risk allowances, not to increase them
- Questioning whether a new risk or new information has emerged that justifies the proposed risk allowance, and how this and other risk allowances are mitigated by the additional shared data made available through joint risk management.[1692]

The FAC-1 Framework Alliance Contract and TAC-1 Term Alliance Contract include a pro forma Risk Register, as considered in Sections 10.6 and 11.7. The members of a framework alliance and term alliance can undertake joint risk management at a strategic level so that the risk management actions they undertake can benefit successive projects or tasks. For example:

- On the £70 million framework alliance created by Harrow Council with Kier to deliver new build and improved schools, one school caught fire during the school holidays in August. Joint risk management enabled the team to agree how renovation work

---

1688 As described in Section 18.2.
1689 Association of Consultant Architects (2010), 38.
1690 Association of Consultants Architects (2010), 12 and 13, as summarised in Section 18.10.
1691 For example, this occurred on the North Wales Prison Trial Project, apparently as part of a standard internal approval process conducted by the main contractor's board.
1692 On the North Wales Prison Trial Project, the main contractor was reminded of its contractual commitments under PPC2000 (2013) clause 12.9 and of the joint risk management work done by the team. It was agreed that the additional risk allowance should be removed from the proposed fixed price.

could be started immediately, with installation of temporary classrooms so that the damaged school could re-open in September[1693]

- The National Change Agent consortia agreed systems of joint change and risk management, 'as a result of which there have been no reported disputes by any of the NCA consortium members with any of their constructors or service providers over the five years of the NCA programme'.[1694]

18.6   How Can a Collaborative Team Manage Ground Risk?

Ground risk is a major consideration in any construction project and the assessment of who is liable will depend on the circumstances affecting the contract, including the nature of the work and the ability of the contractor to discover the actual site conditions before entering into the relevant contract. Under English law, a main contractor is expected to inspect and identify apparent or likely physical matters on site and, under a fixed price construction contract, will generally be responsible for the risk of encountering unforeseen site conditions, with the result that these conditions will not entitle it to additional remuneration.[1695]

However, a main contractor may have a remedy if information provided by the client proves to be incorrect,[1696] or where a professional has actual knowledge of site conditions and does not share it.[1697] Other relevant factors include:

- Unforeseeable site conditions that make it impossible to perform the project or create a legal obstacle to work being performed, which may entitle the contractor to justify non-performance[1698]
- What it is possible for the contractor to do in order to validate the information provided[1699]
- A possible claim by the contractor for misrepresentation or negligent misstatement, for example where ground conditions were described by reference to an 18 feet borehole which in fact had never been sunk.[1700]

The extent that a contractor will be entitled to base its risk assessment on the information provided by the client is far from clear. For example, in an English case by reference to a FIDIC contract, the tender documents had estimated the quantity of contaminated

---

1693 Association of Consultant Architects (2010), 22, where the Head of Property and Infrastructure at Harrow Council reported that 'The complete programme has been delivered on time to provide a great boost to education in Harrow. Relations between team members and the schools' head teachers have remained positive in what could have been stressful circumstances'.

1694 Association of Consultants Architects (2010), 34.

1695 As confirmed in *Worksop Tarmacadam Company v Hannaby* [1995] 66 Con LR 105 (CA).

1696 As considered in *Simplex Concrete Piles v Borough of St Pancras* [1958] 14 BLR 80.

1697 As considered in *Bacal Construction (Midlands) v Northampton Development Corporation* [1975] 8 BLR 88 (CA).

1698 As illustrated in *Havant Borough Council v South Coast Shipping (No. 2)* [1997] 14 Const LJ 430 (CA).

1699 In *Wimpey Construction UK v Poole* [1984] 2 Lloyd's Rep 499 (QB) the client's site investigation report did not reveal artesian pore water pressures. Webster J concluded that the design and build contractor could rely on this report and held that 'If the data contained in the T report were incapable of verification by Wimpey, then in my view Wimpey did not warrant such part of the design as was consequent upon that data'.

1700 *Munro v Bognor UDC* [1913] 3 KB 167.

materials to be taken into account, but the judge held that bidders also had to make their own independent assessments because:

- 'The contractor must draw upon its own expertise and its experience of previous civil engineering projects'
- 'The contractor must make a reasonable assessment of the physical conditions which it may encounter'
- 'The contractor cannot simply accept someone else's interpretation of the data and say that is all that was foreseeable'.[1701]

FIDIC 2017 provides for the client to make available to the contractor all relevant data 'in the Employer's possession on the topography of the Site and on sub-surface, hydrological, climatic and environmental conditions at the Site',[1702] but makes the contractor 'responsible for interpreting all data referred to'.[1703] FIDIC 2017 also provides that the contractor (to the extent practical taking account of cost and time) is 'deemed to have obtained all necessary information as to risks, contingencies and other circumstances which may influence or affect the Tender or Works' and is 'deemed to have inspected and examined the Site, access to the Site, its surroundings, the above data and other available information'.[1704] These provisions require all risk assessments made by the contractor to be undertaken in isolation from the client and consultants, and for them to be completed not only prior to entering into contract but also prior to the contractor submitting its tender.

When assessing the time and cost effects of a claim by the contractor under FIDIC 2017 for 'unforeseeable physical conditions', the Engineer 'may take account of any evidence of the physical conditions foreseen by the Contractor' when submitting its tender, 'but shall not be bound by any such evidence'.[1705] The apparent freedom of the Engineer to ignore the reliance placed by the contractor on data provided by the client underlines the weakness of the contractor's position in the absence of joint risk management.

ICC 2014 provides for the client to warrant 'that he has provided with the Site Information all such data in his possession or control relating to the Site which is relevant to the Works or their design',[1706] but also treats the contractor as 'deemed to have inspected the Site…and to have obtained for himself all necessary information as to risks, contingencies and other circumstances which may affect his tender'.[1707]

NEC4 provides for a compensation event if 'the Contractor encounters physical conditions which are within the Site, are not weather conditions, and [which] an experienced

---

1701 Jackson L.J. in *Obrascon v H M Attorney General of Gibraltar* [2015] EWCA Civ 712. The Court of Appeal in this case also noted that the historical material provided to the contractor made it clear that very extensive contamination was foreseeable across the site. The contractor needed to make provision for a possible worst-case scenario; it should have made allowance for a proper investigation and removal of all contaminated material.

1702 FIDIC (2017) clause 2.5.

1703 FIDIC (2017) clause 4.10.

1704 FIDIC (2017) clause 4.10.

1705 FIDIC (2017) clause 4.12.5.

1706 ICC (2014) clause 4.4.

1707 ICC (2014) clause 4.5.

contractor would have judged at the Contract Date to have such a small chance of occurring that it would have been unreasonable for him to have allowed for them'.[1708] NEC4 states that 'in judging the physical conditions for the purpose of assessing a compensation event, the Contractor is assumed to have taken into account:

- The Site Information,
- Publicly available information referred to in the Site Information,
- Information available from a visual inspection of the Site and
- Other information which an experienced contractor could reasonably be expected to have or to obtain'.[1709]

However, NEC4 also states that 'if there is an ambiguity or inconsistency within the Site Information (including the information referred to in it), the Contractor is assumed to have taken into account the physical conditions more favourable to doing the work'.[1710]

With the exception of JCT MPCC (2016) clause 14.1, JCT 2016 contracts remain silent on ground risk and do not include provisions equivalent to those considered above. The JCT contracts therefore rely on the English common law position described above, when interpreting liability for the condition of the project site. All the common law authorities and contract provisions referred to in this section seek to create a reasonable balance of liability for ground risk, but they do not assist the parties in managing risk by actions designed to mitigate its effects. Instead, the complexity of the standard form contract clauses and the continued assumption that contractors will price a worst-case scenario[1711] suggest that early joint risk management is a safer bet for all parties than the limited reliance that can be placed on tender information combined with formulaic risk allocation under a construction phase contract.

Under a single stage selection process each bidder is expected to undertake its own site investigations at risk and to satisfy itself as to the nature of the site conditions before submitting its bid. It is not realistic to expect each bidding contractor to assess site conditions during the tender period by expending money on its own investigations, and also somehow to find time to interpret the results of those investigations before submitting its bid. In order to improve the quality of bid submissions, Latham recommended 'a two-stage tender process for more complex and substantial projects'.[1712]

PPC2000 and ConsensusDocs 300 take a different approach to other standard form contracts, creating conditional preconstruction phase appointments during which joint risk management can include additional site investigations. As a result, under PPC2000, any exceptions to the main contractor's liability for 'the state and condition of the soil and rock strata and any structures and Environment comprising the site' are set out in the Commencement Agreement under which the alliance members agree to proceed

---

1708 NEC4 (2017) clause 60.1(12) states that 'Only the difference between the physical conditions encountered and those for which it would have been reasonable to have allowed is taken into account in assessing a compensation event'.
1709 NEC4 (2017) clause 60.2.
1710 NEC4 (2017) clause 60.3.
1711 As in *Obrascon v H M Attorney General for Gibraltar* [2015] EWCA Civ 712.
1712 Latham, M. (1994) Section 6.32.3.b. He also suggested that clients should allow bidders to save money by pooling the costs of ground investigations undertaken by one consultant who is retained by all of them', Latham, M. (1994) Section 6.32.3.c and d.

with the construction phase of the project.[1713] Similarly, ConsensusDocs 300 provides that 'the Core Group shall review whether additional information or joint investigations are needed concerning the Worksite or reasonably required to validate Owner's Program'.[1714]

A 2001 Institution of Civil Engineers report recognised that 'ground-related factors are a common cause of lengthy delays and large increases in building and construction costs', and encouraged new working practices including creation of 'a risk management system to reduce and, if possible, avoid these problems and to exploit any opportunities for improvement that may arise'.[1715] It recommended that 'to provide more certainty of outcome in an increasingly fast-track and fragmented construction environment, the following are required:

- Good communication
- A team approach to problem-solving
- An integrated total project process
- A risk-based approach to construction management and design'.[1716]

Ground risk offers a clear example of the way in which joint risk management can be incorporated into early project processes. Clients and their advisers can share better knowledge of ground conditions by commissioning suitable investigations below ground prior to inviting main contractor bids. Early alliance appointments then allow additional time during the preconstruction phase for the client, consultants, contractor and relevant subcontractors and suppliers together to examine:

- The contractor's perception of the ground risk
- Whether the client has additional information available to alter that perception
- Whether additional site investigations would alter that perception
- Whether early works packages, such as demolition, would alter that perception
- Whether any aspect of the risk should be covered by insurance
- Whether the client should assume all or part of the risk
- Whether any of these courses of action would give rise to the removal of, or significant reduction in, any amount of money allowed for the risk in the contractor's price for the project.

On the Hampstead Heath Ponds project alliance, 'risk management workshops took place throughout the project'. 'It had been agreed that clay from around the site would be used as part of the dam construction works but ground investigation suggested that the top layers of clay would be weathered and therefore not fit for this use. The parties agreed to assume a level of 1 metre and then agreed a 50% pain/gain share to the extent that the useable clay was above or below that level. This avoided re-measurement'.[1717]

On the Ministry of Justice North Wales Prison project 'the team faced significant early challenges caused by ground conditions on the extensive project site, including asbestos, soft spots, and remnants of the concrete foundations of an old munitions factory. Joint

---

1713 PPC2000 (2013) clause 18.9.
1714 ConsensusDocs 300 (2016) Article 5.1.2 (Joint Worksite Investigation).
1715 ICE/DETR (2001), 20, 21.
1716 ICE/DETR (2001), 20, 21.
1717 Case study summarised in Section 3.10.

working in co-located offices enabled the team to agree a strong response to these issues in finding ways to minimise any delays and cost overruns'.[1718]

18.7   How Can Risk Be Managed Through No Blame Clauses?

We have considered in Section 7.7 the impact of no blame clauses and in Sections 16.4 and 16.8 how cost reimbursement, target costs and pain/gain incentives enable costs to be shared among team members in ways that cultivate greater mutual trust and cooperation. As a result, and in varying degrees, collaborative cost models can redistribute and share among team members the risk of cost overruns in ways that depart from traditional rights and obligations. No blame clauses attempt to take the redistribution and sharing of risk further by excluding or limiting certain legal remedies irrespective of the cause of loss or damage.

The concept of risk management through a no blame clause is based on concerns that:

- Not all risk issues can be ironed out during preconstruction phase collaborative working
- Any allocation of construction phase risks will encourage main contractors and their supply chain members to create hidden risk allowances and will therefore undermine a collaborative culture
- A no blame clause is necessary to preserve a collaborative culture and to sustain joint risk management during the construction phase of a project.

However, these concerns appear less compelling in the absence of claims and disputes arising on collaborative projects supported by alliance models that do not contain no blame clauses. It is possible that the promotion of no blame clauses fails to take account of:

- The contractual commitment of alliance members to declare all risk allowances[1719]
- The preconstruction phase opportunities for alliance members to undertake detailed risk costing before agreeing a fixed price or target price
- The strength of the collaborative relationships that can be built up during the preconstruction phase, that can continue to thrive during the construction phase and that can respect the agreed allocation of project risks.

We also need to examine where liability actually falls under a no blame clause when considered alongside other contract provisions. For example, no blame clauses are usually combined with the sharing of cost savings and cost overruns among alliance members, except those arising from a limited range of compensation events, but they do not usually eliminate compensation events altogether. The remaining compensation events could, therefore, lead to the risk of additional time and cost being incurred by the client and not shared by other alliance members. For example:

- Compensation events under the NEC4 Alliance Contract include an event which stops or delays the alliance in completing the works, which 'the Alliance could not prevent' and 'which people or organisations experienced in works similar to the works

---

1718 North Wales Prison Trial Project case study.
1719 For example, in PPC2000 (2013) clause 12.9.

would have judged at the Contract Date to have such a small chance of occurring that it would have been unreasonable to have allowed for it'[1720]

- Such an event entitles alliance members to quote for a change to the project budget and the Performance Table,[1721] which resets the basis for incentivisation and increases the amount below which the client shares cost savings and above which it shares over-runs[1722]
- Such an event also entitles alliance members to quote for a change to the project completion date which leaves the risk of delay with the client.[1723]

Any lack of clarity in a no blame clause is likely to lead to disputes, for example:

- The NEC4 Alliance Contract sets out the liabilities of the client and the partners (namely the other alliance members), and states that 'Liabilities other than those stated as being Client's or Partners' liabilities are an Alliance Cost unless recovered from an insurer'[1724]
- This appears to be a very broad exclusion and, as considered in Section 7.7, the Client's and Partners' liabilities include third party claims arising from intentional acts or omissions but do not appear to include any claims between the alliance members themselves, suggesting that all claims between alliance members are an alliance cost payable in agreed shares unless recovered from insurers
- Beyond their intentional acts and omissions, the liabilities of each Partner are very limited indeed, covering only breach of intellectual property rights and death or injury to its own employees, although the liabilities of the client mention neither of these categories.[1725]

The potential for disputes as to the meaning and scope of no blame clauses, like those in relation to non-binding protocols and good faith, reminds us that well-intentioned words can be counter-productive if they are not clearly defined. Where a badly drafted no blame clause blocks the enforceability of a team member's rights, its effect may be more adversarial than collaborative if it transfers risk to a team member because that team member is deprived of its ability to rely on its contract. For a no blame clause to be effective it should clarify its scope and impact rather than use vague or cryptic wording that, on closer analysis, may be revealed to be no more than wishful thinking.

18.8   How Can Risk Be Managed Through New Insurance?

Risk management can include the insurance of risks where this is affordable and appropriate, but insurances are subject to the terms and limitations of each policy.[1726]

1720 NEC4 Alliance Contract (2018) clause 60.1(7).
1721 NEC4 Alliance Contract (2018) clause 62.
1722 NEC4 Alliance Contract (2018) clause 53.3.
1723 NEC4 Alliance Contract (2018) clause 62.1.
1724 NEC4 Alliance Contract (2018) clause 82.1.
1725 NEC4 Alliance Contract (2018) clauses 80.1 and 81.1 as considered in Section 7.7.
1726 For example, in *Shepherd Homes v Encia Remediation* [2007] EWHC 70 (TCC), risk was passed down the supply chain to a party with limited insurance cover and limited liability, thereby in effect transferring the uninsured risk back up the supply chain.

Insurance is a risk management option that safeguards the insured against the financial consequences of risks rather than attempting to avoid or reduce those risks through agreed actions. An insurance policy is therefore usually seen as a support system where unavoidable risks arise, rather than as a component of a procurement and contract model.

The Integrated Project Insurance ('IPI') collaborative procurement model introduces a new insurance policy to cover areas where there is no contractual recourse and is intended in part to overcome concerns caused by the exclusions and limitations of liability in no blame clauses. The IPI model suggests a new type of relationship with insurers who take a more active role in the project through their own review of technical and financial proposals as preconditions to offering cover. Collaborative procurement supported by joint insurance raises new questions as to the extent of the cover, who is covered and what they have to do to obtain this cover.

An IPI policy is unusual in that it covers:

- All usual construction project risks
- Design errors or omissions of any alliance member without a right of subrogation under which the insurer can pursue the party at fault
- Cost overruns caused by any alliance member without a right of subrogation under which the insurer can pursue the party at fault.

The benefits of whole project insurance cover similar to IPI were tested on the University Hospital Dubai project summarised in Section 18.10, where the insurance provided a stabilising influence on intensive work undertaken by a transnational team working simultaneously out of Australia, UAE, UK and USA. This insurance was combined with collaborative working and early contractor involvement in a multi-party project alliance contract.

The first Trial Project exploring the potential of IPI cover was a new build education facility at Dudley College, as summarised in Section 18.10.[1727] The insurer under the Dudley College IPI policy relied on the advice of its 'technical independent risk assessor' and 'financial independent risk assessor',[1728] both of whom were engaged separately by the insurer and whose views were influential. The Dudley College case study illustrates how closely these assessors worked with the alliance team, and the evidence from the Dudley College project shows how an IPI policy can act as an enabler of alliancing and BIM. However, to embed an insurance policy in a procurement model, with the insurer as an active participant, raises new questions as to:

- The appropriate role of an insurer in the risk management, quality management and financial management of a project
- The extent to which an insurer is likely to support innovation
- Whether an insurer, in the absence of rights of subrogation, will be more or less willing to pay when claims are made.

As a market for IPI develops, alliance members will need to consider how their work is likely to be viewed by the insurer and its advisers who may, for example, require a more conservative approach to design in order to minimise the insured risks. Also, the

---

1727 As summarised in Section 18.10 and as considered in University of Reading (2018).
1728 As described in the Dudley College Trial Project case study and in University of Reading (2018).

additional savings and improved value achieved through IPI will need to exceed the IPI premium, which on the Dudley College Trial Project was 4% of project value.[1729] It will be easier to assess the potential of IPI cover if it can be offered in the marketplace without the current requirement to use a confidential, bespoke form of project contract.[1730]

Unlike a professional indemnity policy, an IPI insurance policy is not connected to the legal liability of any team members, and the cover it provides is not linked to an assessment of whether a team member has fulfilled a duty of reasonable skill and care.[1731] This raises the question of whether the latent defects cover that forms part of IPI cover[1732] goes as far as insuring the fitness for purpose of the project.[1733] If so, this would be a bold move by insurers, whose professional indemnity policies currently exclude fitness for purpose commitments made by their clients.

An IPI insurance policy is designed to support a more collaborative culture by reassuring team members that they are less likely to face claims from other alliance members. However, IPI needs to be viewed not only by comparison to the cautious behaviour and unnecessary disputes that are encouraged by traditional procurement models. It also needs to be compared to other collaborative approaches such as Two Stage Open Book, Cost Led Procurement and Supply Chain Collaboration where well-structured teams have avoided disputes and have achieved impressive results without a new insurance policy.

18.9 What Is the Role of Project Bank Accounts?

Project bank accounts are a UK Government policy driven by the poor payment record of main contractors and by risk of supply chain insolvencies, and they supplement several statutory remedies for late payment.[1734] All team members should receive prompt payment for their works, services and supplies as a legal and contractual right, as a basic demonstration of good practice and as a simple way to build trust at all levels of the supply chain. Fair payment practices make an obvious difference to the credibility of a collaborative team, and prompt payment can even earn discounts on previously agreed rates.[1735]

A project bank account ('PBA') is 'a ring-fenced bank account from which payments are made directly and simultaneously by a client to members of his supply chain:

- A PBA has trust status which secures the funds which can only be paid to the beneficiaries, being the supply chain members named in the account
- Payments out of the PBA are made simultaneously to all parties

---

1729 University of Reading (2018), 116 and comprising 2.5% plus consultant fees.
1730 University of Reading (2018), 97–99.
1731 As considered in Sections 5.3 and 14.2.
1732 As described in University of Reading (2018), 114.
1733 As considered in Sections 5.3 and 14.2.
1734 For example, under the Housing Grants, Construction and Regeneration Act 1996 as amended, the Late Payment of Commercial Debts (Interest) Act 2008 and Section 113 of the Public Contract Regulations 2015.
1735 As shown in the Surrey Project Horizon Trial Project case study summarised in Section 5.10 and in the Futures Housing Group FAC-1 case study summarised in Section 10.10.

- The PBA is held in the names of trustees who are likely to be the client and lead contractor but could also be members of the supply chain
- The advantage of trust status is that, in the case of insolvency, monies in the account due for payment to the supply chain members are secure and can only be paid to them'.[1736]

PBAs have been strongly recommended for use on all UK Government projects.[1737] It may seem strange that the Government encourages collaborative procurement through early appointment of main contractors and at the same time supports a payment system that can take significant amounts of money out of the main contractor's direct control. Detailed provisions governing PBAs appear in a number of standard form contracts such as NEC4[1738] and PPC2000.[1739]

PBAs are intended to address concerns that main contractors who seek collaborative relationships with clients and consultants may not recognise the importance of paying their subcontracted supply chain members on time. Late payment demotivates supply chain members and reduces the likelihood of them contributing to improved project value. Therefore, to avoid the risk of contractors not passing on the amounts due to supply chain members in each interim payment, a PBA enables the client to ensure those amounts are paid directly out of funds lodged with a bank. A PBA also safeguards funds due to subcontracted supply chain members in the event that the main contractor becomes insolvent, a risk highlighted in early 2018 following the collapse of the major UK contractor Carillion.

Trial Projects that provided for PBAs include the North Wales Prison project using the option at PPC2000 Appendix 9 and the Dudley College project using a bespoke alliance contract. The Reading University report on the Dudley College Trial Project describes the intended PBA arrangements in detail, including the problems encountered with Barclays Bank in the context of other banking arrangements. They reported that 'establishment of a PBA took considerably more time than was anticipated', with the result that ultimately the team decided not to create a PBA at all on this project.[1740]

Collaborative procurement is designed to achieve a stable supply chain, with no temptation for a main contractor to seek additional profit by covertly replacing approved supply chain members with cheaper alternatives. An incidental effect of a PBA is to make the client aware of any changes in those members of a main contractor's supply chain who are paid through the PBA.

1736 PBA Briefing (2012).
1737 'The Government Construction Board decided in September 2009 that Central Government Departments, their Agencies and Non-Departmental Public Bodies would move to a position where PBAs are adopted unless there are compelling reasons not to do so'. PBA Guidance (2012). 'Cabinet Office is working with public sector construction procurers in pioneering a new way of paying supply chain members in construction projects through Project Bank Accounts (PBAs). PBAs will see construction SMEs working in government projects receiving payment in five days or less from the due date. Government Construction Board members have committed, over the next three years, to deliver £4 bn worth of construction projects using PBAs'. PBA Briefing (2012).
1738 NEC4 (2017) Option Y (UK) 1.
1739 PPC2000 (2013) Appendix 9.
1740 University of Reading (2018), 121.

18.10   Collaborative Risk Management Case Studies

The following case studies illustrate how a collaborative team can undertake joint risk management as part of the project planning processes, and show how three teams benefited from joint risk management activities during the preconstruction phase.

In the first case study the Bermondsey Academy project alliance integrated the roles and responsibilities of all team members who agreed to develop designs, risk management and accurate costing at an early stage so as to ensure that the project could be built within the available budget. They undertook preconstruction phase joint risk management activities that enabled them to agree a fixed price while also developing innovative designs that were affordable and achievable.

Through agreement of key activities under an integrated preconstruction phase risk register, the team managed significant risks such as limited site access, removal of on-site contamination, dealing with the presence of Japanese knotweed and, most significantly, the last-minute need to change the site of a temporary school.

In the second case study the University Hospital Dubai project alliance also undertook preconstruction phase joint risk management activities that enabled them to agree a price within a previously agreed budget. Each risk was described in a risk register that stated agreed risk management actions, an anticipated value of each risk and the chance of it occurring, with a total of these risk values added to the project budget. If actual expenditure on the listed risk items was lower than the total allowed, then the team members shared the benefit unless it was offset by other project overruns, and if actual expenditure on the listed risk items was higher than the total allowed, then the team members shared the pain unless it was offset by other project gains.

The alliance also took out whole project insurance in respect of design liability, without rights of subrogation, in order to enhance the strength of their collaborative relationships. The whole project insurance was seen as encouraging integrated collaborative working without anyone being held back by their professional indemnity insurers.

In the third case study the Dudley College Trial Project alliance undertook preconstruction phase joint risk management activities but did not fix a price. Instead they relied on pain/gain shares to motivate cost control and cost savings. The team also developed a new collaborative relationship with their insurers who covered not only design liability but also cost overruns, in both cases without the usual rights of subrogation that would have enabled them to seek recovery from other team members. The Dudley College case study is supplemented by a University of Reading report which describes in detail the project procurement and delivery process, the IPI insurance cover and the collaborative approach to risk management.

Bermondsey Academy

The Bermondsey Academy project comprised a £22 million new build academy school on a restricted and contaminated site. The project alliance comprised City of London Academy (Southwark) (client) Willmott Dixon Construction (main contractor) Northcroft (cost consultant), Studio E (architect), Dewhurst MacFarlane (structural engineer), Max Fordham (services engineer) and FIRA Landscape (landscape architect).

City of London Corporation wanted to integrate the roles and responsibilities of all team members to progress design, risk management and accurate costing early so as to ensure that the project could be built within the available budget. The main contractor and design consultants were appointed simultaneously under PPC2000 and worked to an agreed timetable of preconstruction activities to develop innovative designs (creating 90% natural lighting and ventilation) that were affordable and achievable.

Through agreement of key activities under the shared Partnering Timetable and Risk Register, the team managed significant risks such as limited site access, removal of on-site contamination, dealing with the presence of Japanese knotweed and, most significantly, the last-minute need to change the site of the required temporary school. They utilised:

- Preconstruction design development and finalisation of open-book costs to achieve robust fixed price, supported by detailed information to assist change and risk management
- Successful management of on-site contamination and site changes, implementing agreed actions set out in a Risk Register
- Performance measures through KPIs that focused on the client's priorities
- Early agreement of commercial incentives by way of shared savings to motivate improved design and risk solutions.

The project achieved the status of an exemplar project (Department for Education and Skills), and won the 'Quality in Construction Medium-Sized Project of the Year Award 2006' (Department for Trade and Industry and Constructing Excellence). It also won the Prime Minister's Better Public Building Award 2006, which expressly recognised how 'partnering between the contractor and design team produced a strong building which benefited from nearly two years of preparation'.

John Frankiewicz, Chief Operating Officer of Willmott Dixon reported: 'When commissioned at an early stage we involve groundwork contractors who may be able to identify potential risks that could be avoided through consideration in regard to orientation of the building or the location/ availability of drainage services'.

Association of Consultant Architects (2010), 12, 13.

University Hospital Dubai

University Hospital Project, Dubai was a project alliance led by Kathleen Gilroy and Roger Bridgeman, and comprised the UAE University Hospital FZ (client), Al Naboodah Laing O'Rourke (main constructor), BK Gulf (M&E constructor) Ellerbe Becket USA (lead designer) GHD Australia (structural subconsultant), SKM Australia (M&E subconsultant) Heery International (client representative) and Davis Langdon (cost consultant).

Risk management had to cover the design and construction, the operational side of running the hospital and the business that controlled and funded the hospital, including corporate and clinical governance. These were all linked through joint risk management

(Continued)

so that changes in any one of them could be examined to see if it affected any of the others, and so that the agreed course of action to overcome the problem could be reviewed to ensure that it did not cause a problem elsewhere.

Each risk needed to be accurately described and a positive but achievable course of action planned, implemented and progressively checked. Each was also given an anticipated value and chance of occurring, so that all risks could be put into a Monte Carlo simulation to see the overall effect. That final value was included in the project budget. If reduced against that budget, then the constructors and consultants benefited in the pain/gain scheme and if it was greater, then it was their shared pain. The risk log was reviewed monthly and formed part of the Core Group report.

It was decided during the preconstruction period to take out Whole Project Insurance as the client wanted the designers and constructors to undertake integrated collaborative working without anyone being held back by their professional indemnity insurers. Under the policy:

- The Whole Project Insurance covered the first level of $10 million for each and every claim with $30 million in the aggregate, with an option of a further $30 million in the aggregate if the first aggregate level of claims was reached
- The next level, if and when the Whole Project Insurance cover had been used, was the insurance of each alliance member to their normal limits
- The third level was the client taking the risk above all of these levels under their Group (Dubai Holding) insurance.

The Whole Project Insurance policy had an excess of $100 000 for each claim on CAR and PI; a waiver of non-vitiation clause; a retroactive date for any design, site supervision, construction supervision or project management carried out prior to the placing of the insurance; cover for infringement of any trademark or patent with its own limit of $5 million; and cover for failure of computer software/hardware with its own limit of $5 million.

When the credit crunch first hit Dubai the other 89 projects being undertaken for Dubai Holding were immediately suspended or terminated. The University Hospital Project kept going for a further 18 months, the team members having met and agreed a plan of action to use unamortised advance payments to continue the project and pay all parties from those funds.

The extended period of the credit crunch and its effects on Dubai meant that eventually the client was unable to keep the project going and it was brought to an amicable termination with sufficient funds to pay all parties the monies that they were owed.

Reported to the ACA by Roger Bridgeman of UHD, at www.ppc2000.co.uk.

Dudley College

Dudley College was the first IPI Trial Project and comprised a £11.685 million new build education facility. The alliance members were Dudley College (client), Derry Building Services (specialist), Fulcro (engineering services and project coordinator), Metz (architect),

Pick Everard (structural), and Speller Metcalfe (constructor). The independent facilitator ('IF') was Integrated Project Initiatives, the technical independent risk assurer ('TIRA') was SECO (Belgium) and the financial independent risk assurer ('FIRA') was Rider Levett Bucknall.

The designers, specialist contractors, contractors and project coordinator were appointed as an integrated project team at the outset, in accordance with the Public Contracts Regulations (current at the time) and under a bespoke alliance contract. The project timeline ran from OJEU advertisement in September 2014, tender returns in December 2014, contract award in February 2015, alliance contract signature in May 2015 to project completion in late 2017.

With IPI and its new partnership with insurers, the open culture prevented risk from being hidden until a claim notification appeared, and a transparent culture was embedded with the IF, TIRA and FIRA sharing information and advice between the alliance/IPT and the insurers, with the objective of avoiding surprises and instead collaborating in managing both opportunities and risks. IPI policy inception was a crucial milestone, when the design solution, programme and target cost were agreed, and:

- The client gained the security of the policy's cost overrun cover
- Each alliance member agreed the benefits of gain share and the risk of pain share (subject to the limits)
- The insurers committed to the risk of overspend above the pain share (subject to their limit of indemnity).

Policy inception did, however, inevitably require a leap of faith, which was eventually taken due to commitment to the future of IPI and the belief that risks could be contained and opportunities realised under the motivation of the no blame/no claim mechanism. The client's primary motivation for policy inception was the security of the cost overrun cover – in contrast to the risk of final account escalation. The interaction with insurers did not stop with policy inception. Griffiths & Armour were instrumental in promoting regular contact with the IF/TIRA/FIRA so that status of opportunities and risks was openly discussed.

Under waiver of claims other than for wilful default, BIM was used without the liability constraints under traditional contracts. BIM lessons learned included concerns that suppliers did not readily adopt the information management processes and that proactive delivery and management of information was a challenge.

Dudley College Trial Project case study. Further detailed analysis in University of Reading (2018).

19

How Can Collaborative Procurement Reduce Disputes?

19.1 Overview

It is always possible that potential disputes will arise between the members of a collaborative team, for example by reason of an unforeseeable event or by reason of the failure of a team member to perform its agreed obligations. Successful collaborative procurement depends on the individuals representing different organisations seeking to avoid or resolve problems. Where these problems might otherwise escalate into disputes, team members can use the additional information built up and shared among them, combined with the support of collaborative governance systems, to provide a more stable basis for non-adversarial dispute resolution.

However, there is a deep sense of uncertainty and mistrust in an industry where:

- 'The adversarial relationship established by the traditional contractual framework does not stop with the completion of the project
- Claims and counter-claims continue often for years afterwards, exhausting the industry from energy, resource and cost aspects'.[1741]

Some industry professionals have developed a mindset of always preparing for conflict. For example, I once represented a Danish subcontractor in negotiations with a major UK contractor, whose representative literally banged his own head on the table in order to emphasise his point of view on any disputed issue. This tactic shocked my client at the time, but the head-banger explained later that he saw every day of his working life as a relentless search for inevitable battles in which he must fight to protect his company's interests.

Usually, a dispute is a painful reminder of why construction team members are dependent on each other and why no one engaged on a project is wholly immune from the effects of it failing. We will consider in this chapter the causes of disputes and the ways in which these disputes can be reduced as a result of early contractor involvement and collaborative relationships. We will also consider in this chapter the means by which an integrated team can seek to avoid or resolve an impending dispute utilising:

- Early warning between team members
- Decisions by a Core Group
- Support from an Independent Adviser
- Other collaborative options.

1741 NEDC (1991), 9.

Collaborative Construction Procurement and Improved Value, First Edition. David Mosey.
© 2019 John Wiley & Sons Ltd. Published 2019 by John Wiley & Sons Ltd.

19.2   Are Construction Disputes a Bad Thing?

It has been suggested that construction disputes, particularly those arising from claims for additional time and money, are evidence of opportunism, bad faith and incompetence on the part of main contractors.[1742] However, this is not fair or accurate and it is neither straightforward nor reliable for contractors or anyone else to make money through dispute resolution,[1743] and the underlying reasons for disputes are more complicated. For example, we have seen in Chapter 6 how poorly-informed risk transfer can tempt bidders to offer excessively low prices in order to win a tender and then to exploit errors and omissions in the time and money claims that often lead to disputes.

The law evolves in part through the resolution of disputes, and the complex matrix of legal relationships that underpin construction projects generates important judicial decisions. However, no one gains much from being a party in a test case for future industry learning, and disputes are usually a humbling experience that scrutinises in detail the ways that we have failed to fulfil our declared objectives.

There are three ways to approach resolution of a dispute:

- Preventative, by ensuring that a problem does not escalate into a dispute through active joint risk management
- Facilitative, by keeping dispute resolution within the control of the parties until they agree a solution, a system which can lead to mutual recognition of the underlying facts, mutual acceptance of a compromise and preservation of some future goodwill
- Evaluative, by putting dispute resolution outside the control of the parties and accepting a decision made by a third party, a system which is more likely to involve a winner and a loser, considerable expense and loss of most future goodwill.[1744]

Collaborative procurement should offer reliable options for team members to adopt the first and second options. For example, the English courts have upheld a contractual requirement:

- To attempt in good faith to resolve a dispute by negotiation between senior executives with authority to settle
- Failing that to attempt in good faith to resolve the dispute by alternative dispute resolution as recommended by Centre for Effective Dispute Resolution[1745]
- Not without reasonable cause to initiate legal action until these alternative dispute resolution processes were complete.[1746]

As to whether different contract forms can affect the avoidance of disputes, Latham observed that 'no doubt some academic lawyers will say of PPC2000, as they also said of NEC3, that it should not be used because it has not been tested in the courts. The absence of courts is a plus, in my view. If you want a document that is regularly tested in the courts, you can use JCT80'.[1747]

1742 Duncan Wallace referred to 'the reputation for aggressive claimsmanship of United Kingdom contractors generally' Duncan Wallace, I.N. (1996), Vol. 2, 506.
1743 In R.J. Smith's view, when disputes arise, the 'cost of victory' is considerable. Smith, R.J. (1995), 44.
1744 Facilitative and evaluative options are considered as 'negotiation', 'third party intervention' and 'an adjudicative process' in Gould et al. (2010), 6, 7.
1745 https://www.cedr.com.
1746 *Cable and Wireless v IBM UK* [2002] EWHC 2059 (Comm).
1747 Latham, M. (2002).

19.3 What Are the Causes of Construction Disputes?

A collaborative approach to procurement can only help to prevent construction disputes if it can address the likely causes of those disputes. The top five reasons for disputes reported in a 2018 UK survey were extensions of time, valuation of final account, valuation of variations, defective work, and loss and expense.[1748] However, all except defective work are not actually reasons for disputes, only references to the contract mechanisms being called into question.

A 2018 global survey did not find that the primary causes of disputes are negligence or unforeseeable ground conditions or unpredictable cost inflation. Instead, it reported that they are poor contract administration, poorly drafted, incomplete and unsubstantiated claims, and failure of an employer or contractor or subcontractor to understand or comply with contract obligations.[1749] All these causes of disputes could be avoided or at least mitigated by a more integrated and collaborative approach to project procurement:

- Poor contract administration
  We might think that this is a problem only for the project manager, but it is not that simple. Contract administration is not just a matter of issuing instructions and evaluating payment claims but also includes sensitive contractual decisions made by the client, the design consultants, the main contractor and its supply chain. For all parties to participate in collaborative contract administration, rather than backing away and hoping it is someone else's fault, depends on a governance system that involves them and gives them accurate information to work with.
- Poorly drafted, incomplete and unsubstantiated claims
  These sound like a contractor's problem, and certainly the habit of holding back detailed evidence in the opportunistic hope of pushing up the total claim does little to help an early settlement or to maintain good relationships. Yet disputes can also arise from tactical games played by a project manager who objects automatically to proposed changes and other compensation events in order to minimise or delay payment as much as possible, an adversarial approach which the main contractor then has to impose on its own supply chain. These games could be avoided if the project manager and contractor together create intelligent foundations for their relationship through the build-up of more accurate design, cost and time data that is reviewed and agreed with the supply chain ahead of start on site.
- Failure of an employer or contractor or subcontractor to understand or comply with contract obligations
  Non-compliance with contract obligations can take many forms, ranging from inadvertent errors to wilful default, but a lack of understanding raises different questions. For example, team members should to be trained as to the effect of the contracts they sign and administer, and contract drafters should use standard forms as a way to establish consistent terms that are well understood rather than muddy the waters with extensive amendments. Procurement advisers and lawyers store up trouble when they create complex contract terms that transfer risks without also creating the means to review and mitigate those risks.

1748 NBS (2018), 25.
1749 Arcadis Global Construction Disputes Report (2018), 10.

A 1995 UK study identified the following main reasons for unanticipated increases in the construction cost of projects:

- 'The objectives were unrealistic or changed during the course of the project;
- Estimates for project approval were too optimistic;
- The project brief was incomplete, unclear or inconsistent;
- The design did not meet planning or statutory requirements;
- The design was incomplete at the time of tender;
- The design lacked coordination, buildability or maintainability;
- Risk allocation was ambiguous; and
- Management control was inadequate'.[1750]

All but the last of the listed reasons relate to project planning and other preconstruction phase activities, and nearly all of them could be examined more closely and possibly avoided if there was early contractor participation alongside the client and consultants in an integrated team.

For example:

- The robustness of client estimates and of contractor tender prices can be tested through joint examination of actual costs in advance of start on site
- The suitability of objectives, any lack of clarity or omissions or inconsistencies in the project brief, any incompleteness or lack of coordination, buildability or maintainability in the designs, and any non-compliance with planning or statutory requirements can all be challenged and rectified by the client and consultants working in conjunction with the contractor and its supply chain as preconditions to start on site
- Preconstruction phase joint working can in turn reduce the ability of any party to ignore or exploit any ambiguities in agreed risk allocation.

19.4 How Can Early Contractor Involvement Reduce Disputes?

Detailed research as to the causes of disputes, as perceived by contractors, clients and consultants in Hong Kong, revealed the following in descending order of overall perceived significance:

    (i) 'Inaccurate design information
   (ii) Inadequate design information
  (iii) Inadequate site investigations
  (iv) Slow client response (decisions)
   (v) Poor communications
  (vi) Unrealistic time targets
 (vii) Inadequate contract administration
(viii) Uncontrollable external events

1750 Efficiency Unit (1995), Section 145, 53.

(ix) Incomplete tender information
(x) Unclear risk allocation'.[1751]

The majority of these causes of disputes are directly linked to activities undertaken during the preconstruction phase of the project and the risk of these causes arising can be avoided or reduced by collaborative planning activities undertaken through early contractor involvement:

- Inaccurate design information is more likely to be identified if, during a period prior to the construction phase, the main contractor and the specialist supply chain members review, comment on and contribute to the designs
- Inadequate design information is more likely to be revealed if the main contractor and the specialist supply chain members make design contributions prior to commencement of the construction phase. In addition, where further detailed design information is required during the construction phase, earlier contractor and supply chain engagement reduces the risk of misunderstandings as to what information is required from whom and as to when it becomes time-critical
- Inadequate site investigations are less likely to be accepted if the client and its consultants exchange views with the main contractor as to the nature and detail of the investigations that are appropriate, and if joint working during the preconstruction phase provides an additional opportunity to agree what site investigations are necessary and how to deal with their results, avoiding the need for the main contractor simply to price the client's investigations and argue about inadequacies later
- Slow client responses and decisions can be avoided if timescales are agreed as part of an integrated timetable through which team members can together establish and agree the key dates and periods that include those governing client responses and decisions
- Poor communications can be addressed by early identification of individuals whose authority and methods of communication are agreed and who can build up mutual trust and respect while finalising project details
- Unrealistic time targets can be challenged if team members have an opportunity for joint programming during the preconstruction phase, agreeing deadlines and interfaces in respect of construction phase dates rather than leaving these to be established by unilateral client or consultant decisions
- Inadequate contract administration can be addressed where team members agree the contract administration resources, establish that the project manager is a team member working in collaboration with all other team members and avoid wasteful duplication of roles
- Uncontrollable external events may be unavoidable despite the joint work of team members but these are less likely to be disputed if the scope for disagreement has been narrowed by the build-up of agreed preconstruction phase design, cost, time and risk data
- Incomplete tender information is less likely where the client and consultants prepare tender information appropriate for early main contractor selection and then work with the selected contractor in preparing and issuing further tender information for

---

1751 Kumaraswamy, M. (1997), 5.

second-stage selection of subcontracted supply chain members. This two-stage collaborative approach also enables the client to seek an unambiguous commitment from the main contractor and its supply chain members that information is sufficiently detailed for the project to proceed on site, and to reduce the risk of later claims that any information was inadequate[1752]

- Unclear risk allocation can be rectified through a joint preconstruction phase risk management process that clarifies any points of doubt, for example resolving any gaps or duplications in the team members' roles and responsibilities.

It is therefore arguable that the likelihood of disputes arising is reduced by the agreed data and joint commitments that can be developed during the preconstruction planning processes that should be inherent in early contractor involvement.

19.5 How Can a Collaborative Team Avoid Disputes?

We have considered in Sections 7.2, 7.3, 7.4 and 7.5 a range of case law illustrating how informal collaborative arrangements and declarations of good faith do not automatically lead to avoidance of disputes. We have considered in successive chapters the strength of more robust collaborative commitments, including the influence of collaborative leadership, management and project governance and the potential for dispute avoidance through joint risk management.

One approach to the avoidance of disputes is to exclude legal rights under the contract itself, for example under NEC4 Alliance Contract 'no claims' clause whereby 'the members of the Alliance agree that any failure by a member of the Alliance to comply with their obligations stated in the conditions of contract does not give rise to any enforceable right or obligation at law except for an event which is a Client's or Partner's liability. Any disputes between the members of the Alliance arising out of or in connection with the contract are only resolved in accordance with these conditions of contract'.[1753] The first part of this clause restricts the types of permissible claims among alliance members, and we have considered this in Sections 7.7 and 18.7.

The interpretation and enforceability of the second part of the NEC4 Alliance Contract 'no claims' clause remains to be tested at the time of writing. It cross-refers to the agreed dispute resolution roles of the Alliance Board, expert determination, senior representatives and the support of a mediator but it does not attempt to exclude a statutory right of adjudication, as acknowledged in Option Y(UK) 2.6. It is, therefore, difficult to see how it adds anything to the separate clauses that set out these agreed dispute resolution procedures.[1754] To state restrictions on the permitted methods of dispute resolution in a 'no claims' clause appears to confuse two unrelated matters, firstly the restriction of the grounds for claims and also the means by which a dispute is resolved.

An effective support system for collaborative dispute resolution is to create more reliable channels for communication of the issues between team members so that they can

---

1752 For example, the PPC2000 (2013) Commencement Agreement includes an undertaking whereby 'The Partnering Team members…agree…that…to the best of their knowledge the Project is ready to commence on Site'.
1753 NEC4 Alliance Contract (2018) clause 94.1.
1754 NEC4 Alliance Contract (2018) clauses 95, 96 and 97.

resolve their differences by direct negotiation. Research undertaken into a sample of 102 disputes found that:

- 'By creating channels through which differences in perspective will be resolved, coordination provisions help mitigate misunderstandings of the kind that raise questions about the intent of another party; this promotes – or at least minimizes damage to – attributions of goodwill during a conflict'[1755]
- Contractual 'coordination provisions increase the likelihood of continued collaboration after a dispute and perceptions of competence mediate this effect'.[1756]

Where alliance members seek to resolve a dispute by direct negotiation, they need to ensure that the dispute is notified by early warning so as to maximise the chances of settlement, and this is considered in Section 19.6. Alliance members also need to agree how direct negotiations will be undertaken, either through the members of a Core Group or Alliance Board as considered in Section 19.7 or through other authorised individuals, for example by means of:

- 'Direct, good faith negotiations' between 'senior executives nominated in the Contract Particulars (or if either is not available, a colleague of similar standing)', as provided for in JCT 2016[1757]
- Meetings between 'Senior Representatives' for up to four weeks if a dispute is referred to them by the Alliance Board under the NEC4 Alliance Contract[1758]
- Meetings between individuals named in a 'Problem-Solving Hierarchy', as provided for in PPC2000.[1759]

Surrey County Council reported on their highways term alliance that 'with the size of the task (delivering over 500 major schemes in the first 3 years) this commitment was always tested in regard to disputes. Through the cultural workshops face to face discussions showed the team shared the same goals and desires to succeed and issues were discussed and agreed through open challenge and honest conversations'.[1760]

To provide the best chance of preserving relationships, an agreed system for the negotiated resolution of disputes should precede any other dispute resolution options. For example, FIDIC 2017 provides for a 28-day period during which 'both Parties shall attempt to settle the dispute amicably before the commencement of arbitration', but this obligation arises only after reference of that dispute to a Dispute Avoidance/Adjudication Board ('DAAB') and only after one party has given notice of its dissatisfaction with the DAAB's decision.[1761]

1755 Malhotra & Lumineau (2011), 984. This research involved access to the files of 102 disputes advised on a law firm in Western Europe between 1991 and 2005.
1756 Malhotra & Lumineau (2011), 993.
1757 JCT SBC/Q (2016) Schedule 8 clause 6.
1758 NEC4 Alliance Contract (2018) clause 97.
1759 PPC2000 (2013) clause 27.2.
1760 Extracts from May 2018 Trial Project report prepared by Surrey County Council for KCL Centre of Construction Law. See also Section 11.5.
1761 FIDIC (2017) clause 21.5. The period for amicable settlement has been reduced from the 56-day period provided for in clause 20.5 of FIDIC (1999).

19.6 How Can Early Warning Avoid Disputes?

Potential disputes can be averted if a warning of an emerging problem is issued as soon as the problem arises, and if the warning is issued to the correct party on the understanding that notification will lead to timely decisions and actions. For example, on the ConnectPlus M25 DBFO 'joint ownership of the Model by the client and supply chain members has enabled transparent and timely transmission of information throughout the supply chain and the early identification and solution of potential problems'.[1762]

In order to help in resolving problems or mitigating their effects, a system of communication needs to be planned and monitored. Otherwise there is the risk that:

- 'Information comes too late, or goes to the wrong place for decisions to be made'
- 'The information then becomes a mere record, and is of little value'
- 'The records are then used to allocate blame for problems, rather than to stimulate decisions which will control the problems'.[1763]

The English courts have increasingly taken the view that contractors as well as design consultants and professional advisers have a duty to warn their clients of any design defects of which they become aware.[1764] A contractual communication system can clarify and extend a duty to warn so that it applies to any potential problem and so that it is linked to a decision-making group which can review the problem when notified. Early warning can identify and enable resolution of problems and disputes, and for this purpose warnings need to be referred to a decision-making group, such as a Core Group or Alliance Board as considered in Section 19.7.

FIDIC 2017 includes provision for advance warning of 'any known or probable future events or circumstances' which may 'adversely affect the work of the Contractor's Personnel, adversely affect the performance of the Works when completed, increase the Contract Price and/or delay the execution of the Works or a Section (if any)'.[1765] FIDIC 2017 also provides that 'the Engineer may request the Contractor to submit a proposal to avoid or minimize the effects of such event(s) or circumstance(s)'[1766] but it does not provide for a forum or process for:

- Consultation on the merits of this proposal
- Who will decide on its acceptability and by what deadline
- What the contractor should do if a proposal is rejected or if no decision is made.

ICC 2014 provides for early warning and joint risk management through 'a meeting of appropriately authorised persons to consider actions or measures in response to the matter…notified'.[1767] The ICC Partnering Addendum includes its own provisions for joint risk management, early warning and a core group.[1768]

1762 ConnectPlus M25 DBFO Trial Project case study.

1763 Smith, N.J. (2002), 12.

1764 As considered in, for example, *Tesco Stores v The Norman Hitchcox Partnership* [1997], 56 Con LR 42 (QB).

1765 FIDIC (2017) clause 8.4, although there is no equivalent provision in the FIDIC Consultant Appointment (2017).

1766 FIDIC (2017) clause 8.4.

1767 ICC (2014) clause 6.1.

1768 ICC Partnering Addendum (2011) P4 and P5.

There are no early warning provisions in the JCT 2016 contracts except the requirement under the JCT Framework Agreement 2016 to warn of any matter 'likely to affect the out-turn cost or programme or the quality or performance'.[1769] However, there is no requirement in the JCT 2016 Framework Agreement for the party giving early warning to propose solutions, there are no stated consequences or required actions following an early warning and there is no contractual decision-making forum to review the early warning and agree a solution. There are also no provisions or guidance as to the effect of early warning on time/money claims or on dispute resolution under any project-specific JCT consultant appointment or construction contract.

NEC4 provides for an obligation to give early warning linked to early warning meetings which can act as a means of notifying and resolving prospective disputes.[1770] Early warning under NEC4 clause 15 is not a matter for a Core Group meeting under Option X12, and Option X12 clause X12.3(3) has a separate early warning system by reference to 'any matter that could affect the achievement of another Partner's objectives'.

FAC-1, TAC-1 and PPC2000 each provide for an early warning system linked to Core Group meetings which can act as a means of notifying and resolving disputes.[1771] Early warning requires the PPC2000 Client Representative as alliance manager to call a Core Group meeting in order to agree an appropriate course of action 'unless all the Core Group members agree such course of action without a meeting'.[1772]

The NEC4 and PPC2000 early warning provisions are also considered in the context of risk management in Section 18.4. The success of early warning depends on team members overcoming their instinctive wish to remain silent, and instead recognising that early warning proposals can be acceptable to all parties and may serve their interests better than reliance on other contractual rights.[1773] Early warning 'is clearly more than a mechanism for one party informing the other of the other's faults. It requires confession of the parties' own faults'.[1774]

It is possible that early warning obligations can clash with the constraints of a team member's professional indemnity insurance policy, for example where insurers require that their clients do not acknowledge any information regarding a possible liability without the insurer's consent. Professional indemnity insurers should be consulted on proposed early warning obligations, although in practice they have been willing to cover contracts which contain early warning clauses.[1775]

It is not possible to define rigidly in a contract all the circumstances in which early warning should be given. There is therefore a risk that, if the parties are entitled to notify problems, they will use early warning excessively as a means to seek contractual waivers

---

1769 JCT Framework Agreement (2016) clause 19.
1770 NEC4 (2017) clause 15.
1771 PPC2000 (2013) clauses 3.7 and 27.1, FAC-1 Framework Alliance Contract (2016) clauses 1.8 and 15.1 and TAC-1 Term Alliance Contract (2016) clauses 1.8 and 15.1.
1772 PPC2000 (2013) clauses 3.7 and 27.3.
1773 As illustrated in the Bewick Court Case study summarised in Section 19.10 and in Association of Consultant Architects (2010), 14 and 15.
1774 Eggleston, B. (2006), 117.
1775 None of over 90 projects referred to in Association of Consultant Architects (2010) encountered any reported reluctance of PI insurers to provide cover to any of the team members or any objection to the early warning provisions in the relevant standard form alliance contracts.

or leniency or simply to cause a distraction. At worst this runs the risk of wasted time and can be managed by peer group pressure and common sense.[1776]

19.7 How Can a Core Group or Alliance Board Resolve a Dispute?

In any negotiation the parties in dispute will make tactical and strategic decisions with their respective advisers, unaware of how other parties are approaching the same issue and without access to relevant information held by others that could usefully be shared. This can breed suspicion and can undermine collaborative relationships, and there is a wealth of evidence to show how self-interested decisions conceived in isolation may be self-defeating.[1777]

If nominated individuals cannot resolve a dispute by direct negotiation, then it is possible that a wider contractual forum can encourage a solution. We have considered in Section 12.7 the collaborative project governance provided by a Core Group or Alliance Board of named individuals with agreed terms of reference, and this forum also has a role to play in dispute avoidance and dispute resolution. Questions may arise as to whether an agreement by alliance members empowering a Core Group or Alliance Board to negotiate is unenforceable as an agreement to agree, but English case law supports the enforceability of a commitment to follow a clear negotiating process.[1778]

Notification of a problem to the Core Group or Alliance Board provides the basis for consultation and agreement with other interested parties. This is preferable to the notification of a problem only to the client or project manager, which may lead to a private review process without consultation and may result in a unilateral decision that is not accepted by other team members. Examples of the work of a Core Group include:

- On the Royal Borough of Greenwich term alliance the 'Early Warning and Core Group systems encouraged collective resolution of problems: Greenwich Council suffered no claims under its TPC contracts'[1779]
- The Macclesfield Station project alliance established a 'clear communications strategy under PPC "Core Group" and "Early Warning" systems to create a culture of trust and cooperation so that the team could overcome problems on site'.[1780]

A Core Group or Alliance Board can enable the open review of data, can seek to resolve issues of conflict or confusion and can encourage disclosure. However, in order to succeed in dispute resolution, a Core Group or its equivalent requires access to reliable sources of information in relation to, for example, design, risk, cost and time. For example, Building Information Modelling (BIM) has the potential to help ensure

---

1776 Eggleston noted in relation to NEC3 that 'Some degree of common sense and some tests of reasonableness and seriousness must be applied to avoid trivial matters obscuring the true purpose of the provisions', Eggleston, B. (2006), 117.
1777 Such as the 'Prisoner's Dilemma', an exercise described, for example, in Gerber & Ong (2013), 2.13.
1778 For example, in *Cable and Wireless v IBM UK* [2002] EWHC 2059.
1779 Association of Consultant Architects (2010), 19.
1780 Association of Consultant Architects (2010), 31.

that a Core Group has access to relevant data, and in this way to help avoid and resolve disputes.

The FAC-1 and TAC-1 Core Group is required to contribute to dispute avoidance and non-adversarial dispute resolution as follows:

- By seeking an agreed course of action following early warning, to be given by an alliance member with its proposals for avoiding or remedying the matter as soon as it is aware of any matter that adversely affects or threatens the alliance or the programme of work or any alliance member's performance[1781]
- By proposing actions to achieve agreed targets or revised targets in order to avoid termination of an alliance member's appointment[1782]
- By making recommendations in the event of notified breach of contract in order to avoid termination of an alliance member's appointment[1783]
- By meeting with all involved alliance members when notified of a dispute and seeking to achieve an agreed solution.[1784]

Using a Core Group has proved very effective in resolving any problems or misunderstandings. For example, the Supply Chain Management Group (SCMG) framework alliance members used a Core Group to resolve 'potential disputes with the benefit of full cost and time information plus the motive to retain long-term relationships'.[1785]

The Arup report observed in its commentary on PPC2000 that 'the creation of a Core Group to guide the project also has a dispute resolution function. This ensures the visibility of problems and any impact of those problems upon the project irrespective of the point in the supply chain at which they are found…In providing these processes it is expected that the parties will find that the terms of the contract provide a swifter and more cost-effective way of resolving points of difference than they might obtain from other dispute resolution mechanisms available such as adjudication or litigation'.[1786]

The NEC4 Alliance Contract provides that:

- 'A dispute between members or past members of the Alliance arising under or in connection with the contract is referred to the Alliance Board'
- 'Unless and until the Alliance Board finally resolves the dispute, the members of the Alliance proceed as if the matter disputed was not disputed'.[1787]

It is important that a Core Group or Alliance Board makes its decisions by a unanimous vote rather than a majority vote.[1788] If any party involved in a dispute fears that it will be obligated to compromise its interests by losing a majority vote, rather than persuaded to accept a unanimously agreed solution, it will be unlikely to participate in the Core Group meetings. This exposes a potential weakness in ConsensusDocs 300 which

---

1781 FAC-1 Framework Alliance Contract (2016) clause 1.8.
1782 FAC-1 Framework Alliance Contract (2016) clause 14.2.1.
1783 FAC-1 Framework Alliance Contract (2016) clause 14.4.
1784 FAC-1 Framework Alliance Contract (2016) clause 15.1.
1785 SCMG Trial Project case study.
1786 Arup (2008), 38 and 39.
1787 NEC4 Alliance Contract (2018) clauses 95.1 and 95.3.
1788 As illustrated in the requirement for unanimous decisions to be made by the Core Group under PPC2000 clause 3.6 and by the Alliance Board under NEC4 Alliance Contract clause 21.3.

firstly permits a majority decision and secondly, in the event of failure to achieve a decision, refers the matter to a Senior Executive Team with a decisive client vote.[1789] The ConsensusDocs 300 approach creates the risk of a majority decision that a team member does not accept, and also raises the question of whether a Senior Executive Team would have sufficient knowledge of the project to make an informed decision, bearing in mind that the Senior Executive Team have not built up collaborative relationships through previous joint activities.

In order to be equipped for dispute avoidance, a Core Group or Alliance Board should be active from the inception of the project so that its members can develop the mutual trust and detailed data that will enable them to make informed, consensus-based decisions. One problem experienced on a number of collaborative projects was a failure to hold any Core Group meetings until a problem arose. If a Core Group only meets to consider potential disputes, it cannot contribute actively to the success of collaborative project procurement[1790] and will have a limited role similar to that of a dispute board.[1791]

A Core Group or its equivalent is less effective if its client member is reluctant to attend meetings, particularly where there is a likelihood of conflicting views. For example, a client may seek to delegate difficult meetings to a consultant to attend on its behalf, and this does not create or sustain a credible client commitment to seeking an agreed solution. By contrast, commentators have noted a beneficial change in the management and organisation of the supply chain under PPC2000 where 'the client takes an active part in the problem-solving processes'.[1792]

On the Bewick Court project summarised in Section 19.10, the appointed alliance manager had difficulty in persuading the client to attend a critical Core Group meeting following early warning of problems arising from the insolvency of the cladding subcontractor. However, having attended and participated in a valuable risk management exercise, the client was aware of what the Core Group could achieve and used the same forum to resolve a problem that the client encountered later in the project.

On another troubled alliance project, the contractual requirement for all members to attend Core Group meetings, or to risk a decision being made in their absence, ensured that 'the core group members step by step acknowledged the parties' respective contributions to the problems and ultimately agreed a compromise'.[1793]

19.8 How Can an Independent Adviser Resolve a Dispute?

We have considered in Section 12.9 the role of the Independent Adviser in supporting a collaborative culture, and some alliance contract forms also provide for advice from an Independent Adviser in support of collaborative dispute resolution. For example, PPC2000 provides that the search for an agreed solution by direct negotiation between

---

1789 ConsensusDocs 300 (2016) Articles 3.10 (Failure to reach unanimity) and 3.11.2 (Decisions): 'If the Senior Executive Team is unable to reach a unanimous decision or action, Owner shall make a decision in the best interest of the Project'.
1790 In the ways described in Section 12.7.
1791 As considered briefly in Section 19.9.
1792 Gruneberg & Hughes (2006), 47.
1793 Mosey, D. (2009), 248.

the individuals named in a Problem-Solving Hierarchy shall be 'guided as necessary by the Partnering Adviser'.[1794] Where the individuals named in the Problem-Solving Hierarchy cannot agree a solution bilaterally, PPC2000 also provides for the Partnering Adviser to attend a Core Group meeting called for the purpose of seeking an agreed solution to the dispute.[1795]

Similar provisions appear in FAC-1 and TAC-1 where an appointed Independent Adviser supports the Core Group members in their search for agreed resolution of a dispute.[1796] There is no Independent Adviser role in the FIDIC 2017, ICC 2014 or JCT 2016 contracts.

The NEC4 Alliance Contract introduces the role of an 'Independent Expert' to whom the Alliance Board may refer a dispute in order to obtain an opinion.[1797] However, being available to provide an opinion is a limited role and appears to suggest that an Independent Expert would have no previous involvement in the project. This is not the same as the ongoing relationship and shared understanding that can be built up with an Independent Adviser.

The role of the Independent Adviser is comparable to that of the dispute resolution adviser ('DRA') which has been developed in Hong Kong and where:

- 'The DRA effectively acts as a project facilitator and mediator, becoming familiar with the project and key personnel and is thus able to identify potential problems before they become formal disputes'
- 'If formal disputes do arise, the DRA is able to adopt an adviser role and assist the parties in choosing the most appropriate, speedy and economical dispute resolution technique for the particular dispute, whilst avoiding the all too common adversarial attitudes that develop between disputants'.[1798]

The rules governing the role of the DRA provide that:

- 'The DRA is an independent expert jointly chosen, appointed and remunerated by the contracting parties
- The DRA becomes familiar with the project and participants by regular site visits and meetings
- The DRA assists the parties in dispute prevention, including individual communications/meetings with each party
- The DRA undertakes processes similar to facilitative mediation but at an early stage in any conflict
- The DRA does not undertake dispute evaluation or make decisions'.[1799]

The DRA system also recognises the benefit of independent advice to bidders so as to secure buy-in to the role of the DRA but sees this a role for a separate pre-contract consultant facilitator appointed to report on equitable risk allocation.[1800]

1794 PPC2000 (2013) clause 27.2.
1795 PPC2000 (2013) clause 27.3.
1796 FAC-1 Framework Alliance Contract (2016) clause 15.1 and TAC-1 Term Alliance Contract (2016) clause 15.1.
1797 NEC4 Alliance Contract (2018) clauses 95.2 and 96.
1798 Gerber & Ong (2013), 166–188, describing the Dispute Resolution Adviser role.
1799 Gerber & Ong (2013), 166–188.
1800 Gerber & Ong (2013), 166–188.

An Independent Adviser should be appointed when an alliance contract is entered into so that he or she can become familiar with the project and can be available at any time. If an Independent Adviser is appointed at a later date, it may be harder for the parties to agree on who is suitable, and doubts may arise as to the neutrality of an Independent Adviser recommended at a later stage by the client or another alliance member. Whoever is selected and at whatever time, the alliance contract should provide for:

- Whether the Independent Adviser is paid and instructed by the client or by all parties
- Whether the Independent Adviser has a facilitative or evaluative role.

A decision as to who pays the Independent Adviser is a practical issue that should not affect how he or she performs the agreed functions. However, psychologically it can be important for the Independent Adviser to be seen to be neutral by reason of not being paid by only one party. For example, the NEC4 Alliance Contract states that 'the members of the Alliance that appoint the independent expert share its costs equally'.[1801]

There are no formal qualifications for an Independent Adviser and no rules governing his or her appointment.[1802] As general guidance the requirements of independence and impartiality for an Independent Adviser could reflect those for the members of a FIDIC Dispute Avoidance/Adjudication Board ('DAAB') in terms of:

- Having no other financial interest in the project or any party
- Having no previous appointment by any party in the last 10 years
- Having no previous judicial or arbitral role in relation to the contract
- Disclosing any professional or personal relationships with any party
- Not seeking future business or employment from any party
- Not seeking any gift or gratuity.[1803]

The role of an Independent Adviser is strongest where he or she is supporting the parties in seeking to negotiate a solution, either directly or through a Core Group or Alliance Board, as illustrated in the Bewick Court case study in Section 19.10.

19.9 Are There Other Collaborative Ways to Resolve a Dispute?

Macneil noted that a 'relational response to the breakdown of cooperation…tends to be defined in terms of what is necessary or desirable to restore present and future cooperation'.[1804] He referred to negotiation and also to mediation and arbitration as 'processes fostering cooperation'.[1805] As mediation and arbitration require involvement

1801 NEC4 Alliance Contract (2018) clause 96.5.
1802 Although the Association of Consultant Architects developed an examination and minimum level of experience for Partnering Advisers and a related Code of Conduct, Mosey, D. (2009), Appendix F, 300.
1803 FIDIC (2017) Appendix: General Conditions of Dispute Avoidance/Adjudication Agreement, clause 4.1.
1804 Macneil, I.R. (1974), 741.
1805 Macneil, I.R. (1974), 741.

of a third party, it is arguable that in varying degrees they signal the failure of relationships and processes to provide a cooperative solution.

However, where a dispute cannot be resolved by the team members themselves, working through structured negotiation or a Core Group, and with or without the support of an Independent Adviser, it will be necessary to refer the dispute for resolution by a third party. Dependent on the agreed terms of reference, and on the approach adopted by the third party, this move from facilitative to evaluative dispute resolution in varying degrees deprives the team members of direct control over the way in which a solution may be achieved, although collaborative contract terms can influence the way in which dispute resolution should be conducted.[1806]

The following standard forms provide for third party collaborative dispute resolution as follows:

- ICC 2014 provides the option for the parties to agree that a dispute shall be referred to conciliation or mediation[1807]
- FIDIC 2017, NEC4, FAC-1, and TAC-1 all provide the option to appoint a dispute board[1808]
- JCT 2016 provides that 'each Party shall give serious consideration to any request by the other to refer the matter to mediation'[1809]
- PPC2000, FAC-1 and TAC-1 all provide the option to appoint a conciliator.[1810]

FIDIC 2017 includes a full set of rules governing the work of its DAAB. FIDIC 2017 also suggests that a DAAB can be active throughout the project rather than only in response to the reference of a specific dispute, with an informal advisory role whereby:

- 'If the Parties so agree they may jointly request...the DAAB to provide assistance and/or informally discuss and attempt to resolve any issue or disagreement'
- 'Such informal assistance may take place during any meeting, Site visit or otherwise. However, unless the Parties agree otherwise, both Parties shall be present at such discussions'.[1811]

The rules governing the roles of the mediator, conciliator and dispute board vary in the provisions of different standard forms but in all cases the principles include the requirement that the parties approve any decision before it becomes enforceable. This maintains a degree of control by the parties themselves, and leaves open the potential for any party to reject a decision and to insist that the dispute is referred to adjudication, arbitration or the courts according to the applicable law and relevant contract provisions.

1806 In *Willmott Dixon Housing v Newlon Housing Trust* [2013] EWHC 798 (TCC), it was held that a contractual commitment to 'transparent and cooperative exchange of information' should guide the conduct of the parties during the adjudication of a dispute.
1807 ICC (2014) clause 19.3.
1808 FIDIC (2017) clauses 21.1–21.4, General Conditions of Dispute Avoidance/Adjudication Agreement, and DAAB Procedural Rules; NEC4 Option W3; FAC-1 Framework Alliance Contract (2016) clause 15.2; TAC-1 Term Alliance Contract (2016) clause 15.2. Standard form dispute board provisions were first set out in FIDIC (1999) and this approach was trialled in the UK in an adapted version of NEC3 on the London 2012 Olympics.
1809 JCT SBC/Q (2016) clause 9.1.
1810 PPC2000 (2013) clause 27.4 and Appendix 5 Part 1; FAC-1 Framework Alliance Contract (2016) clause 15.2 and Appendix 4 Part 1; TAC-1 Term Alliance Contract (2016) clause 15.2 and Appendix 4 Part 1.
1811 FIDIC (2017) clause 21.3.

Questions may arise as to the enforceability of an agreement to refer a dispute for collaborative dispute resolution as a precondition to adjudication or arbitration.[1812] For example, in jurisdictions such as the UK where adjudication is a statutory right that can be exercised at short notice, it may be difficult to ensure that team members firstly undertake other more collaborative processes.[1813]

Alternative procedures will be easier to uphold if clearly set out in the contract terms. A breach of contract in bypassing agreed dispute resolution procedures may be actionable in itself and may also influence the award of further work, for example under an overarching framework alliance. Where a collaborative dispute resolution procedure is agreed, the English courts will be unlikely to allow any party to bypass it unless that party is exercising a statutory right to refer the dispute to adjudication.[1814]

19.10 Collaborative Dispute Resolution Case Studies

Where disputes arise, collaborative teams have found negotiated solutions through direct engagement. The following case studies illustrate how this has been achieved on Bewick Court, a UK housing refurbishment project in Newcastle, and on a bank headquarters project in Bahrain. In both case studies, the agreement of solutions to major disputes was enabled by:

- The relationships and additional information developed through early contractor involvement, through clear commitments to collaborative working and through the contract provisions establishing a Core Group

---

1812 In *Channel Tunnel Group v Balfour Beatty Construction* [1993] AC 334, the House of Lords upheld agreement for the reference of a dispute firstly to panel of three 'acting as independent experts but not as arbitrators' for a unanimous decision, failing which, as a last resort, reference of the dispute to ICC arbitration. Lord Mustill stated that 'those who make agreements for the resolution of disputes must show good reasons for departing from them'.

In *Wah (Aka Alan Tang) v Grant Thornton International* [2012] EWHC 3198 (Ch), the court held that for a mediation clause to be enforceable:

- Either the mediation clause must provide reasonable certainty as to the beginning, the ingredients and the end of the mediation process
- Or the subject matter of the mediation must be determinable by fairness or reasonableness so that the court can infer the necessary procedural ingredients.

Finding that the mediation clause satisfied neither criteria, the court upheld the arbitrator's decision that the mediation clause was ineffective.

1813 On a project where I was appointed as Independent Adviser but rarely, if ever, consulted, I discovered that the parties had initiated an adjudication regardless of their commitment to seek a negotiated solution. I arranged and facilitated an urgent meeting of chief executives who successfully agreed a compromise and halted the adjudication. Their solution owed less to a collaborative culture than to the approaching financial year end and their wish to avoid reserving large sums as contingent liabilities.

1814 In *Peterborough City Council v Enterprise Managed Services* [2014] EWHC 3193 (TCC), Peterborough engaged EMS to design and install a solar energy plant. FIDIC sub-clauses 20.2–20.7 set out the procedure for dispute resolution by a Dispute Adjudication Board ('DAB') to be appointed on an ad hoc basis after any dispute had arisen, provided that if at the time a dispute arose there was no DAB in place 'whether by reason of the expiry of the DAB's appointment or otherwise' then either party could go to court. Following completion Peterborough alleged that the plant had failed to achieve the required power output and argued that, because the dispute raised complex questions relating to legislation, mandatory codes and standard industry practice, the 'rough and ready' process of DAB adjudication was inappropriate. However, the judge noted that the complexity of a potential dispute was foreseeable from the outset, and upheld the agreement by the parties to adopt the FIDIC DAB machinery.

- The roles of client and other leading team members, and the support proved by an Independent Adviser.

On the Bewick Court project all team members signed the multi-party PPC2000 contract and were committed to the agreed roles of the Core Group and an Independent Adviser in the non-adversarial resolution of the problems that arose. There was some initial client reluctance to attend a Core Group meeting when a problem arose because it perceived this to be a matter only for the main contractor, namely the insolvency of a domestic subcontractor after start on site. However, the 'Client Representative' as alliance manager reminded the client that under the PPC2000 project alliance contract all team members are committed to attend Core Group meetings in order to seek solutions to problems that affect any of them.

The Bewick Court Core Group had previously approved the design, cost and supply chain data presented to it by team members in relation to the selection of the specialist cladding subcontractor. The Core Group members could rely on this data in considering and agreeing constructive proposals presented by the main contractor following the cladding subcontractor's insolvency. The value of the Core Group in facilitating an agreed collaborative solution was recalled by the client when it encountered a problem later in the project and depended on cooperation and support from the main contractor.

On the Bahrain bank headquarters project a FIDIC 1999 Red Book contract was adapted by adding a conditional preconstruction phase contract, providing for the build-up of agreed design, cost and supply chain data through early contractor involvement. However, this adaptation did not go so far as to create a multi-party alliance contract with direct contractual links between team members. The FIDIC 'Engineer' was engaged under a separate appointment and refused to include amendments that formalised the roles of a Core Group and Independent Adviser.

Nevertheless, the client persuaded the Engineer to participate in Core Group meetings seeking to resolve a potential dispute and for these meetings to be supported by the Independent Adviser under specific terms of reference. Agreement of these terms of reference was a difficult process in itself which used up valuable time and owed a lot to the influence of the client. In addition, the cynicism displayed by the Engineer's representative throughout the process placed strain on the negotiated resolution of the dispute.

Bewick Court Newcastle Residential Tower Block Refurbishment

The Bewick Court project comprised the recladding and other refurbishment of a residential tower block by a team comprising North British Housing, now Places for People Group (client), Kendal Cross Holdings (main contractor), Elliott Associates (client representative), Red Box Design Group (architect), Gilwood Engineering Services/BES Consulting Engineers/WSP (engineers).

The preconstruction phase build-up of supply chain and cost included joint client/main contractor selection of a cladding specialist subcontractor whose package accounted for a significant element of the project (£1.5 million).

After start on site, the cladding specialist (Allscott) went into administrative receivership. The PPC2000 Client Representative was aware that there were no comparable cladding specialists available within a wide geographical radius, and by an Early Warning

(Continued)

notice under PPC2000, the main contractor was invited to put forward alternative proposals to the Core Group.

The main contractor's proposed solution was to take on direct liability for cladding, recruiting the insolvent specialist's workforce and buying the required materials cheaply from the administrative receiver: a consequent four week delay and additional costs of £7352 were approved by the client.

Later in the project, when mobile phone companies delayed shutting down their aerials on the roof of the building, the Client Representative gave Early Warning of the need for suspension of work. The main contractor agreed to identify savings elsewhere in the project that could fund the costs of suspension, and the client agreed that any compensation received from the mobile phone companies would be spent on the project: the client recovered £40 000 of compensation and spent it on an improved window-cleaning system. Dispute avoidance benefited from:

- Involvement of the client and Client Representative with the main contractor in early selection and appointment of the cladding specialist, creating clear cost information with which to analyse the cost consequences of replacing that specialist
- Early establishment of a communications strategy, utilising the Core Group and Early Warning for joint risk management
- A clear role for the client participating in Core Group problem-solving activities
- Avoidance of claims and cross-claims through early establishment of key individuals authorised to achieve commercially sensibly solutions.

David Pearson of Elliott Associates, Client Representative, commented: 'We could have seen contractual claims against both the client and the contractor and worst of all a project not yet concluded, resulting in another cold winter for Bewick Court residents. Instead, the project finished on time and within its maximum price and the team remains firmly on speaking terms'.

Association of Consultant Architects (2010), 14, 15.

Bahrain bank headquarters

A new build bank headquarters project in Bahrain used an adapted FIDIC (1999) Red Book contract to implement early contractor involvement and to appoint an Independent Adviser. Early selection of a main contractor was supported by a bespoke preconstruction phase contract added to FIDIC, but without substantially changing the other FIDIC conditions. The pre-construction phase contract provided for:

- Early contractor involvement under a conditional contract
- Build-up of shared designs and related supply chain, cost and programming information
- Agreed preconditions to award of the unconditional construction phase FIDIC contract
- Creation of a Core Group to resolve problems with support from the Independent Adviser.

The appointed FIDIC Engineer refused to agree corresponding amendments to its own appointment and was concerned that the Core Group and Independent Adviser would reduce its authority.

A dispute arose between the main contractor, steel specialist and Engineer as to whether steel had been inspected and accepted by the Engineer prior to installation. This was a crucial issue as a suspended steel structure was a central feature of the design. All parties in dispute (including the Engineer) agreed the terms of reference and timescales for negotiated resolution, including 'without prejudice' negotiation meetings facilitated by the Independent Adviser.

Although the parties were preparing to suspend work and enter into arbitration, the Core Group met regularly with advice from the Independent Adviser and the parties were persuaded to exchange additional design and supply chain data rather than hold it back for use as later evidence. The client agreed an extension of time to allow for further steelwork tests to be completed. These actions provided the basis for negotiated resolution of the dispute.

Reported to KCL Centre of Construction Law by the author, based on his work as an Independent Adviser to the project team.

In 2009 the Bahrain Government commissioned a new suite of standard form construction and engineering contracts, incorporating provisions for early contractor involvement through a 'Pre-Construction Appointment' governing services such as additional site investigations, contractor design contributions, value engineering, subcontractor procurement and joint risk management, all of which are governed by a 'Key Dates Schedule'.

20

How Does Collaborative Procurement Operate in Australia?

Professor Paula Gerber and Marko Misko

20.1 What Is the Approach to Alliances in Australia?

The Australian construction industry has an unenviable reputation for being highly adversarial, with the direct costs of resolving construction disputes estimated at being A\$560–840 million annually.[1815] The non-financial costs of disputes, such as fractured relationships, and wasted time, energy and resources devoted to resolving disputes, are of course much harder to measure.

Given the Australian construction industry's track record with protracted disputes, it is perhaps not surprising that a former Victorian Supreme Court judge went so far as to say that 'It is *inevitable* that the commercial activities of those engaged in the construction industry will always produce disputes' [emphasis added].[1816] The authors of this chapter do not share such a pessimistic view. Given the multitude of parties, the risky nature of construction and the complex and lengthy nature of designing and building structures, *conflicts* are inevitable, but *disputes* are not. A key factor in ensuring that conflicts do not escalate into disputes is to get the contractual relationship between the parties right. This is an area where there is significant scope for improvement in Australia.

This chapter considers the impact that the construction industry's approach to contracts has on the uptake of alliances and Building Information Modelling (BIM) on major projects in Australia. It concludes that the time is right for a significant shift in the industry's approach to construction contracts in order to ensure that Australia is keeping up with global best practices when it comes to contractual frameworks and digital technology.

In 1990, the Construction Industry Development Authority issued its seminal report 'No Dispute',[1817] which – in a nutshell – identified the causes of the high incidence of claims and disputes in the Australian construction industry and proposed measures for avoiding those claims and improving productivity.

In a partial response, in the 25 years or so since then, the construction industry and its advisers have progressively developed (or adapted from international experience) a suite of collaborative contracting delivery models. Those models include partnering,

1815 Cooperative Research Centre for Construction Innovation (2009).
1816 Byrne, D. (2007), 405.
1817 Construction Industry Development Authority (1990).

Collaborative Construction Procurement and Improved Value, First Edition. David Mosey.
© 2019 John Wiley & Sons Ltd. Published 2019 by John Wiley & Sons Ltd.

managing contractor ('MC'), early contractor involvement ('ECI'), integrated project delivery ('IPD'), delivery partner model ('DPM') and project alliancing, to name but a few.

All these models share, to a greater or lesser degree, an increased focus on people, relationships and alignment of commercial interests together with a decreased focus on blackletter contractual rights and obligations. However, each model also has its unique features, which need to be fully understood in order to avoid the risk of imposing an entirely inappropriate delivery model and risk allocation on a project.

In Australia, the project owner typically selects the delivery model and form of contract. Once the owner makes the choice that it wants to adopt a collaborative approach, the next question is: how far?

If the owner needs a degree of legal certainty as to time, cost and quality outcomes – but still wants to be collaborative – it may choose MC, ECI or IPD. In particular, the Department of Defence Managing Contractor Contract[1818] – Australia's first and only true industry standard form collaborative contract since 1992 – has stood the test of time and provides the most frequently used collaborative alternative to alliancing.

However, if a project requires a more radical legal approach, then alliancing comes to the fore. This is most likely where the risk profile is high, where there are significant technical challenges, where the works share an interface with existing operations and systems or where there is a need for significant flexibility as to scope, budget or programme.

Why radical? Well, from a legal perspective, alliancing in Australia has been underpinned by several distinguishing features that are not shared by the other collaborative models and that are designed to further its precise commercial intent. These include:

- A 'no blame/no dispute' mechanism, under which the participants release each other from all liability, except for 'wilful default' (which is carefully defined and does not include mere breach of contract or negligence); this enables the alliance to shoot for the stars without the distracting spectre of legal liability
- Project governance requiring unanimous decision-making by an alliance board (except for decisions reserved to the owner) in order to drive maximum collaboration
- A cost-plus payment mechanism under which the non-owner participants are reimbursed the direct cost of all work they perform, including rework of defective design and construction, subject to margin being at risk depending on performance against agreed key performance indicators.

This legal approach – enshrined in the alliance contract – means that all participants (especially the owner) need to focus on practical risk mitigation. This is often done by having genuine in-house technical expertise; investing significantly in project management; ensuring that the capability, key people and culture of the other participants are appropriate; carefully structured subcontracting; tailored insurances and – perhaps most importantly – a commercial gain share/pain share regime which aligns the interests of all participants.

The rub is, of course, that a failure to implement these practical risk mitigation strategies effectively could result in project failure and heavy losses for one or more participants.

Alliancing was first used in Australia on the Wandoo Offshore Gas Platform, in 1995. Since then, it has had a mercurial history, which can be broadly split into two phases.

1818 MCC-1 (2003).

In the period from 1995 to 2009, the model became like the 'emperor's new clothes'; so much so that – by the early 2000s – it was being extensively used by both the public and private sectors (primarily the former) to deliver both building and civil engineering projects. However, unlike the emperor's wardrobe, the practical concerns with this new fashion were very real.

First, alliancing was being vigorously recommended by those who had a vested interest in its use – alliancing practitioners (fees) and contractors (softer risk allocation) – regardless of whether it was genuinely best for project.

Second, there was a growing body of evidence that the alliancing hyperbole was not always matched by actual performance. In particular, value for money in the public sector was being increasingly questioned. In part, this was because alliances were being competitively procured on a non-price basis, under which the target out-turn cost ('TOC') for the project was not set until after contract award.

Third, alliancing advocates were rigidly evangelical about how they spread the gospel – this meant that, although there was no official industry standard form in existence, in practice there was little latitude in the way in which alliances could be structured. Paradoxically, this tended to stymie (at least from a legal perspective) flexibility in alliance procurement.

The upshot? The Victorian Department of Treasury and Finance commissioned its report 'In Pursuit of Additional Value: A benchmarking study into alliancing in the public sector' (the 'DTF Report')[1819] which essentially concluded that, although alliancing had the potential to deliver value for money, it was often not doing so and instead was actually resulting in projects significantly exceeding business case estimates.

The DTF Report found that alliancing had become the default model for the delivery of public sector projects, due in part to economic conditions and in part to the need to overcome skills and capacity shortages and programme imperatives. The result was the ill-considered use of alliancing on projects which should have been delivered using other models.

This led to the second phase from 2009, in which alliancing is now used in a more measured fashion. There is greater focus on value for money in the procurement phase, greater freedom to tailor the terms of the alliance contract and a more balanced use of the model as part of the broader church of collaborative delivery models.

As at 2018, this is perhaps best exemplified by the way in which infrastructure mega-projects are being structured. Rather than using a single delivery model, they are now being split according to the unique characteristics of the respective work packages. In that context, alliancing is frequently used to deliver those packages which involve the highest risk, such as interfaces with other new works or existing systems/operations. Other models, such as a public–private partnership or a lump sum design and construct, are then used to deliver the less risky balance of the project.

At the same time, where owners require greater legal certainty as to project outcomes in conjunction with their collaboration, other collaborative models have come to the fore. In particular, there is far greater recognition of the roles that MC, ECI and IPD each have to play in collaborative project delivery.

Ultimately, this freedom of choice can only be good for the industry, and the ongoing quest for increased productivity and innovation and decreased claims and disputes.

1819 Defined in the main bibliography as Department of Treasury and Finance (2009).

20.2   What Is the Approach to BIM in Australia?

BIM has been used by the Australian construction industry for a number of years, on both public and private sector projects. However, despite its acknowledged benefits for both asset design/construction and for operation/maintenance,[1820] BIM is yet to reach anywhere near its full potential in either phase of the asset lifecycle.

This is primarily the direct result of a palpable absence of strong leadership as to how BIM should be implemented. For instance, the strong government mandate in the UK has been lacking in Australia despite the clear recognition of its importance. Instead, the market has been characterised by a fragmented approach, with numerous industry bodies issuing their own guidance and corresponding documentation.

This has resulted in a lack of nationally consistent BIM processes and consequent reluctance by industry players (whether asset owners, designers, constructors or facilities managers) to take the plunge and fully invest in the BIM initiative. The implication for industry has been that the implementation of BIM has been sporadic and episodic, and typically driven by designers and constructors to improve their own outcomes, without these being shared with the owners.

This is notwithstanding some very clear policy guidance on the issue. For example, in 2014, the Productivity Commission Report into Productivity Improvement[1821] recommended that the government should use BIM to lower the cost of delivering its complex infrastructure projects. The Report also recommended that, to facilitate consistent use of BIM by the public sector, governments – at the national, state and territory levels – should develop a common set of standards and protocols (in consultation with industry) and include in their procurement guidelines detailed advice as to the efficient use of BIM.

In 2016, the Australian Infrastructure Plan (released by Infrastructure Australia)[1822] went further. In part, the purpose of that plan was to set out the infrastructure challenges and opportunities faced by Australia, and then identify the potential solutions. In that context, the plan recommended that governments should make BIM mandatory for the design of large-scale complex infrastructure projects. To support this, the plan recommended the development of appropriate guidance for the adoption of BIM and common standards and protocols.

Despite this strong policy imperative, the recommended changes have not yet occurred. Although the Australian Department of Defence is working hard to develop a standard form approach to BIM for the delivery of its significant annual infrastructure programme,[1823] this is only one agency (albeit a major player) which cannot necessarily speak for the balance of industry.

1820 In 2010, Allen Consulting Group published the results of its industry-sponsored study into the impacts of BIM on productivity. The study's findings reported that the accelerated and widespread adoption of BIM on the Australian economy was likely to increase GDP by 0.2 basis points in 2011. Whilst this may not seem much, the study also noted the likelihood of a cumulative effect given the likely increase in the pace of adoption by 2025. This would mean an increase of 5 basis points in comparison to what they describe as a 'business as usual' scenario.

1821 Public Infrastructure Productivity Commission (2014).

1822 Australian Infrastructure Plan (2016).

1823 KCL Centre of Construction Law (2016), Appendix E Part 2.

At the Australian Government level, procurement (including construction) is governed by the Commonwealth Procurement Rules ('CPR').[1824] They set out numerous mandatory rules which must be observed by Australian government agencies in conducting their procurements, with a view to ensuring efficient, effective, economical and ethical procurement[1825] – at the end of the day, the paramount objective is maximising value for money.[1826]

This is the most logical avenue through which the Australian Government could drive broader BIM reform. It has certainly used the CPR to do so in other procurement policy respects, for example in relation to free trade agreement issues and the specifications for goods and services.[1827] Despite this opportunity, no CPR amendments have yet been proposed with a view to mandating BIM.

The risks are that, without government leadership to establish a consistent national approach to BIM, its adoption (and the realisation of its manifest benefits) will continue to be limited and there will be a divergent approach from project to project and client to client. In particular, this will deprive those owners with significant asset portfolios of the opportunity to realise benefits throughout the operation and maintenance phase of their assets.

To address these very issues, the Australasian BIM Advisory Board ('ABAB') was established in May 2017, representing various state and territory governments, government agencies, industry bodies, universities and consultants.[1828] It is focused on a more consistent approach to the adoption of BIM across jurisdictional boundaries; it also recognises that – without central principal coordination – there is a significant risk that the development of protocols and guidelines will be fragmented, leading to wasted effort and inefficiencies.

The major areas in which BIM requires strong national coordination are process consistency, consistent information requirements and deliverables, bespoke project delivery models and standard form contracts. If achieved, then they will result in more widespread national adoption of BIM, reduced BIM implementation costs (for all industry participants), decreased waste and increased productivity, greater ability to benchmark, greater market understanding and confidence, a more consistent approach to risk allocation and improved asset lifecycle management.

In particular, in the context of an industry characterised by standard form contracts, the optimal implementation of BIM by a project owner will depend heavily on choosing the right delivery model, and then documenting it accordingly. For instance, Defence – in its ongoing search for the BIM 'holy grail' – has found that the delivery model is the key.[1829]

1824 Commonwealth Procurement Rules (2018).
1825 Commonwealth Procurement Rules (2018), r 6.
1826 Commonwealth Procurement Rules (2018), r 4.
1827 Commonwealth Procurement Rules (2018), r 10.9–10.13.
1828 See www.abab.net.au.
1829 'We started our BIM journey 3 years ago by consulting extensively with industry ahead of developing our third generation suite of infrastructure delivery models. We found that there was a strong link between BIM and the underlying contract models and that – before we could fully implement BIM – we had to get those contract models right. It soon led us to the conclusion that – firstly – integrated project delivery (IPD) was required to enable BIM and – secondly – the level of integration required from project to project would vary' Bob Baird AM and Jolanta Skawinski, Australian Department of Defence, in KCL Centre of Construction Law (2016).

Defence has been developing its own suite of IPD models, which seek to optimise BIM implementation, and have been trialled on pilot projects since 2015.[1830] These models incorporate:

- Early contractor involvement
- Collaboration through an integrated project team involving all key project contractors and stakeholders
- A deconstructed supply chain to facilitate access to buildability
- Two-phase delivery to enable the project to be de-risked in the initial design phase and to provide maximum flexibility as to scope, programme and budget
- Removal of traditional fixed time/fixed price tensions in the risk allocation
- Heavy focus on whole lifecycle and asset information
- Positive financial incentives through a 'project performance dividend'[1831]
- Dispute avoidance through a 'project risk and issue settlement model' (the PRISM).

With the new Defence IPD suite released formally to market in 2018, together with the coordinated initiatives now being driven by ABAB and the Australian Government's call for submissions for the Australian Government's Digital Economy Strategy,[1832] there are encouraging signs of the strong national leadership needed to ensure that BIM reaches its full potential within the Australian construction industry.

20.3 What Is the Approach to Construction Contracts in Australia?

The construction project, with its linked set of construction contracts between many participants and its complexity, could not survive today without standard construction contracts. Yet we know so little about them.[1833]

The standard form construction contracts commonly used in Australia all follow the traditional contract theory, which sees a contract as 'a discrete exchange involving rational and utility-minded parties with more-or-less equal bargaining power, who seek to maximise their *self-interest* in the transaction, with little concern for the interests of the other party' [emphasis in original].[1834]

A move away from traditional construction contracts to those based on relational contract theory would be a good starting point in changing the adversarial culture of the construction industry to a culture that emphasises and values teamwork and collaboration. Contracts based on relational contract theory recognise the interdependence of

---

1830 KCL Centre of Construction Law (2016), Appendix E Part 2.
1831 Although subject to a final review prior to formal release to market, the current intention is that the 'project performance dividend' will be an incentive pool determined during the initial design phase and payable at the end of the subsequent construction phase – to all relevant project parties (including the contractor, the key design team, key trades and suppliers and the facility manager) – depending on the achievement of agreed key performance indicators and the project parties' respective agreed contributions to project success.
1832 Department of Industry, Innovation and Science (2018).
1833 Sweet, J. (2011).
1834 Gerber & Ong (2013), 71.

different parties on a construction project and are designed to facilitate positive working relationships between the parties that are based on trust and cooperation.

Whilst many countries have standard form construction contracts that are informed by relational contract theory, such as NEC4 and PPC2000 in the UK and ConsensusDocs in the USA, Australia has no such contracts generally available.[1835] All the popular standard form construction contracts in use in Australia are based on the traditional contract theory, and thus contain no incentives for the parties to act in good faith, collaborate or make decisions that are informed by what is best for the project.

Thus, whilst the Australian construction industry is often an early adopter of new technologies, processes and materials, it has been reluctant to embrace contemporary contracts that encapsulate best practice when it comes to contract theory. This may, in part, be due to a lack of research and understanding in this area.

In response to 'a dearth of publicly-available data as to the Australian construction industry's use of, and attitude towards, standard forms of contract',[1836] the Melbourne Law School undertook a much-needed study into the use of standard form contracts in Australia. It found that 'there is broad support in principle for the industry having available to it standard forms of contract which are capable of being used without substantial amendment. However, a majority (54%) of respondents believe that there is no such form currently available'.[1837]

This suggests that the time is ripe for the introduction of a new standard form construction contract that meets the needs of key stakeholders, yet any optimism that change is imminent must be tempered. The reason for pessimism regarding the introduction of a new standard form contract is the recent experience with AS11000. In 2013, Standards Australia released a draft new standard form construction contract intended to replace AS4000 which was published in 1997. Public comments on this draft were invited and many submissions received. However, four years later, in 2017, Standards Australia announced that it was not proceeding further with AS11000. This was disappointing to the many who were looking forward to seeing a new contract that included features hitherto unknown in Australian standard form construction contracts, including the introduction of an obligation of good faith, early warning procedures and dispute avoidance processes such as contract facilitation and dispute resolution boards.

Regardless of the difficulties associated with convincing key stakeholders of the need for more modern standard form construction contracts, we must not give up on our efforts to modernise our standard form contracts. This is because 'the contract remains the first port of call for legal resolution of disputes arising on a project'.[1838] It is therefore incumbent on all those working in this space, including construction lawyers, to ensure that the parties have a contract that assists the parties to manage conflict so that it does not escalate into disputes, and to resolve any disputes that do arise in a fair, timely and efficient manner.

1835 The exception being the Department of Defence contracts referenced in earlier sections.
1836 Sharkey et al. (2014), 10.
1837 Sharkey et al. (2014), 4.
1838 Bell, M. (2009).

20.4   What Is the Potential for a Framework Alliance in Australia?

The potential for introducing a framework alliance delivery model into Australia needs to be assessed at two levels: first, from the perspective of a new standard form construction contract, generally accepted by industry; second, from the perspective of a model used episodically from project to project, without it being promulgated as a generally accepted standard form from the outset. Both perspectives are discussed below.

Relevant to the first perspective, one of the surprising findings of the study into the use of standard form construction contracts in Australia was that the number one reason, by a significant margin, why a particular standard form contract was chosen, was 'familiarity with the form of the party choosing it'.[1839] This explains why 'the four main Australian Standards forms[1840] represent close to 70% of the standard forms which are used'.[1841] It is concerning that those responsible for choosing which standard form contract is used on a project, predominantly the principal or the principal's lawyer, are more comfortable using wholly outdated contracts – the four most popular contracts were published in 1992, 1995, 1997 and 2000 – rather than try a more contemporary standard form.

This finding suggests that any attempt to introduce a new standard form contract into the Australian market is likely to be met with a great deal of resistance, or what the authors of the Melbourne Law School report call 'familiarity-generated inertia'.[1842] Such an attitude is confounding, given that the industry appears to be sick and tired of the high number of disputes and the amount of time and money that is expended every year on attempting to resolve these disputes. This disconnect is possibly due to a lack of understanding of the causal link between the contract and disputes. Or to put it another way, parties do not see the contract as having a role to play in reducing the incidence of disputes, or in ensuring that any disputes that do emerge are resolved in a fair, efficient and timely manner.

What does this mean for the potential use of a framework alliance in Australia? It would seem to suggest that getting the Australian construction industry to explore the greater use of such a structure will depend on the extent to which project owners might see it as a viable alternative for programmes of work for which they might otherwise use DPM, MC or even a strategic alliance.[1843] In the current market context – with public sector delivery agencies being tasked with procuring increased volumes of infrastructure work – that appetite should exist. For example, the DPM in essence puts in place a framework for the delivery of a programme of works by a delivery partner who then

---

1839 Bell, M. (2009), 33. Other reasons included, suitability of the risk profile, ease of contract administration, minimising costs, reflects the deal, form well-drafted, form recommended and commercial advantage.

1840 AS4300, AS4000, AS2124 and AS4902.

1841 Baird and Skawinski, Australian Department of Defence, in KCL Centre of Construction Law (2016), 4.

1842 Baird and Skawinski, Australian Department of Defence, in KCL Centre of Construction Law (2016), 34.

1843 Strategic alliances have been used in Australia for the delivery of programs of work (often involving a combination of repairs and maintenance, upgrades and other works) over an agreed contract term. The risk allocation is often similar to a project alliance, but can vary, especially as to the no blame/no dispute liability regime.

contracts separately designers and contractors to deliver various packages of design and construction work. The DPM model has been increasingly used since its introduction to the Australian industry on the $4.2 billion Woolgoolga to Ballina Upgrade by New South Wales Roads and Maritime Services in 2014.

So, in a nutshell, even if there may be less of an immediate likelihood of a framework alliance standard form being introduced into the Australian market in its own right, there is no reason why it should not be considered by owners (and their legal advisers) as one of the viable collaborative contracting models available for use. This is particularly so given that:

- Although they share common ground, a project alliance and a framework alliance are different delivery models; as discussed above, the latter is closer to DPM in structure, and therefore it provides a real alternative to project alliancing for extensive programmes of works, for which the Australian market has already shown an appetite
- The framework alliance is not itself a project contract, and as such there may be less resistance to it, since it is not replacing any existing contracts, but instead supporting and assisting the parties with the project contracts
- As discussed above, BIM requires a project delivery model which provides for optimal integration and collaboration of all key project participants; to the extent that this is what a framework alliance is designed to do, this provides owners with a model which can assist in driving a more integrated approach to the implementation of BIM on their projects.

Given that the framework alliance approach (at least as specifically structured and drafted in the UK) will be largely unfamiliar to the Australian construction industry, there will be the need for a comprehensive awareness raising and education programme to ensure that key players understand the purpose and potential benefits of a framework alliance, including increased efficiencies, stronger collaborations, improvements in time and cost savings and a reduction in risk, including the risk of disputes. It may be that this awareness programme should (at least initially) focus on owners and projects which might otherwise be considering the use of similar collaborative models such as DPM. At the same time, many of the concepts underpinning a framework alliance are likely to be very similar to the 'integrated project delivery framework' which is a core feature of the Australian Department of Defence standard form IPD model discussed above, which is designed to facilitate optimal BIM implementation and which has been trialled on various pilot projects in readiness for market release.

20.5 What Are the Legal Issues Affecting an Alliance in Australia?

The answer to this question is now a very different thing from what it would have been in the early pioneering days following the establishment of the first alliance in 1995. When alliances were in their infancy in Australia, the parties contemplating the use of such a structure on their project, and their advisers, had to understand the true legal and commercial dynamics of the delivery model and then had to candidly question whether it was suitable for them and whether they were suitable for it.

More particularly, project owners and the other participants needed to get comfortable with:

- The no blame/no dispute regime which meant the release of all legal rights of redress, other than for 'wilful default'
- The requirement for unanimous decision-making by the alliance board (giving rise to the prospect that the alliance 'contract' may in fact be a non-binding agreement to agree)
- A cost-plus payment mechanism entitling participants to the direct cost of rework of defective design and construction
- The need to have a bespoke insurance regime which responded to the radically altered legal liability landscape.

In essence, an owner had and still has three options when confronted by the implications of the legal risk allocation under an alliance:

- Use another collaborative delivery model which provides greater contractual certainty of outcome
- Use an alliance but seek to amend the 'business as usual' risk allocation
- Use a 'pure' alliance and then use the procurement process to genuinely select the best non-owner participants, together with a carefully structured gain share/pain share regime which drives those participants to achieve exceptional outcomes.

These days, the pure alliance risk allocation is better understood by the market, so that there is less anxiety about the traditional alliancing legal issues. However, it is still important for all project participants to consider the peculiar features of alliancing, and how they could affect their legal and commercial positions. Five of these key features are now discussed briefly.

First, although the no blame/no dispute regime is accepted as part and parcel of an alliance, it is the owner who needs to understand that it is the party most likely to be affected by this release of legal remedies for defective work, cost overruns or delays. Even if the gain share/pain share mechanism is robust and genuinely drives performance, it will not fully compensate the owner for its losses when things go wrong: it will just result in the owner paying less margin to the other participants.

Secondly, the early legal concern about the requirement for unanimity being an agreement to agree has been largely overcome by a deadlock-breaking mechanism (such as independent expert) as a last resort for issues which cannot be resolved unanimously by the alliance board, or by reserving certain unilateral powers to the owner with the implications of the decisions (such as the effect on the payment mechanism) still being agreed in the usual way.

Thirdly, given the no blame/no dispute legal structure, there has always been the risk that traditional project insurances will not respond, whether because they are fault-based, such as professional indemnity in the case of defective design, or because the insurer might lose its rights of subrogation. It is therefore critical for all participants that they put in place bespoke insurance products, which are tailored and respond to the alliancing risk allocation.

Fourthly, termination has always been an important issue in the context of alliancing. In particular, the owner typically preserves a power to terminate for convenience subject to full reimbursement of the other participants for both direct costs and a proportional amount of agreed gain share/pain share. This right can be important to the owner if the alliance is not proceeding as originally anticipated. It is particularly important under two-phase delivery, involving the refinement of the TOC and other aspects of the project in the initial phase, if it becomes clear that the alliance is unlikely to deliver the outcomes required by the owner.[1844]

Finally, value for money or a failure to demonstrate it was one of the key findings of the DTF Report.[1845] The procurement process underpinning the original alliancing approach was founded on selection of non-owner participants on the basis of non-price criteria because this was thought most consistent with the whole collaborative alliancing dynamic. Although there were various checks and balances which could be used to support value for money under that original approach, more recently there has been a tendency to move to:

- Competitive alliances under which the preferred participants are not selected until the TOC has been bid by proponents
- A phased structure under which the owner does not proceed to the subsequent full delivery phase until the project proposal has been finalised to its satisfaction in the initial phase.

As discussed above, the alliancing delivery model is now in its mature product lifecycle phase. It is no longer a creature of fashion, applied on occasion to projects without a discriminating consideration of the legal and commercial implications of its bespoke structure, and regardless of whether they are truly best for project. It is has also outgrown its churlish adolescent tendency towards rigidity in approach, marked by the reluctance of alliance practitioners to deviate from the original model.

At the end of the day, this maturity can only be good for the ongoing resilience, sustainability, and usefulness of the alliancing delivery model as part of the broader range of collaborative contracting options in the construction industry.

1844 The termination for convenience right is typically reserved to the owner. However, where a non-owner participant commits a wilful default or suffers financial default, then the other participants together usually have the ability to exclude the defaulting participant from further participation until the default is remedied and ultimately to engage a replacement third party.
1845 Department of Treasury and Finance, Victoria (2009).

21

How Does Collaborative Procurement Operate in Brazil?

Dr Alexandre Aroeira Salles, Mariana Miraglia and Matatias Parente

21.1 What Is the Approach to Alliances in Brazil?

Some Brazilian legal scholars and practitioners contend that a culture of alliancing has not yet been consolidated in Brazil. Since the 1990s, there have been few reported cases of alliances entered into by Brazil's largest construction companies, and information as to the development or success of such projects is very limited. Moreover, as most alliances have been formed in the private sector, little industry learning has been publicly exchanged in relation to such experiences.

Brazilian law does not provide for a general legal definition of alliance contracts.[1846] However, pursuant to Article 425 of the Brazilian Civil Code[1847] ('BCC'), parties are free to enter into contract forms which are not specifically provided for in the BCC or in any other statute, as long as compliance with the general rules on contracts set out in the BCC is verified.[1848] There are few such general rules, and, on the whole, they are highly abstract and allow for great freedom in establishing contract terms and forms. The mandatory general rules include:

- The requirements for the validity of contracts (i.e. parties' capacity, lawful, possible and determined – or determinable – object and compliance with formal requirements set out by the law, whenever applicable)[1849]
- The requirements of good faith when entering into the contract and throughout contract performance[1850]

1846 There is a definition of alliances in Article 2.3 ('i') of Decree 2,745/1998, governing the simplified bidding procedures applicable to Petróleo Brasileiro S.A. ('Petrobras', the Brazilian National Oil Company). The definition, however, is limited to the context where it is used, and is not capable of immediate general application.

1847 Brazilian Federal Law 10.406, 2002 as amended.

1848 Title V of the BCC – Contracts in general – (Articles 421–480) sets out rules applicable to all private contracts under Brazilian law. Article 54 of the Brazilian Public Procurement Act ('PPA' – Law 8,666/1993) states that, where the PPA does not provide specific rules on a matter, the general theory of contract law, and the provisions applicable to private contracts, are applicable to the administrative contracts referred to in the PPA.

1849 BCC, Article 104.

1850 BCC, Article 422.

Collaborative Construction Procurement and Improved Value, First Edition. David Mosey.
© 2019 John Wiley & Sons Ltd. Published 2019 by John Wiley & Sons Ltd.

- The exercise of freedom of contract by virtue and within the limits of the social function of the contract[1851]
- General rules on contract formation,[1852] defects,[1853] and termination,[1854] among others.

Cases concerning alliances are scarce in Brazilian courts and, with the exception of some material produced by legal professionals with practical experience in the field, little has been written about this form of contracting under Brazilian law.[1855]

It is important to note that where contracts with the state are considered, the contracting parties' freedom to negotiate contract terms and to decide on contract forms is limited by provisions in a number of statutes applicable to such contracts, including:

- Federal Law n. 8,666/1993 (the Brazilian Public Procurement Act – 'PPA')
- Federal Law n. 12,462/2011 which created the Differential Public Procurement Regime ('RDC' in Portuguese) enacted initially to apply to works performed in relation to the 2013 Confederations Cup, the 2014 FIFA World Cup and the 2016 Olympics and Paralympics
- Federal Law n. 13,303/2016, which establishes the legal regime and procurement procedures for Public Companies and Mixed Capital Corporations
- Federal Law n. 8,987/1995 (the Concession and Permission for Public Services Act – 'CPPSA')
- Federal Law n. 11,079/2004 (the Public-Private Partnerships Act – 'PPPA').

Such federal laws on public procurement and contracting procedures must be complied with at all three levels of the federation (federal, state, and municipal) and they impose restrictions on the use of any contract form that is not explicitly set out in the statutes. Considering that there is no legal provision for alliance contracts, such a contractual arrangement is not permitted for projects in the public sector.

A current exception to this restriction is Petrobras, Brazil's National Oil Company, a mixed joint-stock corporation controlled by the Federal Government. Prior to the enactment of Federal Law n. 13,303/2016 (which now governs the legal regime and procurement procedures for all public companies and mixed capital corporations), Article 67 of Federal Law n. 9,478/1997 (the Brazilian Petroleum Law) granted special permission for the creation, by presidential decree, of a simplified bidding procedure for Petrobras. The procedure was then established by decree n. 2,745/1998.

Federal Law n. 13,303/2016, however, revoked the authorisation granted by the Petroleum Law, giving public companies and mixed capital corporations a two-year period from the enactment of the law to adapt their internal processes to provisions of the new law (Article 91). Hence, Petrobras was required to adapt its bidding process and to cease to apply the decree by July 2018. The Decree contained a specific provision allowing Petrobras to enter into alliance contracts on the basis that 'Bidding is not

---

1851 BCC, Article 421.
1852 BCC, Articles 427.
1853 BCC, Articles 441–446.
1854 BCC, Articles 472–480.
1855 Silva (2017), 181 defines such alliance contracts as: 'the contractual arrangement through which two or more companies are hired by the project owner to, with him, as a single integrated team, in return for compensation, deliver a specific project owned by the contracting party, in a collaborative system based on the sharing of risks and rewards, on the principle of good faith and on broad transparency between the parties'.

required when competition is not viable from a factual or legal perspective, especially: i) for the execution of "alliance contracts", considered as being those contracts which have as their aim the combination of efforts between legal entities, for the joint management of enterprises, including the planning, administration, procurement services, civil construction works, assembly, pre-operation, commissioning and start-up of units, by establishing "target" and "ceiling" prices, for the purposes of calculating bonuses and penalties, in relation to such prices, time limits and the verified performance'.[1856]

In addition to providing a definition of 'alliance contracts' for the purposes of Petrobras' contracting procedures, the provision allowed Petrobras to enter into such contracts in special circumstances where competition would not be possible. Over the period the decree has been in force, Petrobras entered into a number of alliance contracts, some of which were subject to audit procedures by the Federal Court of Accounts (TCU, in Portuguese).[1857]

In the course of its auditing functions, the TCU has put forward some relevant considerations on what the court understood to be the core elements of an alliance contract. Drawing heavily on international literature on alliancing,[1858] the TCU contended that the defining characteristics of an alliance include:

- Equitable sharing of risks and responsibilities
- Joint management of the project
- Commitment to the agreed target cost and target price
- Remuneration based on an established system of bonuses and penalties in connection with performance indicators.[1859]

As the Petrobras Decree has ceased to apply to the company's bidding and contracting procedures, the state-controlled entity is no longer able to enter into new alliance contracts. Such forms of contract thus only remain possible in the context of private sector construction projects.

21.2 What Is the Approach to BIM in Brazil?

A report published as part of the study 'Experiences Exchange in BIM – Building Information Modelling'[1860] shows that, since 2007, a number of BIM initiatives have been reported in Brazil in the public and private sectors as well as in academia.[1861] However, despite the growing debate concerning the use, implementation and diffusion of BIM for both public and private construction projects in Brazil, the report demonstrated that the use of BIM is still incipient in this country.[1862]

1856 Section 2.3, Item 'i' of Decree 2,745/1998.
1857 The Federal Court of Accounts (TCU) is the federal government institution responsible for assisting Congress in its constitutional incumbency to exercise external control over the executive branch. In broad terms, the TCU's duty is to ensure the effective and regular management of public resources to the benefit of Brazilian society.
1858 For instance, in its Audit Report (TCU (2013)), the TCU specifically references the Australian Government's National Alliance Contracting Guidelines.
1859 TCU (2013), item 18.
1860 The study is a product of a cooperation project between Brazil's Ministry of Development, Industry and International Trade ('MDIC') and the European Union.
1861 Kassem & Amorim (2015), 22–39.
1862 Kassem & Amorim (2015).

From a regulatory perspective, in 2010, the Brazilian Technical Standards Organisation ('ABNT' in Portuguese)[1863] created a special study committee with a view to establishing technical norms and standards on the use of BIM. As a result of the committee's efforts, ABNT published in 2010 'ABNT NBR ISO 12006-2:2010', a translation of ISO 12006. The same organisation has also been working on a classification system of building information, 'ABNT NBR 15965', a norm consisting of seven different parts of which four have already been published.

From a legislative perspective, an amendment to the Brazilian Public Procurement Act (Law 8,666/1993 – 'PPA') was proposed under Bill 6,619/2016. The proposed legislative change aimed to establish the use of BIM as a mandatory requirement in public construction projects governed by the PPA. In justifying the amendment, the proposing Federal Representative stated that the compulsory use of BIM would make potential bidders more qualified to execute the projects, while also reducing errors and omissions and boosting productivity. The amendment would change Article 7 of the PPA, which would then provide that 'The bids for the execution of works and for the provision of services shall comply with this article, and, in particular, with the following sequence:

I – basic design
II – executive design
III – execution of works and services.

The execution of each phase shall necessarily be preceded by the completion and approval, by the competent authority, of the works relating to the previous phase, with the exception of the executive design, which may be developed simultaneously with the execution of the works and services, provided that authorisation is granted by the Government in that regard, and compliance with the applicable parameters of the system of modelling of building information, identified by the acronym BIM – Building Information Model is verified.'[1864]

A broader reform of the PPA was proposed under Bill 6,814/2017, which aims to repeal and replace the current Federal Law 8,666/1993 in its entirety. The latest draft did not, however, include any provision in relation to the use and implementation of BIM for public construction projects.

An important step towards the diffusion of BIM in Brazil was taken on 5th June 2017, when a presidential decree created the Strategic Committee on the Implementation of Building Information Modelling ('CE-BIM') with a specific mandate of proposing a national strategy on the dissemination of BIM. The initiative suggests a possible future intention that BIM be implemented for construction projects at a federal level. There have also been actions at a state level, a prominent example of which is the publication, by the State of Santa Catarina, in 2014, of 'Guidelines for the Presentation of Projects in BIM'.[1865] Compliance with the guidelines was already included as a requirement

---

1863 The Brazilian Technical Standards Organisation (Associação Brasileira de Normas Técnicas [ABNT]) is a non-profit organisation engaged in the preparation of national technical standards. It is recognised by law as a Public Utility Organisation (Federal Law 4,150/1962) and serves as the unified national forum for standardisation.
1864 Bill n. 6,619/2016, viewed 5 January 2018, http://www.camara.gov.br/sileg/integras/1517799.pdf.
1865 Caderno de Apresentação de Projetos em BIM.

in bidding procedures in projects for the construction of hospitals and centres for innovation in the state.[1866]

Even though there is still no legal requirement for the use and implementation of BIM, there are examples of state organs (such as the Brazilian Army) and state-owned enterprises (such as Banco do Brasil) setting out the use of BIM as a prerequisite for competitors in bidding procedures. Cases of voluntary use of BIM by construction companies have also been reported in public projects related to the 2014 FIFA World Cup and the 2016 Olympic Games in Brazil.[1867] Despite the absence of information on the extent to which BIM has contributed to these projects, it has been suggested that the use of BIM was possibly made easier due to the fact that such projects were procured under the Differential Public Procurement Regime ('RDC'), governed by law n. 12,462/2011, which established a regime similar to a design-build (DB) process – which is different to the position under the PPA which establishes a design-bid-build (DBB) system.[1868] The report contended that the possibility of having early engagement of the companies responsible for the design and those in charge of the construction may have contributed to the decision to use BIM in such context.[1869]

Research published by the Brazilian Chamber of the Construction Industry ('CBIC') indicated that current difficulties associated with the implementation and use of BIM in Brazil have mainly resulted from behavioural factors, such as resistance to change, lack of professional training in BIM and cultural elements of the construction sector in Brazil – such as its highly adversarial nature.[1870] The current laws concerning public procurement may also limit the use of BIM, as most of the projects are procured under a design-bid-build system, hence restricting the engagement of all parties involved in the construction project from its outset.

There has been significant involvement of Brazilian governmental agencies in discussions and initiatives on the diffusion of BIM as a tool to improve productivity, sustainability, transparency, control and optimisation of resources in the construction industry.[1871] Initiatives such as the BIM Platform – a free national repository of BIM templates – and the efforts of the ABNT to provide standardisation of norms concerning the use of BIM are also indispensable for the diffusion of BIM in the country. The inclusion of BIM as a requirement for public construction projects and the development of a regulatory framework for its implementation and use also seem to be desirable steps for the consolidation of BIM in the Brazilian construction industry.

21.3 What Is the Approach to Construction Contracts in Brazil?

The public–private divide described in relation to alliance contracts in Brazil also applies to construction contracts. Not only is such a dichotomy relevant for a

---

1866 For further information, see http://www.spg.sc.gov.br/noticias/1475-caderno-de-projetos-em-bim-orienta-uso-da-tecnologia-em-obras-publicas, accessed 7 January 2018.
1867 Kassem & Amorim (2015), 26.
1868 Kassem & Amorim (2015), 26; Design-bid-build is equivalent to the 'traditional' procurement model in the UK where design is procured separate from construction.
1869 Kassem & Amorim (2015), 26.
1870 CBIC (2016).
1871 ABDI (2017).

theoretical understanding of the approach to construction contracts in Brazil, but it is also pivotal in determining practical aspects of the contracts, such as risk allocation and the parties' main obligations.

From a private law perspective, the regulatory framework for construction contracts under Brazilian law is comprised of relevant provisions from the BCC as well as provisions from other statutes capable of applying to construction contracts.[1872] Contracts entered into with the state are governed by different statutes, depending on the legal nature of the governmental entity, as illustrated by the statutes mentioned in Section 21.1. Contracts entered into with public companies and mixed capital corporations are subject to a somewhat mixed regime: Article 173, Paragraph 1, Item 'II' of the Brazilian Constitution establishes that these entities will be subject to the legal regime of private enterprises – including in relation to their civil, commercial, labour and tax obligations. However, the Brazilian Constitution also imposes certain special obligations on all government entities (for example, state organs, state-owned enterprises, and state-controlled companies) which derive from their public law nature.

Contracts entered into with state organs for the procurement of services and works are usually governed by the provisions of the PPA.[1873] As a general rule, the PPA establishes a design-bid-build system ('DBB'), whereby the design process (basic and executive design) determines, but is usually kept separate from, the construction process, making it difficult to adopt a more integrated approach to the construction project.

The RDC a regime initially introduced to expedite procurement of the construction projects related to the sports events held in Brazil in 2013, 2014 and 2016, institutes the possibility of further integration, bearing close resemblance to a design-build process ('DB').[1874] Further to the perceived success of the RDC, its scope of application was considerably widened to include projects within the Federal Government Growth Acceleration Programme ('PAC'), engineering works and services required within the Unified Healthcare Services ('SUS'), as well as those related to improvements in urban mobility and logistics infrastructure among others.[1875]

The law governing the legal regime and procurement procedures for Public Companies and Mixed Capital Corporations[1876] allows for both DBB and DB procurement processes, having due regard to the specifics of the relevant construction project.[1877]

1872 Federal Law 8,245/1991, which regulates urban property lease agreements, provides an example of statutory provisions other than the BCC which are capable of applying to construction contracts. Article 54 A of the law provides specific rules to regulate 'built-to-suit' contracts, a type of transaction which merges elements from constructions contracts and lease agreements.

1873 As briefly summarised by Pereira (2016), 233: The scope of application of the PPA 'extends to all forms and levels of government. Its scope of application comprises the "direct administration", which corresponds to those government departments directly linked to political entities of the federal, state, and municipal administrations, and the "indirect administration", which comprises autonomous government agencies, foundations, government-owned corporations (in which the government holds 100% of the shares), mixed-capital companies (in which government has a controlling stake), public consortia and all possible forms of government-controlled entities'.

1874 The requirement for an outline of the basic design ('*anteprojeto de engenharia*') for bidding works under the RDC is one of the differences from a traditional design-build system.

1875 The scope of application is indicated in Article 1 of Federal Law 12,462/2011.

1876 Federal Law n. 13,303/2016.

1877 Article 43 of Federal Law 13,303/2016 law provides a list of the procurement regimes and their designated applications.

The content of a contract entered into between a public contracting authority and a private party is determined by the conditions set out in the bid documents issued by the contracting authority and the proposal made by the party and accepted under the bidding procedure.[1878] Although both parties are bound by the express terms of the contract, the bid and the proposal, the public law regime (governing what are known as 'administrative contracts') grants a contracting authority the power unilaterally to modify several aspects of the contract entered into with a private party (for example, to terminate the contract or to modify its scope)[1879] in order to achieve specific goals, namely the fulfilment of supra-individual needs relating to the public interest.

The power of the state unilaterally to change the contract is balanced by its obligation to maintain the original economic-financial balance of the contract in the course of its performance.[1880,1881] Furthermore, supervening occurrences altering the original balance of the bargain made by the parties at contract formation must lead to a corresponding amendment to the contract terms so as to restore the original contractual balance. Such situations include unforeseeable circumstances, foreseeable circumstances with unforeseeable economic consequences, circumstances that make performance more difficult or impossible, force majeure, or even the so-called *factum principis* (acts of state unrelated to the contractual relationship that incidentally interfere with the contract, e.g. the creation of a new tax).[1882] Thus, under Brazilian law, it is a matter of principle that in contracts with the state the risks related to such events are borne by the contracting authority.

Private construction contracts, as with any other contract type, are subject to the general rules governing private contracts set out in the BCC.[1883] In addition to general rules, the BCC also sets out a general legal framework for a number of contract types (for example sale of goods, supply of services, lease agreements and contracts for construction works), through fairly abstract provisions on key aspects of each type of contract, implying terms either in a mandatory manner or where the parties have not agreed otherwise. Under the system established by the BCC, the provisions relating to contracts permit the parties, in most cases, to deviate from such provisions and to determine contract terms as they deem fit.

'Construction contracts' are not listed among the contract types set out in the BCC. The variety of arrangements which could possibly fit under the description of a 'construction contract' arguably makes it difficult to provide a definition for such a contract

---

1878 Article 54(1) of the PPA.

1879 Article 58(I) of the PPA.

1880 Article 37 (XXI) of the Constitution of the Federative Republic of Brazil 1988 states that 'with the exception of the cases specified by law, public works, services, purchases, and sales shall be contracted by means of a public bidding procedure that ensures equal conditions to all bidders, with clauses that establish payment obligations, maintaining the effective conditions of the bid, pursuant to the law, whose procedure shall only allow the requirements of technical and economic qualifications that are indispensable to guarantee the performance of the obligations'.

1881 Under a private law regime, such as that governing public companies and mixed capital corporations, such prerogatives (i.e. unilateral termination and modification of the contract) are not guaranteed to the contracting party. On the other hand, the obligation to maintain the original economic-financial balance of the contract in the course of the performance of the contract is arguably applicable, as Article 37 of the Constitution expressly applies to all government entities (including public companies and mixed-capital corporations).

1882 L8666 1993 (BR) article 65 (II) (d).

1883 For an overview of the general provisions in the BCC, see Section 21.1 and notes 2–8 of this chapter.

type or to determine rules applicable to it in general terms. However, the BCC regulates a 'contract of construction works' ('*contrato de empreitada*')[1884] as a specific contract type. This is a relatively simple kind of construction contract, through which a contractor may provide only workmanship or both workmanship and materials.[1885] Hence, it does not capture the complexities of intricate construction projects, where a number of connected services may be provided such as engineering, procurement, operation, commissioning and start-up.

For complex contracts, which may present elements of a variety of contract types, questions may be raised as to which are the applicable rules. Although no clear answer is provided by statute or previous court decisions, at least two relevant approaches can be identified. The first is to consider that a construction contract falls within the category of an 'uncodified agreement', as provided for under Article 425 of the BCC. This means that only the general rules of contract set out in the BCC are applicable, placing a heavier burden on the parties to draft the contract in greater detail in order thoroughly to document their intentions, given the absence of any default rules applicable in case of omissions. The second approach is to consider whether a construction contract displays elements of a variety of contract types, a so-called 'mixed contract', for example an EPC[1886] contract which is considered to contain obligations of a contract of construction works and of a services contract. In these cases, the provisions governing different contract types apply to the parts of the transaction which could be identified as falling within the category of each contract type. Taking the EPC example, the provisions relating to a services contract in the BCC would apply to the obligations concerning engineering and procurement services, whereas the provisions relating to a contract of construction works in the BCC would apply to the construction obligations.

As demonstrated, the public–private divide described in this chapter determines fundamental differences in relation to the approach to contracts governing public and private construction projects in Brazil. From the perspective of the contractor, perhaps one of the most obvious differences relate to the level of party autonomy in negotiating and establishing contract terms and contractual arrangements. Whereas in the private sphere party autonomy prevails, in public construction projects tendered under a public law regime the parties may only enter into contract types expressly provided for under the applicable public procurement laws. This is also reflected in risk allocation provisions where the parties can freely allocate most risks in private contracts, as compared to public contracts where the guarantee of maintaining the original economic-financial balance of the contract sets the main parameters for risk apportionment.

21.4   What Is the Potential for a Framework Alliance in Brazil?

Taking into account the legal considerations put forward in this chapter, in relation to the approaches to alliances and to construction contracts in Brazil, the potential for a framework alliance is currently concentrated on private sector construction projects.

From a legal perspective, the general framework for contracts set out in the BCC provides the parties with great freedom to define the terms of an alliance. Furthermore, the

---

1884 BCC, Articles 610–626.
1885 BCC, Article 610.
1886 An engineering, procurement, and construction contract.

introduction and diffusion of a framework alliance may serve the purpose of promoting non-adversarial relationships and may represent significant cost and time savings, reduction and better allocation of risks, increased transparency and sustainability as well as other efficiencies, as experience has proven in other jurisdictions.

Introducing a standard form framework alliance contract in Brazil would also contribute to standardisation in a country where bespoke contracts are predominantly used. A standard form successfully used in a number of countries and designed to adapt well to normative requirements across various jurisdictions may also benefit from evidence that its provisions are balanced and have been tested elsewhere. This may encourage clients, contractors and consultants to consider a framework alliance as a possible contractual arrangement for their projects, as well as a way to reduce transactional costs.

The current efforts to foster the implementation and use of BIM can also be supported by a framework alliance in Brazil. The development of a culture of collaborative working and the early engagement of contractors involved in the design, construction and management of a project provide favourable conditions for the successful use of BIM and enable the maximisation of the benefits associated with framework alliances. The early involvement of all parties in a more integrated project approach, combined with early BIM model contributions from the relevant parties, are also key to avoiding delays and common disputes verified in projects where the design and construction phases are fragmented.

The lack of a statutory provision allowing the state (its organs, entities, and controlled enterprises) to enter into alliances currently poses a limitation on the potential for a framework alliance in Brazil. This means that public construction projects awarded under the PPA, the RDC or the legal regime applicable to state-owned enterprises cannot be structured under any other legal arrangement other than those expressly set out in the respective statutes.

The successful use of framework alliances in private projects, however, may drive significant changes in the construction sector in Brazil, leading to amendments in legislation so as to include the possibility of this type of contractual arrangement for public construction projects. The implementation and use of BIM in Brazil may also lead to the conclusion that, in certain circumstances, a more integrated approach and the use of relational contracts provide for improved results, transparency and cost-efficiency. The success in private projects may, therefore, provide normative justification for deviations from the currently prevailing contracting system for public construction.

Considering the many benefits of introducing a framework alliance in Brazil, the Brazilian Institute of Construction Law (IBDIC) and the Brazilian law firm Aroeira Salles Advogados worked together with King's College London and the Association of Consultant Architects to produce a version of the FAC-1 framework alliance contract in Portuguese and adapted for use in Brazil. The amendments to the English version are summarised in Diagram 15. The Brazilian version of the contract was presented at an international conference organised by IBDIC held in September 2018 in São Paulo, and attracted a significant amount of interest from clients, professionals and academics across the construction sector. Given the high level of engagement, plans are in place to hold more focused events with key industry representatives regarding FAC-1, and alliancing methods more broadly, with a view to encouraging their adoption on construction projects in Brazil.

21.5   What Are the Legal Issues Affecting an Alliance in Brazil?

The prevalence of party autonomy has been described as one of the advantages of Brazilian contract law. The general provisions set out in the BCC allow for a measure of certainty in relation to key issues, while also maintaining great freedom for the parties in determining contract terms as they deem appropriate. When alliance contracts are considered, the absence of specific provisions in relation to this type of contract (for example, absence of a legal definition or of terms specifically implied in this contract type) may be construed as allowing parties to adapt the alliance contract conditions to their specific requirements. However, this advantage may also constitute one of the main legal issues affecting an alliance in Brazil. The absence of statutory provisions governing this type of contract, in addition to the lack of practical experience and the absence of a body of cases judged in Brazilian courts, may exacerbate the uncertainties to which the parties are subject when opting for this new arrangement.

A fundamental point of uncertainty surrounding alliance contracts is whether pursuant to Article 425 of the BCC only the general rules on contracts are applicable to such contracts, or whether there may be contract types which are capable of applying to certain aspects of an alliance, based on the assumption that alliances are 'mixed contracts'.[1887]

This matter was discussed during the 1st Company Law Journey promoted by the Brazilian Federal Council of Justice in 2013, under the coordination of the Superior Court of Justice Minister Ruy Rosado.[1888] As a result of the discussions, a formulation was put forward that 'With the exception of the warranty established in article 618 of the Civil Code, all other articles pursuant specifically to a contract of construction works (arts. 610 to 626) shall apply in a subsidiary manner to the contractual conditions agreed upon by the parties to complex construction and engineering contracts, such as EPC, EPCM and Alliances'.

The conclusion seems to suggest that the participants in the 1st Company Law Journey took the view that an alliance is a 'mixed contract'. The wording of the statement transcribed above also seems to suggest that scholars attending the event viewed alliance contracts as being part of the category of 'complex construction and engineering contracts', a stance that is not appropriate to all the types of alliance mentioned in this chapter.

The opinions of the participants in the 1st Company Law Journey as expressed in the approved statements do not constitute authoritative statements of the law, although they may carry persuasive weight. The matter remains, therefore, open to discussion. If the issue is raised in litigation, for instance, substantially different outcomes may result from the court adopting the 'mixed contract' approach or solely applying the general rules of contract law.

Another matter posing overarching issues relates to the fact that basic concepts of the Brazilian law of obligations such as 'reciprocal obligations', 'breach' and 'excessive

---

1887 As referred to in Section 21.3.
1888 The 1st Company Law Journey was an event promoted by the Center for Judicial Studies of the Federal Justice Council in 2013, with the aim of gathering commercial law specialists to discuss and promote the interpretation of commercial law in Brazil. After the work of the committees, the propositions were subject to vote, approved and published, with the purpose of guiding the interpretation of the current state of the law. Federal Justice Council, Center for Judicial Studies (2013).

burden' seem to have been developed to apply to transactional, rather than relational, contracts. Therefore, in analysing and construing such concepts in light of the facts of a case concerning an alliance contract, the courts would have to interpret and apply them through the lens of the cooperative system that the alliance attempts to establish. Such an exercise may result in further uncertainties for the parties to an alliance contract in Brazil.

Despite the legal issues raised, the benefits that alliance contracts may bring – as the experience in other jurisdictions suggests – can outweigh the legal hurdles that the initial uncertainty is likely to cause. A paradigm-shift from a highly adversarial contract culture to more collaborative, result-oriented, efficient and transparent practices would indeed be a welcome change in the Brazilian construction sector.

Reference	English version	Brazilian version	Explanations
Signature blocks	Executed as a deed	The option to execute the contract as a deed was not included in the Brazilian version.	The concept of a deed does not exist under Brazilian law in the same way that it does under English law. It should be noted that the parties have the option to register the contract with a notary public (*cartório de registro de títulos e documentos*) if they would like the contract to be disclosed publicly.
Clause 10: Duty of Care	'Use reasonable skill and care appropriate to their respective roles, expertise and responsibilities' (clause 10.1)	'Atuar com as habilidades e diligências adequadas às suas respectivas funções, competências e responsabilidades' (Free translation: 'Act with the abilities and diligence appropriate to their respective roles, expertise and responsibilities').	The word 'reasonable' was not translated directly as under Brazilian law there is an implicit duty to act prudently. Furthermore, including the word 'reasonable' could weaken the contractual obligation.
Clause 10: Duty of care	Clauses 10.2–10.6	In translating clauses 10.2– 10.6, a literal approach was adopted.	In Brazil, the duty of care owed to all parties under a contract is determined by statute and cannot be derogated from by the parties. Therefore, the effect of clauses 10.2–10.6 is different under Brazilian and English Law. In Brazil, these provisions are likely to be construed as reinforcing pre-existing legal obligations in relation to the parties under statute.

Diagram 15 Table of the main differences between the English and the Brazilian versions of FAC-1 (version dated 27 June 2017).

Reference	English version	Brazilian version	Explanations
Clause 14: Duration and termination		New clause added to deal with the provisions under Brazilian law relating to termination where contractual obligations become excessively onerous.	Under the Brazilian Civil Code where, during the term of a contract, an unpredictable event occurs which makes the obligations of one party excessively onerous to the benefit of another party, the adversely effected party may make a claim for termination of the contract. The party benefiting from this event may offer a contractual amendment to rebalance the contractual rights and obligations with a view to keeping the contract effective.
Definition: Insolvency Event	Insolvency Event	The events constituting an Insolvency Event were amended to reflect the position under Brazilian law.	It is important to note that the definition covers equivalent events in jurisdictions outside Brazil.
Definition: Improved Value	Improved Value	Agregação de Valor (Free translation: 'Addition of Value').	A direct translation of Improved Value was not included due to concern that it would not be widely understood in Brazil. The alternative term was used as it is well established in the marketplace in Brazil.
Appendix 4, Part 2: Adjudication		The Appendix was translated for use in Brazilian projects with one minor amendment (the addition of the words 'Ato jurídico perfeito fazedor de título executivo' to the end of Paragraph 7 of Appendix 4, Part 2).	It should be noted that although adjudication can be used in Brazil for the interim settlement of disputes, the mechanism is not provided for by law. This means that there is substantial room for questioning the enforceability of adjudicators' decisions. The minor amendment made to Paragraph 7 just clarifies that in the event of an agreement between the parties in relation to the adjudicator's decision, such agreement can be enforced through execution proceedings in Brazil.
Appendix 4, Part 3: Arbitration		Small adjustments were made to Appendix 4, Part 3 relating to arbitration to reflect the position under Brazilian law.	The adjustments were made to reflect the specific language relating to arbitration under the Brazilian Civil Procedure Code.

Diagram 15 (*Continued*)

22

How Does Collaborative Procurement Operate in Bulgaria?
Adriana Spassova

22.1   What Is the Approach to Alliances in Bulgaria?

Bulgaria is a 1300-year-old, relatively small European state, with a population of over seven million, situated on the Balkan Peninsula. After the Second World War Bulgaria became a one-party socialist state and most of the private property was nationalised. After the collapse of socialism in 1989, Bulgaria started privatisation, transition into a democracy and development of a market-based economy. The country joined the European Union in 2007, but remains the poorest among the EU members,[1889] with a bureaucratic administration[1890] and problems with the rule of law.[1891]

During the era of socialism, the state-owned design, construction and manufacturing companies were united in national or regional structures, called State Economic Unions.[1892] They were created to improve the efficiency of the industry. Each state-owned company preserved its legal and economic autonomy within the union and competed for national projects, but national and regional administration developed centralised planning, tendered for international projects and managed social and welfare programmes. For example, the State Economic Union 'Metalni Construczii' (Steel Structures) united design companies, scientific research laboratories, testing laboratories, specialist contractors, several plants for steel cold bended profiles, steel structures and prefabricated elements. Early contractor involvement was being

---

1889 In 2016 Bulgaria was the EU Member State with the lowest per-capita GDP, at 51% below the EU average; the price levels at less than 60% below the EU-28 average. http://ec.europa.eu/eurostat/statistics-explained/index.php/GDP_per_capita,_consumption_per_capita_and_price_level_indices, accessed on 26 December 2017.
1890 Spassova, A. (2014): The Multi-Criteria Analysis ranking the level of bureaucracy in design approval for a construction permit shows that the administrative burden in Bulgaria is six times greater than that in England, and more than double compared to Germany, France and Italy.
1891 The Parliamentary Assembly of the Council of Europe PACE (2017) pointed to 'serious problems' with the rule of law in Bulgaria, related to 'independence of the judiciary and the principle of the separation of powers, mainly due to attempts to politicise judicial councils and courts, dismiss judges and prosecutors (or try to) and limit the legislative power of parliaments, as well as the effects of corruption'. http://assembly.coe.int/nw/xml/News/News-View-EN.asp?newsid=6828&lang=2&cat=8, accessed 16 December 2017.
1892 Chavdarov et al. (1983), 157: Central construction unions: Industrial Construction ('Промишлено строителство'), Hydro construction (' Хидрострой'), Construction Mechanisation ('Строителна механизация'), Steel Structures ('Метални конструкции'), Construction Linings ('Строителни облицовки') and the territorial union for the residential and administrative buildings in Sofia ('Софстрой').

Collaborative Construction Procurement and Improved Value, First Edition. David Mosey.
© 2019 John Wiley & Sons Ltd. Published 2019 by John Wiley & Sons Ltd.

practised and designers and contractors collaborated with research laboratories.[1893] The entities had planning departments and goals for improvement of efficiency, time and cost savings.

There were also many state initiatives supporting value engineering and research into new technologies in the fabrication of construction materials and preassembly of construction. For example, different methods for mounting/pushing steel structures with hydraulic jacks were tested, developed and used for buildings and bridges. Young engineers strived for innovations and cost savings. A collaborative culture facilitated design and construction: the designers were aware of the capabilities of the fabrication plants; the contractors had long-term arrangements with the producers, which made the logistics easier and the construction process more efficient. However, quality often suffered due to the drive to overcome all unforeseeable obstacles and fulfil the five-year plan within three years.

From 1989 we saw the privatisation and decline of the state-owned organisations. The change from state property to private property, the difficult economic and political situation, the loss of previous markets and the lack of resources led to the fragmentation of the construction industry. At the same time, Computer-Aided Design (CAD) gradually became a standard, enabling engineers to design from home offices without technicians. Only a few of the big design organisations, construction companies and scientific research centres survived after privatisation and with much reduced staff. Hard competition for survival gradually erased the former collaborative culture.

Sadly, in the last 30 years, research and innovation in construction are rare phenomena in Bulgaria. International investors, contractors and consultants bring know-how and good practices. At the same time, they have introduced an adversarial culture in construction contracts, which in some cases is even in conflict with mandatory 'good faith' obligations by which:

- Under the Obligations and Contracts Act (OCA)[1894] Article 12 the parties must act in good faith in the course of negotiations and when entering into contracts
- Article 63 of OCA stipulates, that 'Each of the parties under a contract must perform its obligations arising from it accurately and in good faith'.

Many foreign investors and international contractors and consultants pulled out of Bulgaria, due to the bureaucracy, uncertain investment environment and high perception of corruption. The lack of private investors focuses the construction business on public procurement. This creates further distortion of the market, and some companies win many tenders with political protection because 'Among the main risks of corruption are the concentration of market power in the hands of the public administration, the lack of effectiveness of procurement procedures, the lack of integrity and control of procurement officials and bodies, political interference in the work of the Public Financial Inspection Agency (PFIA), structural governance deficiencies, and too many frequent changes to the legal framework of public procurement'.[1895]

1893 Chavdarov et al. (1983) 105 '…the state organizations in the sector are active participants at the stage of the elaboration of the technical design and the Bills of Quantities. Later during the construction stage the engineering-implementation organisations participate in the organization and execution of construction'.
Note: some design organisations were transformed into engineering-implementation organisations, with the goal to develop design nomenclatures, standard technical specifications, model designs and calculations for more efficient prefabrication and construction of specific buildings and structures.
1894 Obligations and Contracts Act (1950) 275 SG, last amended (2008) 50 SG.
1895 Bulgaria Corruption Report (2017).

Construction professionals are discouraged and strive to adapt themselves to the existing market conditions by bidding very low prices. In this situation, the incentives and innovations come mostly from the few international investors, or from companies working internationally.

Presently some NGOs that unite members through sharing common goals call themselves alliances, for example, the Digital National Alliance[1896] and the International Business Alliance (IBA) Bulgaria.[1897] However, alliances in the sense of integrated teams, sharing common goals aligned with customer outcomes, are not known in legal theory or in the construction industry in Bulgaria.

There are also some quasi-alliances such as clients who have a short list of designers, contractors and suppliers working for them on many projects. They use open-book supply chain management[1898] but the main target is to be invited to work on the next project. An example of a quasi-alliance is the organisation of the Regional Waste Management Centres whose EU co-financed projects were designed in a central location in each region, with transfer loading stations handling the waste and spread in different municipalities within the region. Representatives of these municipalities were united in a Regional Waste Management Association, which was the project beneficiary[1899] and supported the project activities by permitting connections to utilities and risk management. Several contractors worked on the sites, interfacing with each other. The representatives of the Regional Waste Management Association participated in coordination meetings together with the contractors, employer, designers and supervisor, but their involvement did not include strict obligations relating to the achievement of the project objectives. On some projects the management structure was inefficient and led to delays and cost increases, since the decision-makers were not present at the coordination meetings and rarely helped with risk management.

The Bulgarian mandatory good faith provisions are a good start for the reinstatement of forgotten collaborative traditions, but generally the UK concept of an alliance[1900] is presently unknown in Bulgaria. There are some similar relationships, but a more focused and structured approach is still to be developed. The leading construction professionals understand the benefits of alliances but are not informed as to how to achieve them in a market economy.

1896 Established in 2013, in Bulgarian named 'Digital National Coalition' ('Дигитална Национална Коалиция'), part of the EC initiative Digital Skills and Jobs Coalition. 'The main goal of the organization is to promote digital technologies, to achieve more efficient implementation of digital opportunities and to assist the development of the ICT sector in Bulgaria'. https://www.digitalalliance.bg/en-about, accessed on 11 December 2017.

1897 'IBA's mission is to develop and support Bulgarian entrepreneurship globally and to nurture the sustainable development of Bulgarian owned businesses around the world'. https://www.linkedin.com/company/bulgarianbusinessalliance, accessed on 11 December 2017.

1898 For example, BLD, sell about 200 apartments annually in Sofia, and work in parallel on several high quality residential projects. They are competing with other investors through innovations and improvements in building materials and facades, thus achieving more sales than their competitors. For them commercial risk management is very important and closely related with schedule management. www.bld.bg, accessed 23 December 2017.

1899 For example, the Kardjali Regional Waste Management Centre, where the client was the Ministry of Environment and Water and the Final Beneficiary was the Regional Association 'For Cleaner Rhodopi', ISPA Measure No.2003 BG PPE019. https://www.google.bg/url?sa=t&rct=j&q=&esrc=s&source=web&cd=1&ved=0ahUKEwjMlZOB-anYAhWCwxQKHQ8jBw8QFggrMAA&url=http%3A%2F%2Fwww.minfin.bg%2Fupload%2F2002%2F%2509kardjali_fm.pdf%2509&usg=AOvVaw3mpiab7xCE-B9pQvW6zJ0k, accessed 23 December 2017.

1900 Infrastructure Client Group (2015).

22.2   What Is the Approach to BIM in Bulgaria?

The construction industry in Bulgaria needs Building Information Modelling (BIM), and BIM needs framework alliance contracts. A report on the state of BIM in Bulgaria in 2017 concluded that the practical use of BIM in the Bulgarian construction industry lags behind the global trend.[1901] CAD has been widely used for many years in Bulgaria,[1902] usually for calculations and 2D drawings. In recent years the leading designers (architects for buildings, engineers for infrastructure and plants) often prepare 3D models, especially for more complicated projects. The 3D model in some cases is used for clash detection. In most cases the other design disciplines submit 2D drawings and the clients do not know the benefits of BIM and do not request or use 3D models. BIM is often confused with the software for digital 3D modelling.[1903]

Commentators have suggested that: 'Today, most architects use BIM software products, but they are using them just to draw their models – they do not utilize the full benefits of BIM. This is just like having an amphibious car and using it as a normal vehicle. Yes, architects start the 3D model part of BIM, but if it is not populated with information by the structural and mechanical engineers and the model is not used by the contractor and the owner, BIM loses its purpose'.[1904]

Complex projects use a common data environment, organised on FTP (File Transfer Protocol) servers. Some have more sophisticated online document management and project management systems.[1905] Bulgaria has excellent coverage with high-speed internet,[1906] but ranked 27th out of 28 EU members for integration of digital technology in 2017; this is due to the very low scores for e-commerce, while the adoption of e-business is in the middle EU range.[1907] E-business includes BIM in construction where: 'High use of BIM on construction projects is small in Bulgaria (11%) when compared to UK (33%)'.[1908] The acceptance of the ISO BIM standards as Bulgarian State Standards and the future translation into Bulgarian are included in the activities of the Task Group 101 'Sustainable Development' at the Bulgarian Standardisation Institute. Four of the ISO BIM Standards were already active in Bulgaria at the end of 2017.[1909]

1901 Georgiev, V. (2017b)

1902 The answers of Bulgarian architects in a Facebook forum in November 2017 show that only 4% draw by hand, 45% use ArchiCAD, 39% use AutoCAD, 13% use Revit.

1903 Zheleva & Ivanova (2016).

1904 Georgiev, V. (2017a).

1905 For example, the Document Retrieval System (DRS), elaborated by the EPC Contractor ALSTOM in 2006 on the 1.5 billion EUR '600 MW net TPP AES Galabovo' project. The design was developed in many countries simultaneously and incorporated via the leading designer in Germany into the 3-D model, processed in India. The clash detection, the visibility of the design revisions, the easier design management on a concise industrial site, were among the main DRS benefits. The project parties (Owner AES, designers, contractors, design approval and construction supervision consultant) had access to the digital information; all design documents for permits, stamped and approved in hard copies in compliance with the building control regulations, were scanned and uploaded in the DRS. Regrettably, the AES attempt to deliver the first paperless project in Bulgaria was blocked by the bureaucracy of the public administration.

1906 Bulgaria ranks 5th in the world for quality of internet broadband speed, according to the results of research through Cisco System Inc., conducted by Oxford University. http://www.dospeedtest.com/speedtest-result/country-statistics/Bulgaria, accessed 26 December 2017.

1907 Europe's Digital Progress Report (2017).

1908 Eadie et al. (2017).

1909 http://www.bds-bg.org, accessed on 26 December 2017.

A research project of the University of Architecture, Civil Engineering and Geodesy (UACEG) 195/16 questioned 113 professionals, including 57 designers, investors, suppliers and contractors regarding the use of BIM. A total of 47% knew what BIM is, but only 23% used separate specialised BIM instruments, such as 3D modelling (26%), specifications (9.8%), execution plans (7.8%), etc. Only 7% used the universal file format IFC. In all, 61% intended to use BIM in the future, including 23% who will continue to use BIM, 18% who will start to use it in the next three years and 20% who would use it only if BIM is accepted as a standard.[1910]

The UACEG research project 195/16 concluded that 'BIM introduction and implementation culture is to be built up in Bulgaria, based on international standards and already tested and validated good practices, in relation to BIM introduction policies and procedures, applied by trained experts in accordance with the specifics of the Bulgarian market. A lot of focused work has to be done in different directions: (1) acquaintance of the professionals with the BIM concept, with the benefits and challenges, connected with the use of this instrument; (2) elaboration of national policy for the development of the use of BIM and (3) education of experts in appropriate formats for team work'.[1911]

At the end of 2017 Bulgaria had no road map for BIM implementation. Neither the public authorities nor the professional bodies have made any visible efforts towards the digitalisation of construction. In practice, about 20% of designers have been using BIM Level 1 in the last two years, and many more are being trained and intend to start working in a BIM environment. Only a few use BIM Level 2 and Level 3, mostly for international projects.

The lack of communication regarding pilot projects leads to fragmented BIM efforts: the architects using Level 2 for buildings think that the engineers are lagging behind; the transport engineers leading the design in the infrastructure think that all other disciplines are pulling them back; at the same time some HVAC engineers are working in Level 3 for projects outside Bulgaria. However, the BIM implementation process is gradually developing, even without a national strategy and it is expected that 'In a natural way, the digital generation shall grow up in the profession and push out the analogue generation'.[1912]

The present BIM drivers are listed in Diagram 16 and may be grouped as follows:

- Market drivers:
  - Architects prepare 3D visualisations to satisfy their clients and to be more competitive
  - Developers request 3D visualisations to sell/rent apartments/offices more easily
  - Educated clients request BIM for better project and asset management
  - Designers want to work internationally

1910 Kouteva-Guentcheva, M. (2017).
1911 Kouteva-Guentcheva & Boshnakov (2018), 9.
1912 Tzvetan Georgiev, Manager of Ircon Ltd.: the engineering company developed 11 years ago a CAD/BIM department, specialised in designing steel work, platework, mechanical and piping in BIM, by a multidisciplinary team of architects, structural, water and sewage, and electrical engineers. They also deliver output from 3D steelwork and platework models for fabricators equipped with CNC machines. http://www.irconltd.com/en/cadcam-proektirane-tekla.html, accessed 27 December 2017.

- Foreign companies outsource BIM services.[1913]
- Organisation culture:
  - Bulgarian companies apply BIM as a corporate standard[1914]
  - Progressive managers understand the benefits of BIM and start implementation[1915]
  - Young professionals from the digital generation are becoming BIM managers[1916]
  - Companies working on international projects are educated to use BIM.[1917]
- IT developments:
  - BIM software is developing and becoming user-friendly (some of it is translated)
  - The hardware and cloud services are enabling easy digitalisation, processing, storage and exchange of data
  - BIM software developers are continuously educating, training and supporting users and cooperating with universities
  - The IT sector has doubled its income in Bulgaria in the period 2011–2016 and is reported to have sustainable development, with average salaries four times higher than the average in Bulgaria and 90% of the employees are younger than 35[1918]
  - BIM applications are successfully being developed for foreign clients.[1919]

Diagram 16 BIM drivers in Bulgaria in 2017.

1913 The UK leading architectural practice Stephen George established its largest international office in Sofia with more than 40 architects and engineers, http://stephengeorgeint.com/stephen-george-international-sofia, accessed 27 December 2017.
1914 Sweco Energoproekt. http://www.sweco.bg/en/our-offer/architecture/?service=Building+information+modelling+(BIM), accessed 27 December 2017.
1915 In 1997 the founding partners of Transgeo Ltd created their own software for railway design and cadaster maps, aiming to digitalise the terrain information, and they continue to design in 3D models. http://en.transgeo-bg.com/1990/01/software-development.html, accessed 27 December 2017.
1916 Tzvetan Petrov, son of the founder of Ivo Petrov Architects, manages the biggest BIM team in Bulgaria: 86 experts, including 4 BIM Managers, 12 engineers and 50 architects are using BIM Level 2 and Level 3. He introduced BIM to the company in 2012, when the staff rose to more than 30 people for easier management and clash detection. http://www.ip-arch.com, accessed 27 December 2017.
1917 The innovative design bureau A&A Architects worked in partnership with Americans for a complicated project. http://aaa.bg/en/projects/details/muzeiko, accessed 27 December 2017.
1918 https://technews.bg/article-104015.html, accessed 27 December 2017.
1919 Co-builder started its Bulgarian office in 2013 with several employees and in 2017 had 105 experts, including. 25 BIM experts in the R&D Department – architects, engineers and consultants – generating ideas and planning future solutions, further developed by 50 IT experts. The products invented in Bulgaria are mainly for UK, Scandinavia, France and Germany, http://cobuilder.com/cobuilder-bulgaria-granted-funding, accessed on 27 December 2017.

Professionals, including BIM software providers and users, take different positions in relation to the possibility of top down implementation of BIM in Bulgaria:

- About 50% consider that public leadership is the only way to develop BIM as a national standard; most of them think that this should be planned at least five years ahead and recommend a positive approach with a system of awards for the pioneers, rather than regulation for mandatory implementation
- The other 50%, most of whom have used BIM for years, consider that the top down approach may overload the process with bureaucracy and waste. They recommend education of clients and joint activities of professional practices. They think that when professionals taste BIM, they will love it.[1920]

Experts recommend the following activities for BIM implementation:

- Formation of a Task Force with representatives from the Ministry of Regional Development and Public Works (MRDPW) and other public clients, the National Association of the Municipalities in Bulgaria, the Bulgarian Institute for Standardisation and a range of professionals
- Task Force implementation of the EU BIM Task Group strategic actions and a BIM Implementation Plan or Road Map, continuing the development of national standards based on the adopted ISO EN standards
- Pilot projects using BIM to be executed and shared in the next two to three years
- EU funds available in the 'Operational Programme Innovation and Competitiveness' 2014–2020 to be used to support SMEs for the supply of BIM software and related hardware.[1921]
- Continuous BIM education in high schools, universities and professional bodies
- Construction e-permitting to be implemented, enabling checking and approval of designs using BIM
- Regulations to be amended, providing a set of incentives for the next five years, so that in future the mandatory 'Technical Passports' and as-built documents for the completed construction projects are extracted from BIM models.[1922]

22.3 What Is the Approach to Construction Contracts in Bulgaria?

22.3.1 Legal Framework

In Bulgaria, the term 'construction contract' is not defined in legislation. All contracts for works and services in construction are generally subject to the rules for 'the contract for manufacture' ('договор за изработка'), governed by the OCA Articles 258–269

1920 Adriana Spassova conducted the interviews in December 2017; among the interviewees were designers (architects; structural, hydro technical, road, railroad, HVAC engineers), university professors, BIM software distributors and trainers, engineering companies' managers, Chamber of Architects board members, Bulgarian Construction Chamber President, public employers, project managers; a written record of the interviews is available.
1921 Eadie et al. (2017); http://www.bds-bg.org, accessed on 26 December 2017; Kouteva-Guentcheva, M. (2017).
1922 The Spatial Development Act (2001) 1 SG, last amended (2017) 13 SG and its regulations have to be amended to accommodate the changes.

whereby 'Under a contract for manufacture, a contractor undertakes, at his risk, to manufacture something in accordance with the other party's order and the latter undertakes to pay remuneration'.[1923]

This broad definition makes it clear that construction contracts as 'contracts for manufacture' are not limited to building projects, but also cover contracts that involve repair and maintenance works.[1924] In addition, 'The special regulation for construction activity is contained in the Spatial Development Act (SDA)[1925] and the ordinances for its application. The requirements for the persons, executing construction activities, their registration, etc. are governed by the Construction Chamber Act'.[1926,1927]

The SDA defines construction as 'over-ground, semi-underground, underground and under-water buildings, constructions, additional constructions, superstructures, strengthening, rehabilitation works, conservation, restoration, reconstruction per authentic data in the meaning of Art. 74, par. 1 of the Law for Cultural Heritage, and adaptation of immovable cultural valuables, fences, networks and facilities of technical infrastructure, public works and sport facilities as well as their major repairs, reconstruction and restructuring, with or without change of the designation'.[1928]

SDA Article 160(1) lists the participants in construction as the employer; the builder (contractor); the designer; the building control consultant;[1929] the person exercising technical control; the site manager; and the supplier of machinery, equipment and plant. Further, Article 160 refers to the 'Ordinance for minimum warranty periods', introducing strict liability for latent defects.[1930] In addition, 'Separate provisions in other special acts find application in the relations between the contractor and employer depending on the specifics of the particular subject of the construction contract (the type of project to be built – building, energy project, road structure, water project) or the procedure for its conclusion. These are the Energy Act, the Roads Act, the Water Act, the Cultural Heritage Act, the Public Procurement Act, etc'.[1931]

The Public Procurement Act (PPA)[1932] imposes restrictions on those construction contracts within its ambit in relation to variations and subcontracting.

22.3.2 Essential Terms

It has been stated that 'The construction contract as a contract for manufacture has two essential terms namely the subject to be constructed or improved and the remuneration of the contractor':[1933]

1923 OCA, Article 258.
1924 Bulgarian Chapter developed by the BSCL in 2016 and to be included in the next edition of Studies in European Construction Law (ESCL 2015),
1925 Part III 'Construction' of the Spatial Development Act.
1926 Construction Chamber Act (2006) 108 SG, last amended (2017) 92 SG.
1927 Tormanov, Z. (2012), 119.
1928 Section 5, 38 of the additional provisions of the Spatial Development Act.
1929 In 1999 the State transferred some of its building control duties to private consulting companies, then licensed and presently registered by the Government. To qualify for registration, they must have architects, engineers, and lawyers in the mandatory 'Lists of qualified staff', meeting the minimum criteria for experience.
1930 10 years for structures, 5 years for finishing works, etc., pursuant to Ordinance 2 of 31 July 2003 on Commissioning of Construction in the Republic of Bulgaria and minimum warranty periods for executed construction works (Ministry of Regional Development and Public Works) (2003) 72 SG, last amended (2016) 65 SG. Longer periods may be agreed in the contract.
1931 Dimitrov, M. (2012), 25.
1932 Public Procurement Act (2016) 13 SG, last amended (2017) 102.
1933 Dimitrov, M. (2012), 98.

- The subject is usually defined by reference to a concept design or developed design
- The remuneration to which the contractor is entitled does not have to be determined in detail,[1934] and it is enough to be clear that the contractor is entitled to payment: OCA Article 258 includes only the obligation 'to pay remuneration' but does not mention the amount of the 'agreed' remuneration.

Specific characteristics of construction contracts include:

- The construction contract is in most cases a commercial transaction, in the ambit of the Commercial Act, thus the contractor must 'exercise the care which a good merchant would'[1935]
- A specific requirement of the SDA is for a construction contract to be concluded in writing[1936]
- Mandatory administrative provisions related to building control govern the responsibilities of the parties, including the formal documenting of the execution of the works.[1937]

22.3.3 Design Responsibilities

The civil law principle of freedom of contract allows the parties to agree that responsibility for elaborating and approving designs and securing necessary permits is transferred to the contractor. If so, the contractor will act on behalf of the employer in the relevant administrative procedures but the employer will continue to be considered the party responsible for, and required and entitled to participate in, all these administrative procedures. Normally, where responsibility for securing the approval of designs and permits is shifted to the contractor, the contract will require the employer to grant the contractor authority to act on its behalf.

In any case, regardless of which party is contractually responsible for design, its elaboration may be done only by a designer authorised in accordance with Bulgarian legislation. The designer is also required to perform designer's supervision during the construction process.[1922]

22.3.4 A Duty to Cooperate

The duty to cooperate follows from the general principles of civil law. The parties are obliged to perform their obligations arising from a contract accurately and in good faith, in accordance with the legal requirements, and each must not impede the other from performing its obligations in the same manner.[1938] Hence, in contractual matters the employer has a general duty to cooperate with the contractor and vice versa.

Article 95 of the OCA provides that a creditor is in default if upon being requested by the debtor, the creditor fails to provide the debtor with the necessary cooperation, where as a result the latter is not in a position to perform (the 'creditor's default rule'). The OCA provides another express obligation for cooperation by which the employer is obliged to take over the works, once executed in accordance with the contract.[1939]

1934 Supreme Court of Cassation Decision 404-2000-V Civil College.
1935 Commercial Act (1991) 48 SG, last amended (2017) 102 SG, Article 302.
1936 SDA, Article 160(2).
1937 SDA, Article 170(1).
1938 OCA, Article 63.
1939 OCA, Article 264.

The SDA also imposes specific cooperation obligations on the employer at each stage in the execution of the project. During the earliest stage, this cooperation covers the conditions required for the commencement of the works such as securing zoning and legal rights, access to the site and the necessary permits. The employer's cooperation is also needed to secure the mandatory construction supervision and designer's supervision, and in signing agreements with the relevant participants in construction. The employer is also obliged to cooperate in signing the statutory protocols which record different stages of construction, and upon completion to take over the works, following a specific procedure.[1922]

22.3.5 Early Warning

Under Bulgarian law, the contractor should notify the employer immediately if the design provided or materials supplied are unfit for the works. If the contractor fails to give notice, it is liable for any costs and losses caused.[1940] The limits of the contractor's liability are laid down in the OCA. It covers those losses, including loss of profit, which are a direct and immediate consequence of the breach and are foreseeable when the duty to warn arises. However, if the contractor has acted in bad faith, it will be liable for all direct and immediate losses.[1941]

The duty to warn may also be inferred from the general good faith principle. The duty to warn must be performed with due diligence, except when the law provides for some other degree of diligence.[1942] As far as commercial relations are concerned, the standard for the duty of care is that of the good professional.[1933] In addition to the above, the construction contract may contain express provisions on the contractor's duty to warn.

Current Bulgarian law contains no provision for an equivalent duty on the employer. However, taking into consideration the general requirements for good faith and due diligence in contractual relations,[1943] we can conclude that the employer must also warn the contractor of any defects in the design which may affect performance of the assigned works.

Bulgarian law does not provide expressly for a designer's or consultant's duty to warn, nor is this duty included in any professional codes. However, the duty to warn is implied in the obligations of a designer performing designer's supervision and in the general good faith and due diligence obligations.[1922]

22.3.6 Use of Standard Forms

There are no standard forms of construction contract in general use in Bulgaria. The international financing institutions (for example EBRD and KfW) usually require FIDIC Conditions of Contract to be used. In 2016, at the request of the Bulgarian Construction Chamber, the Bulgarian Society of Construction Law (BSCL) developed Contract Agreements and Particular Conditions for the FIDIC Red[1944] and Yellow[1945]

1940 OCA, Article 260.
1941 OCA, Article 82.
1942 OCA, Article 63.
1943 OCA, Article 63 and CA, Article 302.
1944 FIDIC 1999 Conditions of Contract for Construction.
1945 FIDIC 1999 Conditions of Contract for Plant and Design Build.

Books, harmonised with the national legislation. At present, FIDIC forms are used only by the Ministry of Environment and Water, on projects financed by the EBRD, and on highway projects with EU co-financing and private projects with foreign clients.

On 4th September 2018, the Minister of Finance approved tender documents for mandatory use in the water sector Programme 'Environment 2014–2020' and Programme 'Development of Rural Regions 2014–2020', using the FIDIC Red and Yellow Book 1999 General Conditions.[1946]

22.4 What Is the Potential for a Framework Alliance in Bulgaria?

22.4.1 Use of Framework Contracts

Framework contracts are used in Bulgaria for repair and rehabilitation of schools, roads, railways, public utilities and energy efficiency projects, financed with public funds. Some examples are set out below:

- *Sofia.* For emergency and repair works of the water supply and sewage mains, the concessioner Sofiyska Voda AD is the client; the assets are public municipality property; the capital Sofia is divided into several regions and one framework contractor is selected for each region. The in-house team works in one of the regions. According to the Executive Director of the Concessioner 'the distribution of the works in this way helps to optimise the costs and make comparative analysis between different options for maintenance of the mains'.[1947] The client supplies the materials for the repair works.
- *Burgas.* Another water supply and sewage company Vodosnabdiavane i Kanalizazia EAD[1948] uses framework contracts, selecting separate contractors for two years for repairs, including materials, of water mains, sewage and buildings.
- *Electricity networks.* The electricity distribution company CEZ Razpredelenie Bulgaria AD,[1949] responsible for the maintenance of the electricity mains in Western Bulgaria, signs framework agreements, using one contractor for a region for one type of works (the works are divided into overhead lines, transformers, cable lines and new customer connections). Also, they sign framework agreements with design companies and others with construction supervision companies. The electricity system operator in Sofia signs framework agreements for repair of substations and planned reconstruction of electrical cable lines.

In addition to the above case studies, the National Railway Infrastructure Company[1950] spent about €200 million on repairs in the period 2012–2016 via four framework agreements.

One commentator on framework agreements in Bulgaria concluded that 'perhaps the insufficient knowledge of the essence and the advantages of the framework agreements

---

1946 Published by the PPA, http://www.aop.bg/fckedit2/user/File/bg/obraztzi/razpechatvane2/Zapoved_04092018.pdf, visited 6 September 2018.
1947 http://www.capital.bg/biznes/kompanii/2013/09/29/2150417_sofiiska_voda_podpisa_nov_dogovor_s_firmi_ot_galchev 29 сеп 2013, accessed 27 December 2017.
1948 http://www.vik-burgas.com, accessed 27 December 2017.
1949 www.cez.bg, accessed 27 December 2017.
1950 http://www.rail-infra.bg, accessed 27 December 2017.

is a reason they tend to be concluded relatively rarely for public procurement in Bulgaria'. She notes the benefits of framework agreements for taxpayers, contractors, and especially for the innovation and competitiveness of SMEs: 'It is beneficial for the state and for taxpayers to save money spent conducting multiple uniform procedures, while ensuring transparency and publicity...From the contractor's point of view, the framework agreement is more favourable than several short-term public procurement contracts. The framework agreement can secure the contractor work for his staff and revenue over a longer period of time, save him time and money to apply for multiple projects, and give him more certainty that his investment will be recovered. This is especially important for smaller enterprises that cannot afford to invest in the modernisation of facilities, and therefore cannot maintain their competitiveness, unless the term of the contract is long enough to recover the funds invested. A framework agreement would give more security and economic stability to such contractors'.[1951]

In order to create and sustain a collaborative working environment, all participants in the construction process[1952] should sign a framework alliance contract. This will ensure easier project management and provide tools for mandatory early warning. Presently, the framework agreements for public works in Bulgaria are only for construction works or only for design or supervision services, which cannot bring the full benefits of integrated framework alliance agreements. Also, at present public clients do not use their frameworks as a means to achieve innovations and cost savings.

Framework alliance contracts could be used in the following cases:

- Railway and underground projects: billions of euros are to be invested in high-speed railway projects and in the Sofia metro lines in the next few years. The projects are complicated and many stakeholders have to be managed due to overlaps with other infrastructure and cultural heritage. International and national contractors, consultants and suppliers usually win the tenders for construction and supervision in joint ventures; the clients are educated and open to innovations; due to the huge investments, the works are divided into different lots or separate projects, so the transfer of data is essential
- The Sofia central heating operator Toploficazia Sofia EAD[1953] has an investment programme for the rehabilitation of the network and heating stations
- The biggest municipalities, Sofia and Burgas, invest millions of euros in roads, social infrastructure and energy efficiency. They are open to innovations and may try framework alliance contracts if the provisions of PPA permit their use
- Other operators of public infrastructure may use framework alliance contracts for planned (not emergency) repairs and rehabilitations, as well as for new build projects.

Framework alliance contracts have potential in Bulgaria. They can support BIM and can facilitate the transfer of data and know-how on projects constructed with public and private funds. In March 2017 the BSCL, full member of the European Society of Construction Law, signed a licence agreement with the Association of Consultant Architects (ACA) and the author of FAC-1, Professor David Mosey, for:

1951 Moskova, N. (2013), 230.
1952 As defined in the SDA, see Section 22.3.
1953 https://toplo.bg, accessed 27 December 2017.

- the creation and use of amendments necessary for the FAC-1 Framework Alliance Contract to comply with Bulgarian law and
- the creation and use of a Bulgarian language version of FAC-1.

A BSCL FAC-1 Task Group, led by the lawyer Boyana Milcheva, created the Bulgarian language version of FAC-1 and also Schedule 6 Part 1 Legal Requirements in English, stating the amendments necessary for the English language version of FAC-1 to comply with Bulgarian law. These amendments are summarised in Diagram 17 and the Bulgarian FAC-1 was launched in Sofia on 8 February 2019.

22.5 What Are the Legal Issues Affecting an Alliance in Bulgaria?

As stated above, the OCA good faith provisions enable collaborative work in construction. Since framework alliance contracts have significant potential for public works, the interpretation of PPA restrictions is seen as the main hindrance for their implementation in Bulgaria:

- Article 81(2) of the PPA stipulates that a framework contract is concluded for a period up to four years with the purpose of establishing the terms of multiple contracts, including with regard to prices. Many public procurement employers interpret the provision for agreement of prices as requiring tendering with fixed rates and prices. Some consider that framework agreements make projects more expensive, despite the evidence of how savings are achieved. In addition, many public employers are so cautious that they are not particularly interested in the reduction of public spending
- According to Article 81(6) of the PPA additional members cannot join the framework agreement and the substitution of a member is possible only under the conditions of Article 116 (4) where:
  - An existing member can no longer perform under the framework and the possibility of a substitution has been foreseen in the tender documentation and in the framework contract, with clear clauses related to specific conditions
  - Novation to a new member is possible only if the new member complies with the tender selection criteria and if the substitution does not lead to essential variations or circumvention of the law
- The restrictive procedures for subcontracting do not allow re-negotiation of the terms and conditions which is seen as limiting the possibility for Supply Chain Collaboration.

Other legal issues to consider arise from:

- The Bulgarian Protection of Competition Act under which the exchange of specific business information between competitors can be treated as a violation[1954]
- The provision of Bulgarian law that, where several persons undertake obligations as part of a commercial transaction, they are jointly liable unless otherwise expressly agreed.[1955]

1954 Protection of Competition Act (2008), 28 SG, last amended (2015) 56 SG, Article 37(1): It is forbidden to learn, to use or to disclose a business or trade secret, contrary to good faith commercial practice.
1955 Commercial Act (1991), Article 304.

Alliances and BIM need successful pilot projects in Bulgaria. It is of utmost importance that public procurement policy is refocused from the spending of more public money using restrictive procedures, to the support of systems that improve efficiency and innovation.

Reference	English version	Bulgarian version	Explanations
FAA, clauses 10.1 and 10.2	The following amendments apply to the duties of care under clauses 10.1 and/or 10.2	Added: 'but in any case, the duty to act in good faith and the minimum standard of duty of care as provided for in the legislation should not be affected'.	According to Bulgarian law, a party to a contract should act in good faith towards the other parties to the contract, as well as to any third parties which may be affected by the performance of the contractual obligations. The duty of bonus pater familias is the minimum mandatory standard of duty of care.
FAA, Clause 15.2	Conciliation Procedure	Added after 'The Model Conciliation Procedure shall be': 'the Mediation Procedure provided in the Mediation Act'.	The Bulgarian Mediation Act explicitly provides for a mediation procedure which is appropriate for this purpose.
FAA, Clause 15.3	Adjudication	Deleted	Bulgarian law does not provide for statutory adjudication.
FAA, Clause 15.4	Arbitration	Dots substitute 'English' in the fourth sentence. Added: 'In case that both parties to the dispute are persons and/or entities with registered addresses, places of actual management or domicile in the Republic of Bulgaria, the location of an arbitration will be Sofia, Bulgaria'.	Users are enabled to select the language. According to the Bulgarian Civil Procedure Code, the parties can agree on a location of an arbitration abroad only in cases where the registered address, the place of actual management or the domicile of one of the parties is outside of the Republic of Bulgaria.
FAA, Signatures		Deleted: 'as a DEED'.	In Bulgarian law the deed does not exist. The typical means of executing a contract are (in ascending order of formality) signature, authenticated signature by a notary, and a public act drafted by a notary public. For this type of contract, no notarial form or authentication of signatures is required.

Diagram 17 Table of the agreed amendments between the English and Bulgarian versions of FAC-1.

Reference	English version	Bulgarian version	Explanations
Schedule 4	Guidance note	Added new bullet point at the start: 'The procedures are compliant with the requirements of the applicable mandatory provisions of the legislation in the field of the public procurement'.	The PPA provides for limited applicability of the *Direct Award Procedure* and clarifies procedural rules for other award procedures.
Contract Terms, Clause 1.11.6		Added new clause 1.11.6: 'For avoidance of doubt, the *Alliance Members* shall not be jointly liable, unless otherwise explicitly agreed'.	According to Bulgarian law, in cases where several persons and/or entities undertake obligations under a commercial transaction, they would be jointly liable unless otherwise explicitly agreed. The added Clause 1.11.6 is intended to exclude the joint liability of the *Alliance Members*.
Clause 3.2	Representation of *Client*	Added: 'by an explicit notarized power of attorney'.	Bulgarian law requires a notarised power of attorney for representation of most governmental and municipal authorities.
Clause 6.3.1	Exchange of information between the *Alliance Members*	Added: 'as far as it is allowed by the applicable legislation in the field of protection of competition'.	The exchange of specific business information between competitors could be treated as a violation under the Bulgarian Protection of Competition Act.
Clause 7.5	Termination of an *Order* by the *Client* or *Additional Client*	Added: 'In this event the *Alliance Member* to whom the *Order* has been issued shall be entitled only to be paid the agreed costs for that part of the *Alliance Activities* or *Pre-Contract Activities* subject to the *Order* which were performed by that *Alliance Member* prior to the receipt of the notice of termination. For avoidance of doubt, no *Alliance Member* shall be entitled to compensation for profit lost due to the termination of any *Order*'.	Under Bulgarian law in cases of termination of a works contract by the employer, the contractor shall be entitled to receive the incurred costs, the price of the works/services performed prior to the termination and compensation for the profit which the contractor would have received if the contract had not been terminated. The added text is intended to exclude the right of the contractor to seek compensation for loss of profit.

Diagram 17 (*Continued*)

Reference	English version	Bulgarian version	Explanations
Contract Terms, Clause 8.12		Added new clause 8.12 'Tax legislation': 'The *Payment Notice* shall not contradict Bulgarian tax legislation. The stipulations of clauses 8.3–8.8 represent a payment mechanism. The *Final Date for Payment* shall be date when the payment has become due and the *Alliance Members* shall proceed in accordance with the applicable tax legislation'.	The provision is added in order to make the entire contract compliant with the requirements of the Bulgarian VAT Act.
Clause 10.3 and Clause 10.4	Duty of care in respect of specific documents and the *Alliance Manager*	Added in the beginning of both clauses 'Without prejudice to those duties of other *Alliance Members* implied by law'	According to Bulgarian law, a party to a contract owes a duty of care (to act in good faith) to the other parties to the contract, as well as to any third parties which may be affected by the performance of the contractual obligations.
Clause 13.2	Assignment and subcontracting	Added new paragraph: 'In case of a legal succession, all rights and obligations of an *Alliance Member* shall pass on to its legal successor(s)'.	The new clause was added to avoid any uncertainty in case of legal succession of an *Alliance Member*.
Clause 15.3	Adjudication	Deleted Clause 15.3 and reference to Clause 15.3	Adjudication is not regulated by Bulgarian law and the decision of a *Dispute Board* could not be recognised and enforced.
Clause 15.5	Law and jurisdiction	Added at the end of the clause: 'and it is subject to the jurisdiction of its courts for disputes which are not arbitrable'.	The provision is adapted to reflect the Civil Procedural Code rules regarding non-exclusive jurisdiction.
Definition: Partnership	Partnership – a business entity creating joint and several liability between its members	Revised as follows: 'a company under the Commercial Act, a consortium, a civil partnership, or any other entity or unpersonified formation'.	The definition is adapted to reflect entities provided for under Bulgarian law.
Definition: insolvency event	Insolvency Event	Deleted the text after the word 'including' and added: • Any event where a court decision establishing the insolvency and/or the overindebtedness of an *Alliance Member* is rendered, regardless of its entry into force;	The definition has been adapted to reflect the concepts of insolvency and winding up of a company under Bulgarian law.

Diagram 17 (*Continued*)

Reference	English version	Bulgarian version	Explanations
		• Any event where a resolution for winding up of an *Alliance Member* is passed; • Any event where the entire commercial enterprise of an *Alliance Member* is subject to enforcement procedure under the Registered Pledges Act; • Any event equivalent to any of the above in any jurisdiction outside Bulgaria.	
		Added: 'Mediation Act'	Act published in State Gazette 110 of 17 December 2004 as amended from time to time.
Appendix 4		• Added: 'The *Conciliator* should meet the requirements under the Mediation Act'. • Does not consider adjudication • **Added: 'The disputes which are not arbitrable as per the** *Legal Requirements* **shall be referred to the jurisdiction determined in accordance with the** *Legal Requirements*'.	

Diagram 17 (*Continued*)

23

How Does Collaborative Procurement Operate in Germany?

Dr Wolfgang Breyer and Professor Stefan Leupertz

23.1 What Is the Approach to Construction Contracts in Germany?

Traditionally, the vast majority of German construction contracts can be divided into two different models, namely unit-price contracts and lump sum contracts, and lump sum contracts can be further divided into subcategories. Cost-plus and guaranteed maximum price models are not commonly used in Germany.

23.1.1 Einheitspreisvertrag (Unit Price Contract)[1956]

The unit price contract is very similar to a remeasurement contract. In an 'Einheitspreisvertrag' or unit price contract, the employer usually takes on full responsibility for designing the project. On this basis the employer prepares and submits to the contractor a bill of quantities that itemises and describes the individual works. The contractor then inserts prices for each unit of the listed items. To ascertain a position price for each item, the unit price is multiplied by the estimated quantities required for the item and the sum of the position prices then constitutes a pre-estimated contract price. After completion, the actual works that have been completed are measured and the contract price is determined in the same manner as the pre-estimate, but on the basis of the actual measured quantities instead of the estimated figures derived from the bill of quantities. In this way, both parties avoid the risks arising from inaccurate predictions of quantities and the employer accepts the uncertainty that arises from the fact that the exact price for the tendered works cannot be reliably estimated at the point in time when the contract is concluded.

23.1.2 Pauschalpreisvertrag (Lump Sum Contract)[1957]

In contrast to the structure of unit price contracts, the parties can agree on a 'Pauschalpreisvertrag' or volume-independent lump sum for the tendered works. This lump sum price remains effective even if the quantities calculated for the tendered construction work are actually reduced or exceeded. Therefore, both parties bear the risk of there

---

1956 Kapellmann & Messerschmidt (2018), Section 4 VOB/A, para 10.
1957 Leupertz, in: Messerschmidt & Voit (2012), I. Teil K III 3. c), d), para 18.

Collaborative Construction Procurement and Improved Value, First Edition. David Mosey.
© 2019 John Wiley & Sons Ltd. Published 2019 by John Wiley & Sons Ltd.

being an incorrect estimation of quantities. Lump sum contracts are further divided into subcategories with significant differences between them. The distinctions are essentially based on the allocation of design responsibility and the question of which works are covered by the contractually agreed lump sum. Against this background, two basic types of lump sum contracts have emerged, commonly referred to as the detailed lump sum contract ('Detailpauschalvertrag') and as the global lump sum contract ('Globalpauschalvertrag'). Global lump sum contracts can be further subdivided into basic global lump sum contracts ('Einfache Globalpauschalverträge') and complex global lump sum contracts ('Komplexe Globalpauschlverträge'). In practice, hybrid contracts are often agreed to that cannot be definitively assigned to any of the aforementioned contract types.

23.1.3 Detailpauschalvertrag (Detailed Lump Sum Contract)[1958]

The Detailpauschalvertrag or detailed lump sum contract differs from the unit price contract only on account of the fact that the contract price is determined by the parties as a specific fixed amount (the lump sum) and is thus decoupled from the actual quantities required for the construction works. The contractual agreement of the parties as to performance and price are, similarly to unit price contracts, based on a detailed specification of the required works, which in practice is usually produced by the employer using its project design.[1959] The contractor then provides its prices for the individual items. On this basis, in distinction to unit price contracts, the parties agree on a lump sum for the works as a whole. This is the total contract price, which is payable for all works required to achieve the completion of the works in respect of functionality ('funktionaler Werkerfolg'). The only effect of the lump sum nature of the contract on the amount of the contractually owed remuneration is that the contracting parties base the agreed price on the quantities determined in the specifications (bill of quantities) and not, as under a unit price contract, on the quantities that are actually required as measured after the completion of the construction project. This means that the contractor bears the risk of the correctness of the measurement of the necessary quantities, i.e. the quantities in addition to the amount specified, and which are necessary for the completion of the project, are not compensated for.[1960] If lower quantities are actually required than the employer estimated to have been necessary, the contractor benefits if the employer's estimated quantities are the basis of its calculation. Additional work, that has not been identified in the bill of quantities but that has to be undertaken at the demand of the employer for the works to be completed in accordance with the contract, is not covered by the lump sum price and must be remunerated separately pursuant to Section 650b of the German Civil Code ('BGB'), even if the definition of the functional completion of the works has remained unchanged.

23.1.4 Globalpauschalpreisvertrag (Global Lump Sum Contract)[1961]

The responsibilities and risks of the parties under a global lump sum contract are significantly different to those under detailed lump sum contracts. Although the contractor has to provide all the works that are required for the achievement of the functional completion of the works, it does not receive any extra remuneration in addition to the

---

1958 Leupertz, in Messerschmidt & Voit (2012), I. Teil K III 3. c), d), para 19–21.

1959 Leupertz, in Messerschmidt & Voit (2012), I. Teil K III 3. c), d), para 577.

1960 Kleine-Möller et al. (2014) 140; Kapellmann & Messerschmidt (2018), Section 4 VOB/A, para 34.

1961 Leupertz, in: Messerschmidt & Voit (2012), I. Teil K III 3. c), d) OB/A, para 22–26.

agreed lump sum for additional works insofar as the employer does not request or order a change of the definition of the functional completion of the works.[1962] Hence, the contractor not only bears the monetary risk for the correct estimation of the quantities required but may also have to undertake certain work that was not specified by the employer at all and that, therefore, was not part of the contractor's price-calculation. There are generally two ways to organise this risk allocation contractually.

A basic global lump sum contract is created by adding an individually agreed and thus effective 'clause of completeness' to a detailed lump sum contract.[1963] By virtue of this legal agreement, the contractor assumes the calculation of design risk, but only to the extent that it would reasonably agree to it, taking into account the monetary consequences of divergences between the specification of the works and what is actually required for fully functional works.[1964] This does not apply to works which, outside the recognisable design specifications, result from individual user or site-specific requirements. Also, in accordance with general principles of German contract law, the contractor is generally not liable for evident errors in the design provided by the employer. The additional works required as a result of such errors are therefore not covered by the scope of works any more than additional expenses arising from change orders of the employer.[1965]

In a complex global lump sum contract, the employer outlines only the function/purpose of the works and does not devise a detailed bill of quantities. The specifications consist merely of an explanation of the objectives of the project and a basic description of the works tendered for. Quantities, detailed specifications of the works and designs of any kind are not provided by the employer. The contractor takes on full design responsibility and it therefore has to ensure the functional completion of the works and do all that is necessary to achieve it without receiving any extra remuneration.

23.1.5 Hybrids

In Germany, lump sum contracts are commonly used for turnkey projects.[1966] Those projects are characterised by split planning responsibilities, which often lead to severe problems. Under hybrid lump sum contracts, the lump sum is intended to account for the cost of the construction work in total and for the greater part of the design work, especially if a design and build approach is used. However, unlike with 'pure' complex global lump sum contracts, the contractor generally takes on full design responsibility only after the building permit has been submitted and prepares the outstanding drawings starting with the detailed design ('Ausführungsplanung'). This quite frequently gives rise to difficult questions of contractual interpretation in respect of the precise delineation of design responsibilities and thus also in respect of how much work is covered by the contractually agreed remuneration.

1962 Busche, J. (2018), Section 631, para 94.
1963 Langen in Nomos Kommentar BGB (n.d.), Appendix IV to Section 631–651: *Vertragstypen im Baurecht*, para 46, 49.
1964 This method of contracting has been seen to be vulnerable to exploitation by employers, who on occasion will use a bill of quantities as a basis for a global lump sum contract, allocating all risks related to quantities and incomplete Employer's Requirements (including defective design) to the contractor by referring to the detailed description of the works provided by the bill of quantities. As might be expected, this approach has been a source of many disputes.
1965 Kapellmann & Messerschmidt (2018), Section 2 VOB/B, para 265.
1966 Kleine-Möller et al. (2014), 576.

The existence of such hybrid contracts has contributed to the emergence of a highly adversarial contracting culture in Germany, which precludes the efficient and financially successful execution of construction projects, and can only be effectively overcome by the introduction of functional alliance models.

23.1.6 Failures of Traditional German Contracting Methods

In recent years, a number of highly complex and very large public construction projects have encountered severe cost overruns and have drastically failed to meet original cost estimates. The failures of such projects to meet targets has spurred various organisations in Germany to create alternatives to traditional contracting models for large, complex construction projects, particularly (but not limited to) public projects, where it has sometimes been suggested that political figures and administrators are motivated to favour time and cost estimates at the lower end of the scale.[1967] The advantages of alternative contract models are considered to derive from early contractor involvement, fewer claims due to carefully thought-out risk management, increased cooperation and the existence of mutual incentives of the contract parties to achieve success. What follows is a summary of the most controversial recent German public projects.

23.1.7 Stuttgart 21

Stuttgart 21 is intended to be a part of a high-speed rail corridor stretching from Paris to Bucharest (including Munich),[1968] and is defined by three elements: creating a 60-km high-speed rail line between Stuttgart and Ulm, creating a high-speed link to Stuttgart Airport and converting Stuttgart main rail station (Stuttgart Hauptbahnhof) into an underground through station (as opposed to a terminus). Stuttgart 21 was originally expected to cost €2.6 billion, but that projection has increased nearly threefold to €7.6 billion.[1969] Furthermore, while the project was originally expected to be completed in 2019,[1970] completion was deferred until December 2024.[1971]

23.1.8 Berlin Brandenburg Airport

Berlin Brandenburg Airport was promoted as one of the largest airports in Germany, and for that matter Europe. The new airport is intended to see approximately 34 million passengers annually and replace the German capital's ageing Cold War-era Tegel Airport.[1972] The problems with this project emerged in dramatic fashion after the high-profile June 2012 opening ceremony was postponed when it was discovered that the fire protection system was faulty. Subsequently, numerous other defects were discovered, including overheated wiring, escalators that were too short and serious structural faults in the ceiling. Incredibly, the expected opening of the airport was deferred until at least 2020.[1973]

1967 Diekmann et al. (2013).
1968 Ward, J. (2010).
1969 Knight, B. (2017).
1970 Stuttgarter Zeitung (2016).
1971 Zeit Online (2017).
1972 Euronews (2017).
1973 Euronews (2017).

23.1.9 Elbphilharmonie

The Elbphilharmonie is a large complex in Hamburg consisting of two concert halls, a hotel and luxury apartments that was built on top of the site of Kaispeicher A, a cocoa, tobacco and tea warehouse built in 1966. The Elbphilharmonie was designed by Herzog & de Meuron, famous for designing the Tate Modern in London and the Allianz Arena in Munich. Although the Elbphilharmonie has been completed, the project suffered from massive delays and cost overruns. Construction began in 2007 and was intended to be finished in 2010. However, construction only concluded on 31 October 2016 at a cost of €789 million, compared to an original cost estimate of €241 million.[1974]

23.2 What Is the Approach to Alliances in Germany?

Thus far, alliances have seen relatively little uptake in Germany.[1975] However, the collapse of several large German construction companies resulted in the first movements in the German construction market towards alliance contracts. In recent years, the success and increasing acceptance in the UK of alternative contracting forms such as PPC2000 and FAC-1 has led to a renewed push in Germany to develop alliance contracting models. These initiatives have come predominantly from four different directions: contractors' initiatives, Initiative TeamBuilding, research undertaken by the Federal Ministry for the Environment, Nature Conservation, Building, and Nuclear Safety, and the Deutscher Baugerichtstag. PPC2000 has been recently translated into German with the translation available at https://shop.bundesanzeiger-verlag .de/bau-und-architektenrecht-hoai/ppc-international-e-book. Additionally there are plans in place to utilise the German FAC-1 version for a mega project (combined office/residential) in Germany as an overarching umbrella to create harmonised collaboration between parties to a multi-party contract and subcontractors who are not part of the main contract. Most recently FAC-1 has been implemented in conjunction with the FIDIC Silver Book contract on a trial project with the reconstruction of the German embassy in Sofia. The goal is to utilise FAC-1 as an 'under-arching' umbrella to encourage value engineering and the use of BIM amongst all subcontractors. Thus far the use of FAC-1 has been met with enthusiasm from the participants.

The collapse of construction companies in the 1980s resulted from a phenomenon similar to what took place in England prior to the release of the Latham Report, variously described as 'cut-throat bidding' and 'suicidal practices'.[1976] As a result, some of the largest German contractors became aware of the need for reform with regard to contracting models for large complex projects and took it upon themselves to design alternative contracting methods in order to minimise conflict at the development and management stages of these projects and to implement these projects more effectively and more amicably. Three examples of these models are outlined here:

23.2.1 PreFair by Hochtief

Hochtief AG is one of the largest German contractors, founded in 1875, and has approximately 51 000 employees. This company developed the 'PreFair' procurement

---

1974 Diekmann et al. (2013).
1975 Messerschmidt & Voit (2012), I. D. para 258.
1976 Gardiner, J. (2014). For the discussion of the current situation in Germany, see Bosch & Zühlke-Robinet (2000), 49–57.

model in 2002. PreFair was designed so that the architect, the employer and the contractor (Hochtief) work together from the earliest possible stage. PreFair is structured according to two main phases: (i) Preconstruction and (ii) Construction. During Preconstruction, Hochtief provides the employer with critical information in order to optimise the execution of the project for all participating parties. Because Hochtief is involved from the very beginning, it can make a binding offer to the employer at the end of the preconstruction phase. Then it is up to the employer to instruct Hochtief to execute the construction or procure another contractor. Hochtief is remunerated for its work during the preconstruction phase even if the employer declines to instruct Hochtief to undertake the construction.

23.2.2   i.volution by Bilfinger SE (Former: Bilfinger Berger)[1977]

Bilfinger SE, a listed contractor, was founded in 1880, has approximately 35 000 employees and developed 'i.volution' (formerly: 'Gemeinsam Miteinander Partnerschaflich (GMP)') in 2000. This splits the project into five phases:

1. Conception and Development
2. General Design
3. Consulting Preconstruction
4. Construction
5. Operation.

Phases 3 and 4 are mandatory for the client to be involved in and the client has a discretion whether or not to employ Bilfinger for the earlier or later stages. Generally, the contract takes the form of a guaranteed maximum price contract in which the price is set during the early project stages.

23.2.3   Teamconcept by Züblin[1978]

Ed. Züblin AG was founded in 1898, has approximately 13 000 employees, and developed 'teamconcept' in 1994. This divides the project into five phases:

1. Design
2. Planning
3. Calculation
4. Construction
5. Operation.

Züblin begins its involvement in the design phase and is obliged to provide three stages of expense budgeting: the first as a simple estimate, the second as a calculation and the third as an offer with a guaranteed maximum price. The contract is then concluded as an agreed maximum price, cost-plus-fee or lump sum contract.

All three of the above contract models rely on early contractor involvement as the contractor must contribute to the planning in the preconstruction phase and must try to prevent the employer from incurring additional claims which it could not foresee. In different ways the models use open-book principles, duties to cooperate and communicate

---

1977 Agthe et al. (2016), 28; Heilfort & Strich (2003), 14.
1978 Agthe et al. (2016), 28; Heilfort & Strich (2003), 16.

and key performance indicators as tools to foster the success of the project. However, all these approaches remain based on traditional bilateral relationships, rather than the multi-party relationships that alliances are designed to foster.

23.3 What Is the Potential for a Framework Alliance in Germany?

The most prominent and hopeful attempts to formulate a framework alliance have arisen as a result of the efforts of Initiative TeamBuilding. Initiative TeamBuilding is a group that is led by lawyers, academics and representatives of leading employers, contractors, designers and other consultants with the aim of achieving the following goals:

- The development of models for integrated project development
- The drafting of contract structures and contract clauses for multi-party contracts to implement the project development models that are being developed
- The creation of organisational and procedural guidelines for project parties.

Initiative TeamBuilding has developed four working groups, each with its own topic: Working Group 1 deals with organisation structures and communication during the project; Working Group 2 deals with methods for procuring and selecting parties; Working Group 3 concerns itself with matters of remuneration, including how to determine target prices; and Working Group 4 investigates risk management, including working out processes for the joint determination, evaluation and distribution of risks.

Many of Germany's most important employers, contractors and construction managers are initiators and supporters of Initiative TeamBuilding, including BMW, DB Netz AG, ECE Project Management, Fraport (Frankfurt Airport), Hochtief, Porr, Züblin, Arcadis, Drees and Sommer, and others. The fundamental objective of Initiative TeamBuilding is to establish parameters for the implementation of alternative contracting models into the future based on a multi-party or alliance approach.

23.4 What Is the Approach to BIM in Germany?

Although the use of Building Information Modelling (BIM) is allowed by the German Regulation on the Award of Public Contracts,[1979] the uptake of BIM in Germany remains nascent for the time being.[1980] Several public pilot projects have used BIM and the will to implement BIM is strong, both in the private and the public sectors.[1981] Nevertheless, in comparison to several neighbouring countries, Germany has been relatively slow in adopting BIM.[1982]

1979 Section 12, Paragraph 2, Sentence 1 VgV.
1980 Eschenbruch, K. (2016), 375; Eschenbruch & Grüner (2014), 403; Boldt, A. (2015), 393–394; ARGE BIM Guide (2013), 24.
1981 Boldt, A. (2015), 394; Eschenbruch, K. (2016), 375; Cf. Federal Ministry of Transport and Digital Infrastructure, 'Stufenplan für digitales Planen und Bauen', 2015. https://www.bmvi.de/SharedDocs/DE/Publikationen/DG/stufenplan-digitales-bauen.pdf.
1982 Eschenbruch & Grüner (2014), 402.

The root cause of the slow rate of uptake of BIM may be because thus far multi-party contracts have rarely been used in Germany.[1983] In large-scale projects in Germany, participants are often obliged to provide others with data. However, the usage of data servers and Computer-Aided Design (CAD)[1984] and CAFM[1985] have been much more prevalent than the use of BIM.[1986]

A controversial issue that has emerged is whether the usage of BIM is compatible with the HOAI (the Act of Fees for Services by Architects and Engineers). Since BIM requires a much higher standard of design in advance than under the HOAI, the HOAI might not be compatible[1987] because it necessitates a phase-based model. If BIM is used, such a model is not practicable as the parties develop the project jointly at an earlier phase using BIM. Under certain circumstances, the question may then arise as to how much the architect can bill under the HOAI.

Many in the German construction industry are still attempting to implement BIM within traditional contract structures. As a result, the actual potential of BIM, which is directed towards facilitation of the early exchange of information and creating the technical possibility of networking among construction partners beyond contractual boundaries, is largely being lost. Therefore, more than a few in the German construction industry have recognised the power of BIM and pledge to use it to full effect together with alliance contracting models. They argue that the uptake of BIM in Germany will not only continue to rise, but that it will also be a driving force in the adoption of multi-party contracts and alliances,[1988] and that the drive towards BIM will likely happen sooner rather than later and will likely be driven largely by the private sector.

23.5 What Are the Legal Issues Affecting an Alliance in Germany?

Two of the most prominent initiatives that are investigating the legal issues around alliances in a German context are the Deutsche Baugerichtstag and a research project commissioned by the Federal Ministry for the Environment, Nature Conservation, Building, and Nuclear Safety, project 17.

23.5.1 The Deutsche Baugerichtstag

The Deutsche Baugerichtstag is an eingetragener Verein ('registered non-profit association') founded by Prof Dr Rolf Kniffka and currently presided over by Prof Stefan Leupertz, which organises a conference every two years dedicated to addressing current construction law issues in the German market and presenting recommendations to legislators. Recommendations given by the Deutsche Baugerichtstag were the basis of the first ever construction contract law provisions that have been inserted into the

---

1983 Ritter (2017), 85; Boldt, A. (2015), 394.
1984 Computer-Aided Design.
1985 Computer-Aided Facilities Management.
1986 Eschenbruch & Leupertz (2016), 14; Eschenbruch & Grüner (2014), 403; ARGE BIM Guide (2013), 28.
1987 Cf. Kemper (2016), 426–428.
1988 ARGE BIM Guide (2013), 22–23; Ritter (2017), 87; Boldt, A. (2015), 394. Whether BIM works most efficiently in conjunction with multi-party-contracts is disputed, see Eschenbruch & Leupertz (2016), 12–14.

German Civil Code, which came into force on 1 January 2018 after decades of impasse and uncertainty.

The thesis papers for the 7th Deutsche Baugerichtstag give an insight into the legal and practical issues that are of concern. The main issues with regard to traditional procurement models as applied in large complex projects, as identified by the Deutsche Baugerichtstag, include:

- Insufficient needs assessment or definition of project goals and, accordingly, insufficient budgeting and estimation of the deadline timeframe
- Insufficient risk management
- Insufficient process integration of critical project participants
- Lack of trust among the project participants and a culture of distrust around the project
- Lack of rules for decision-making and insufficient conflict-resolution mechanisms.

One of the conundrums that has been identified by the Baugerichtstag is whether multi-party contracts should be classified in the same way as other construction contracts, i.e. as a 'Werkvertrag' (works contract), as a 'Dienstvertrag' (contract for services), as a 'BGB-Gesellschaft' (a partnership agreement), or as a unique form of construction contract. The last idea stems from the fact that under Section 311, paragraph 1 of the German Civil Code (BGB), contractual parties have a broad-ranging freedom with regards to contracting.[1989] The Baugerichtstag has, however, identified that if multi-party contracts are classified as unique, such contracts will need to be made compliant with German statutory law, and most controversially, German public procurement law.

One of the procurement methods that is being discussed as a potential fit is 'Wettbewerblicher Dialog', which is very similar to the EU concept of 'Competitive Dialogue'.[1990] This is a procurement method that public sector agencies can currently elect only if the project is too complex to create a precise description of works required.[1991] The procurement process consists of four phases:

- Disclosure phase: The procurement documents are published Europe-wide and contain certain information, such as the employer's requirements, deadlines for tenders, participation criteria and selection criteria.[1992]
- Selection phase: The future employer selects bidders to be invited to the dialogue phase. Only those bidders that fulfil the participation criteria can be invited[1993]
- Dialogue phase: The selected bidders (not less than three)[1994] develop and present information on the realisation of the project, may request further information from the employer and make (non-binding)[1995] offers. The employer informs the participants if their offer accords with the employer's expectations (this is also

---

1989 Stadler, A. (2016), Section 311, para 2ff.; Emmerich, V. (2016), Section 311, para 1.
1990 As considered in the Bank Station case study in Section 6.10.
1991 Section 101, para 5, sentence 1 GWB ('Act against Restraints on Competition'), Section 18, para 1, sentence 1 ('Public Procurement Contract Tender Award Regulations'); Drömann, D. (2008).
1992 Section 18, para 1, para 2 VgV.
1993 Section 18, para 2, sentence 2, para 4 sentence 1 VgV.
1994 Section 18, para 4, sentence 2, 51 para 2 sentence 1 VgV; Schröder, H. (2007), 222.
1995 Section 18, para 8, sentence 1 VgV.

non-binding).[1996] The fundamental principles of the dialogue are confidentiality and equal treatment of the participants[1997]

- Valuation phase: Only at this stage does the employer decide on who should become the contractual partner according to the disclosed selection criteria.[1998]

A drawback of Wettbewerblicher Dialog is that bidders have to reveal their ideas to others, which, regardless of the principle of confidentiality as generally applies, creates the risk that competitors could use these ideas in their own projects.[1999] Wettbewerblicher Dialog is, however, not the only potential solution to the problem of fitting alternative contracting models within the confines of public procurement law, and is arguably not a perfect solution in other respects. One of the reservations noted with regard to Wettbewerblicher Dialog is that this procurement method restricts contractors' proposals to the period when they are still in competition, which has been identified as one of the features of traditional contracting models that alliances are aimed at improving.

Another question at the forefront of discussion at the Baugerichtstag is whether alternative contracting models require the extension of the principle of joint and several liability. This is a question that the increasing uptake of BIM promises to bring to bear on projects regardless of the uptake of alternative contracting models.

23.5.2 Research of the Federal Ministry for the Environment, Nature Conservation, Building, and Nuclear Safety (Project 17)

In June 2017, the Federal Ministry for the Environment, Nature Conservation, Building and Nuclear Safety chose Dr Wolfgang Breyer, Prof Dr-Ing Dipl-Kfm Shervin Haghsheno and Prof Dr Antje Boldt to undertake a research project into alternative contracting models.[2000] The aim of the research project is to consider alternative contract models from abroad that are already in use, to analyse if and to what extent they can be applied under German and EU law and regulations, especially in terms of public procurement and tendering issues, and to analyse what would need to be changed with regard to particular regulations in order to clear the way for the application of alternative contracting models. Another objective is to make a recommendation to the Ministry as to which model is most suitable.

The tasks are:

- To analyse advantages and disadvantages of the unit price contract in terms of time, cost and quality, and to show the typical difficulties and risks in procurement and execution of construction projects
- To examine the most effective alternative contract model in terms of procurement, segregated procurement of planning and execution, and staff costs of the employer
- If possible, to create a suitable standard form contract.

1996 Section 18, para 7, sentence 2 VgV.
1997 Section 18, para 5, para 7 VgV; Heiermann, W. (2005), 774.
1998 Section 18, para 9, VgV; Heiermann, W. (2005), 775.
1999 As considered in the Bank Station case study in Section 6.10.
2000 *Alternative Vertragsmodelle zum Einheitspreisvertrag für die Vergabe von Bauleistungen durch die öffentliche Hand*; transl.: alternative contract models to unit price contracts in terms of procurement of building works by the public sector.

The focus of the research lies on integrated project delivery, project alliancing and project partnering, and relevant questions in this context include:

- Would no-blame-no-claim-clauses, as used in Australian alliance models to avoid disputes, be appropriate or valid?
- How can the emphasis on the lack of price competition, as present in some of the alternative models, be reconciled with the fact that it is impossible under German public procurement law to take price out of the procurement process?
- Which internal dispute settlement mechanisms can be applied?
- What are the issues with respect to liability among the parties if a multi-party contract is used, including with regard to other non-member contractors?

Acknowledgements

Dr Wolfgang Breyer and Professor Stefan Leupertz would like to thank Konrad Anderson for his assistance in the drafting of this chapter.

24

How Does Collaborative Procurement Operate in Italy?
Professor Sara Valaguzza

24.1 What Is the Approach to Alliances in Italy?

Historically, there has been a negative attitude in Italy towards alliance contracts and other collaborative contracts. This can be traced to a number of reasons, for example:

- These contracts are not included among the specific contract types described in the Civil Code[2001]
- Italian operators are sceptical of adopting standard contracts because they believe that they are too complex and it is common practice, even for major contracts, to come to an agreement with a handshake
- Legal advisers still prefer to draft tailor-made contracts setting out the details case by case.

In order to describe how Italian academics and practitioners look at collaborative contracts, it is necessary to consider how scholars in Italy as a civil law jurisdiction have developed contract theory by reference to the Civil Code.[2002] Despite the Code recognising the principle of party authority,[2003] it is a common perception that the Code offers a framework that covers most of the issues arising from the implementation of any contract. This creates a reluctance to introduce contractual models designed to support and expand on the law, and a preference to trust the law rather than the contract.[2004]

2001 This provides a general framework that regulates all kind of contracts; the fact that alliances and collaborative contracts are not included in the Civil Code is very relevant because in Italy the law is entrusted with the task of authorising private and public organisations to undertake transactions with a legal value. There is a famous saying used a lot by Italian legal experts: *Ubi lex voluit dixit, ubi non voluit non dixit* – meaning 'when the law wants something to happen it says it, when the law does not want something to happen, it remains silent'. So, while parties are free to create collaborative contracts, the absence of a legal category may suggest to some parties that collaborative contracts are not needed.

2002 For general guidance on the theory of contracts according to the Italian literature: De Nova, G. (2014); Galgano, F. (2011); Sacco & De Nova (2004). On atypical contracts in the Italian system: De Nova, G. (2010). On public contracts in the light of Italian scholars, see, *ex multis*: Racca et al. (2011); Racca, G. M. (2012); Racca and Yukins (2014); Torchia, L. (2016); Ramajoli and Galli (2017); Valaguzza, S. (2018).

2003 Article 1322(1) of the Civil Code.

2004 A detailed description of the civil law system that highlights the differences with the common law can be found, in particular, in Monateri & Somma (2016).

Collaborative Construction Procurement and Improved Value, First Edition. David Mosey.
© 2019 John Wiley & Sons Ltd. Published 2019 by John Wiley & Sons Ltd.

The contract is perceived as a bespoke tool to govern only particular elements of a legal relationship. As a consequence, with the exception of specific sectors,[2005] the potential good practice of standard model contracts is underestimated and the contract as a tool governing the issues that can affect a specific relationship is undervalued. Limited attempts to create model contracts have been pursued by professional or trade associations but superficially drafted, tailor-made contracts are still used even in relation to complex projects. The commercial assumption persists that model contracts cannot solve 'real' problems, underpinned by the culture of the Italian legal profession which is more focused on litigation than on contract engineering.

Italian construction operators, especially in the private sector, do not attempt to draft inclusive, clear and consistent contracts, but instead base the success of a legal relationship on the selection of their counterparts. Reliance on people replaces the faith in the contract which, while it might appear consistent with collaborative relationships, is a risky approach in a litigious environment.

In Italy, there is no special legislation governing the construction market, and regulation of the sector is divided between provisions applicable to:

- Construction involving only private sector organisations, contained in the Civil Code and mainly regulating procurement[2006]
- Provisions applicable to construction by a public sector client, contained in the Code of Public Contracts (now the Legislative Decree no. 50/2016).[2007]

In addition, Italian construction contracts are often:

- Fragmented, in that they are divided among many contracts awarded to each of the parties involved in each specific project
- Static, in that they focus on every individual negotiated relationship, without taking into account wider interactions
- Occasional, in that the contents of contracts are often the result of decisions by the most powerful contracting party, not necessarily consistent with the principles of the legal system or with the overall sector.

A lack of specialisation in construction law is a weakness which could be addressed by adopting guidelines and model contracts designed to simplify the issues most commonly faced and to improve contractual performance through clauses based on studying the needs of the sector.[2008] Among these needs, a pivotal role can be played by contracts that improve the coordination between the professionals involved in a specific project from its initial conception.

2005 In particular, when one contracting party is dominant, for example in the case of contracts stipulated by banks and insurance companies.

2006 Luminoso, A. (ed) (2010); Costanza, M. (ed). (2000); Cuffaro, V. (2011); De Tilla, M. (2007); Mangini & Iacuaniello Bruggi (1997); Russo & Criaco (2005); Miglietta, M. & Miglietta, A. (2006); Di Gregorio, V. (2013); Panetta, R. (2016).

2007 Cianflone & Giovannini (2002). Lasalvia, M. (2017); Cabiddu & Colombo (2016); Carullo & Iudica (2011); Garella & Mariani (2016); Ferrari and Morbidelli (2013); Clarich, M. (ed) (2010).

2008 For example, Italian case law shows that, with regard to the execution of procurement contracts for design and works, the issues most commonly examined by judges are those regarding: supposed errors of one of the parties; the admissibility of requests from the public administration to the contractor to execute variations to the project; price revisions; interruption of works; the promptness of supplies; the interpretation of specific contract terms.

The economic crisis in Europe, the complexity of Italian national rules, the reluctance of the legal system to depart from more traditional tools and the high level of litigation are all elements that demand a vigorous and innovative intervention, both in the private and in the public sectors, aimed at greater cost and time control, optimisation of processes and exchanges of information, improvement of efficiency and promotion of healthy competition.

In practice, the lack of exposure to collaborative contracts means that all the relationships between the parties involved in a project are regulated by separate contracts that form a chain, without the integration offered by an alliance. The consequence is a lack of cooperation that often creates cost overruns and delays and that greatly increases the risk of disputes. Health and safety records also show that something is not working correctly in the relationships between contractors, subcontractors, designers and managing bodies:[2009] The adoption of alliance contract models, and their regular use as a new contractual norm, could provide the answer to real and unavoidable needs.

To overcome reactionary attitudes, it is very important to offer a clear picture of the advantages that alliance contracts can provide, in particular as to the cost and value of any project and more generally to the economy of the country. A first step towards the promotion of collaborative procurement was taken by the University of Milan, when in December 2016 it received from King's College London and the Association of Consultant Architects (ACA) Council a licence for the translation and adaptation of the Framework Alliance Contract 'FAC-1'. This contract promotes the collaborative environment that is needed in Italy to overcome the adversarial attitude that dominates the construction sector and that is responsible for extra costs and delays. The application of FAC-1 is also consistent with the general principles of Italian public law, namely transparency, efficiency and control over quality.

The Italian version of FAC-1 was subject to consultation among Italian academics, public and private sector clients, consultants and contractors and has been adapted to make it compatible with the Italian legal framework. Consultees included the Municipality of Milan, the Ministry of Infrastructure and Transport, the Administrative Court of Milan, Associations of Constructing Companies and the most important public players in railways and in highways infrastructure. The reactions were very positive.

FAC-1 is increasingly attracting the interest of economic operators, public authorities and the public contracts regulator. FAC-1 would give both public authorities and private clients the opportunity to act strategically, while taking advantage of the opportunities offered by new digital technologies and modern contracts.

FAC-1 represents the first example of an alliance model contract in Italy, designed for and directly applicable to the construction sector.[2010] Diagram 19 at the end of this chapter summarises changes made to FAC-1 for the purpose of its translation into Italian. In October 2017, the Italian version of FAC-1 was launched and, since then it has been adopted on several important projects.

2009 For example, in the period 2013–2017 we counted seven deaths and many people injured because of collapsed bridges and viaducts in the north and in the south of Italy.
2010 The Italian version of FAC-1 can be purchased at the following link: www.ebuildingcontracts.co.uk/italian-fac-1. For info and technical support is possible to email the following address: fac-1@unimi.it.

In July 2018, the Italian Centre of Construction Law and Management[2011] (CCLM) launched the first two trials of FAC-1 in Italy in the context of the public sector, both of which are summarised in Section 10.10:

- FAC-1 Liscate School – a €5 million works contract for a new build public school in the Municipality of Liscate
- FAC-1 Science for Citizens Project – a €335 million project for the Private Finance Initiative (PFI) design, construction and operation of a campus for University of Milan.

The very first trial of FAC-1 in Italy concerned the construction of a school in Liscate, a municipality in the area of Milan, for an amount of approximately €5 million. The idea of applying FAC-1 in this context resulted from the project being modelled in Building Information Modelling (BIM) by a design team supported by the Polytechnic of Milan. They saw that the application of FAC-1 would enable a deeper synergy between the client, the designers, the main contractor and the supply chain.

Two crucial factors had a positive impact on the adoption of FAC-1 in this case. Firstly, the management of the public tender was entrusted to a central purchasing body acting on behalf of several municipalities. It had wide experience of public contracts and was willing to trial something new. Secondly, the accountability of the universities involved as consultants to the client, as well as the role of CCLM, strongly influenced the decision to trial FAC-1.

FAC-1 was explained during a meeting with prospective contractors, following an on-site inspection for the purpose of assessing the condition of the site. The Polytechnic's representative explained the criteria for evaluation of offers and the representative of CCLM outlined the features of FAC-1. It was specified that the goal of the contracting authority was to secure the commitment of participants to being members of a coordinated team with a shared objective, namely to respect the timeline and costs associated with construction.

The agreed objectives of the alliance are:

- To monitor the timeline and costs of the works
- To avoid modifications necessitated by design errors or insufficiently detailed study of the project
- To deal with potential unforeseen events in the most efficient way possible, both by means of information modelling and collaborative contracting techniques.

The alliance members are the client, the contractor, the design team, the construction manager and the safety coordinator. Subcontractors and suppliers will be invited to join the alliance but not the FAC-1 Core Group. The FAC-1 Independent Adviser is CCLM.

Among the FAC-1 Alliance Activities are weekly meetings of the alliance members, the preliminary development of a supplies calendar and development of a timeline for the appointment of subcontractors in order to secure Supply Chain Collaboration.

For the FAC-1 Success Measures and Targets, mathematical formulae have been developed that will allow mapping of compliance with the FAC-1 Timetable and with agreed cost estimates for the project. If application of the formulae indicates that one or more Objectives have not been met, the alliance member identified will be required

---

2011 www.cclm.eu.

to provide explanations as to how such failures occurred and to propose remedial measures to the FAC-1 Core Group.

Regarding Incentives, it was agreed that the achievement of Objectives would result in alliance members being officially recognised as eligible to carry out alliance activities and experienced in using innovative contract forms. These measures are being considered as the basis for award of a corporate credit rating by the National Anti-Bribery Authority.

The second and most noteworthy test of FAC-1 in Italy concerns the design, construction and operation of a new university campus in Milan in the area where it hosted the international exhibition Expo 2015. This privately funded project will involve Lendlease as the concessionaire, noting their experience of collaborative projects such as North Wales Prison in the UK.

The campus project involves urban regeneration in an important part of the city, planned with the intention of placing research centres of excellence (such as the Human Technopole) side by side with the University of Milan's scientific, educational and research faculties, and with a range of recreational and other facilities.

The specific circumstances of the campus project have made the University's request to use FAC-1 especially compelling. In particular, the promoter raised risk issues that may obstruct the success of the project, depending on the activities and interests of third parties such as:

- The public entities involved in administrative procedures and in the issue of building permits in compliance with town-planning legislation and environmental law
- The owners of surrounding properties
- The developer of the area surrounding the campus
- The university's professors and researchers.

FAC-1, with the aim of coordinating and controlling a wide range of activities, has been recognised as a useful instrument to deal with the above complexities.

Alliance members will include the client and the publicly-owned company responsible for urban regeneration (Arexpo), and also the concessionaire, the construction manager and the safety coordinator. Stakeholders will include project funders and public bodies involved in permits and authorisations.

FAC-1 will govern the complex interactions between the alliance members and stakeholders, during the design and construction phases, in order to ensure that the design responds to the needs of the client and in order to measure the time and cost of the works, also to avoid modifications deriving from design errors or insufficient study of the project. The Italian National Anti-Bribery Authority will monitor the use of FAC-1 on this project in order to assess the potential for improved transparency and improved value through collaborative construction procurement.

The objectives of the alliance are:

- To coordinate the agreed activities of each alliance member
- To minimise the negative effects of unforeseen events
- To address promptly, and in a cooperative manner, the needs of the design and construction teams
- To point out promptly, in accordance with the Early Warning system, each and every issue that emerges during the design and execution of the works.

In the FAC-1 Risk Register relating to Alliance Activities, we have included a 0–5 index for the assessment of specific risks, the impacts of those risks on the Framework Programme and the agreed risk management strategies. For example, the risk of insufficient flexibility in the project is considered very high and could negatively affect compliance with the client's needs. Consequently, the information and coordination obligations relating to this risk have been reinforced.

In this context, there is relevant innovation in a new move towards model contracts and alliances in the public sector. Legislative Decree no. 50/2016 has delegated the National Anti-Bribery Authority to adopt and to impose on contracting authorities new model contracts, model tender documents and model public notices, with the intention of promoting efficiency in the public contracts marketplace. This is a unique initiative and the model contracts, if accepted, could also be drafted by an organisation that has the specific task of preventing corruption and that is entrusted with regulatory functions. Therefore, new contracts will emerge as a regulatory tool created by a third party for use in the Italian public sector and not as a spontaneous product of self-regulation. Standardisation resulting from regulation implies that model contracts will be imposed rather than selected on the basis of their merits and consistency. It is possible that the future of alliance contracts in Italy may in part depend on the foresight of the National Anti-Bribery Authority, to whom the Italian translation of FAC-1 has been presented as a possible way to better orient the industry in relation to public contracts.

The adoption of FAC-1 on two pilot projects in Italy has triggered a keen interest in this contract form. The monitoring of these pilot projects will be crucial.

24.2   What Is the Approach to BIM in Italy?

In Italy, in the ambit of private construction, BIM represents a medium for improved value that currently only the best architecture and engineering firms are able to offer. Working with BIM represents an important competitive advantage, as well as improving the ability to answer the client's needs for accurate timing, cost and design solutions.

In Italy, there are no specific regulations applicable to BIM in relation to private sector works, and the use of BIM[2012] is not yet compulsory. Nevertheless, the 2016 Public Contracts Regulations[2013] introduced some significant provisions, applicable to the construction of public works, that enhance a digital transition and indicate a strategy to promote digital modelling tools. This is of great importance since, in Italy, the public sector provides the most significant quantity and quality of work for the construction industry.

In particular, the new regulations established that the planning and design of public works should ensure the 'incremental use of specific digital methods and tools such as those for modelling building and infrastructure'.[2014] The regulations provide that public authorities such as Government, public enterprises and local administrations may require the use of digital methods and processes in a tender.[2015] The regulations also

---

2012   We rely here on the definition given by Eastman et al. (2016). With regard to the Italian legal system, the most complete analysis of BIM as an innovative method of planning, especially with reference to its practical application in the construction sector, is made by Ciribini, A. (2013).

2013   The Code of Public Contracts, Legislative Decree no. 50/2016, implementing EU Directives 23, 24 and 25/2014.

2014   Article 23, para. 1.

2015   Article 23, para. 13.

entrusted a commission established within the Ministry of Infrastructure and Transport with definition of the methods and timeframe for the progressive mandatory implementation of BIM.

In December 2017 the Minister of Infrastructure and Transport published Decree no. 560/2017, which provides a gradual introduction of BIM in the design of public works, making it mandatory from 2019 for complex projects with a tendered value equal to or higher than €100 million. By 2025, it is proposed that this mandate will apply to all projects commissioned by contracting authorities, with the intention of promoting the evolution of the construction sector through the application of advanced technological processes. In deciding on this incremental approach the public sector recognises that in a 'fragmented landscape of actors and responsibilities' in the markets of services, architecture, and engineering 'only a slow but gradual evolution of a strong public customer and its function seems to be able to determine the necessary change'.[2016]

The intended path and established deadlines show that, in the public sector, BIM will gradually be made mandatory with a timescale that is not aligned with other more technologically developed countries.[2017] The delays that have characterised the switch from traditional design to digital modelling may derive from the fear of reduced competition between operators in the sector. In Italy, the fragmented and diverse market for design services appears inward-looking and driven by the interests of professional associations that are often not able adequately to keep pace with new developments. This tendency can frustrate the promotion of good practice, precision and quality in public works and can protect those enterprises that are not in line with current developments.

That said, we are witnessing the use of BIM in some public tenders, with some inevitable difficulty for contracting authorities, especially in the arrangement of the tender documents. The requests to offer a BIM model in a public tender mainly follow these alternatives:

- To specify what is expected from all the participants, without evaluating the different BIM offers
- To make BIM one of the aspects that must be evaluated as part of the technical offer.

In the latter case, it is essential that the contracting authority specifies in detail the expected BIM modelling in order to make the technical offers comparable. It should also be taken into consideration that in Italy there are 36 000 contracting authorities.[2018] Therefore, it is unrealistic to imagine that they all have the adequate professional skills to set up a tender procedure for BIM or to understand how BIM models work.

The first uses of BIM in public tenders have been followed by disputes arising from the adjudication of the contract.[2019] In May 2017, the Administrative Courthouse of Milan had to deal with a judicial claim related to the request for BIM in a public tender for the design and construction of a school in Milan. The plaintiffs challenged the evaluation of

2016 Ermolli & De Toro (2017).
2017 For example, data provided by the European Commission shows that in Italy the internet and other informative tools have very low rate of application in the ordinary relationships between citizens and public administrations: Eurostat report (2017). This report states that in the last 12 months only 25% of Italians used the internet to interact with public authorities.
2018 According to data collected in November 2017.
2019 Regional Administrative Court for Lombardy, Milan, 29 May 2017 no. 1210, in http://www.giustizia-amministrativa.it.

the offer presented by another bidder, who supposedly had presented an incomplete and inconsistent BIM model containing a two-dimensional (2D) representation of the entire electrical system of the building, instead of a three-dimensional (3D) one as required by the tender documents.

The case was dismissed by the administrative judges on the basis of a technical assessment carried out by the Director of the Department of Structural, Geotechnical, and Building Engineering of the Polytechnic of Turin, on behalf of the court. The judgment established that:

- A 'BIM format' does not exist, since BIM is a method of work and not an instrument; therefore, the digital representation of the model is irrelevant; this is consistent with the technical literature, which has affirmed that 'BIM is not a thing nor a type of software but rather a human activity that involves, ultimately, extensive modifications to processes in the construction sector'[2020]
- The goal of BIM is to introduce 'a more efficient process of planning, design, construction, management and maintenance [by using] a standard model of information in a digital format for each building, new or existing, containing all the information created or gathered on that building in a format usable by all the parties involved in its lifecycle'
- It is not mandatory to use BIM guidelines when requesting BIM in a public tender.

As the cited case shows, the path towards the acceptance of BIM in Italy is rough and it will take time before the necessary knowledge and good practice will be shared by all the operators in both the public and the private sectors. In this regard, the use of FAC-1 could increase the spontaneous and efficient use of BIM in construction contracts in Italy: indeed, a modern collaborative contract could be used as a platform to facilitate the coordination of each party's competences and to control the multiple activities of the alliance members during the design and the execution phases.

When presenting FAC-1 to public authorities, the CCLM emphasised the fact that BIM is going to become compulsory for public contracts over a certain threshold. We proposed collaborative contracts as a necessary corollary to the efficient use of BIM. It was easy to explain to public clients and economic operators that, if a project is BIM-based, there is a need to coordinate different participants with different roles and responsibilities.

According to this logic, FAC-1 is likely to become the most effective instrument to govern the interactions between the different players involved in a project using BIM, while also regulating the responsibilities and activities of each alliance member.

The use of digital tools and methods is changing the way in which design activity is conceived, and it is also affecting team-working habits which now tend to be more and more integrated at all levels. This significant change in contractual relationships requires a re-thinking of the contractual framework, for example to deal with issues of shared responsibility among team members and to integrate the intellectual property rights connected with digital projects.

In Italy, as a consequence of the attention to so-called 'Legal BIM', the level of interest in alliancing and collaborative contracting is now very high. The use of new technolo-

---

2020 Eastman et al. (2016), 361.

gies is bringing about a radical modernisation of the traditional contractual approaches hitherto used in the construction world. Professionals, businesses and scholars are now beginning to appreciate how beneficial this evolution can be. It represents a shift from adversarial contract types to collaborative contracts, in which all involved parties can move in the same direction, with an awareness of being key players in the success or failure of a project or programme of work. In this way, all parties are winners or all are losers.

In summary, the public sector is particularly suited to collaborative procurement because the compulsory use of BIM requires a contractual framework that imposes discipline on the relationships between the team members involved in the realisation of a project.

24.3 What Is the Approach to Construction Contracts in Italy?

The approach to construction contracts in Italy is fragmented, driven by the regulations applicable to public contracts, not adequately specialised and highly adversarial. Construction contracts are not well understood and the current law does not address the real needs of the sector. The approach based on general rules in the Civil Code weakens the construction sector, which has suffered a significant crisis in recent years.

In Italy, there is no general regulation of construction law applicable to both public and private contracts. Therefore, the approach varies according to the private or public nature of the client. The most interesting and innovative rules and provisions come from the field of public works and, more specifically, the regulatory activity of the National Anti-Bribery Authority. This is an example of strategic regulation[2021] aimed at enhancing the medium and long-term development of the country. Therefore, we can say that currently in Italy the public sector is the driver of the positive values of progress and development. It is the public sector that is promoting model contracts and digital methods, along with other objectives such as sustainable development, the preference for quality over price, new technologies for innovation and public–private partnerships.

It must be noted that the very detailed public contract regulations in the Code of Public Contracts are also taken into account by private operators in order to regulate their businesses, but that the very limited provisions of the Civil Code applicable to private procurement are unable to deal with the issues concerning the realisation of complex works. Other activities of contracting authorities also provide useful material for the private sector. For example, the price list based on estimates of market prices, used for public tenders, has become an easy point of reference for private sector clients.

Due to a lack of non-adversarial norms, a lack of trust in contractual modelling and the low costs of litigation, most issues continue to be resolved in courts of law. Diagram 18 shows the high number of disputes concerning public procurement claims brought before the Regional Administrative Courts of First Instance in 2016 and divided into subtopics. The data has been collected with the support of the

---

2021 The expression 'strategic regulation', in the field of public contracts, has been coined in Valaguzza, S. (2016a). See also Valaguzza, S. (2016b), 1; La Chimia & Valaguzza (2017).

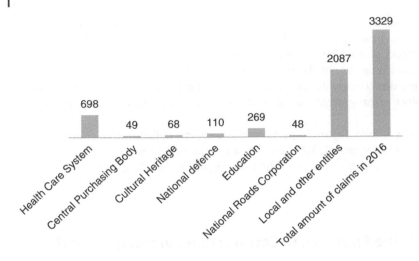

Diagram 18 Number of public procurement claim disputes, before the Regional Administrative Courts of First Instance in 2016 and divided into subtopics.

National Anti-Bribery Authority's database and elaborated by the Council of State's Press and Communication Office (Diagram 18).[2022]

Disputes in the construction sector discourage foreign enterprises and cause the isolation of the Italian market.[2023] Claims and disputes derive from the fact that bespoke contracts, usually drafted on behalf of clients, are less able to express and balance the different needs and attitudes of the parties involved in a project. In this context, disputes often arise to avoid compliance with the requests of the client. It will be possible to overcome this situation only with the adoption of model contracts and good practice equivalent to those used in other countries.

The attempt to reduce litigation in the construction sector is one of the courses of action that the Italian Minister of Infrastructure has declared a willingness to promote, and is one where the public sector can play a leading role.[2024] The reduction of excessive disputes in this sector is one of the current goals of Italian politics, also in the public sector, and the new Code of Public Contracts proposes alternative resolution methods that favour extra-judicial conciliation.

24.4 What Is the Potential for a Framework Alliance in Italy?

In the context described in the previous sections, the introduction of alliance contract models has very significant potential. Alliance contracts would allow us to deal with the

---

2022 According to the research, 1023 public procurements challenged were below the €200 000 threshold, 689 were between the €200 000 and €1 million thresholds and 1617 were above €1 million. All data is retrieved from *Giorn. dir. amm.*, issue 2, 2018, 249.

2023 According to the report of PWC/European Commission (2016) 'Stock-taking of administrative capacity, systems and practices across the EU to ensure the compliance and quality of public procurement involving European Structural and Investment (ESI) Funds', only 1% of procurements in Italy are won by foreign firms.

2024 A system is under consideration that rewards enterprises that complete projects without bringing any conflict to trial.

issue of shared responsibility using a clear and appropriate discipline for each of the complex relationships existing between all the parties involved in any project, including those with subcontractors.

In the private sector, the use of alliance contracts would give clients the chance to involve construction companies and professionals from the initial stage of the project, thus reducing errors in the execution of the works and improving compliance with the timescales and costs of the operation. Furthermore, an alliance contract could benefit businesses in the construction sector by enabling them to regulate their mutual and shared responsibilities. This is very important in civil law jurisprudence which assumes the joint responsibility of all the subjects involved in a project.

In the public sector, alliance contracts would allow the application of general principles established by the public law, such as:

- Transparency (in line with anti-corruption policy)
- Efficiency (avoiding or reducing extra costs)
- Control over the quality of execution (with reduction of mistakes in the design of public works and consequently with reduction of contractual variations necessary to correct errors in project development).

Where BIM is applied, it is also useful to use an alliance contract. If the parties agree to share relevant information related to a given project, it is necessary for them to coordinate each other's competencies and to control the multiple activities of team members during execution using appropriate and specific contractual systems.

In short, the use of alliance contracts in Italy would enable alliance members to:

- Reduce costs during implementation and verify the timing of execution
- Coordinate the activities of alliance members with more guaranteed results and reduced unforeseen interferences
- Manage complex works in an efficient way, integrating the contributions of different expertise and integrating the work of multiple clients
- Achieve improved value in terms of sustainability of the works or services, improved organisation of sites, improved employment and working conditions and improved cooperation with the supply chain
- Involve all professionals in cooperation that could prevent or reduce the errors made by other members of the alliance
- Promote transparency in relation to the purposes and specific goals of cooperation, and define the methods for measurement of performance
- Prevent risks during the construction phase and establish procedures to manage them
- Reduce litigation by preventing and managing potential claims.

In addition, alliance contracts would enrich the system with solid legal support for the use of BIM in the construction sector, through:

- The development of positive interaction between different members of a design team
- The coordination of the different phases of work
- Agreement of all critical aspects of the multilateral relationships between the parties, such as intellectual property rights and responsibilities for managing the project.

The alliance contract appears to be an advantageous model for all members of a team involved in any project. All professionals can benefit from greater control over the realisation of the project, from the saving of time and costs through the alliance, from the anticipation of issues that could arise during the construction phase and from the elimination of barriers between the design and the construction phases, with a consequent reduction of variations as well as of litigation. All team members can be urged to demonstrate their expertise, for example by exploiting economies of scale, and to earn additional payments or other incentives.

Furthermore, all members of the alliance can enjoy the benefits deriving from the collaborative culture in which their work is executed, through clear allocation of responsibilities, simplification and coordination of those activities and reduced likelihood of errors during execution. All members of the alliance would also benefit from the increased level of transparency and from shared data, so that the objectives of each party are more easily achieved and are combined with general benefits for the alliance members as a whole.

24.5 What Are the Legal Issues Affecting an Alliance in Italy?

The first obstacle to the adoption of alliances in Italy is the mentality with which construction sector businesses sometimes approach contracts, not in order to clarify and develop legal relationships, but as a source of privileges and advantages to the strongest party in the negotiation. In both the public and private sectors, operators attempt to maintain competitive advantages deriving from information asymmetries. This creates reluctance to be part of an alliance that gives responsibilities to all the parties to share information, to connect their responsibilities and to disclose to the client the risks and potential obstacles to the success of the project.

However, if we examine Italian case law more thoroughly, the attitudes described above should be overcome. The most recent decisions of the civil courts recognise the existence of 'a substantial cohesion of the relationship' between the designer, the construction manager, the contractor and also the client, thus recognising in distinct contracts an alliance in substantial terms.[2025] Therefore, it is becoming crucial to agree the interactions and responsibilities of the different parties involved in a specific project. This should lead to greater adoption of the new approach offered by collaborative contracts, where the risks and the tasks of each professional are regulated in a rational and consistent manner in order to avoid the imposition or assumption of inappropriate responsibilities.

For example, if all the team members involved in a given project were aware of the legal risks of being responsible for another party's mistake when they are engaged under different contracts[2026] then, in order to avoid being trapped in the negative effects of

---

2025 Court of Monza, Sect. II, 1 September 2016, no. 2364. On the topic also cf. Cass. Civ., sect. II, 27 August 2012, no. 14650; Cass. Civ., Sect. II, 2 February 2008, no. 2800; Cass. Civ., Sect. II, 26 September 2016, n. 18831; Cass. Civ., Sect. II, 27 A 2012, no. 14650.

2026 It has been recently stated, for example, that 'the contractor, when he realises another party's projects, must always respect the rules and is subject to responsibilities despite interference from the client, so that the responsibility of the contractor, with the consequent compensatory obligation, does not fail even in the case of faults attributable to planning errors or defective supervision of works, if having observed the presence of a fault the contractor did not promptly report it to the client and notify his objection', Cass. Sez. Civ., ord. no. 20214/2017.

another party's act or omission, they would agree to a more coordinated approach through alliance activities. What we lack in Italy, especially in public and private construction projects, is education as to the benefits of cooperation and of building contractual relationships inspired by the successful resolution of conflicts. FAC-1 could be the solution to reduce litigation, inefficiencies, high expense and fragmentation in the construction sector.

There are no real obstacles, in the Italian legal system, to overt recognition of the model contracts that establish and discipline an alliance.[2027] The illusion of the self-sufficiency of the law and of the tailor-made single bespoke contracts could be overcome if economic operators had a clear picture of the risks they run on a daily basis and of the advantages they could obtain by adopting and implementing alliance contracts. The legal assessment of the pros and cons of alliance contracts can be highlighted to operators in the sector, in order to convey the economic and legal advantages of cooperation, sharing information and implementing collaborative procurement in ways that cross the borders between separate contracts.

In the pilot projects described in Section 24.2 we have assessed the use of FAC-1 in the public sector, using tender procedures carried out according to traditional methods and applying the Italian regulations on public procurement. A primary benefit of this approach will be to encourage contracting authorities to think not only about the award procedure but also about the substance of what is expected from a given project in terms of added value.

Furthermore, to reward an economic operator's willingness to cooperate and to evaluate its ability to propose solutions to issues arising – instead of taking advantage of them – encourages strategic re-thinking by the construction sector in Italy. With the aim of introducing a new construction strategy, it is very important that the National Anti-Bribery Authority has shown interest in the legal framework of alliance contracts. For this purpose, CCLM was recommended to trial FAC-1 as a collaboration tool, without at this stage offering additional economic incentives to alliance members but instead offering formal recognition that enhances the reputation of alliance members who adopt collaborative behaviours.

Formal acknowledgement of the benefits of collaborative contracts, as instruments that can create added value in public procurement, would help in spreading adoption of this model. It would assist all who believe in alliancing if a soft regulation at a supranational level (such as a green paper of the European Commission) described the benefits of collaboration in the field of public sector contracts and offered suggestions as to its wider adoption. Therefore, discussion and circulation of evidence from the use of the alliance contracts in practice is very important.

2027 Rolfi, F. (2006); La Rocca, D. (2006); Angeloni, F. (2002), 41; Cannata, C.A. (1999), 12; Camardi, C. (1997), 74; Barcellona, M. (1987), 677; Donisi, C. (1980), 649; Cian, G. (1968), 253; Giorgianni, M. (1951), 29; De Vincentiis, Q. (1906), 249.

Reference	English version	Italian version (translated into English)	Explanations
Title	Framework Alliance Contract	Framework dell'Accordo Collaborativo	The acronym of both definitions is 'FAC-1'.
Inside cover	Description of the nature and origin of FAC-1 and details about intellectual property rights on the document and publication	A preface was added, explaining the contents and the structure of FAC-1.	The explanatory preface is intended to bring the operators closer to the model contract, which is far from the way that contracts are structured in Italy.
FAA, reference to clause 1.6	The Core Group members are:	Added footnote: 'Every Alliance Member should appoint a delegate to be part of the Core Group'.	The footnote was added – answering the request of the sector to clarify that individual members should not fear the decisions of the Core Group, since they are all represented in the same body.
FAA, clauses 10.1 and 10.2	The following amendments apply to the duties of care under clauses 10.1 and/or 10.2	Added in italics: '(only to provide a higher, additional or more specific duty than the one established by the Contract Terms)'.	Italian law does not permit us to derogate from the duty of care.
FAA, reference to clause 15.3	Adjudication	Deleted	Adjudication does not exist in Italian law and is arguably not compatible with the approaches to alternative dispute resolution permitted in the Italian system.
FAA, signatures	Or executed as a deed by…	Signed and authenticated by…	In Italian law execution as a deed does not exist. The typical approaches are signature, authenticated signature and a public act drafted by a notary public.
Clause 10	Duty of care	10.1 makes reference to a suitable duty of care, according to the roles, expertise and responsibilities of the Alliance Members. 10.2 makes reference to a duty of 'fairness and bona fide'. 10.3 and 10.4 make reference to the duty owed in respect of contractual obligations.	The clause has been adapted to the Italian contractual duty, which cannot be derogated from by the parties.

Diagram 19 Table of differences between the English and the Italian versions of FAC-1.

Reference	English version	Italian version (translated into English)	Explanations
Clause 14	Termination	**14.2.2.1** In case an Alliance Member doesn't meet an Objective or a revised Objective, the Client or the Additional Client will have the power to terminate the contract or (only in the case in which the performance of said Alliance Member is not considered essential) the appointment of that Alliance Member, only after the notification of an invitation to remedy within 20 Working Days. If the Objective is not reached after that period, the contract or the Alliance Manager's appointment will be automatically terminated. **14.3.1.** The other Alliance Members shall have the right to cease to be a party, to require immediate execution of the Client's performance and to suspend their own performance unless adequate warranty is provided. **14.3.2.** The norms contained in the bankruptcy law shall apply. **14.4** In case of breach, every Alliance Member shall notify it to the Core Group, which shall provide guidance and recommendations; in parallel, the Alliance Member shall invite the breaching member to remedy within 20 Working Days. If the breach is not resolved within that period, the FAC-1 contract or (only in the case in which the performance of said Alliance Member is not considered essential) the appointment of the breaching Alliance Member will be automatically terminated.	**The clause has been** adapted to the Italian rules on termination which cannot be derogated from by the parties. **The Italian bankruptcy law** forbids termination in case of insolvency events.
Clause 15	• Adjudication • Conciliation or Dispute Board • Arbitration	• No adjudication • Conciliation or amicable agreement (i.e. a contract by which the parties, through mutual agreement, define an imminent or actual difference) • Arbitration, as it is described in the English version	Adjudication does not exist in Italian law. Amicable agreement is a common tool to solve disputes in Italy. Arbitration may not be applicable to public administrations.

Diagram 19 (*Continued*)

Reference	English version	Italian version (translated into English)	Explanations
Definition: Independent Adviser	An independent adviser who may be identified in the Framework Alliance Agreement and appointed on terms agreed by the Alliance Members to provide fair and constructive advice to the Alliance Members on the implementation of the Framework Alliance Contract and the avoidance or resolution of any dispute.	There can be more than one independent adviser, of different expertise.	This reflects the fact that differences and disputes could rise both on technical issues and on legal issues.
Definition: insolvency event		• any case in which (where the Alliance Member is a legal person) events or other exterior facts demonstrate that it is not able to regularly satisfy its obligations; or: • any case in which (where the Alliance Member is a natural person) he/she has filed a restructuring plan in court; • any other event equivalent to those indicated above, in any other jurisdiction outside Italy;	The definition has been adapted to the Italian insolvency rules which cannot be derogated from by the parties.
Appendix 4		Does not consider adjudication.	
Final page	Signatures	A double signature is required for clauses 1.3, 1.4, 5.6, 7.5, 8.10, 13.1, 13.2, 14 and 15.	According to the Italian Civil Code, such clauses – where inserted in a document such as the Contract Terms – should be expressly double-signed for acceptance.

Diagram 19 (*Continued*)

25

How Does Collaborative Procurement Operate in the USA?

Howard W. Ashcraft

25.1   What Is the Approach to Alliances in the USA?

Project alliancing, as a term, has not found favour in the United States, although there have been some projects in Canada that have used this terminology. Integrated Project Delivery ('IPD'), which is, in some ways, similar to alliancing, is the predominant method for contractual, collaborative projects.

The most commonly cited definition of IPD is from the American Institute of Architects/AIA California Council guide. It is aspirational and broadly inclusive, potentially applying to a range of collaborative project delivery systems: 'Integrated Project Delivery (IPD) is a project delivery approach that integrates people, systems, business structures and practices into a process that collaboratively harnesses the talents and insights of all participants to optimize project results, increase value to the owner, reduce waste and maximize efficiency through all phases of design, fabrication, and construction'.[2028]

The generality of this definition has made it difficult to differentiate IPD from other collaborative methods. The AIA California Council narrowed the IPD definition by additionally requiring five specific elements:[2029]

- Continuous involvement of owner and key designers and builders from early design to project completion
- Business interests aligned through shared risk/reward, including financial gain that is at risk dependent upon project outcomes
- Joint project control by owner and key designers and builders
- A multi-party agreement or equal interlocking agreements
- Limited liability among owner and key designers and builders.

In an optimal IPD project, these elements are further developed and separated into an integrated business model, an integrated contract structure and enabling behaviours (Diagram 20).[2030]

As can be seen from these requirements, many of the characteristics of alliances, such as joint control, waivers of liability, and painshare/gainshare, are also contained within IPD. In addition to these structural characteristics, many IPD projects use lean tools and

---

2028 IPD: A Guide (2007), preface.
2029 IPD: An Updated Working Definition (2014), 4.
2030 IPD: An Updated Working Definition (2014).

Collaborative Construction Procurement and Improved Value, First Edition. David Mosey.
© 2019 John Wiley & Sons Ltd. Published 2019 by John Wiley & Sons Ltd.

Business model	Contract structure	Enabling behaviours
Separate profit from cost	Early involvement of key participants	Optimise the whole, not the parts
Guaranteeing costs to perform the work	Joint project control and decision-making	Trust
Limited entitlement to change orders	Shared/Risk reward based on project outcomes	Integration (information, people, and systems)
Profit based on agreed project outcomes	Jointly developed validated targets/goals	Continuous improvement/learning
	Reduced liability among risk/reward members	Appropriate technology
		True collaboration

Source: IPD: An Updated Working Definition, AIACC 2014, 11.
Diagram 20 Optimal IPD elements.

principles,[2031] sometimes referred to as Lean IPD or Integrated Lean Project Delivery ('ILPD'), and the use of Building Information Modelling (BIM) within IPD projects is commonplace.

Standard form suites published by the AIA and ConsensusDocs include IPD contracts although many, perhaps most, IPD projects are executed under proprietary or manuscript contracts. The American Institute of Architects has issued a three-party IPD agreement,[2032] and an agreement utilising a special-purpose entity.[2033] The AIA, in conjunction with the AIACC, has issued an IPD guide.[2034] ConsensusDocs, a coalition of owners, contractors, and others, has issued a multi-party IPD agreement that has been recently updated,[2035] and the Canadian Contract Documents Committee is near release of its IPD agreement.[2036] Thus there is substantial support for IPD among the professional associations.[2037]

Although IPD has been used in a wide variety of private projects including healthcare, entertainment, biotech/pharmaceutical, manufacturing, commercial, higher education, data centres, semi-conductor manufacturing, research laboratories and institutional work, it has not been adopted – in its fullest form – by public agencies in the United States due to statutory procurement regulations.[2038] However, public projects have

---

2031 For a general introduction to lean construction, review the public resources at http://www .leanconstruction.org.
2032 AIA Document C191-2008.
2033 AIA Document C195-2008. Although this document preceded the C195, it is used infrequently due to its more complicated structure and increased legal requirements.
2034 IPD: A Guide (2007).
2035 ConsensusDocs (2016), CD-300.
2036 Canadian Contract Documents Committee CCDC-30.
2037 See, also, *Integrated Project Delivery for Public and Private Owners*, A joint effort of the National Association of State Facilities Administrators (NASFA); Construction Owners Association of America; APPA: The Association of Higher Education Facilities Officers; Associated General Contractors of American (AGC) and American Institute of Architects (AIA) (2010). https://www.coaa.org/Documents/Owner-Resources/Industry-Resources/IPD-for-Public-and-Private-Owners.aspx.
2038 The author has structured pure IPD public projects in Canada, has assisted a public hospital project in Norway and is aware of public sector IPD projects in Finland.

been successfully accomplished using many of the principles set out above, albeit without the single multi-party agreement and with reduced use of contractual waivers.

Researchers have studied the effectiveness of IPD. The American Institute of Architects[2039] and the University of Minnesota[2040] have published case studies and there have been surveys[2041] and case studies of specific projects published in seminars and similar papers. Although not specifically studying IPD, a comparative study of 360 projects by researchers at Pennsylvania State University and the University of Colorado found a favourable statistical correlation between project outcomes and team integration and cohesion – both characteristics of IPD.[2042] IPD has also found favour with prominent international agencies, such as the World Economic Forum.[2043]

The Lean Construction Institute recently published a study of 160 projects from 80 owners that investigated the use of lean tools and their effect on project outcomes, but also compared the success of projects based on the project delivery type used.[2044] The study compared 'typical' projects that performed poorly on criteria of cost, schedule, quality, safety, and performance with 'best' projects that met or exceeded owner expectations. In this study, IPD significantly outperformed other project delivery approaches. Twenty-two percent of the 'best' projects, but only 1% of the 'typical' projects, were IPD projects.[2045] Recently, a University of Minnesota research team concluded that IPD effectiveness has been adequately shown and has turned to investigating what aspects of IPD most affect project outcomes.[2046]

There is no reliable data on the number of IPD projects in North America, due mostly to a lack of centralised tracking. This chapter author's team has structured over 130 pure IPD projects, but as these are only a portion of the market, the full extent is unknown. Another difficulty derives from inconsistent use of IPD terminology. Some use the term generally to describe highly collaborative projects, while others require that IPD projects meet specific characteristics, such as a multi-party collaborative contract. The number of IPD projects is, thus, dependent on the strictness of the definition used.

It is clear, however, that use of IPD is expanding into new project types and is being adopted by multi-national firms with extensive construction portfolios as well as significant national and regional organisations. In part, the issue facing these new adopters is not whether IPD is beneficial, but how they should begin their journey and adapt IPD to their specific requirements. To address these concerns the Charles F. Pankow Foundation,[2047] in collaboration with the Center for Innovation in the Design and Construction Industry[2048] and the Integrated Project Delivery Alliance,[2049] sponsored a guide to commencing IPD projects.

2039 Integrated Project Delivery Case Studies (2010).
2040 Cheng et al. (2012).
2041 Cheng et al. (2015).
2042 Molenaar et al. (2014).
2043 World Economic Forum (2016).
2044 This aspect of the study compared 'typical' project outcomes to 'best' project outcomes for design-bid-build, construction management at risk, design-build and IPD.
2045 Lean Construction Institute (2016), 14.
2046 Cheng et al. (2016).
2047 www.pankowfoundation.org.
2048 www.cidci.org.
2049 www.ipda.ca.

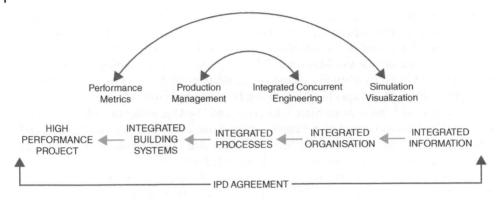

Diagram 21 A simple framework for IPD.

Integrating Project Delivery,[2050] the most recent and comprehensive text on IPD, uses an outcome-based approach to create a theoretical framework for IPD. Starting with the outcomes of a high-performance project – one that is buildable, usable, operable and sustainable – it deconstructs what is necessary to achieve these outcomes and then reassembles them into an integrated whole. The Simple Framework for IPD shown in Diagram 21 represents graphically the key requirements for a high-performance project, each of which is detailed in the text with relevant case studies.

25.2   What Is the Approach to BIM in the USA?

BIM is not mandated in the United States and Canada. Nonetheless, a recent comparative study concluded that North America led the world in years of BIM use, depth, and proficiency (Diagram 22).[2051]

Although there is no broad government mandate, many US governmental institutions support the use of BIM for their projects. The General Services Administration (GSA)[2052] has comprehensive standards and maintains an often-cited internet resource. Similarly, the Veteran's Administration,[2053] the Department of Defense[2054] and National Aeronautics and Space Administration (NASA)[2055] have published manuals and standards. Many state and local governmental entities have similar information or have adopted standards from the federal level. The upshot is that BIM may not be mandated but it is definitely used.

2050 Fischer et al. (2017).
2051 Jung & Lee (2015).
2052 See, resources at https://www.gsa.gov/real-estate/design-construction/3d4d-building-information-modeling.
2053 US Dept of Veterans Affairs (2017).
2054 *MHS Facility Life Cycle Management (FLCM) Building Information Modeling (BIM) Minimum Requirements*, v. 2.1, (2014) https://facilities.health.mil/repository/getfile/1463; and materials generally at http://www.wbdg.org/ffc/dod.
2055 National Aeronautics and Space Administration (2011).

Region	Years of use[a]	Depth[b]	Proficiency[c] (%)
North America	8.5 (5.3)	73.02% (29.4)	82.1
Europe	5.3 (3.2)	55.9% (35.0)	75.0
Asia	4.9 (2.9)	46.4% (33.2)	46.3
Oceania	7.7 (3.5)	65.5% (34.6)	81.8
The Middle East/Africa	5.9 (3.7)	60.0% (36.7)	80.0
South America	3.4 (1.0)	55.7% (33.1)	71.4

a) Years of BIM use. Average and standard deviation.
b) Depth of implementation. Average and standard deviation.
c) Ratio of users who are 'advanced' or 'expert' in their use of BIM.
Source: Jung & Lee (2015), 408.

Diagram 22 Regional maturity of BIM use.

The National Building Information Modelling Standard[2056] ('NBIMS') is supported by the National Institute of Building Sciences (NIBS). NIBS is congressionally chartered and is the US affiliate of the Building Smart Alliance. NBIMS is a broad-based, ongoing effort that is currently in its third version (NBIMS v.3.0).[2057] NBIMS 3.0 draws heavily from the work by the General Service Administration as well as the BIM execution guide[2058] and owner's guide[2059] developed at Pennsylvania State University. In addition to the NBIMS materials, professional associations, such as the American Institute of Architects[2060] and ConsensusDocs,[2061] have standard contract documents for use of electronic information and BIM.

One of the differences between North American and worldwide BIM practice relates to the support of Industry Foundation Classes (IFC). The US governmental organisations require IFC compliance in support of open source standards but, in practice, compliance is achieved through software that can translate to IFC standards, rather than software that is built on IFC. IFC-based interoperability, which was a major discussion topic a decade or more ago, generates much less interest because true IFC interoperability has proved more difficult to achieve and has provided fewer benefits than anticipated.

In practice, there is variance in the depth of BIM adoption. BIM is virtually always used for the development of the design drawings and coordination/conflict checking, and is often used for scheduling and cost analysis. Use for construction simulation and

---

2056 National Building Information Standard – United States (NBIMS) https://www.nationalbimstandard.org/home.
2057 NBIMS v. 3.0, https://www.nationalbimstandard.org/buildingSMART-alliance-Releases-NBIMS-US-Version-3.
2058 *BIM Project Execution and Planning Guide*, v. 2, Charles Pankow Foundation/Pennsylvania State University, http://bim.psu.edu.
2059 *BIM Planning Guide of Facility Owners*, v. 2, Building SMART Alliance/Pennsylvania State University, 2013, http://bim.psu.edu/resources/owner/bim_planning_guide_for_facility_owners-version_2.0.pdf.
2060 Building Information Modeling and Digital Data Exhibit. AIA Document E203-2013; Project Digital Data Protocol Form, AIA Document G201–2013; Project Building Information Modeling Protocol Form, AIA Document G202–2013; Digital Data Licensing Agreement, AIA Document C106–2013.
2061 Building Information Modelling (BIM) Addendum, ConsensusDocs B301-2015, see, also, Hurtado, K. (2015).

design optimisation is seen less often but is not unusual. BIM information is also used for facilities management, but the information is often extracted and imported into a Computerised Maintenance Management System (CMSS) or Building Automation System (BAS), rather than maintained in the 3D model. 3D scanning to BIM is quite common, especially in retrofit projects. Direct BIM to fabrication occurs, and is a developing area, but is most often seen in specific trade work such as mechanical fabrication. Leading firms are pioneering robotic framing[2062] and similar automated construction.

The depth of BIM use is correlated to the type of project. Complex projects with detailed structural systems, complex mechanical systems or sophisticated building envelopes will generally use more detailed and interrelated BIM models. The depth of BIM use is also correlated to the project delivery approach. Traditional 'design-bid-build'[2063] cannot incorporate trade information into the design BIM because trades are not procured until the design is essentially complete. 'Construction management at risk'[2064] allows earlier trade involvement, and design-build,[2065] engineer-procure-construct[2066] and IPD offer even more. Thus, one tends to see deeper BIM interactions on highly collaborative projects.

25.3 What Is the Approach to Construction Contracts in the USA?

Construction contracts in the United States are divided between public and private contracts. In general, procurement of public construction is constrained by the enabling legislation applicable to a specific jurisdiction or agency. The default approach is design-bid-build with a complete separation of design and construction. Work is awarded to the responsive (filled out the paperwork properly) and responsible (meets financial and experience requirements) contractor that submits the lowest bid. All other forms of procurement are considered 'alternative' and are only allowed if there is supporting procurement authority. Design-build, generally on a 'bridging' basis,[2067] is available to many agencies, although it may be limited in the types of project that can be procured. Construction management at risk is available in some jurisdictions.

IPD and alliances are not generally permitted for public agencies, although there may be possibilities if intertwined in a public–private partnership (P3)[2068] because the enabling statutes for P3 may be broad enough to encompass IPD or an alliance. P3 is

---

2062 Examples are Digital Building Components https://www.digitalbuilding.com and Blueprint Robotics, http://www.blueprint-robotics.com.
2063 Equivalent to the 'traditional' procurement model in the UK where design is procured separate from construction.
2064 Equivalent to the 'construction management' procurement model in the UK where construction is procured by the client appointing multiple specialists or trade contractors.
2065 Equivalent to the 'design and build' model in the UK where a contractor is appointed to design and construct.
2066 Engineer-procure-construct (or 'EPC') is a variant of design and build by which contractual commitments establish the performance-based responsibilities of a contractor for all aspects of design and construction.
2067 As described in Section 25.5.
2068 Collaborative procurement and public–private partnerships are considered in Section 8.9.

permitted on a state-by-state basis, with varying requirements. P3 is less common in the United States than in Europe or Canada.

Private construction projects have few constraints and a wide variety of project delivery types are used. In the total number of contracts, design-bid-build still leads but, as project size increases, projects quickly move to construction management at risk, design-build or IPD. Organisations that build repeatedly for their own use often utilise master service agreements as described in Section 25.4, executed with a limited number of preferred providers.

25.4 What Is the Potential for a Framework Alliance in the USA?

Framework agreements, as generally known in Europe, are not a common American practice. Public agencies, in general, do not use long-term contracting for procurement, although there is some use of 'Indefinite Duration/Indefinite Quantity' (IDIQ)[2069] or task order contracting. Most significant infrastructure is still procured project by project on a design-bid-build or design-bid basis. It is unlikely that this will change in the near future as public procurement is based on statute or regulation and there are differing interests that make it difficult to reform.

Private organisations are free to enter framework agreements, but they generally approach long-term procurement through use of 'master service agreements'[2070] or preferred provider arrangements. The master service agreements contain the bulk of the legal and pricing information with project-specific scope and cost contained in a task or work order. This creates a standing relationship between the parties, reduces the transactional cost of contracting, but allows flexible specification of the project cost, schedule and scope. Many major corporations use master service agreements for most of their procurement requirements. Preferred provider agreements are often used with suppliers to obtain discounted pricing, and sometimes preferred delivery, from materials or equipment suppliers. These tools obviate the need for framework agreements that could effectively achieve the same ends.

Framework alliance agreements are unlikely to gain traction because firms interested in this approach use one of two different methods to achieve the same goals. Several firms that use IPD consistently have developed 'master IPD agreements'[2071] that are executed by their preferred designers and builders. In one instance, developed by the chapter author, approximately 85 firms have signed a single master IPD agreement that contains the legal terms applicable to all, including how costs are calculated and paid. The terms of the agreement were jointly negotiated among all parties in three web conferences.

The master IPD agreement does not authorise or guarantee any specific work. Projects are implemented by issuing a project authorisation, signed by a subset of those signing the master IPD agreement. The project authorisation specifies the commercial terms

---

2069 The US equivalent of the procurement model known as a 'term contract' in the UK.
2070 A master service agreement has certain features that are equivalent to those of a framework agreement and demonstrates how US parallel initiatives have evolved in the USA and Europe.
2071 A master IPD agreement is designed to fulfil the same functions as a framework alliance contract.

and scope, plus any modifications made necessary to accommodate local legal requirements.

Other firms have developed standard IPD agreements that they have pre-negotiated with their principal designers, builders and vendors. In this approach, a draft IPD agreement is circulated for comment and the comments are then correlated and negotiated in a series of joint meetings or web conferences. Based on the comments and negotiations, a final IPD agreement is created that addresses consistent concerns. Comments or concerns that are only raised by a single organisation are generally not incorporated into the final agreement. Responding firms may, or may not, be permitted to list exceptions to the base IPD agreement. It is presumed, however, that very few exceptions will be permitted to the pre-negotiated IPD agreement.

The master IPD agreement fits owners that build very similar projects, but in many different geographic regions. It provides contracting stability and long-term relationships. Moreover, because the project types are somewhat similar, the participants can be chosen for this project type. The pre-negotiated agreement fits owners that use IPD regularly, but have widely different project types. One biotech firm, for example, uses this approach to execute headquarters buildings, office buildings, research laboratories and pharmaceutical manufacturing plants – with their associated support structures. Because of this variation, the owner prefers to choose from teams of differing capabilities, depending on the needs of the project.

If framework agreements develop traction in the United States, it is possible that a framework alliance agreement would do so as well. But as IPD is the predominant collaborative form, and IPD alternatives exist that reduce commercial contracting effort and support long-term relationships, it is expected that these will satisfy the niche where framework alliances would fit.

25.5 What Are the Legal Issues Affecting an Alliance in the USA?

As noted previously, there is a sharp divide between private and public design and construction procurement. Public procurement is strictly prescribed by statutes and regulations. If a procurement approach is not permitted, then the contract can be contested and is likely to be void. Depending on the jurisdiction involved, it may be impossible to obtain payment for services performed under a void agreement and it is possible that any payments that were made would have to be disgorged. Needless to say, these potential outcomes chill the willingness to stray outside the procurement regulations.

One of the principal difficulties is the difference between procuring design and construction. Design firms, in particular, have strongly supported legislation that mandates Quality Based Selection ('QBS'), such as the Brooks Act at the Federal level and mini-Brooks acts enacted by states. Under QBS, the public agency interviews firms that have responded to a Request for Qualifications and selects the firm based on the best technical qualifications. It then attempts to negotiate commercial terms with the selected firm. If acceptable commercial terms cannot be reached, the agency can then commence negotiation with the next most qualified firm until an agreement is reached.

In contrast, builders are generally procured through competitive bid, with the contract being awarded to the lowest cost, responsive and responsible bidder. This creates two major difficulties. First, because there is nothing to bid on until the design is complete (or at least well-developed), procurement of designers must precede procurement of builders. Second, because the procurement methods are quite different (you can't bid designers and you must bid builders), they are inherently incompatible.

In some state jurisdictions, it is possible to procure builders on a 'best-value' basis. Best value is a hybrid between QBS and lowest responsible bid. A best-value procurement usually uses a formula that weights commercial terms and technical qualifications to select the firm that provides the owner with the 'best-value'. It is possible, although difficult, to create a QBS procured design appointment and a best-value procured construction contract that have interrelated terms. But many of the terms are often different. It is difficult to flow through key provisions to specialist contractors and trade contractors and to correctly build the risk/reward model that would exist in IPD. Moreover, because the procurements are separate, the participating firms do not get to choose who they will agree to work with – a key component of IPD. Nonetheless, this approach has been used in some instances. UCSF Medical Center Women's and Children's hospital is a very successful example of this hybrid approach.

One might believe that an IPD or alliance outcome could be nearly achieved through design-build procurement, which is permitted at the Federal level and in most states. But design-build is allowed by specific procurement statutes that must be followed. In general, most of these statutes call for 'bridging' design-build on a best-value basis. In 'bridging', the public agency retains a designer through a QBS procurement who prepares design documents to a specific level of detail (usually 30% design), which are then used as a basis for cost comparisons among responding builders and perhaps the setting of a fixed price. The selected builder then must generally hire another designer to complete the design, which might be reviewed by the first designer. This is less efficient and less flexible than IPD.

There are difficulties in some jurisdictions, where to include designers in a 'shared savings risk/reward program' would be viewed as a conflict of interest. Many jurisdictions, including the Federal government, have institutional conflict of interest rules that prohibit a party that was involved in developing design or contract requirements from bidding to perform the work. Having designers and builders jointly engaged in developing the detailed design – and then being incentivised jointly to achieve these ends – is disquieting to public officials who have expected designers to guard the public agency against poor construction or dilution of program requirements.[2072]

In addition, the liability waiver and risk/reward provisions of alliances and IPD agreements may exceed the authority of public agencies. Often, the doctrine of sovereign immunity is waived by specific statute or constitutional provisions, but these establish specific rules for governmental liability that may not be waived. Moreover, some jurisdictions prohibit a 'gift of public funds', which occurs if a party receives something of value for work it was already required to perform. In these jurisdictions, bonuses or shared savings would be deemed a gift of public funds.

In summary, it will be difficult to have true public sector alliances or true public sector IPD agreements without enabling legislation that resolves the difficulties discussed above.

2072 Although this concern is often raised, the experience of those that have studied IPD is that quality is higher, even without specific quality incentives. Thus, this argument is a make weight, not a justified concern.

In the private sector the legal issues are more easily addressed and none preclude alliances or IPD. A few of these issues are discussed below:

- *Conflation of design and construction.* In all states, the lead designer ('the designer in responsible charge') must be appropriately licensed in that jurisdiction. Many jurisdictions also license builders at the general level and often at the specialty contractor level. Designers and contractors are separately regulated and governed – their scope of their respective obligations has little overlap. Thus, it is important to maintain some separation of design and construction within the IPD or alliance agreement. Moreover, because the respective licensing statutes may limit the type of legal entity used for design or construction, and may limit who may be an owner of a design or construction firm, it is important that the IPD agreement is not a joint venture, partnership or corporation unless that form is specifically permitted. Given the additional difficulties involved in having a larger partnership of designers and trades (taxation, corporate compliance, etc.), it is almost always preferable to have a contractual, rather than an organisational, basis for an alliance or IPD project.
- *Insurance provisions.* Insurance policies tend to mirror the requirements of traditional project delivery and are not designed specifically for IPD or alliances. Professional liability ('PL') insurance often excludes liability for means and methods (construction) and loss of fee. In addition, the PL policy may limit coverage for joint ventures, which reinforces the concern of an entity-based alliance or IPD project. PL policies generally prohibit waiving the subrogation of losses. The commercial general liability ('CGL') policies obtained by builders exclude liability for design errors. Thus, unless the IPD agreement carefully maintains these distinctions, there is a possibility that insurance will not be available when needed.
- *PL insurance and CGL insurance are also different temporally.* The PL policy that is in force when a claim arises is the policy that will respond. This may be years after the error occurred or was manifest. The CGL policy that is in place when the damage occurred is the policy that will respond. Thus, if a loss occurs that involves both design and construction, the relevant policies may be from entirely different policy years. And in the case of a latent defect, the policies – if any – that exist may be different from those that are specified.
  These issues can be overcome through careful drafting and consultation with insurance experts. There are insurance products that solve some of these difficulties and, for larger projects, fully integrated insurance programs can be created that reflect the risk/reward intent of alliances and IPD.
- *Intellectual property.* Under copyright law, work that is jointly prepared is jointly owned. Thus, if many parties participate in the design, there are potentially many joint owners. Ownership of created and pre-existing intellectual property should be clarified in the IPD or alliance agreement. In addition, rights to use, and responsibility for use, of intellectual property need to be contractually delineated.
- *Termination.* It is quite possible that a party may need to be terminated for poor performance, lack of collaboration or other reasons. Because the IPD agreement is a multi-party agreement, there should be a procedure and appropriate payment provisions, that allow removing a recalcitrant participant.
- *Contractual waivers.* There are limits to what liabilities can and should be waived. In most jurisdictions, liability for fraud or intentional misconduct may not be waived.

Moreover, some rights that are granted by statute, such as mechanic's liens or prompt payment obligations, are not waivable. Similarly, some safety responsibilities cannot be changed by contract. In creating an IPD or alliance project, it is important to consider what can and should be waived. A blanket 'no sue' clause will not work.

- *Statutory provisions.* Design and construction are closely regulated and the contract drafter must understand which statutes might apply to an IPD or alliance project. For example, many states have 'anti-indemnity' laws that are designed to limit the ability of one party to transfer liability by contractual indemnification. These provisions were not drafted with IPD or alliancing in mind and may be broadly read to interact with the liability waivers. These, and similar statutes, should be reviewed to ensure that the contract language does not inadvertently activate a statutory limitation.

These are only a few of the legal issues to consider when creating an IPD or alliance agreement. Because many of our legal institutions and insurers assume traditional delivery models, the drafter must accommodate these limitations while maintaining the fundamental principles of IPD or alliancing. In summary, unless there is supporting procurement legislation, in the United States it will be difficult to have governmental pure IPD or alliance projects. In contrast, there are better opportunities for public IPD projects in Canada. Private projects in North America are almost completely unrestricted, but the alliance or IPD agreement must consider the legal and insurance environment and accommodate specific issues as necessary.

Appendix A

Research Timelines

A.1   How Has Improved Value Been Delivered Using Early Contractor Involvement, Collaborative Working and BIM?

2011: Publication of UK Government Construction Strategy including commitment to early contractor involvement, collaborative working and Building Information Modelling (BIM), to be tested through a programme of Trial Projects.

2012: Trial Projects Delivery Group agreed methodology (summarised in Appendix B).

2012: KCL Centre of Construction Law appointed as mentor on four Trial Projects and as academic partner on a fifth Trial Project.

2012–2014: Monitoring, collection and analysis of evidence of Trial Projects through documentation review and meetings with Trial Project teams.

2014: Case study reports on first round of Trial Projects included five reports prepared by KCL Centre of Construction Law, reviewed and agreed by academic partners and by Trial Project teams for approval by Constructing Excellence, Trial Projects Delivery Group and Cabinet Office.

2014: Publication by UK Cabinet Office of guidance emerging from analysis of first round of Trial Projects, including 'Guidance on Two Stage Open Book and Supply Chain Collaboration' drafted by KCL Centre of Construction Law.

2014: KCL Centre of Construction Law appointed as mentor on two additional Trial Projects.

2014–2018: Continued monitoring, collection, and analysis of evidence from Trial Projects.

2014–2018: Case study reports on continuing and additional Trial Projects prepared by KCL Centre of Construction Law, reviewed and agreed by academic partners and by Trial Project teams for approval by Constructing Excellence, Trial Projects Delivery Group and Cabinet Office.

A.2   What Is the Relationship Between Procurement, Contracts and BIM?

2014: Ministry of Justice Trial Project case study at Cookham Wood illustrated links between procurement, contracts and BIM.

Collaborative Construction Procurement and Improved Value, First Edition. David Mosey.
© 2019 John Wiley & Sons Ltd. Published 2019 by John Wiley & Sons Ltd.

2014: Society of Construction Law and Association of Consultant Architects awarded grant to KCL Centre of Construction Law for research into procurement, contracts and BIM.

2015: Formation of KCL BIM Research Group.

2015: BIM Research Group members agreed terms of reference and identified 12 private sector and public sector BIM projects for analysis and 40 BIM practitioners for confidential structured interviews.

May 2016: Draft BIM Research Group report shared with 20 participants in full day workshop and with 115 participants in full day industry conference, whose comments influenced the final version of the report signed off by the BIM Research Group.

July 2016: 'Enabling BIM Through Procurement and Contracts' published on-line by KCL Centre of Construction Law.

March 2018: Construction Industry Council published Second Edition BIM Protocol incorporating amendments recommended in 'Enabling BIM Through Procurement and Contracts'.

October 2018: Signature of FAC-1 integrating BIM contributions on school project in Milan.

A.3   What Are the Benefits of a Standard Form Framework Alliance Contract?

2012: UK Government 'Effectiveness of Frameworks' report identified need for new standard form framework contract.

2014: Trial Project case studies revealed economic and social value attributable to collaborative frameworks and alliances.

January 2015: Association of Consultant Architects invited KCL Centre of Construction Law to draft new forms of alliance contract.

March 2015: KCL Centre of Construction Law conference on frameworks and alliances.

2015–2016: Draft alliance contract forms circulated with invitation for comments from 120 organisations in 14 jurisdictions (full list in **Appendix E**).

November 2016: Six consultation sub-groups reviewed aspects of draft alliance contract forms at industry conference hosted by Association of Consultant Architects (published feedback summary in **Appendix F**).

January 2016: Futures Housing Group agreed to trial draft FAC-1 Framework Alliance Contract.

June 2016: FAC-1 published and launched at industry conference chaired by Sir John Armitt.

August 2016: Futures Housing Group launched first FAC-1 entered into with 23 local and regional contractors and national builders' merchant.

October 2016: Association of Consultant Architects and King's College London created www.allianceforms.co.uk website providing guidance and news on FAC-1.

June 2017: FAC-1 used for procurements worth a total over £9 billion in the first 12 months.

August 2017: Futures Housing Group recorded savings in excess of 9% under FAC-1 framework alliance, benchmarked against their previous equivalent frameworks.

September 2017: Department for Business, Energy and Industrial Strategy awarded grant to KCL Centre of Construction Law to develop FAC-1 and TAC-1 as the basis for model long-term contracts to improve efficiency of new build housing, particularly through off-site manufacture.

October 2017: Kier Services Highways adopted FAC-1 for first contractor-led supply chain alliance, with commitment to early contractor involvement and improved social value through support for local supply chains and development of local employment and skills.

March 2018: Kier reported agreed savings of 8% from FAC-1 supply chain alliance.

October 2018: Formation of Transnational Alliancing Group.

January 2019: Following guidance and support from King's College London, Crown Commercial Service advertised procurement of their £30 billion collaborative contractor framework alliances using FAC-1.

Appendix B

Trial Projects Process

Don Ward of Constructing Excellence led the Trial Projects programme, in collaboration with Cabinet Office/Infrastructure and Projects Authority and a Trial Projects Delivery Group comprising representatives of Government and the construction industry, whose most recent representatives are listed below.

The Trial Projects process required each team to commit to sign-off at each of the following four stages by a mentor and by an academic partner:

- Kick off Meeting
- Brief/Team Engagement
- Decision to Build
- Build and Occupy.

On the six Trial Projects supported by King's College London as mentor, the academic partners with whom we worked were Alison Ahearn at Imperial College London, Rob Garvey and Martin Ball at University of Westminster, Professor Peter McDermott at University of Salford and Ed Evans at Constructing Excellence. The mentor on the Property Services Cluster project, where King's was academic partner, was Professor Peter McDermott.

Both the mentor and the academic partner were provided by each Trial Project team with all requested documentation and were invited to attend project team meetings. Any improved value reported by Trial Project teams was subject to verification by a cost consultant team member and also subject to public sector audit.

Each Trial Project case study was written up by the mentor, approved by the academic partner, by Constructing Excellence and by the Trial Projects Delivery Group before being published online. The Trial Project case studies illustrate how collaborative working, early contractor involvement and BIM contributed to improved value. They comprise:

- Trial Project summary (with project time line and key features)
- Client objectives and vision
- New procurement techniques and processes
- Cost targets and savings
- Additional benefits
- Overall assessment
- Key lessons
- Matrix of how the reported savings were achieved including:

Collaborative Construction Procurement and Improved Value, First Edition. David Mosey.
© 2019 John Wiley & Sons Ltd. Published 2019 by John Wiley & Sons Ltd.

- Strategic context
- Common new delivery model characteristics deployed
- Two Stage Open Book/Cost Led Procurement/Integrated Project Insurance characteristics deployed
- Other cross-cutting initiatives deployed (Building Information Modelling, Procurement Routemap, Government Soft Landings, Project Bank Accounts).

The Government Construction Strategy 2016–2020 recorded that:

- 'New models of construction procurement were trialled under GCS 2011–2015 to explore the potential to drive better value and affordability in the procurement process. The new models include the principles of early supplier engagement, transparency of cost, and collaborative working to deliver a value for money outcome. Alongside the potential for efficiencies, the models can support improved relationships across clients and the supply chain, increased supply chain innovation, and reduced risk. Case studies of the trial projects demonstrate that the new models can help realise efficiencies. The Trial Projects Working Group will continue to encourage the use of new models of construction procurement and the Project Initiation Routemap'
- 'The Government, with support from industry, will continue to establish trials of the new models of construction procurement outlined in GCS 2011–2015 (Cost Led Procurement; Integrated Project Insurance; Two Stage Open Book). The trials will be monitored by the Trial Projects Working Group and will form part of the growing evidence base to inform their adoption'.

The Trial Projects Delivery Group currently comprises:

Role	Name	Job Title	Organisation
Chair	Don Ward	Chief Executive	Constructing Excellence
Mentor	David Mosey	Director, Centre of Construction Law and Dispute Resolution	King's College London
Mentor	Vaughan Burnand	Managing Director	VEB Lean Ltd
Mentor	Martin Davis		Integrated Project Initiatives
Client	Rob Taylor	Commercial Services Manager	Environment Agency
Client	David Wharton		Crown Commercial Service
Client	Charles Mills		London Underground
Client	Keith Heard	Chair	National Association of Construction Frameworks
Industry	Mark Taylor	Group Commercial Director	Osborne/Build UK Main Contractor rep
Industry	Alasdair Henderson	Operations Manager	BAM Ritchies/Build UK Specialist Contractor rep

Industry	Alan Muse	Global Director of Built Environment Professional Groups	RICS (and Construction Industry Council)
Industry	Peter Caplehorn	Deputy Chief Executive and Policy Director	Construction Products Association
Industry	Trevor Hursthouse	Chairman, Specialist Engineering Contractors' Group	Lingwood Management Services
Academic	Rob Garvey	Senior Lecturer	University of Westminster

The 2017 report submitted by the Trial Projects Delivery Group to the UK Infrastructure and Projects Authority recorded that:

- Levels of awareness of the new models are low, let alone of the benefits they offer. This is the major barrier prohibiting greater take-up
- Other barriers to change need to be addressed before wider roll-out can be achieved. The group considers key ones to be as follows:
  - *Clearer government policy*
    More robust, unequivocal and overt recommendations are required in the Government Construction Strategy and from Government Construction Board and the Construction Leadership Council for these better modern models.
  - *Lack of private sector take-up*
    Ministers often look to the private sector for validation. We have identified many informed clients in the private sector who use the models or variants with many of the same features, e.g. water utilities, rail, Heathrow Airport.
  - *Client capability*
    This requires engagement in the 'demand chain' long before the procurement phase, starting with client leadership, a focus on outcomes, aligning policy, and departmental procurement teams.
  - *Procurement policies, processes, and capability*
    Identify barriers to formation of an integrated project team (IPT) at the earliest possible stage, including when to make key decisions. The Gateway Process, Digital Plan of Work, BS8534, BS11000 inter alia all need to be aligned and consistent with each other. An unequivocal statement is required with high-level government endorsement that the models are fully compliant with OJEU and other procurement law – a fact which was confirmed by advisers to all the Trial Projects.
  - *Existing frameworks*
    If these do not explicitly refer to use of the new models, advisers may feel that they cannot adopt their use. Crown Commercial Service (CCS) have shown the way in including them in their latest frameworks, and this should be emulated by all other frameworks as they renew. Other tools to consider, which will further facilitate the new models, are the Association of Consultant Architects FAC-1 and TAC-1 Framework and Term Alliance Contracts, launched in 2016 and adopted already on procurements with values ranging from £7.5 million to £2.8 billion in the public and private sectors.

- *Professional Institutions*
  Better engagement is required with the professional institutions, consultants and professional advisers who can appear to have an agenda which favours old models.
- *Evidence*
  Further evidence and clarity of the benefits of the IPI model is required.

Appendix C

Case Studies

A total of over 50 case studies have been researched in preparing this book. They can be categorised by client, location, value, sector, works and contract type as follows, with cross-reference to the sections where they are mentioned, whether in the text or in footnotes:

- Education
  - Archbishop Beck Sports College, Liverpool – £15.9 million new build school – Bespoke framework contract with NEC3 (Sections 3.8, 4.8, 15.3, 15.4, 15.6 and 16.10)
  - Bermondsey Academy, London – £22 million new build school – PPC2000 (Sections 12.4, 18.5 and 18.10)
  - Dudley College, West Midlands – £11.685 million new build college – Bespoke project contract (Sections 2.8, 4.8, 6.8, 12.9, 14.6, 15.3, 18.8, 18.9 and 18.10)
  - Hackney City Academy, London – £30 million new build school – PPC2000 (Sections 15.9 and 18.4)
  - Harrow Schools Programme, London – £70 million framework of new and improved schools – Bespoke framework contract with PPC2000 (Sections 4.8, 12.5 and 18.5)
  - LHC central purchasing programmes for schools, community buildings and housing, nationwide – £5 billion and £800 million national programmes of new build and refurbished schools and community buildings (separate national programme of modular and other new build housing) – FAC-1 with JCT 2016, NEC4 and PPC2000 (and with housing development agreements) (Sections 8.8, 10.9 and 10.10)
  - Milan Union of Municipalities Liscate School, Italy – €5 million new build school, using BIM – FAC-1 with bespoke project contracts (Sections 10.9, 10.10, 14.8 and 24.1)
  - Portslade/Downs Park School Improvement Programme, East Sussex – £1.66 million extension to two existing primary schools – PPC2000 (Section 15.8)
  - Property Services Cluster Education Basic Needs Programme, Hampshire, West Sussex, Surrey, Reading – £119 million new build and refurbished schools – Bespoke framework agreement with JCT 2011 and NEC3 (Sections 4.6, 4.8, 6.6, 8.8, 8.10, 12.3, 13.7, 15.3 and 15.5)

Collaborative Construction Procurement and Improved Value, First Edition. David Mosey.
© 2019 John Wiley & Sons Ltd. Published 2019 by John Wiley & Sons Ltd.

- University of Milan Campus PFI project, Italy – €335 million privately-funded campus regeneration project, using BIM – FAC-1 with bespoke project contracts and PFI contract (Sections 8.9, 10.9, 10.10 and 24.1)
- University of Westminster, London – £2 million corporate services relocation and fit out – PPC2000 (Section 17.8)
- Environment
  - Hampstead Heath Ponds, London – redesign and improvement of reservoirs – PPC2000 (Sections 3.5, 3.10, 6.7, 12.5, 17.8 and 18.6)
  - Rye Harbour, East Sussex – £9.6 million improvement of flood defences – Bespoke framework contract with NEC3 (Sections 2.8, 4.8, 6.6, 9.3, 12.5, 12.6, 15.3 and 17.10)
- Health
  - Bath and North East Somerset Council, Somerset – £12 million new build and refurbished residential care homes – PPC2000 (Sections 3.5, 12.4, 12.5 and 12.10)
  - ORTUS Learning and Events Centre, King's College Hospital, London – new build mental health facility – PPC2000 adapted for construction management (Sections 3.5, 3.7, 6.5, and 7.10)
  - Poole Hospital – £2.5 million operating theatres refurbishment – PPC2000 (Sections 17.2 and 17.10)
  - St. George's Hospital, London – £37.5 million key worker accommodation comprising flats, nursery, retail and management facilities – PPC2000 (Sections 16.8, 17.9 and 18.5)
  - University Hospital Dubai – US$900 million new build hospital – PPC International (Sections 18.8 and 18.10)
- Highways
  - A30 Bodmin/Indian Queens, Devon and Cornwall – £43 million new road construction – NEC3 (Section 15.9)
  - Kier Highways Services supply chain alliance, Surrey – £54 million highway repairs and improvements – FAC-1 with SPTC 2005 subcontracts (Sections 2.9, 5.8, 6.3, 7.6, 10.9, 11.5, 11.6, 11.10, and 15.3.)
  - Connect Plus M25, outer London – £350 million motorway operation and improvements – Bespoke framework contract with NEC3 and overarching PFI contract (Sections 4.7, 4.8, 8.9, 12.6, 12.8, 12.10, 15.3, 15.4, and 19.6)
  - Surrey County Council Project Horizon term alliance, Surrey – £100 million highway repairs and improvements – TPC2005 with bespoke alliance contract (Sections 2.9, 4.8, 5.6, 5.7, 5.8, 5.10, 6.3, 6.9, 7.1, 7.6, 8.5, 10.1, 11.2, 11.5, 11.6, 11.10, 12.3, 12.5, 15.3, 15.7, 15.9, 17.7 and 19.5)
- Housing
  - Amicus Horizon (now Optivo) – £27 million responsive and void repairs – TPC2005 (Sections 5.4 and 7.8)
  - Anchor Property – £3.4 million housing reactive and planned maintenance – NEC3 (Sections 15.3 and 16.8)
  - Bewick Court, Newcastle – £3.5 million residential tower block improvements – PPC2000 (Sections 19.6, 19.7, 19.8 and 19.10)
  - Epping Forest District Council – £22 million house building programme, FAC-1 with JCT project contracts (Section 10.5)

- Erimus Housing decent homes alliance, Middlesbrough – £105 million housing improvement programme – Bespoke framework alliance contract with PPC2000 (Sections 16.6, 16.8 and 16.10)
- Futures Housing Group framework alliance, East Midlands – £30 million housing improvements – FAC-1 framework alliance contract with TPC2005 (Sections 10.2, 10.9, 10.10, 15.3, 15.5 and 15.6)
- Glasgow Housing Association, Glasgow – £1 billion housing stock refurbishment and new build – Bespoke framework contract with PPC2000 (Sections 4.8, 12.7 and 15.7)
- Havelok Homes, London – £600 000 gas maintenance – TPC2005 (Sections 12.5, 12.8 and 12.9)
- Maidstone Housing Trust, Kent £20 million decent homes improvement programme – TPC2005 (Sections 5.8, 15.7 and 16.7)
- National Change Agent procurement consortia, nationwide – £1.6 billion national programme of housing improvements – Consortium agreement and bespoke framework contracts with JCT, NEC3 and PPC2000 (Sections 8.8, 8.10, 15.9 and 18.5)
- Rogate House, London Borough of Hackney – £13 million housing improvements – PPC2000 prototype (Sections 3.10, 15.4, 15.8, 17.5 and 17.8)
- Royal Borough of Greenwich – £200 million housing repairs and maintenance – TPC2005 prototype (Sections 5.8, 5.10, 11.2, 12.7, 15.1 and 19.7)
- SCMG (Hackney Homes and Homes for Haringey), London – £200 million housing improvement programme – Bespoke framework alliance contract with PPC2000 (Sections 2.9, 4.2, 4.4, 4.6, 4.8, 4.10, 5.7, 6.3, 6.6, 6.9, 8.7, 8.8, 9.3, 10.1, 12.8, 15.3, 15.4, 15.5, 15.6, 15.7, 15.8 15.9, 16.6, 17.7 and 19.7)
- Southern Housing Group term alliance, London and South East England – £23 million per annum housing repair and maintenance – TAC-1 term alliance contract (Sections 11.4, 11.10 and 16.7)
- Welwyn Hatfield District Council, Hertfordshire – £160 million housing asset management – TPC2005 (Sections 5.7, 11.2 and 17.7)
- Whitefriars Housing Group, Coventry – £240 million housing refurbishment – Bespoke framework alliance with PPC2000 (Sections 4.8, 4.10, 12.7 and 15.7)
- Leisure
  - Eden Project Core, Cornwall – new build education centre – Bespoke framework alliance with NEC3 (Sections 4.8 and 15.4)
  - Football Foundation (with Football Association and Sport England), nationwide – £150 million programme of new build changing rooms and £60 million programme of mini-pitches – each using FAC-1 with JCT 2016 (Sections 10.4, 10.9 and 10.10)
  - Jarvis Hotels, Manchester Piccadilly – £17 million hotel refurbishment – PPC2000 adapted for construction management (Sections 3.5, 7.10 and 17.9)
- Offices
  - Bahrain bank headquarters, Manama, Bahrain – New build bank headquarters – FIDIC adapted for early contractor involvement (Sections 9.4, 12.4 and 19.10)

- – Department for Work and Pensions and Land Securities Trillium Job Centre Plus programme, nationwide – £737 million programme of adapted and refurbished offices – Bespoke framework alliance with PPC2000 (Sections 4.8, 4.10 and 16.4)
- – Tottenham Court Road, London – new build office development – PPC2000 (Sections 6.5, 6.7, 6.10 and 17.1)
- Public buildings
  - – Central Criminal Court (Old Bailey), London – mechanical and electrical refurbishment programme, using BIM – TPC2005 (Section 13.6).
  - – Cookham Wood Young Offenders Institution, Kent – £20 million new build young offenders institution, using BIM – Bespoke framework alliance contract with PPC2000 (Sections 4.6, 4.8, 4.10, 8.8, 9.3, 12.3, 12.4, 13.4, 13.5, 13.6, 13.7, 14.3, 14.7, 14.10, 15.3, 15.4, 15.5, 15.9, 16.3, 17.8 and 18.3)
  - – Crown Commercial Service consultant alliance – £2.8 billion project management and design services – FAC-1 with NEC3 and bespoke consultant appointments (Sections 8.8, 10.5, 10.9 and 10.10)
  - – HMP Bure – £64 million new build prison – Bespoke framework alliance contract with PPC2000 (Section 17.8)
  - – Ministry of Justice new build alliance frameworks – £1.2 billion national programme of prisons, other custodial facilities and courts – Bespoke framework alliance contract with PPC2000 (Sections 4.6, 4.8, 4.10, 6.6, 10.1, 12.7, 15.3, 15.4, 15.6 and 16.6)
  - – North Wales Prison (HMP Berwyn) – £157 million new build prison, using BIM – Bespoke framework alliance contract with PPC2000 (Sections 3.9, 4.8, 8.8, 12.4, 13.7, 14.10, 15.3, 15.6, 15.7, 15.9, 16.3, 17.8, 18.5, 18.6 and 18.9)
  - – Sheffield City Council/Kier joint venture term alliance, South Yorkshire – £65 million per annum building repairs and improvements to libraries, schools, social services centres and housing – TPC2005 with bespoke joint venture agreement (Sections 8.8, 8.10 and 11.2)
- Rail
  - – Bank Station Capacity Upgrade, London – £564 million London Underground station upgrade – NEC3 (Sections 6.9 and 6.10)
  - – Macclesfield Station refurbishment – £500 000 railway station refurbishment – PPC2000 (Sections 12.5, 17.10 and 19.7)
- Utilities
  - – Anglian @One Alliance – £4.5 billion (over a five-year period) programme of water network maintenance and improvement – Bespoke framework alliance contracts with NEC3 (Sections 4.10, 7.6, 14.8, 15.9 and 16.8).

Trial Project Case Studies

The case studies listed above include the following 10 Trial Project case studies:

- Anchor Property, housing reactive and planned maintenance using Cost Led Procurement
  http://constructingexcellence.org.uk/wp-content/uploads/2015/12/Trial-Projects-Anchor-case-study_Final.pdf
- Archbishop Beck Sports College, Liverpool City Council, new build school using Two Stage Open Book

https://www.gov.uk/government/uploads/system/uploads/attachment_data/file/ 325945/Archbishop_Beck_Trial_Project_Case_Study__CE_Format__120614.pdf

- ConnectPlus M25, collaborative framework underpinning a highways public private partnership using Two Stage Open Book
  http://constructingexcellence.org.uk/wp-content/uploads/2015/12/Trial-Projects-Connect-Plus-Case-Study_Final.pdf

- Cookham Wood, Ministry of Justice new build, young offenders institution using Two Stage Open Book with BIM
  https://www.gov.uk/government/publications/procurement-trial-case-study-cookham-wood-prison

- Dudley College, new build education facility using Integrated Project Insurance with BIM
  http://constructingexcellence.org.uk/wp-content/uploads/2015/12/Trial-Projects-Dudley-College-Advance-II-Case-Study_Final.pdf

- North Wales Prison, Ministry of Justice new build prison using Two Stage Open Book with BIM
  http://constructingexcellence.org.uk/wp-content/uploads/2015/12/Trial-Projects-North-Wales-Prison-Case-Study_Final.pdf

- Project Horizon, Surrey County Council, highways term alliance using Two Stage Open Book with Supply Chain Collaboration
  https://www.gov.uk/government/publications/procurement-trial-case-study-report-highways-maintenance

- Property Services Cluster, Hampshire County Council, Surrey County Council, Reading Borough Council and West Sussex County Council schools framework using Two Stage Open Book, partly with BIM
  https://www.gov.uk/government/uploads/system/uploads/attachment_data/file/ 325955/Property_Services_Cluster_-_Case_Study__CE_format__120614.pdf

- Rye Harbour, Environment Agency infrastructure project using Cost Led Procurement
  https://www.gov.uk/government/uploads/system/uploads/attachment_data/file/ 248614/Procuremen.Trial_Project_Case_Study_EA_Rye_Harbour_0_2.pdf

- SCMG, Hackney Homes and Homes for Haringey housing framework alliance using Two Stage Open Book with Supply Chain Collaboration
  https://www.gov.uk/government/publications/procurement-trial-case-study-social-housing-refurbishment

Appendix D

BIM Research Projects and Interviewees

The King's College London BIM research ran from 2014 to 2016 and the BIM Research Group comprised:

- Professor David Mosey, Director, Centre of Construction Law, King's College London (Chair)
- Darya Bahram, Architect and Project Manager, Research Associate, Centre of Construction Law, King's College London
- Richard Dartnell, Legal Director, Pinsent Masons
- Chris Hallam, Partner, Nabarro
- Christopher Howard, Director of Professional Legal Education, King's College London
- Assad Maqbool, Partner, Trowers & Hamlins
- Kevin Murray, Deputy Director – Construction & Property, Government Property Agency
- Simon Rawlinson, Head of Strategic Research and Insight, Arcadis, member of UK Government Construction Leadership Council, and member of UK Government BIM Task Force
- May Winfield, Senior Legal Counsel, ENGIE.

The research methodology was agreed by the multi-disciplinary BIM Research Group. It comprised reviews of 12 leading BIM projects and confidential interviews with 40 leading practitioners. The resulting Research Report was also informed by a full-day workshop (20 participants) and a full-day conference (115 participants), following which all those present were issued with a draft of the Research Report and were invited to comment. The work leading to the Research Report was partly grant funded by the Society of Construction Law and the Association of Consultant Architects.

Projects

- City of London Corporation, Central Criminal Court
- Crossrail
- Francis Crick Institute, London
- East Riding Leisure Centre, East Riding
- Ministry of Justice, Cookham Wood
- Ministry of Justice, North Wales Prison
- University College London Hospital
- UBS, 5 Broadgate

Collaborative Construction Procurement and Improved Value, First Edition. David Mosey.
© 2019 John Wiley & Sons Ltd. Published 2019 by John Wiley & Sons Ltd.

- Foxwood & Highview School, Kent
- Champion Hill, King's College London
- Birmingham City University
- Northern Line Extension.

Interviewees

- AECOM, Dale Sinclair – Director of Technical Practice
- AHR Architects, Ian Rye – Regional Director
- Allies and Morrison, Indu Ramaswamy – Director
- ArcDox Architecture (Dublin), Ralph Montague – Managing Partner
- Association of Consultant Architects, Alison Low – Secretary General
- Balfour Beatty, Peter Trebilcock – Director of BIM
- BAM Design, Andrew Pryke – Managing Director
- Constructing Excellence, Don Ward – Chief Executive; John Lorimer – Chair of BIM
- Construction Products Association, Noble Francis – Economic Director
- City of London Corporation, Richard Chamberlain – Project Manager
- Crossrail, Malcolm Taylor – Head of Technical Information
- East Riding of Yorkshire Council, Anne Chamberlain – Group Manager Building Design
- Faithful+Gould, Terry Stocks – Head of UK Public Sector
- Freeform, James Bowles – 4D BIM Consultant
- GRFN BIM Consultancy, Stephen Griffin – Director
- Hampshire County Council, Nigel Midmer – Framework Manager
- Hoare Lea, Ewan MacGregor – Director Legal and Risk Management; Ben Roberts – Associate BIM Delivery Leader
- HS2, Sonia Zahiroddiny – BIM Strategy Manager
- ISG Group, Steve Bagland – Pre-Construction Project Manager; Richard Oldfield – Framework Director (Retail); Andrew Stanford – BIM Manager
- King's College London, Kevin Little – Director, Capital Projects and Development
- Laing O'Rourke, Neil Smith – Project Director; Lucas Cusack – Lead Digital Engineer; Tom Inglis – Senior Digital Engineer; Philip Rowen – Commercial Manager; Ronan Burke – Quantity Surveyor
- Lendlease, Brett Wharton – Executive General Manager, Integrated Projects
- Pick Everard, Owen Cockle – BIM Consultant/Senior Architect
- Pollard Thomas Edwards Architects, Krigh Bachmann – BIM Manager
- PRP Architects, Scott Sanderson – Director; Michael Richardson – BIM Manager
- Specialist Engineering Contractors Group, Rudi Klein – Chief Executive
- Tekla UK, Duncan Reed – Digital Construction Process Manager
- Transport for London, Paul Davis – IMMCP (Information Modelling and Management Capability Programme) Delivery Team
- 3DReid Architects, Neil Sterling – Divisional Director
- Turner & Townsend, Tom Oulton – BIM Manager
- UBS, Jason Clark – Director
- University College London Hospital, Paulina Zakrzewska – BIM Lead
- Westfield, Ian Ogden – Design Director
- Willmott Dixon, Garry Fannon – Head of BIM.

Appendix E

FAC-1 Consultation Group Members

Professional Bodies
- Association of Consultant Architects: *Alison Low**
- Association for Consultancy and Engineering: *Rosemary Beales, Dwight Patten*
- Constructing Excellence: *Charles Mills, Craig Keogh*
- Local Government Association (Construction Steering Group): *Tina Holland*
- RICS (Contract Steering Group): *Roland Finch*

Lawyers (UK)
- Anthony Collins: *Andrew Millross*
- Atkin Chambers: *Andrew Burr, Rupert Choat*
- Beale and Co: *Andrew Croft*
- Brechin Tindal Oatts: *Marion Davis, Claire Mills*
- Brookstreet des Roches: *Joe Bellhouse*
- Devonshires: *Mark London, Susannah Davis*
- Fenwick Elliott: *Nick Gould*
- Fieldfisher: *Cecily Davis*
- Goodman Derrick: *Richard Bailey*
- Keating Chambers: *Calum Lamont*
- Lawrence Stephens: *Anne Wright*
- Macfarlanes: *Angus Dawson, Fiona Bishop*
- McQueens: *Emmanuel Amadi*
- Olswang: *Francis Ho*
- Pinsent Masons: *Nigel Blundell**
- Trowers & Hamlins: *Andrew Vickery*
- White & Case: *Ellis Baker*
- Withers: *Julie Teal*

Universities
- Birmingham City University: *Denise Bowes*
- Nottingham Trent University: *David Long*
- Said Business School, University of Oxford: *Lindsay Henshaw*
- Salford University: *Peter McDermott*
- University College London: *Stephen Pryke*
- University of Westminster: *Rob Garvey*

Clients
- Anglian Water: *Michael Meenaghan*
- BAA Heathrow: *David Ferroussat*
- Bedfordshire County Council: *Paul Meigh*
- Birmingham City Council: *Rob Barker*
- Bolton at Home: *Ian Alker*
- British Gas/Centrica: *James Dickinson*
- Cabinet Office: *Peter Groves*
- Circle Housing: *Christian Carlisle**
- City of London Corporation: *Philip Mirabelli*
- Crown Commercial Service: *Kevin Murray**
- Defence Infrastructure Organisation: *Roy Pouncey*
- Devon County Council: *Jamie Staples*
- East Riding of Yorkshire Council: *Steve Baker*
- Environment Agency: *Veronica Flint-Williams*
- Fusion21: *Gordon Ronald*
- Futures Housing Group: *John Thornhill*
- Hampshire County Council: *Alex Chinn, Keith Heard*
- Highways England: *Ken Simmonds*
- King's College London: *Paul Hartmann*
- London Borough of Barking and Dagenham: *Keith Harriss*

- London Borough of Hackney: *Sean Eratt*
- London Borough of Haringey: *David Mulford, Doreen Manning*
- London Underground: *Lee Grant*
- Ministry of Justice: *Guy Beaumont*
- Moat Homes: *Marian Burke**
- Optivo: *Antony Silk**
- Procure Plus: *Derek Greatorex*
- Solihull MBC: *Tim Robbins, Peter Booth*
- Strategic Investment Board (Belfast): *Ronan O'Hara**
- Surrey County Council: *Jason Russell**
- Transport for London: *Simon Addyman, Olu Morgan*
- Whitbread: *Nigel Graham*

Contractors
- Bilfinger HSG International Facilities Management: *Peter Excell*
- Costain: *Adam Freemantle*
- Kier: *Liam McGirl*
- Lendlease: *Simon Caron*
- Higgins: *Adrian Kelly*
- Mears: *Jane Nelson*
- Mitie: *Tyron Stalberg**
- Morgan Sindall: *Helen Mason*
- Mulalley and Company: *Gareth Stephens*
- OpenView Security Solutions: *Pat Sweeney**
- Osborne: *Phil Clarke*
- Seddon: *Keith Waddington*
- United Living: *Daren Moseley*
- Willmott Dixon: *Danny Corley*
- Vinci: *Garry McInerney**

Consultants
- AECOM: *Peter Roberts, Dale Sinclair*
- Amaben: *Mary Bennell*
- Arcadis: *Tes Adamou*
- Atkins: *Mark Dunne*
- Bridgeman Consulting: *Roger Bridgeman*
- Cameron Consulting: *Neil Thody**
- Champion Pearce: *Ronan Champion*
- Cirrus: *Fiona Moore*
- DGA: *Richard Jenkinson*
- Driver Trett: *Nabeel Khokhar*
- Elementa Consulting: *Carl Collins*
- Faithorn Farrell Timms: *Lee Maskell*
- Greenwood Consulting: *Paul Greenwood*
- IP Initiatives: *Kevin Thomas*

- Jacobs: *Kareen Murray*
- KPMG: *Emma-Jane Houghton**
- Lupton Stellakis: *Sarah Lupton*
- Marshall Aerospace: *Jan Robjohn*
- Navigant: *Sena Gbedemah*
- On-Pole: *Russell Poynter-Brown*
- PML Group: *Kim Newman**
- Resolex: *Stephen Woodward*
- Ridge: *Mike Rumbelow, Peter Frank*
- Rider Levett Bucknall: *Paul King**
- Savills: *Shane Hughes*, John Kiely*
- Stradia: *Nigel Barr*
- Sweett: *Andy Taylor, Chung Sam*
- The Change Business: *Alan Mossman*
- Turner & Townsend: *David Barnsley*

Insurers
- Towergate: via *Alison Low*

Australia
- DLA Piper: *John Gallagher*
- Fenwick Elliott Grace: *Jeanie Elliott*
- Jackson McDonald: *Basil Georgiou*
- Monash University: *Paula Gerber*
- Piper Alderman: *Andrew Robertson*
- University of Melbourne: *Wayne Jocic*

Brazil
- Aroeira Salles: *Alexandre Aroeira Salles*

Bulgaria
- EQE Control: *Adriana Spassova*

China
- JustFaith law firm: *Dongfeng Zhao*

Germany
- Breyer Rechtsanwälte: *Dr Wolfgang Breyer**
- Former Federal Judge: *Stefan Leupertz,*
- Züblin: *Peter Steinhagen*

Hong Kong
- Hong Kong City University: *Professor Sai On Cheung*

Indonesia
- PT.PP: *Tommy Anwar*

Ireland
- Dublin School of Architecture: *Malachy Mathews*
- Dublin Institute of Technology: *Louis Gunnigan*
- Trinity College Dublin: *Deirdre Ní Fhloinn*
- University College Dublin: *Orla Hegarty*

Malaysia
- Zul Rafique: *Thaya Baskaran*

New Zealand
- Department of Education: *Stephen Palfrey*

Singapore
- National University of Singapore: *Professor Philip Chan*
- Rodyk & Davidson: *Paul Wong*

Sweden
- Chalmers University of Technology: *Anna Kadefors*
- Foyen: *John Hane*

USA
- Columbia University Law School: *Professor Christina Tvarno*

* = Also members of Alliance Steering Group.

Appendix F

Amendments to FAC-1 and TAC-1 Consultation Drafts

Contract Terms	Consultation comments on FAC-1	Amendments to the FAC-1 consultation draft
1. Roles, Responsibilities, Communication and Additional Alliance Members		
Collaboration	• Reconsider the phrase 'mutual cooperation'.	• Clause 1.1 – 'Mutual cooperation' has been replaced with 'collaboration'.
Framework Prices and Additional Clients	• Need to ensure that *Framework Prices* bind *Additional Clients* as well.	• Clause 1.11.2 – Clarified that prices and proposals are also binding as regards *Additional Clients*.
Priority of documents	• Suggest clarifying priority between the *Framework Documents* and any *Project Contract*.	• Clause 1.5.3 – Confirmed that a *Project Contract* takes priority in the event of a discrepancy.
Stakeholders	• More explanation of the role of *Stakeholders*. Are they potential stakeholders in *Projects* and not *Additional Clients*? • Do the *Alliance Members* owe *Interested Parties* any duty of care?	• Clause 1.10 and Appendix 1 – *Stakeholders* are not *Alliance Members* but have an interest in the *Framework Programme*, which is to be stated in the *Framework Alliance Agreement*. • Confirmed that no duty of care is owed to *Stakeholders* unless expressly agreed.
Additional Clients, Additional Alliance Members, rights and duties of care	• Should clause 1.11 be widened to include other categories of joining parties?	• Clause 1.11 – Provision widened for *Additional Alliance Members*. • These could be of the same or different disciplines dependent on the agreed FAC-1 structure and the *Objectives* of the *Alliance*.
	• This does not say whether an *Additional Client* shall be responsible for any activities that occurred before its joining.	• Clause 1.11.5 – Clarified that any *Additional Client* or other *Additional Alliance Member* is not responsible for any activities prior to its *Joining Agreement*.

Collaborative Construction Procurement and Improved Value, First Edition. David Mosey.
© 2019 John Wiley & Sons Ltd. Published 2019 by John Wiley & Sons Ltd.

Contract Terms	Consultation comments on FAC-1	Amendments to the FAC-1 consultation draft
	• At what point do *Additional Alliance Members* join? • Could get messy if they don't agree to the *Framework Prices* or other *Framework Documents*.	• Clause 1.11.4 – Agreement of all *Framework Documents* is the starting point, but with new option that the *Framework Documents* will be amended by agreement of all *Alliance Members* to reflect the particular interests of an *Additional Alliance Member*.
	• It should clarify that each *Client's* liability to the other *Alliance Members* (if there are one or more *Additional Clients*) should be several and there is no joint liability deemed?	• Clauses 1.11.3, 10.2, 10.3, 10.5, and 11.2 distinguish when bilateral and multiparty duties of care are owed.
	• If *Additional Clients* are appointed, then it might be useful to clarify that they can act individually in relation to a *Project* if need be?	• Clause 1.11.3 clarifies that each *Additional Client* may act independently unless stated otherwise, including as to implementation of the *Direct Award Procedure* or *Competitive Award Procedure*, as to instruction of *Pre-Contract Activities* and as to award of *Project Contracts*.
Core Group governance	• Not sure how the *Core Group* will work. Who pays for their time? Who nominates? • What is the decision-making process? What authority do they have to implement decisions?	• Clause 1.6.2 clarifies that *Core Group* members are nominated and paid by *Alliance Members*. • Clause 1.7 states the *Consensus* decision-making procedure and requires that *Alliance Members* comply with decisions made within the scope of the agreed *Core Group* functions.
2. Objectives, Incentives, Success Measures, and Targets		
Objectives	• Do we want to limit *Objectives* to the *Framework Programme*?	• Clause 6.1 also links *Objectives* to agreed *Alliance Activities*.
Incentives, Success Measures and Targets	• *Incentives* should not be broken down in terms of individual efforts and should always reward the *Alliance* as a whole. • Need details of the impact of the *Success Measures* and *Targets*.	• The *Incentives* are for agreement by *Alliance Members* and can measure and reward the success of the *Alliance* as a whole. • Clause 14.2 and Schedule 1 set out agreed consequences of meeting/failing to meet agreed *Targets*.

Contract Terms	Consultation comments on FAC-1	Amendments to the FAC-1 consultation draft

3. Alliance Manager and Employees

Role of Alliance Manager	• Will the *Alliance Manager* be a representative of the *Client* or can it be another body? • How is the *Alliance Manager* engaged and remunerated?	• The *Alliance Manager* can be an officer of the *Client* or an independent consultant. • The *Alliance Manager Services Schedule* is part of the *Framework Documents*. • The *Alliance Manager Payment Terms* are part of the *Framework Documents*.

4. Agreed Prices

Agreed Prices	• There is no provision for fluctuations.	• Clause 8.2 now provides that fluctuation provisions may be set out in the *Framework Prices*.
Framework Prices	• Should the *Framework Prices* set out agreed costs/rates?	• *Framework Prices* can include known costs/rates in respect of the *Framework Programme* dependant on the nature of the works/services/supplies.

5. Award of Project Contracts

Award of Project Contracts	• It would be useful to have further details of the *Direct Award Procedure* and the *Competitive Award Procedure*. • *Template Project Documents* – don't understand the requirement for Schedule 5.	• A guidance note has been added in Schedule 4. • A guidance note has been added in Schedule 5, as to *Template Project Documents* which create consistency in the structure of *Project Contracts*.
Pipeline of work	• Contractors and other suppliers will be concerned if there is no clear pipeline of work.	• Clauses 5.6 states any agreed minimum value or type of *Project Contracts* and clause 5.7 states any agreed exclusivity.

6. Alliance Activities, Supply Chain Collaboration and Timetable

Supply Chain Collaboration	• What is *Supply Chain Collaboration*? • How do any savings and other *Improved Value* achieved in accordance with this clause get fed into the prices for future *Project Contracts*? • *Supply Chain Collaboration* may be difficult to implement with *Alliance Members*, as they are also competitors in the market.	• *Supply Chain Collaboration* is now defined. • See also Cabinet Office guidance at https://www.gov.uk/government/publications/two-stage-open-book • Processes in the *Framework Brief,* and additional processes agreed by the *Core Group*, clarify how savings may be shared and/or may feed into *Agreed Prices* on future *Projects*.

Contract Terms	Consultation comments on FAC-1	Amendments to the FAC-1 consultation draft
		• Successful *Supply Chain Collaboration* with sharing of data among competitors is evidenced in SCMG Trial Project case study.
Timetable consents/approvals	• Is it worth adding 'consent/approval' to the headings of the first and third columns of Schedule 2?	• 'Consent/approval' has been added.

7. Orders and Pre-contract Activities

Contract Terms	Consultation comments on FAC-1	Amendments to the FAC-1 consultation draft
Payment to Alliance Members	How does the *Alliance* provide for payment in respect of an *Alliance Member's* work?	• *Project Contracts* will govern payment for *Projects*. • Clause 8.2.2 governs any agreed payments for *Alliance Activities* and *Pre-Contract Activities*, and the Appendix 3 form of *Order* states the agreed activities, the dates for their completion and details of the agreed amounts to be paid.

8. Payment

Contract Terms	Consultation comments on FAC-1	Amendments to the FAC-1 consultation draft
Payment to Alliance Manager	• How is the *Alliance Manager* remunerated?	• Clause 8.2.1 and the *Alliance Manager Payment Terms* provide for any agreed payments to the *Alliance Manager*.

9. Risk Management

Contract Terms	Consultation comments on FAC-1	Amendments to the FAC-1 consultation draft
Risk Management	• The *Risk Register* should be regularly updated at set intervals so that impact and likelihood are evaluated at key points because the likelihood and impact of a particular risk will vary over time.	• Clause 9.3 provides for further *Risk Management* actions to be agreed and clause 9.4 provides for the *Risk Register* to be updated.

10. Duty of Care

Contract Terms	Consultation comments on FAC-1	Amendments to the FAC-1 consultation draft
Skill and care	• Why is it necessary to establish contractual duties of care between *Alliance Members* and does this create the scope for litigation?	• Clauses 1.11.3, 10.2, 10.3, 10.4, 10.5 and 11.2 now create a limited duty of care between specific *Alliance Members* where a multi-party duty of care is not appropriate. • There is no evidence that a multi-party duty of care under alliance contracts has created scope for litigation.

11. Intellectual Property

Contract Terms	Consultation comments on FAC-1	Amendments to the FAC-1 consultation draft
Liability for use of documents	• Is there a conflict in clause 11.4 with the clause 10.1 duty of care?	• Clause 11.4 limits the extent of liability for use by reference to agreed purposes whereas clause 10.1 establishes the level of the duty of care.

Contract Terms	Consultation comments on FAC-1	Amendments to the FAC-1 consultation draft
12. Insurances		
Insured activities	• What activities will be undertaken that require insurance under FAC-1 rather than a *Project Contract*?	• Any agreed *Alliance Activities* and *Pre-Contract Activities*.
13. General		
Exclusion of partnership	• In some countries, partnering or alliance contracts may be regarded as a partnership, which will lead to joint and several liability to third parties outside of the *Framework Alliance Contract*.	• Clause 13.1 excludes any *Partnership*.
Confidentiality	• This is a relatively lean confidentiality provision and may benefit from some further detail. • Should there be limits/restrictions on what information the *Alliance Members* are to share amongst themselves?	• Clause 13.3 provides an option for more detailed confidentiality provisions to be set out in the *Framework Alliance Agreement*. • Clauses 13.3.1 and 13.3.2 state the confidentiality of certain documents, and clause 6.3.1 provides for agreement as to the basis for sharing information in relation to *Supply Chain Collaboration*.
Legal Requirements	• *Legal Requirements* should not only apply to England and Wales.	• Clause 13.4 and the Schedule 6 Part 1 guidance note refer to the laws of the country stated in the *Framework Alliance Agreement*.
Special Terms	• It is not clear what is envisaged by *Special Terms*.	• A guidance note has been added to Schedule 6 Part 2.
14. Duration and Termination		
Extension and termination	• Should there be a provision for the duration of the *Alliance* to be extended?	• Clause 14.1 provides an option to extend by agreement of the *Alliance Members* or as stated in the *Framework Alliance Agreement*
Termination by multiple Clients	• Does termination of the *Framework Alliance Contract* invoke termination of a *Project Contract* that has been entered into? • If there is more than one *Client*, should they all have to act together in order to terminate another *Alliance Member's* appointment?	• Clause 14.5 provides for all *Project Contracts* in existence at the time of termination to remain in force. • Clause 14.4.2 clarifies the termination procedure for multiple *Clients*.

Contract Terms	Consultation comments on FAC-1	Amendments to the FAC-1 consultation draft
Insolvency	• Where there is an *Insolvency Event*, the option to novate with a potential purchaser of an insolvent company could in some circumstances be a benefit to a client.	• Clause 14.3 provides an option for *Alliance Members* to agree that termination does not occur by reason of an *Insolvency Event*.

15. Problem-solving and Dispute Resolution

Notice and Core Group meeting	• A dispute could be commercially or legally sensitive, so why would it be shared with the *Core Group* which could include other commercial organisations? • Is a *Core Group* meeting a pre-condition to terminating for *Alliance Member* breach?	• A *Core Group* provides collaborative governance to avoid the escalation of a dispute and creates opportunities to agree a mutually acceptable solution. • FAC-1 dispute resolution is separate from the agreed approaches to dispute resolution under individual *Project Contracts*. • Yes, under clause 14.4.
Dispute Board	• Consider use of a *Dispute Board*.	• Use of a *Dispute Board* as an alternative to conciliation has been added as an option in clause 15.2.

Equivalent comments were raised by Consultation Group members in relation to the draft TAC-1 Term Alliance Contract, and equivalent amendments were incorporated in the published version.

Additional issues were raised in consultation at a conference organised by the Association of Consultant Architects in October 2015, and extracts are reproduced below:

Consultation on FAC-1 and BS11000 (ISO 44001)

In a consultation session led by Paul Greenwood (Accredited BS11000 Collaborative Business Relationships Facilitator):

• It was recognised that some Consultation Group submissions questioned whether FAC-1 goes far enough in supporting the behaviours required for successful collaborative working
• Paul Greenwood described the features of the draft FAC-1 that had been based upon proven enablers of collaborative working under the PPC2000/TPC2005 contract forms, such as the multi-party hub, the shared Timetables of key dates, the role of the Core Group and the Early Warning system, which between them have ensured only two PPC/TPC court cases in many years of extensive use
• While the strength of these contractual processes was recognised, delegates also suggested the need to train and support new users in collaborative behaviour, not least to ensure that they use FAC-1 in the right ways

- Delegates also commented on the importance of leadership by example among alliance members, and the benefits of collaborative behaviour being emphasised in the procurement process and in FAC-1 or related guidance.

Other Consultation Feedback

In a Consultation Session led by Antony Silk (Director of Procurement, Optivo):

- It was noted that Consultation Group submissions expressed strong support for FAC-1, including its potential for strategic collaborative working and its suitability for use with FIDIC, JCT, NEC, PPC or any other project contract form
- Among a total of over 80 delegates, nearly all confirmed that they use frameworks but were not satisfied with the way they operate
- Antony Silk suggested that this dissatisfaction may be caused by the limitations of 'static' frameworks where the starting point can become outdated, and also where the search for savings depends on mini-competitions that cut margins and demotivate participants
- Antony Silk saw FAC-1 as breaking this mould, enabling new insights, improved value and greater creativity through ongoing multi-party dialogue and development.

Consultation Group members requested guidance on how selection procedures and joint working should operate in practice. It was also suggested that there should be provision for agreed payment in respect of certain Alliance Activities.

Consultation responses suggested that FAC-1 should make clear that it adopts the collaborative processes outlined in the UK Government's recommended procurement models, as Trial Projects using these processes have already delivered significant agreed savings and other improved value.

The Consultation Group supported use of a standard form multi-party timetable in FAC-1, integrating the agreed activities of Alliance Members. This was recognised as a strong basis to embed agreed processes such as Supply Chain Collaboration.

The design of FAC-1 for use in any jurisdiction is of great interest. Consultation Group members recommended an option for arbitration because, in many countries, arbitration remains the preferred forum for dispute resolution.

Consultation Group members recommended ways to ensure that FAC-1 achieves its optimum potential in practice, including proposals in relation to roles and responsibilities, information exchange, pricing, risk management, project management, incentivisation, and measurement of value.

Appendix G

Completing FAC-1

Part 1 – Completing the FAC-1 Framework Alliance Agreement

The details inserted in the FAC-1 *Framework Alliance Agreement* and its schedules set out the specific requirements and expectations of the *Alliance Members* and the processes that they will implement in order to bring their *Alliance* to life. Words and expressions in italics are defined in FAC-1 Appendix 1.

Question	Page number in *Framework Alliance Agreement* and clause number of *Contract Terms*	Details to be inserted
What is the scope of works, services or supplies comprising the *Framework Programme* and the name of the *Client*?	Page (i)	Only a title is required here, and details should be set out in the *Framework Brief* and in the Schedule 1 *Objectives* and the Schedule 5 *Template Project Documents*.
		The name of the *Client* should appear here, with further details at signature page (vi).
What are the names of the other *Alliance Members*?	Page (i) Clause 1.1	In addition to the *Client*, any *Additional Clients* and the *Alliance Manager*, the *Alliance Members* may include one or more consultants, main contractors, sub-contractors, suppliers, manufacturers, operators or other providers in any combination.
		A list of names should appear here with further details at signature pages (vi) and (vii).

Question	Page number in *Framework Alliance Agreement* and clause number of *Contract Terms*	Details to be inserted

Are there any additional or amended *Framework Documents* other than those listed?	Page (i) Clauses 1.3, 1.4 and 1.5	Review the list of *Framework Documents* in the *Framework Alliance Agreement* and ensure that these are completed and signed for identification where necessary, including the *Framework Documents* described in Part 2 of this Appendix G and also the following schedules forming part of the *Framework Alliance Agreement*:
		
		• Schedule 1 – *Objectives, Success Measures, Targets,* and *Incentives,* as considered in Section 10.2.
		• Schedule 2 – *Timetable* which sets out the ways in which the *Alliance Members* agree to seek *Improved Value* through:
		
		  – agreed deadlines, gateways and milestones in respect of *Alliance Activities* and achievement of the *Objectives*
		  – timescales for *Supply Chain Collaboration* and other agreed *Alliance Activities,* including the nature, sequence and duration of the agreed actions of each *Alliance Member* and any consents or approvals (whether required from *Alliance Members* or third parties) that are pre-conditions to subsequent actions
		
		• Schedule 3 – *Risk Register,* as considered in Section 10.6
		• Schedule 4 – *Direct Award Procedure* and/or *Competitive Award Procedure,* as considered in Section 10.3
		• Schedule 5 – *Template Project Documents,* as considered in Section 10.3
		• Schedule 6 – *Legal Requirements* and *Special Terms* that give rise to additions and amendments to the *Contract Terms* as considered in Section 10.7.

What are the names of the *Core Group* members, including any alternates?	Page (ii) Clause 1.6	The *Core Group* is considered in Sections 12.7 and 19.7 as a proven governance system to support *Alliance* relationships, to encourage *Improved Value* and to resolve misunderstandings and avoid disputes. Insert the names of the *Core Group* members, including any alternates.
What are the agreed communications systems?	Page (ii) Clause 1.9.3	These may comprise for example electronic information exchange systems for all communications, for management of the *Alliance,* for the *Project Contract* award procedures and in connection with *BIM.*

Question	Page number in *Framework Alliance Agreement* and clause number of *Contract Terms*	Details to be inserted
What are the names of any *Stakeholders* and what is the nature of their intended involvement?	Page (ii) Clause 1.10	Engagement with *Stakeholders* is considered in Section 12.5. Insert the names of any *Stakeholders* and cross-refer to the *Framework Brief* for their interest in the *Framework Programme* and the ways in which *Alliance Members* need to engage with them. *Alliance Members* will not owe a direct duty of care to *Stakeholders* other than pursuant to clause 10.6 of the *Contract Terms*.
Are there any named *Additional Clients* and/or other *Additional Alliance Members* that it is agreed may join the *Alliance*?	Page (ii) Clause 1.11	Insert the names or descriptions of any *Additional Clients* and/or other *Additional Alliance Members* that it is agreed may join the *Alliance*. The form of *Joining Agreement* is set out in Appendix 2 to the *Contract Terms*.
What is the name of the *Alliance Manager* and are there any restrictions on his/her/its authority?	Page (ii) Clause 3.1	The role of the *Alliance Manager* is considered in Sections 10.6 and 12.4. Insert the name of the *Alliance Manager*, who may be a *Client* officer or an independent consultant. Insert any limits on the authority of the *Alliance Manager* by reference to clause 3.1 of the *Contract Terms*.
In what matters can the *Alliance Manager* act on behalf of the *Client*?	Page (ii) Clause 3.2	In addition to the matters described in clause 3.1, insert any authority delegated by the *Client* to the *Alliance Manager* to act on its behalf.
Is there an *Independent Adviser*?	Page (iii) Clause 3.3	The role of the *Independent Adviser* is considered in Sections 12.9 and 19.9, as an optional source of fair and constructive advice to *Alliance Members* on implementation of the *Framework Alliance Contract,* including implementation of agreed FAC-1 procedures, development of and adherence to the FAC-1 *Timetable* and *Risk Register*, development and implementation of *Supply Chain Collaboration* and other agreed *Alliance Activities* and avoidance or resolution of disputes. Insert details/delete as required.
Will *Framework Prices* state separately an *Alliance Member's* agreed levels of *Profit* and *Overheads*?	Page (iii) Clauses 4.3 and 4.4	This approach to costing is considered in Chapters 6 and 16, and will enable the separate agreement of cost reductions through *Alliance Activities* such as *Supply Chain Collaboration*. Delete if not applicable.

Question	Page number in *Framework Alliance Agreement* and clause number of *Contract Terms*	Details to be inserted
Will any minimum value or type of *Projects* will be awarded to *Alliance Members*?	Page (iii) Clause 5.6	The procedure for award of *Project Contracts* under Schedule 4 is considered in Section 10.3. By reference to this procedure under Schedule 4, insert any minimum value or type of *Projects* that will be awarded to any one or more *Alliance Members*.
Will all or any part of the *Framework Programme* be awarded exclusively to any one or more *Alliance Members*?	Page (iii) Clause 5.7	Insert details of any agreed exclusivity, with or without any minimum under clause 5.6. Insert also in Part 2 of Schedule 1 to the *Framework Alliance Agreement* any provision for adjustment of exclusivity according to achievement of agreed *Targets*.
What *Supply Chain Collaboration* and other *Alliance Activities* will be undertaken by *Alliance Members* in accordance with the *Timetable* or as otherwise agreed?	Page (iii) Clause 6	*Alliance Activities* define how joint working under a framework alliance seeks to achieve *Improved Value*. *Supply Chain Collaboration* is a strong example of effective *Alliance Activities* and is considered, for example, in Sections 2.9, 2.10, 3.6, 4.6, 5.6, 15.3 and 15.10 and in Chapters 6 and 16. Insert details of the agreed joint processes and shared working practices intended to achieve improved efficiency, cost savings and other *Improved Value*. These should be linked to dates and periods of time set out in the *Timetable*.
Are there amended payment periods in respect of amounts agreed to be paid for *Alliance Manager Services*, *Alliance Activities* and *Pre-Contract Services*?	Page (iii) Clause 8.6	Insert any amended payment periods in respect of amounts agreed to be paid for *Alliance Manager Services*, *Alliance Activities* and *Pre-Contract Services*. Public sector *Clients* are subject to the maximum 30-day payment periods under legislation such as Section 113 of the Public Contracts Regulations 2015.
Is there an agreed rate of interest on any late payment?	Page (iii) Clause 8.11	Insert the rate of interest on late payment. UK *Clients* are subject to minimum rates under the Late Payment of Commercial Debts (Interest) Act 1998 and the Late Payment of Commercial Debts Regulations 2013.

Question	Page number in *Framework Alliance Agreement* and clause number of *Contract Terms*	Details to be inserted
Are there any amendments to the duty of reasonable skill and care and to the parties between whom a duty of care is owed?	Page (iv) Clauses 10.1 and 10.2	The duty of reasonable skill and care is described in clauses 10.1 and 10.2 and is owed between limited *Alliance Members* in respect of certain documents and activities under clauses 1.11.3, 10.3, 10.4, 10.5 and 11.2. Insert any amendments in respect of the duty of reasonable skill and care and in respect of the parties between whom a duty of care is owed.
Is a duty of care extended to any other parties?	Page (iv) Clause 10.6	Insert details of any agreed third party warranties or other third party rights.
What types and levels of insurance cover are to be taken out by each *Alliance Member*?	Page (iv) Clause 12	Insert agreed types and levels of insurance cover in respect of obligations under FAC-1, which can include third party liability insurance and professional indemnity insurance in respect of *Alliance Manager Services*, *Alliance Activities* and *Pre-Contract Activities*. Any requirements as to insured risks, exclusions, deductibles, and other terms should be stated in the *Framework Brief*. These insurances are separate from insurances of *Projects* which will be taken out under *Project Contracts* and should be stated in the *Template Project Documents*.
Is any *Alliance Member* entitled to assign or sub-contract its rights or obligations under the *Framework Alliance Contract* to another party?	Page (iv) Clause 13.2	Insert any agreed entitlement of an *Alliance Member* to assign or sub-contract rights and/or obligations under FAC-1.
Are there amended confidentiality provisions?	Page (iv) Clause 13.3	Insert any extension or relaxation of the confidentiality provisions set out in clause 13.3, for example by reason of the identity of the *Client* or the nature of the *Projects* comprising the *Framework Programme*.
What are the commencement date and duration of the *Framework Alliance Contract*, and are there any extension procedures?	Page (v) Clause 14.1	Insert the date of commencement of the *Framework Alliance Contract*, its duration and any extension procedures.

Question	Page number in *Framework Alliance Agreement* and clause number of *Contract Terms*	Details to be inserted
Is there an agreed *Conciliator* and *Conciliation Procedure* or a *Dispute Board?*	Page (v) Clause 15.2	These optional non-adversarial dispute resolution approaches are considered in Section 19.9 and are additional to the role of the *Core Group* and any *Independent Adviser*. Insert details/edit/delete as required and see Part 1 of Appendix 4 to the *Contract Terms* regarding conciliation.
Is there an agreed *Adjudicator* and *Model Adjudication Procedure?*	Page (v) Clause 15.3	Adjudication is a statutory right in respect of the resolution of construction contract disputes in the UK and in certain other jurisdictions. Insert details/edit/delete as required and see Part 2 of Appendix 4 to the *Contract Terms*.
Are there agreed arrangements for appointment of arbitrators, applicable arbitration rules and the location and language of arbitration?	Page (v) Clause 15.4	Arbitration is an optional dispute resolution approach. Insert details/delete as required and see Part 3 of Appendix 4 to the *Contract Terms*.
What are the agreed law of the *Framework Alliance Contract* and the courts with jurisdiction?	Page (v) Clauses 13.4 and 15.5	The selected law will govern all aspects of the *Framework Alliance Contract* and may necessitate amendments to the *Contract Terms* as *Legal Requirements* under Schedule 6 the *Framework Alliance Agreement*. Insert applicable law.

Part 2 – Completing the FAC-1 Framework Documents

In addition to the *Framework Alliance Agreement* and the *Contract Terms*, the FAC-1 *Framework Documents* comprise:

Framework Brief

The *Framework Brief* is defined as describing 'the scope and nature of the *Alliance* and of the *Projects* comprising the *Framework Programme*.' It is signed separately by all *Alliance Members* for identification.

The *Framework Brief* sets out 'the *Client's* and any *Additional Clients'* technical, management and commercial requirements in relation to the *Framework Programme*, including the required approach to design, *Supply Chain* engagement, costing, *Risk Management* and programming and all other relevant procedures and expected outcomes and where appropriate the required approach to *BIM'*.

Any change to the *Framework Brief* is governed by clause 9.1 of the *Contract Terms*. It is important to distinguish between the information set out in the *Framework Brief* and the information set out in:

- the *Direct Award Procedure* and/or *Competitive Award Procedure* as regards the procurement model for individual *Projects*
- the *Template Project Documents* as regards the standard structure, components, and requirements for each *Project Contract*.

Framework Prices and Framework Proposals

The *Framework Prices* and *Framework Proposals* are agreed confidentially by each *Alliance Member* with the *Client*, any *Additional Clients* and the *Alliance Manager* (clause 13.3.2 of the *Contract Terms*). Each set of *Framework Prices* and *Framework Proposals* is signed separately by those parties for identification.

The *Framework Prices* are defined as describing 'the prices of an *Alliance Member* in response to the *Framework Brief*, agreed confidentially between that *Alliance Member*, the *Client*, any *Additional Clients* and the *Alliance Manager*'. Their content and structure will vary according to the selected procurement and delivery model and according to whether the *Framework Alliance Contract* governs the appointment of consultants, contractors, suppliers, manufacturers, operators or other providers, or a combination of different roles and disciplines.

There are options for *Framework Prices* to state the *Profit* and *Overheads* of an *Alliance Member*, and to state separately the net costs comprising *Framework Prices* and the *Agreed Prices* for each *Project* (clauses 4.3 and 4.4 of the *Contract Terms*). This is necessary in order to enable *Alliance Members* to identify savings arising from *Supply Chain Collaboration* and other *Alliance Activities*, and to ensure that agreed savings do not erode the margins of *Alliance Members*.

Profit and *Overheads* are set in accordance with the *Framework Prices* and/or in accordance with a *Direct Award Procedure* or *Competitive Award Procedure* (Appendix A definitions). *Profit* and *Overheads* can be percentages of *Agreed Prices* or agreed lump sums, but it is preferable to agree lump sums so that *Alliance Members* do not receive less *Profit* and *Overheads* if and when other costs are reduced.

The *Framework Brief* or the *Template Project Documents* state any provision for fluctuation in the *Framework Prices* (clause 4.2 of the *Contract Terms*).

The *Framework Proposals* are defined as comprising 'each of the documents describing the proposals of an *Alliance Member* in response to the *Framework Brief*, agreed confidentially between that *Alliance Member*, the *Client*, any *Additional Clients* and the *Alliance Manager*'. It is important to state which, if any, of the *Framework Proposals* are approved for use on any *Project* and which of them may be accepted at the option of the *Client* in relation to particular *Projects*.

Framework Proposals can include:

- drawings and specifications in respect of standard *Project* components and/or model *Projects*
- procedures for implementing *Projects*
- approaches to *Risk Management*
- proposals for achieving *Improved Value*.

Alliance Manager Services Schedule and Alliance Manager Payment Terms

The functions of the *Alliance Manager* are stated in FAC-1 clause 3 and, if the *Alliance Manager* is an officer of the *Client*, there is usually no need for an *Alliance Manager Services Schedule* or *Alliance Manager Payment Terms.*

If the *Alliance Manager* is an independent consultant, its agreed services are set out in the *Alliance Manager Services Schedule* so as to integrate its role and responsibilities as an *Alliance Member* and so as to avoid the need for a separate appointment. The *Alliance Manager Services Schedule* is signed by the *Alliance Members* for identification and should include:

- The individuals comprising the *Alliance Manager*'s team and any agreed constraints on their removal or replacement
- The resources and expertise committed by the *Alliance Manager*
- The *Alliance Manager's* agreed approach to its functions.

If the *Alliance Manager* is a consultant, its fees and expenses are set out in the *Alliance Manager Payment Terms,* so as to enable payment under FAC-1 (clauses 8.2–8.11 of the *Contract Terms*) and so as to avoid the need for a separate appointment. The *Alliance Manager Payment Terms* also state any fee fluctuation provisions (clause 8.2 of the *Contract Terms*). The *Alliance Manager Payment Terms* are confidential as between the *Client,* any *Additional Clients* and the *Alliance Manager* and are signed by them separately for identification (clause 13.3.1 of the *Contract Terms*).

Additional Framework Documents

FAC-1 allows for additional *Framework Documents,* although usually all *Framework Documents* will fall within one of the categories listed in the *Framework Alliance Agreement* at page (iii).

Special Terms can be used to explain the status of any additional *Framework Documents.*

Bibliography

Chapters 1–19

Part 1 Standard Forms

Autodesk Terms of Use (2018). *Autodesk General Terms of Use*, effective 18 May 2018 https://www.autodesk.com/company/terms-of-use/en/general-terms

BS EN ISO 19650 (2019). International standard for organization and digitization of information about buildings and civil engineering works, including building information modelling, comprising BS EN ISO 19650-1&2:2018 and PD19650-0:2019

BS 8354:2001 (2011). Construction procurement policies, strategies and procedures, Code of practice, BSI.

BSI 6079, British Standard 6079 (2002). *BS 6079–1 Project Management. Guide to Project Management*. British Standards Institution.

CIC (2007). *Construction Industry Council Consultants Contract Conditions: Conditions of Contract for the Appointment of Consultants on Major Building Projects: CIC/Conditions (first edition)*. London: RIBA.

CIC BIM Protocol (2013). *Construction Industry Council Building Information Modelling (BIM) Protocol First Edition Standard Protocol*. cic.org.uk/download.php?f=the-bim-protocol.pdf

CIC BIM Protocol (2018). *Construction Industry Council Building Information Modelling (BIM) Protocol Second Edition Standard Protocol*. http://cic.org.uk/admin/resources/bim-protocol2nd-edition-1.pdf

CIOB Time and Cost Management Contract Suite (2015). *CIOB Time and Cost Management Contract*. Bracknell: The Chartered Institute of Building.

ConsensusDocs 300 (2016). *Standard Multi-Party Integrated Project Delivery (IPD) Agreement*, 2007, revised 2016. Arlington, VA: ConsensusDocs.

ConsensusDocs BIM Addendum (2016). *ConsensusDocs 301 Building Information Modelling (BIM) Addendum* 2006, revised 2016. Arlington, VA: ConsensusDocs.

FAC-1 Framework Alliance Contract (2016). *FAC-1 Framework Alliance Contract*. Kent: The Association of Consultant Architects.

FIDIC (1999). *FIDIC Conditions of Contract for Construction First Edition 1999*. Geneva, Switzerland: International Federation of Consulting Engineers.

FIDIC (2017). *FIDIC Conditions of Contract for Construction Second Edition 2017*. Geneva, Switzerland: International Federation of Consulting Engineers.

Collaborative Construction Procurement and Improved Value, First Edition. David Mosey.
© 2019 John Wiley & Sons Ltd. Published 2019 by John Wiley & Sons Ltd.

FIDIC White Book (2017). *FIDIC Client/Consultant Model Services Agreement: White Book*, 5e. Geneva, Switzerland: International Federation of Consulting Engineers.

ICC (2014). *Infrastructure Conditions of Contract With Quantities Version November 2014*. London: Association for Consultancy and Engineering and Civil Engineering Contractors Association.

ICC Measurement Version (2011). *Infrastructure Conditions of Contract Measurement Version*. London: Association for Consultancy and Engineering and Civil Engineering Contractors Association.

ICC Partnering Addendum (2011). *Infrastructure Conditions of Contract Partnering Addendum August 2011*. London: Association for Consultancy and Engineering and Civil Engineering Contractors Association.

ICC Target Cost Version (2011). *Infrastructure Conditions of Contract Target Cost Version*. London: Association for Consultancy and Engineering and Civil Engineering Contractors Association.

ICC Term Version (2011). *Infrastructure Conditions of Contract Term Version*. London: Association for Consultancy and Engineering and Civil Engineering Contractors Association.

ISO 31000, ISO 31000:2018 *Risk Management*. 2018. International Organization for Standardization.

ISO 44001 (2017). ISO 44001:2017 *Collaborative business relationship management systems – Requirements and framework*. The British Standards Institution.

JCT CA (2016). *JCT 2016 Consultancy Agreement (Public Sector) (CA)*. London: Sweet & Maxwell.

JCT CE (2016). *JCT 2016 Constructing Excellence Contract (CE)*. London: Sweet & Maxwell.

JCT CE Project Team Agreement (2016). *JCT 2016 Constructing Excellence Contract, Project Team Agreement (CW/P)*. London: Sweet & Maxwell.

JCT Design and Build (2016). *JCT 2016 Design and Build Contract (DB)*. London: Sweet & Maxwell.

JCT Framework Agreement (2005). *JCT 2005 Framework Agreement (FA)*. London: Sweet & Maxwell.

JCT Framework Agreement (2007). *JCT 2007 Framework Agreement (FA)*. London: Sweet & Maxwell.

JCT Framework Agreement (2016). *JCT 2016 Framework Agreement (FA)*. London: Sweet & Maxwell.

JCT MPCC (2016). *JCT 2016 Major Project Construction Contract*. London: Sweet & Maxwell.

JCT MTC (2016). *JCT 2016 Measured Term Contract*. London: Sweet & Maxwell.

JCT PCSA (2016). *JCT 2016 Pre-Construction Services Agreement (General Contractor) (PCSA)*. London: Sweet & Maxwell.

JCT PCSA (SP) (2016). *JCT 2016 Pre-Construction Services Agreement (Specialist) (PCSA/SP)*. London: Sweet & Maxwell.

JCT SBC/AQ (2016). *JCT 2016 Standard Building Contract with Approximate Quantities*. London: Sweet & Maxwell.

JCT SBC/Q (2011). *JCT 2011 Standard Building Contract with Quantities*. London: Sweet & Maxwell.

JCT SBC/Q (2016). *JCT 2016 Standard Building Contract with Quantities*. London: Sweet & Maxwell.

JCT SBC/XQ (2016). *JCT 2016 Standard Building Contract Without Quantities*. London: Sweet & Maxwell.

JCT SBC/Sub/C (2016). *JCT 2016 Standard Building Sub-Contract Conditions (SBC/Sub/C)*. London: Sweet & Maxwell.

NEC3 (2013). *Engineering and Construction Contract*, 3e. London: Thomas Telford.

NEC4 Alliance Contract (2018). *NEC4 Alliance Contract*, 1e. London: Thomas Telford.

NEC4 (2017). *NEC4 Engineering and Construction Contract*, 4e. London: Thomas Telford.

NEC4 Framework Contract (2017). *NEC4 Framework Contract*, 4e. London: Thomas Telford.

NEC4 Professional Services Contract (2017). *NEC4 Professional Services Contract*, 4e. London: Thomas Telford.

NEC4 TSC (2017). *NEC4 Term Service Contract*, 4e. London: Thomas Telford.

NHS England (2015). *NHS England Standard Contract Template Alliance Agreement*, August 2015. https://www.england.nhs.uk/wp-content/uploads/2017/08/3b.-170802-Alliance-Agreement.pdf

Outline Scope of Services for Information Management (2013). *Construction Industry Council CIC/INF/MAN/S2013*. https://www.landscapeinstitute.org/PDF/Contribute/Outline-Scope-of-Services-for-the-Role-of-Information-Managment.pdf

PPC2000 (2013). *ACA Standard Form of Contract for Project Partnering 2000 (amended 2013)*. Kent: The Association of Consultant Architects.

Post Contract-Award BIM Execution Plan (2013). *Post Contract-Award Building Information Modelling (BIM) Execution Plan (BEP), 2013*. Construction Projects Information Committee. www.cpic.org.uk/wp-content/uploads/2013/06/cpix_post_contract_bim_execution_plan_bep_r1.0.pdf

Pre-Contract Award BIM Execution Plan (2013). *Pre-Contract Award Building Information Modelling (BIM) Execution Plan (BEP), 2013*. Construction Projects Information Committee. www.cpic.org.uk/wp-content/uploads/2013/06/cpix_pre-contract_bim_execution_plan_bep_v2.0.pdf

RIBA (2013). *Royal Institute of British Architects Plan of Work: Multi-Disciplinary Services*. London: RIBA.

SCL Delay and Disruption Protocol (2017). *Society of Construction Law Delay and Disruption Protocol*. Society of Construction Law. https://www.scl.org.uk/sites/default/files/SCL_Delay_Protocol_2nd_Edition.pdf

SoPC4 (2007). *Standardisation of PFI Contracts Version 4*. HM Treasury. https://ppp.worldbank.org/public-private-partnership/sites/ppp.worldbank.org/files/documents/UK_Standardisation%20of%20PFI%20Contracts%20(ver4.2007).pdf

SPC2000 (2008). *ACA Standard Form of Specialist Contract for Project Partnering, 2002 (amended 2008)*. Kent: The Association of Consultant Architects.

SPC2000 Short Form (2010). *Standard Form of Specialist Contract for Project Partnering*. Kent: The Association of Consultant Architects.

TAC-1 Term Alliance Contract (2016). *TAC-1 Term Alliance Contract*. Kent: The Association of Consultant Architects.

TPC2005 (2013). *ACA Standard Form of Contract for Term Partnering, 2005 (amended 2013)*. Kent: The Association of Consultant Architects.

Part 2 Articles, Books and Guidance

Arcadis Global Construction Disputes Report (2017). *Avoiding the same pitfalls*, Arcadis. https://www.arcadis.com/en/united-states/our-perspectives/global-construction-disputes-report-avoiding-the-same-pitfalls

Arcadis Global Construction Disputes Report (2018). *Does the industry learn from its mistakes?* Arcadis. https://www.arcadis.com/en/united-states/our-perspectives/global-construction-disputes-report-2018-does-the-construction-industry-learn-from-its-mistakes

Arrighetti, A., Bachmann, R. and Deakin, S. (1997). Contract law: social norms and inter-firm cooperation. *Cambridge Journal of Economy* 21: 171–195.

Association of Consultant Architects (2003). *Guide to the ACA Project Partnering Contracts PPC2000 and SPC2000*. Kent: The Association of Consultant Architects.

Bailey, J. (2016). *Construction Law*, 2e. Abingdon, Oxon: Informa Law.

Baker, E. (2007). Partnering strategies: the legal dimension. *Construction Law Journal* 23: 344–354.

Barber, J. and Jackson, S. (2010). Pre-construction services agreements – early lessons from experience. *Construction Law Journal*. 26 (3): 168–184.

Barlow, J., Cohen, M., Jashapara, A. and Simpson, Y. (1997). *Towards Positive Partnering: Revealing the Realities for the Construction Industry*. Bristol: The Policy Press.

Bayfield, R. and Roberts, P. (2004). Insights from beyond construction: collaboration – the Honda experience. *Society of Construction Law* Paper 119, September 2004.

BBVA (2015). Digital Economy Outlook *Smart Contract: The ultimate automation of trust?* https://www.bbvaresearch.com/wp content/uploads/2015/10/Digital_Economy_Outlook_Oct15_Cap1.pdf

Bell, M. (2018). How is that even possible? Raising construction regulation from the ashes of Grenfell Tower. *Society of Construction Law* Paper 208, March 2018.

Bennett, J. (2000). *Construction – The Third Way*. Oxford: Butterworths-Heinemann.

Bennett, J. and Jayes, S. (1998). *Seven Pillars of Partnering*. London: Thomas Telford.

Bennett, J. and Pearce, S. (2006). *Partnering in the Construction Industry: A Code of Practice for Strategic Collaborative Working*. Oxford: Butterworth-Heinemann.

BIM+, March 2018. www.bimplus.co.uk/people/uk-procurement-framework-informs-italys-move-manda

Bitcoin Magazine (2016). Smart Contracts Described by Nick Szabo 20 Years Ago Now Becoming Reality. https://bitcoinmagazine.com/articles/smart-contracts-described-by-nick-szabo-years-ago-now-becoming-reality-1461693751

Bresnen, M. and Marshall, N. (2002). Partnering in construction: a critical review of issues, problems and dilemmas. *Construction Management and Economics* 18 (2): 229–237.

Burke, R. (2002). *Project Management Planning and Control Techniques*, 3e 1999 (reprinted 2002). Chichester: Wiley.

Campbell, D. and Harris, D. (2005). Flexibility in long-term contractual relationships: the role of cooperation. *Lean Construction Journal* (www.leanconstructionjournal.org) 2: 5–29.

Chitty (2008). *Chitty on Contracts*, 13e. London: Thomson Reuters (Legal).

Cheung, Y., Rowlinson, S., Simons, R. and Rafferty, A. (2006). Alliancing in Australia—no-litigation contracts: a tautology? *Journal of Professional Issues in Engineering Education and Practice* 132 (1): 77–81.

Colledge, B. (2005). Relational contracting – creating value beyond the project. *Lean Construction Journal* (www.leanconstructionjournal.org) 2 (1): 30–45.

Collins, H. (2016). Is a relational contract a legal concept? In: *Contract in Commercial Law* (ed. S. Degeling, J. Edelman and J. Goudkamp). Sydney: Thompson Reuters.

Commons, J. (1934). *Institutional Economics: Its Place in Political Economy*. New York: MacMillan.

Constructing Excellence Hymn-Sheet (2015). http://constructingexcellence.org.uk/wp-content/uploads/2015/01/CW-Hymn-Sheet-FIN2.pdf

Cooper, W. (2013). Professional appointments and BIM. *Construction Law* 24 (7): 26.

Cox, A. and Thompson, I. (1998). *Contracting for Business Success*. London: Thomas Telford.

Cox, A. and Townsend, M. (1998). *Strategic Procurement in Construction: Towards Better Practice in the Management of Construction Supply Chains*. London: Thomas Telford.

Deutscher Baugerichtstag (2018). am 4/5. Mai 2018 in Hamm/Westf. Thesen der Arbeitskreise I bis XII, 67.

Duncan Wallace, I.N. (1996). *Construction Contracts: Principles and Policies in Tort and Contract*. Vol.1 (1986) and Vol. 2 (1996). London: Sweet & Maxwell.

Eggleston, B. (2006). *The NEC3 Engineering and Construction Contract: A Commentary*, 2e. Oxford: Blackwell Publishing.

Finnie, D., Ali, N.A. and Park, K. (2017). Enhancing off-site manufacturing through early contractor involvement (ECI) in New Zealand, *ICE Proceedings, Management, Procurement and Law*, Paper 1700029, 2.

Fischer, M., Ashcraft, H., Reed, D. and Khanzode, A. (2017). *Integrating Project Delivery*. Hoboken, NJ: Wiley.

Gerber, P. and Ong, B.J. (2013). *Best Practice in Construction Disputes*. LexisNexis Butterworths Australia.

Gibbs, D., Lord, W., Emmitt, S. and Ruikar, K. (2015). Building information modelling. *Construction Law Journal* 31 (3): 167–179.

Glover, J. and Hughes, S. (2018). *Understanding the FIDIC Red and Yellow Books*, 3e. London: Sweet & Maxwell.

Gould, N., King, C. and Britton, P. (2010). *Mediating Construction Disputes: An Evaluation of Existing Practice*. King's College London Centre of Construction Law. www.scl.org.uk/sites/default/files/KCL_Mediating_Construction_Parts%20I-III.pdf

Gruneberg, S. and Hughes, W. (2006). Understanding construction consortia: theory, practice and opinions, RICS research paper series Vol. 6, Number 1, March 2006, 18.

Hancock, D. (2010). *Tame, Messy and Wicked Risk Leadership*. Farnham: Gower.

Hawkins, D. (2017). *Raising the Standard for International Collaboration*. Ely: Melrose Books.

ICC Guidance Notes (2014). *Infrastructure Conditions of Contract With Quantities Version Guidance Notes*. London: Association for Consultancy and Engineering and Civil Engineering Contractors Association.

ICE Guidance Note (2018). *Procurement, Commercial and Contracting Key Principles*. https://ice.org.uk/getattachment/knowledge-and-resources/best-practice/procurement/ICE-Procurement-Guidance-Notes.pdf.aspx

ICE proceedings (2014/1). *BIM and construction contracts – CPC2013's approach*, Institution of Civil Engineers proceedings, Vol. 168, Issue MP6, 2014.

ICE proceedings (2014/2). *Building information modelling and the law*, Institution of Civil Engineers proceedings, Vol. 167 Issue MP3, 2014.

ICC Term Version Guidance Notes (2011). Association for Consultancy and Engineering and Civil Engineering Contractors Association.

Introducing the new NEC4 Alliance Contract (2018). NEC News and Media 2 July 2018. https://www.neccontract.com/About-NEC/News-Media/Introducing-the-new-NEC4-Alliance-Contract

ISO 44001 Implementation Guide (2017). https://www.bsigroup.com/LocalFiles/en-GB/iso-44001/Resources/ISO-44001-Implementation-Guide.pdf

Jackson, R. (2018). Does good faith have any role in construction contracts? *Society of Construction Law* Paper 207, January 2018.

Jackson, S. and Fuchs, B. (2015). Good faith: an Anglo-German comparison. *International Construction Law Review* 32 (4): 405.

JCT BIM Practice Note (2016). *JCT 2016 Building Information Modelling (BIM), Collaborative and Integrated Team Working*. London: Sweet & Maxwell.

JCT CE Guide (2016). *JCT 2016 Constructing Excellence Contract Guide*. London: Sweet & Maxwell.

JCT Practice Notice 4 (2001). *JCT Standard Forms of Building Contract 1998 editions: Series 2, Practice Note 4, Partnering*. London: RIBA Publications.

JCT Tendering (2017). *Practice Note*. London: Sweet & Maxwell.

Jones, D., Savage, D. and Westgate, R. (2003). *Partnering and Collaborative Working Law and Industry Practice*. London: Hammonds LLP.

Kaluarachchi, Y.D. and Jones, K. (2007). Monitoring of a strategic partnering process: the Amphion experience. *Construction Management and Economics* 25 (10): 1053–1061.

KCL Centre of Construction Law (2016). *Enabling BIM Through Procurement and Contracts*. King's College London Centre of Construction Law. www.kcl.ac.uk/law/research/centres/construction/assets/bim-research-report-1-jul-2016.pdf

Kier (2005). *A PPP in Housing/Building Repairs and Maintenance*, 4ps. Kier Group and Sheffield City Council, project information briefing.

King's College London (2015). *The Art of Partnering, Cultural Partnerships Report*. www.kcl.ac.uk/Cultural/culturalenquiries/partnership/index.aspx

Kohler, J. (1909). *The Philosophy of Law*. New York: Macmillan.

Kumaraswamy, M. (1997). Common categories and causes of construction claims. *Construction Law Journal* 13 (1): 21–34.

Lahdenpera, P. (2017). Towards a coherent theory of project alliancing: discovering the system's complex mechanisms yielding value for money. *Construction Economics and Building* 17 (2): 41–61.

Lane, T. (2011). The rise of the BIM consultant. *Building*, 4 November 2011. www.building.co.uk/the-rise-of-the-bim-consultant/5026614.article.

Latham, M. (2002). Just my opinion. *Building* 18: 33.

Latham, M. (2004). The cynic's bestiary. *Building* 4: 33.

Latham, M. (2006). Key Note Address to PPC2000 User Group National Conference, 5 October 2006.

Lavers, A. (2008). Ethics in construction law – a review of the position in eight European countries. *Society of Construction Law* Paper 144, April 2008.

Lewis, S. (2014). Playing games of risk. *Building*, February 2014.

Lock, D. (2000). *Project Management*, 7e. Aldershot: Gower.

London Underground (2014). Project report on the implementation of Innovative Contractor Engagement for the appointment of a Design and Build Contractor for the Bank Station capacity upgrade scheme, September 2014. www.secbe.org.uk/content/panels/Report%20-%20Innovative%20Contractor%20Engagement%20Procurement%20Model%20-%20Bank%20Station%20Capacity%20Upgrade-6d5f2a.pdf

Macneil, I.R. (1974). The many futures of contracts. *Southern California Law Review* 47: 691–816.

Macneil, I.R. (1978). Contracts: adjustment of long-term economic relations under classical, neoclassical and relational contract law. *Northwestern University Law Review* 72 (6): 854–905.

Macneil, I.R. (1981). Economic analysis of contractual relations – its shortfalls and the need for a 'rich classificatory apparatus'. *North Western University Law Review* 75 (6): 1018–1063.

Malhotra, D. and Lumineau, F. (2011). Trust and collaboration in the aftermath of conflict: the effects of contract structure. *Academy of Management Journal* 54 (5): 981–998.

Mason, J. (2016). Intelligent contracts and the construction industry. *Journal of Legal Affairs and Dispute Resolution in Engineering and Construction*. http://eprints.uwe.ac.uk/30522/3/intelligentcontractsfinal.pdf

Matthews, D. (2011). First ever BIM legal case in US. *Building*, 17 June 2011.

Milgrom, P. and Roberts, J. (1992). *Economics Organisation and Management*. Englewood Cliffs, NJ: Prentice-Hall.

Mosey, D. (2001). PPC 2000: The first standard form of contract for project partnering. *Society of Construction Law* Paper 95, September 2001.

Mosey, D. (2003/1). Contract or co-operation? Trends and changes within the UK construction industry – an overview. *Society of Construction Law* Paper D39, October 2003.

Mosey, D. (2003/2). *A Guide to the ACA Project Partnering Contracts PPC2000 and SPC2000*. Kent: The Association of Consultant Architects.

Mosey, D. (2009). *Early Contractor Involvement in Building Procurement*. Chichester: Wiley.

Mosey, D. (2011). The strengths of early contractor procurement. *Society of Construction Law* Paper 171, November 2011.

Mosey, D. (2014). BIM and related revolutions: A review of the Cookham Wood trial project. *Society of Construction Law* Paper D171, July 2014.

Mosey, D. (2015). Good faith in English construction law. *International Construction Law Review* 32 (4): 393.

Mosey, D. (2017). The origins and purposes of the FAC-1 framework alliance contract. *International Construction Law Review* 34 (4): 391–405.

Mosey, D., Howard, C. and Bahram, D. (2016). Enabling building information modelling through procurement and contracts. *Society of Construction Law* Paper D192, September 2016.

Mosey, D., Breyer, W., Leupertz, S. and Boldt, A. (2018). *PPC Deutschland: Einfuhrung in PPC und FAC-1*. https://shop.bundesanzeiger-verlag.de/fileadmin/bundesanzeiger-shop/content/produkte/linked_objects/MEDDA/PPC_FAC.pdf

NBS BIM Toolkit. https://toolkit.thenbs.com

NEC Framework Guide (2017). *NEC4 Framework Contract, 2017 Guidance Notes and Flow Charts*. London: Thomas Telford.

NEC4 *preparing an alliance contract* (2018), volume 2. London: Thomas Telford.

NEC4 *managing an alliance contract* (2018), volume 4. London: Thomas Telford.

NEC4 Procurement and Contract Strategies (2017). *NEC4 Procurement and Contract Strategies*. London: Thomas Telford.

O'Reilly, M. (1995). Risk, construction contracts and construction disputes. *Construction Law Journal* 11 (5): 343–354.

Pearson, A. (2003). Dances with penguins. *Building,* 2 May 2003, 56–59.

Pinsent Masons (2014). BIM Survey. http://www.pinsentmasons.com/en/media/press-releases/2014/governments-bim-target-unachievable-says-pinsent-masons-survey

Pinsent Masons (2016). *Collaborative Construction: More myth than reality?* https://www.pinsentmasons.com/PDF/2016/Collaborative-Construction.pdf

PPC Guide (2003). *Guide to the ACA Project Partnering Contracts PPC2000 and SPC2000.* Kent: The Association of Consultant Architects.

PPC Pricing Guide (2008). *Introduction to Pricing under PPC2000 for use with ACA Project Partnering Contracts PPC2000 and PPC International.* Kent: The Association of Consultant Architects.

P13 Blueprint (2018). *P13 Blueprint,* Institution of Civil Engineers. www.p13.org.uk/wp-content/uploads/2018/06/P13-Blueprint-Web.pdf

Rawlinson, S. (2008). Procurement: single stage tendering. *Building* 46: 68–69.

Rhys-Jones, S. (1994). How constructive is construction law? *Construction Law Journal* 10 (1): 28–38.

RIBA (2012). *BIM Overlay to the RIBA Plan of Work,* May 2012. London: RIBA.

RICS (2015). *International BIM implementation Guide,* 1e. Royal Institution of Chartered Surveyors. http://www.rics.org/uk/knowledge/professional-guidance/guidance-notes/international-bim-implementation-guide-1st-edition

Roadmap 2030. *Financing and implementing Global Goals in Human Settlements and City-Regions.* https://ecosequestrust.org/roadmap2030

Saxon, R. (2013). *Growth through BIM.* Construction Industry Council.

SCALA (2000). SCALA Practice Forum October 2000. www.scala.org.uk

Selznick, P. (1969). *Law, Society and Industrial Justice.* New York: Russell Sage Foundation.

Shepherd, D. (2015). *BIM Management Handbook.* NBS, 19 and 21.

Sinclair, D. (2011). *Leading the Team: An Architect's Guide to Design Management.* RIBA Publishing.

Skeggs, C. (2003). Project partnering in the international construction industry. *International Construction Law Review* 20 (4): 456–482.

Smith, A. (2015). *BIM Toolkit and Verification.* https://toolkit.thenbs.com/articles/verification

Smith, N.J. (2002). *Engineering Project Management,* 2e. Oxford: Blackwell.

Smith, N.J. (2003). *Appraisal, Risk and Uncertainty.* London: Thomas Telford.

Smith, N.J., Merna, T. and Jobling, P. (2006). *Managing Risk in Construction Projects,* 2e. Oxford: Blackwell.

Smith, R.J. (1995). Risk identification and allocation: saving money by improving contracts and contracting practice. *International Construction Law Review* 12 (1): 40–71.

Southern Construction Framework Report (2016). www.southernconstructionframework.org.uk/pages/documents-scf.html

Supply Management (2016). Framework forces competitor collaboration. *Supply Management,* 10 August 2016.

Thomas, D. (2012). *Keating on NEC3.* London: Sweet & Maxwell.

TPC2005 Introduction to Pricing (2012). Kent: The Association of Consultant Architects.

Tvarno, C. (2015). To bind or not to bind: it's in the contract. *Journal of Strategic Contracting and Negotiation* 1 (4): 288–314.

Treitel (2011). *Treitel Law of Contract*, 13e. London: Sweet & Maxwell.

Tyerman, D. (2013). Building information modelling and change management – a single version of the truth. *Construction Law Journal* 29 (4): 295–307.

Uff, J. (2018). Is the construction industry waving or drowning? *Construction Law Journal* 34 (3): 176.

University of Reading (2018). *Delivering More for Less Under the IPI Model*, Final Research Report, February 2018. file:///C:/Users/K1217231/AppData/Local/Microsoft/Windows/INetCache/Content.Outlook/WX3IPVBD/IPI%20UoR%20Final%20Report%20(February%202018).pdf

Water Design-Build Council (2016). *Water and Wastewater Design-Build Handbook*. file:///C:/Users/K1217231/AppData/Local/Microsoft/Windows/INetCache/Content.Outlook/WX3IPVBD/WDBC-Handbook-4th_Edition_2016_Web.pdf

Williamson, O.E. (1979). Transaction-cost economics: the governance of contractual relations. *Journal of Law & Economics* 22 (2): 233.

Williamson, O.E. (1985). *The Economic Institutions of Capitalism*. New York: Collier Macmillan.

Williamson, O.E. (1993). The evolving science of organisations. *Journal of Institutional and Theoretical Economics* 149 (1): 36–63.

Part 3 Government and Industry Reports

AE (1999). *Constructing the Best Government Client*. London: Achieving Excellence, HM Treasury.

All-Party Group for Excellence in the Built Environment (2012). *A Better Deal For Public Building*, http://cic.org.uk/admin/resources/appg-for-ebe-report-1.pdf

Arup (2008). *Partnering Contract Review for Office of Government Commerce*. Arup, London, 25 September 2008. http://ppc2000.co.uk/wp-content/uploads/2016/12/partnering_contract_review.pdf

Association of Consultant Architects (2010). *10 years of Partnering Contracts*. http://ppc2000.wiserhosting.com/wp-content/uploads/2016/12/10-Year-Anniversary-PPC-and-5-Year-TPC.pdf

Banwell Report (1964). *The Placing and Management of Contracts for Building and Civil Engineering Work*. London: HMSO.

BEIS (2018). *Construction Sector Deal*, Department for Business Energy and Industrial Strategy. https://www.gov.uk/government/publications/construction-sector-deal/construction-sector-deal#contents

Be PFI (2003). *Improving PFI through Collaborative Working*. Reading: Be – Collaborating for the Built Environment.

BSI (2015). *City data survey report*. https://www.bsigroup.com/en-GB/smart-cities/Smart-Cities-Standards-and-Publication

Building a Safer Future (2018). *Building a Safer Future – Independent Review of Building Regulations and Fire Safety*, Hackitt, J. https://assets.publishing.service.gov.uk/

government/uploads/system/uploads/attachment_data/file/707785/Building_a_Safer_ Future_-_web.pdf

Built Environment (2014). *A Report on Our Digital Future.* cic.org.uk/download.php? f=be2050-cic-bim2050-2014-1.pdf

CE (2011). *Collaborative working: the principles.* Constructing Excellence. http:// constructingexcellence.org.uk/wp-content/uploads/2015/01/CW-Hymn-Sheet-FIN2 .pdf

CECA (2016). Civil Engineering Contractors Association Procurement Report, *Directions in policy for the UK's infrastructure sector* 2016, 19. www.ceca.co.uk/media/246624/ ceca-procurement-report-2016-master-for-upload.pdf

CIC (2002). *Guide to Project Team Partnering.* London: Construction Industry Council.

CIC (2005). *A Guide to Partnering Workshops.* London: Construction Industry Council.

CIPFA (2013). *Voluntary Code of Conduct for Operational PFI/PPP schemes.* file:///C:/Users/K1217231/Downloads/Private_Finance_Initiative%20(4).pdf

CIRIA (1998). *Selecting Contractors by Value.* London: Construction Industry Research and Information Association.

Construction Industry Training Board *Client-Based Approach.* www.citb.co.uk/national-skills-academy-for-construction/what-is-the-national-skills-academy-for-construction/ client-based-approach

Cost Led Procurement Guidance (2014). *Cost led procurement guidance,* Cabinet Office/Efficiency and Reform Group https://www.gov.uk/government/uploads/system/ uploads/attachment_data/file/325012/Cost_Led_Procurement_Guidance.pdf

CRINE (1994). *Cost Reduction Initiative for the New Era (CRINE) Report,* February 1994. https://www.logic-oil.com/content/crine-report

Department for Culture, Media & Sport (2017). *Policy paper: The wider economy – helping every British business become a digital business.* https://www.gov.uk/government/ publications/uk-digital-strategy/4-the-wider-economy-helping-every-british-business-become-a-digital-business

Department of Treasury and Finance, Victoria (2009). *In Pursuit of Additional Value, A benchmarking study into alliancing in the Australian Public Sector.* Department of Treasury and Finance, Victoria, October 2009, 153. www.pc.gov.au/__data/assets/pdf_ file/0007/131794/sub078-infrastructure-attachment12.pdf

Digital Built Britain (2015). *Level 3 Building Information Modelling – Strategic Plan.* https://assets.publishing.service.gov.uk/government/uploads/system/uploads/ attachment_data/file/410096/bis-15-155-digital-built-britain-level-3-strategy.pdf

Effectiveness of Frameworks (2012). A Report by the Working Group on the Effectiveness of Frameworks of the Procurement and Lean Client Task Group, March 2012. https:// www.gov.uk/government/uploads/system/uploads/attachment_data/file/61157/ Procurement-and-Lean-Client-Group-Final-Report-v2.pdf

Efficiency Unit (1995). *Construction Procurement by Government – An Efficiency Unit Scrutiny.* London: Efficiency Unit, Cabinet Office, HMSO.

Egan, J. (1998). *Rethinking Construction. Report of the Construction Task Force.* London: Department of the Environment, Transport and Regions.

Emmerson, H. (1962). *Survey of Problems before the Construction Industries.* London: Ministry of Works, HMSO.

Farmer, M. (2016). *Modernise or Die, Time to Decide the Industry's Future.* Construction Leadership Council (CLC).

Government Construction Client Group (2011). *BIM Working Party Strategy Paper*, March 2011. www.bimtaskgroup.org

Government Construction Strategy (2011). https://assets.publishing.service.gov.uk/government/uploads/system/uploads/attachment_data/file/61152/Government-Construction-Strategy_0.pdf

Government Construction Strategy (2016–2020). https://assets.publishing.service.gov.uk/government/uploads/system/uploads/attachment_data/file/510354/Government_Construction_Strategy_2016-20.pdf

Government Construction Trial Projects (2012). https://assets.publishing.service.gov.uk/government/uploads/system/uploads/attachment_data/file/62628/Trial-Projects-July-2012.pdf

Government Office for Science (2014). *The internet of things: making the most of the second industrial revolution.* https://www.gov.uk/government/uploads/system/uploads/attachment_data/file/409774/14-1230-internet-of-things-review.pdf

Government Office for Science (2016). *Distributed Ledger Technology: beyond block chain,* report by the UK Government Chief Scientific Advisor. https://www.gov.uk/government/news/distributed-ledger-technology-beyond-block-chain

GSL (2014). *Government Soft Landings.* Cabinet Office 2014. https://www.cdbb.cam.ac.uk/Resources/Bimtaskgroupmaterial/GovernmentSoftLandingsSection1Introduction.pdf

Highways Agency (2005): *Highways Agency Procurement Strategy Review 2005*: www.highways.gov.uk

HMEP Supply Chain Collaboration Toolkit (2014). www.highwaysefficiency.org.uk/efficiency-resources/procurement-contracting-and-standardisation/supply-chain-collaboration-toolkit.html

Housing Forum (2010). *Turning the Corner.* London: The Housing Forum.

Housing Forum (2018). *Stopping Building Failures, how a Collaborative Approach Can Improve Quality and Workmanship.* London: The Housing Forum. www.housingforum.org.uk/publications/housing-forum-reports

ICE/DETR (2001). *Managing Geotechnical Risk: Improving Productivity in UK Building and Construction.* Prepared under the DETR Partners in Technology Programme for the Institution of Civil Engineers. London: Thomas Telford.

Infrastructure Client Group (2014). *Alliancing Best Practice in Infrastructure Delivery.* https://assets.publishing.service.gov.uk/government/uploads/system/uploads/attachment_data/file/359853/Alliancing_Best_Practice.pdf

Infrastructure Client Group (2015). *Improving Infrastructure Delivery: Alliancing Code of Practice.* https://assets.publishing.service.gov.uk/government/uploads/system/uploads/attachment_data/file/487294/alliancing_code_of_practice_18122015.pdf

Integrated Project Insurance Guidance (2014). *Integrated Project Insurance guidance.* Cabinet Office/Efficiency and Reform Group. https://www.gov.uk/government/uploads/system/uploads/attachment_data/file/326716/20140702_IPI_Guidance_3_July_2014.pdf

IPA (2014). *Improving Infrastructure Delivery: Project Initiation Routemap Handbook.* Infrastructure and Projects Authority and HM Treasury, 2014. https://www.gov.uk/government/publications/improving-infrastructure-delivery-project-initiation-routemap

Joint Venture guidance (2010). *Joint Ventures: a guidance note for public sector bodies forming joint ventures with the private sector.* HM Treasury, March 2010. https://assets

.publishing.service.gov.uk/government/uploads/system/uploads/attachment_data/file/225321/06_joint_venture_guidance.pdf

Latham, M. (1993). *Trust and Money – Interim Report of the Joint Government/Industry Review of Procurement and Contractual Arrangements in the United Kingdom Construction Industry*. London: HMSO.

Latham, M. (1994). *Constructing the Team – Final Report of the Government/Industry Review of Procurement and Contractual Arrangements in the United Kingdom Construction Industry*. Constructing Excellence. http://constructingexcellence.org.uk/wp-content/uploads/2014/10/Constructing-the-team-The-Latham-Report.pdf

LGA/NACF (2016). *National Construction Category Strategy for Local Government Effective Construction Frameworks*. www.nacframework.org.uk/docs/LGA%20Effective%20Construction%20Frameworks%20January%202016.pdf

McKinsey Global Institute (2017). *Reinventing Construction Through a Productivity Revolution*. https://www.mckinsey.com/industries/capital-projects-and-infrastructure/our-insights/reinventing-construction-through-a-productivity-revolution

More homes, fewer complaints (2016). Report by the All Party Parliamentary Group for Excellence in the Built Environment, Report from the Commission of Inquiry into the quality and workmanship of new housing in England. file:///C:/Users/k1182652/Downloads/more-homes.-fewer-complaints.pdf

NAO (2001). *Modernising Construction*, Report by Comptroller and Auditor General, The Efficiency Office. London: The Stationery Office.

NAO (2005). *Improving Public Services through Better Construction*. London: The Stationery Office.

NAO (2008). *The Roll-Out of the Jobcentre Plus Office Network*. London: The Stationery Office.

NAO (2018). *Report by the Comptroller and Auditor General HM Treasury PFI and PF2*. www.nao.org.uk/wp-content/uploads/2018/01/PFI-and-PF2.pdf

NBS (2012). *National Construction Contracts and Law Survey 2012*. https://www.thenbs.com/knowledge/nbs-national-construction-contracts-and-law-survey-2012

NBS (2015). *National Construction Contracts and Law Survey 2015*. https://www.thenbs.com/knowledge/national-construction-contracts-and-law-survey-2015

NBS (2018). *National Construction Contracts and Law Report 2018*. https://www.thenbs.com/knowledge/national-construction-contracts-and-law-report-2018

NBS BIM Report (2015). https://www.thenbs.com/knowledge/nbs-national-bim-report-2015

NEDC (1991). *Partnering: Contracting without Conflict*, National Economic Development Council Construction Industry Sector Group. London: HMSO.

NEDO (1975). *The Public Client and the Construction Industry*, National Economic Development Office. London: HMSO.

Never Waste a Good Crisis, (2009). *Never Waste a Good Crisis*, Constructing Excellence

Nichols (2007). *Review of Highways Agency's Major Roads Programme*. London: Nichols Group.

OGC (2007). *Achieving Excellence in Construction*. Construction Projects Pocketbook; *Achieving Excellence in Construction Procurement Guides*, Nos 1–11. London: Office of Government Commerce.

PBA Briefing (2012). *Project Bank Accounts – Briefing document*, Cabinet Office Government Construction 10 February 2012. https://assets.publishing.service.gov.uk/

government/uploads/system/uploads/attachment_data/file/62117/Project-Bank-Accounts-briefing.pdf

PBA Guidance (2012). *A Guide to the implementation of Project Bank Accounts (PBAs) in construction for government clients,* Cabinet Office, Government Construction, 03 July 2012. https://assets.publishing.service.gov.uk/government/uploads/system/uploads/attachment_data/file/62118/A-guide-to-Project-Bank-Accounts-in-construction-for-government-clients-July-2012.pdf

PF2 (2012). *Standardisation of PF2 Contracts,* HM Treasury, December 2012. https://assets.publishing.service.gov.uk/government/uploads/system/uploads/attachment_data/file/207383/infrastructure_standardisation_of_contracts_051212.PDF

Procurement/Lean Client Task Group (2012). *Government Construction Strategy Final Report to Government by the Procurement/Lean Client Task Group,* July 2012. https://www.gov.uk/government/uploads/system/uploads/attachment_data/file/61157/Procurement-and-Lean-Client-Group-Final-Report-v2.pdf

Procuring for Value (2018). *Procuring for Value,* Construction Leadership Council, July 2018. www.constructionleadershipcouncil.co.uk/news/procuring-for-value

RIBA Ten Principles (2017). *RIBA Ten Principles for Procuring Better Outcomes,* August 2017. https://www.architecture.com/knowledge-and-resources/resources-landing-page/ten-principles-for-procuring-better-outcomes

Selecting the Team (2005). Construction Industry Council, 2005.

Two Stage Open Book and Supply Chain Collaboration Guidance (2014). *Project Procurement and Delivery Guidance Using Two Stage Open Book and Supply Chain Collaboration,* Cabinet Office/Efficiency and Reform Group and King's College London. https://www.gov.uk/government/uploads/system/uploads/attachment_data/file/325014/Two_Stage_Open_Book_Guidance.pdf

Part 4 Table of Cases

Allied Maples Group Ltd v Simmons & Simmons [1995] EWCA Civ 17, [1995] 1 WLR 1602.

Almacantar (Centre Point) Ltd v Sir Robert McAlpine Ltd [2018] EWHC 232 (TCC).

Al Nehayan v Kent [2018] EWHC 333 (Comm).

Alstom Signalling Ltd v Jarvis Facilities Ltd [2004] EWHC 1232 (TCC).

AMEC Group Ltd v Secretary of State for Defence [2013] EWHC 110.

Amey Birmingham Highways Ltd v Birmingham City Council [2018] EWCA Civ 264.

Bacal Construction (Midlands) Ltd v Northampton Development Corporation [1975] 8 BLR 88 (CA).

Baird Textiles Holdings Ltd v Marks & Spencer Plc [2001] EWCA Civ 274.

Bedfordshire County Council v Fitzpatrick Contractors Ltd [1998] 62 Con LR 64 (TCC).

Berkeley Community Villages Ltd v Pullen [2007] EWHC 1330 (Ch).

Birse Construction Ltd v St David Ltd (No.1) [2000] BLR 57 (CA).

Blackpool & Fylde Aero Club v Blackpool Borough Council [1990] EWCA Civ 13.

Bluewater Energy Services BV v Mercon Steel Structures BV [2014] EWHC 2132 (TCC).

Bolam v Friern Hospital Management Committee [1957] 1 WLR 582 (QB).

Bristol Groundschool Ltd v Intelligent Data Capture Ltd [2014] EWHC 2145 (Ch).

Cable and Wireless Plc v IBM UK Ltd [2002] EWHC 2059 (Comm).

Channel Tunnel Group Ltd v Balfour Beatty Construction Ltd [1993] AC 334.

Compass Group UK and Ireland Ltd v Mid-Essex Hospital NHS Trust [2012] EWHC 781 (QB).

Computer Associates UK Ltd v The Software Incubator Ltd [2018] EWCA Civ 518.

Consarc Design Ltd v Hutch Investments Ltd [2002] All ER (D) 236 (TCC).

Cooperative Insurance Society Ltd v Henry Boot Scotland Ltd [2002] EWHC 1270 (TCC).

Cooperative Group Ltd v John Allen Associates Ltd [2010] EWHC 2300 (TCC).

Costain Ltd v Bechtel Ltd [2005] EWHC 1018 (TCC).

Costain Ltd v Tarmac Holdings Ltd [2017] EWHC 319 (TCC).

Courtney & Fairbairn Ltd v Tolaini Bros (Hotels) Ltd [1975] 1 WLR 297 (CA).

CPC Group Ltd v Qatari Diar Real Estate Investment Company [2010] EWHC 1535 (Ch).

Davis Contractors Ltd v Fareham Urban District Council [1956] AC 696 (HL).

De Beers UK Ltd v Atos Origin IT Services Ltd [2010] EWHC 3276 (TCC).

Director General of Fair Trading v First National Bank Plc [2001] UKHL 5.

Donnelly v Weybridge Construction Ltd [2006] EWHC 2678 (TCC).

Duncanson v Scottish County Investment Company Ltd [1915] SC 1106.

Eckersley v Binnie & Partners [1988] 18 Con LR 1 (CA).

Equitable Debenture Assets Corporation Ltd v William Moss Group Ltd [1984] 2 Con LR 1 (QB).

Esso Australia Pty Ltd v Southern Pacific Petroleum NL [2005] VSCA 228.

First Energy (UK) Ltd v Hungarian International Bank Ltd [1993] 2 Lloyd's Rep 194 (CA).

Foley v Classique Coaches Ltd [1934] 2 KB 1 (CA).

Francois Abballe v Alstom UK Ltd (2000) LTL 7.8.00 (TCC).

Frank W Clifford Ltd v Garth [1956] 1 WLR 570 (CA).

Fujitsu Services Ltd v IBM United Kingdom Ltd [2014] EWHC 752 (TCC).

George Hawkins v Chrysler (UK) Ltd and Burne Associates (1986) 38 BLR 36 (CA).

GLC v Cleveland Bridge and Engineering Co Ltd [1984] 34 BLR 50 (CA).

Gold Group Properties Ltd v BDW Trading Ltd [2010] EWHC 1632 (TCC).

Greaves & Co (Contractors) Ltd v Baynham Meikle & Partners [1975] 1 WLR 1095 (CA).

H Fairweather & Co Ltd v London Borough of Wandsworth [1987] 39 BLR 106 (QB).

Haden Young Ltd v Laing O'Rourke Midlands Ltd [2008] EWHC 1016 (TCC).

Harmon CFEM Facades (UK) Ltd v House of Commons [1999] EWHC 199 (TCC), [1999] 67 Con LR 1.

Havant Borough Council v South Coast Shipping Co Ltd (No.2) [1997] 14 Const LJ 430 (QB).

Henry Bros Ltd and Scott and Ewing v Department of Education for Northern Ireland [2008] NIQB 105.

Hounslow London Borough Council v Twickenham Gardens Development Ltd [1971] Ch 233.

ING Bank NV v Ros Roca SA [2011] EWCA Civ 353.

Interfoto Picture Library Ltd v Stiletto Visual Programmes Ltd [1987] EWCA Civ 6.

JM Hill & Sons Ltd v London Borough of Camden [1980] 18 BLR 31 (CA).

J Murphy & Sons Ltd v Johnston Precast Ltd [2008] EWHC 3104 (TCC).

Leander Construction Ltd v Mulalley & Co Ltd [2011] EWHC 3449 (TCC).

London Borough of Merton v Stanley Hugh Leach Ltd [1985] 32 BLR 51 (Ch).

M.A. Mortenson Company Inc. v Timberline Software Corporation (1999) 93 Wash. App 819.

Mackay v Dick (1881) 6 App Cas 651 (HL).

McLaughlin and Harvey v Department of Finance and Personnel [2008] NIQB 91.

Mediterranean Salvage & Towage Ltd v Seamar Trading & Commerce Inc (The Reborn) [2009] EWCA Civ 531.

Mediterranean Shipping SA v Cottonex Anstalt [2016] EWCA Civ 789.

Miller v Cannon Hill Estates Ltd [1931] 2 KB 113.

Mirant Asia-Pacific Construction (Hong Kong) Ltd v Ove Arup & Partners International Ltd [2004] EWHC 1750 (TCC).

Mowlem Plc v Stena Line Ports Ltd [2004] EWHC 2206 (TCC).

Multiplex Constructions (UK) Ltd v Cleveland Bridge UK Ltd [2006] EWHC 1341 (TCC).

Munro v Bognor UDC [1913] 3 KB 167.

Northern Ireland Housing Executive v Healthy Buildings (Ireland) Ltd [2017] NIQB 43.

Obrascon Huarte Lain SA v HM Attorney General of Gibraltar [2015] EWCA Civ 712.

OLG Brandenburg, (2011) judgment dated 16.03.2011–13 U 5/10.

Pacific Associates v Baxter [1990] 1 QB 993 (CA).

Pagnan v Feed Products [1987] 2 Lloyd's Rep 601 (CA).

Peterborough City Council v Enterprise Managed Services Ltd [2014] EWHC 3193 (TCC).

Pigott Foundations Ltd v Shepherd Construction Ltd [1993] 67 BLR 48 (QB).

Plymouth and South West Co-operative Society Ltd v Architecture Structure and Management Ltd [2006] EWHC 5 (TCC).

Portsmouth City Council v Ensign Highways Ltd [2015] EWHC 1969 (TCC).

Queensland Electricity v New Hope Collieries [1984] UKPC 39, [1989] 1 Lloyd's Rep 205.

Rainy Sky SA v Kookmin Bank [2011] UKSC 50.

Riva Properties Ltd v Foster + Partners Ltd [2017] EWHC 2574 (TCC).

Ross River Ltd v Waveley Commercial Ltd [2013] EWCA Civ 910.

RTS Flexible Systems Ltd v Molkerei Alois Muller GmbH & Co KG [2010] UKSC 14, [2010] BLR 337.

Samuel Payne v John Setchell Ltd [2002] BLR 489 (TCC).

Scottish Power Plc v Kvaerner Construction (Regions) Ltd [1999] SLT 721.

Secretary of State for Defence v Turner Estate Solutions Ltd [2015] EWHC 1150 (TCC).

Shepherd Homes Ltd v Encia Remediation Ltd [2007] EWHC 70 (TCC).

Simplex Concrete Piles v Borough of St Pancras [1958] 14 BLR 80.

Socimer International Bank Ltd v Standard Bank London Ltd [2008] EWCA Civ 116, 1 Lloyd's Rep 558.

Software Solutions Partners v HM Customs and Excise [2007] EWHC 971.

Stephenson Blake (Holdings) Ltd v Streets Heaver Ltd [2001] Lloyd's Rep PN44 (QB).

Sudbrooke Trading Ltd v Eggleton [1983] 1 AC 444 (HL).

Supamarl Ltd v Federated Homes Ltd [1981] 9 Con LR 25.

Sutcliffe v Thackrah [1974] AC 727 (HL).

TA Bickerton & Son Ltd v North West Metropolitan Regional Hospital Board [1970] 1 WLR 607 (HL).

Tesco Stores Ltd v Costain Construction Ltd [2003] EWHC 1487 (TCC).

Tesco Stores Ltd v The Norman Hitchcox Partnership Ltd [1997] 56 Con LR 42 (QB).

Tramtrack Croydon Ltd v London Bus Services Ltd [2007] EWHC 107 (Comm).

Trant Engineering Ltd v Mott MacDonald Ltd [2017] EWHC 2061 (TCC).

TSG Building Services Plc v South Anglia Housing Ltd [2013] EWHC 1151 (TCC).

Trustees of Ampleforth Abbey Trust v Turner & Townsend Project Management Ltd [2012] EWHC 2137 (TCC).

Turiff Construction Ltd v Regalia Knitting Mills Ltd [1971] 9 BLR 20 (QB).

Via Sanantonio Pty Ltd v Walker Corporation [2009] NSWSC 679.

Viking Grain Storage Ltd v TH White Installations Ltd [1985] 33 BLR 103 (QB).

Vranicki v Architects Registration Board [2007] EWHC 506 (QB).

Wah (Aka Alan Tang) v Grant Thornton International Ltd [2012] EWHC 3198 (Ch).

Walford v Miles [1992] 2 AC 128 (HL).

West Faulkner Associates v London Borough of Newham [1994] 71 BLR 1 (CA).

Willmott Dixon Housing Ltd v Newlon Housing Trust [2013] EWHC 798 (TCC).

Wimpey Construction UK Ltd v Poole [1984] 2 Lloyd's Rep 499 (QB).

Worksop Tarmacadam Co Ltd v Hannaby [1995] 66 Con LR 105 (CA).

Yam Seng Pte Ltd v International Trade Corp Ltd [2013] EWHC 111 (QB), 146 Con LR 39.

Yorkshire Water Authority v Sir Alfred McAlpine & Son (Northern) Ltd [1985] 32 BLR 114 (QB).

Chapter 20 (Australia)

Australian Infrastructure Plan (2016). *Priorities and reforms for our nation's future*, Report February 2016. Infrastructure Australia.

Bell, M. (2009). Standard form construction contracts in Australia: are our reinvented wheels carrying us forward? *Building and Construction Law Journal* 25: 79.

Byrne, D. (2007). The future of construction litigation disputes. *Building and Construction Law Journal* 23: 398.

Commonwealth Procurement Rules (2018). www.finance.gov.au/procurement/procurement-policy-and-guidance/commonwealth-procurement-rules

Construction Industry Development Authority (1990). *No Dispute – Strategies for Improvement in the Australian Building and Construction Industry*, May 1990.

Cooperative Research Centre for Construction Innovation (2009). *Guide to Leading Practice for Dispute Avoidance and Resolution: An Overview*, 2009. http://www.construction-innovation.info/images/pdfs/DAR_Overview.pdf

Department of Industry, Innovation and Science (2018). *The Digital Economy: Opening Up the Conversation*.

MCC-1 (2003). *Department of Defence Managing Contractor Contract*. www.defence.gov.au/estatemanagement/Support/SuiteContracts/mccontract.asp

Public Infrastructure Productivity Commission (2014). Inquiry Report Vol. 2, No. 71, 27 May 2014.

Sharkey, J., Bell, M., Jocic, W. and Marginean, R. (2014). *Standard Forms of Contract in the Australian Construction Industry Research Report*, June 2014. http://law.unimelb.edu.au/__data/assets/pdf_file/0007/1686265/Research-Report-Standard-forms-of-contract-in-the-Australian-construction-industry.pdf

Sweet, J. (2011). Standard construction contracts in the USA. *International Construction Law Review* 103: 129.

Chapter 21 (Brazil)

ABDI (2017). Agência Brasileira de Desenvolvimento Industrial 2017, 7. *Guia 2 – Classificação da informação no BIM – Coletânea Guias BIM ABDI-MDIC*, ABDI,

Brasília, DF, accessed 10 December 2017. www.abdi.com.br/Paginas/bim_construcao_
download.aspx

CBIC (2016). Câmara Brasileira da Indústria da Construção 2016, 02, 24–27, *Coletânea
Implementação do BIM para Construtoras e Incorporadoras*, vol. 2, *Implementação BIM*,
CBIC, Brasília, DF. http://cbic.org.br/inovacao/2017/10/18/coletanea-bim

Federal Justice Council, Center for Judicial Studies (2013). I *Company Law Journey*, Brasília,
DF, accessed 5 January 2018. http://www.cjf.jus.br/cjf/corregedoria-da-justica-federal/
centro-de-estudos-judiciarios-1/publicacoes-1/jornadas-de-direito-comercial/livreto-i-
jornada-de-direito-comercial.pdf/view

Kassem, M. and Amorim, S.R.L. (2015). *BIM Building Information Modeling no Brasil e na
União Europeia*. Brasília: Ministry of Planning, Budget and Management.

Pereira, C. (2016). Public Procurement in Brazil: General Rules. In: *Brazil Infrastructure
Law*, International Commerce and Arbitration series, vol. 20 (ed. J.F. Marçal, C. Pereira
and M.A. Rost), 229–255. Eleven International Publishing.

Silva, L.T. (2017). *Contrato de aliança: projetos colaborativos em infraestrutura e
construção*. São Paulo: Almedina.

TCU (2013). Federal Court of Accounts (TCU) 2013, Technical Report in audit procedures
TC 014.754/2011–8 and decision AC-2805-40/13-P rendered in 16 October 2013. www
.tcu.gov.br/Consultas/Juris/Docs/judoc/Acord/.../AC_2805_40_13_P.doc

Chapter 22 (Bulgaria)

Bulgaria Corruption Report (2017). http://www.business-anti-corruption.com/country-
profiles/bulgaria, accessed 16 December 2017.

Chavdarov, T. et al. (1983). *Organisation and Intensification in Construction*. Partizdat.

Dimitrov, M. (2012). *The construction contract* (SIBI 2012 ISBN 978–954–730-771-1).

Eadie, R., Stankov, N., Ivanov, Y. and Perera, S. (2017). *State of e-Business in the Bulgarian
Construction Industry Report for CIB TG83* (Western Sydney University ISBN
978–1–74108-450-4) 22. http://www.irbnet.de/daten/iconda/CIB_DC30028.pdf,
accessed 26 December 2017.

Europe's Digital Progress Report (2017). *Integration of Digital Technology*. https://ec.europa
.eu/digital-single-market/download-scoreboard-reports, accessed 26 December 2017.

Georgiev, B. (2017a). BIM in Bulgaria: an interview with Associate Professor Boyan
Georgiev. http://cobuilder.com/bim-in-bulgaria-an-interview-with-associate-
professor-boyan-georgiev, accessed 26 December 2017.

Georgiev, V. (2017b). *Building Information Modelling in Bulgaria* XVII International
Scientific Conference VSU 2017. https://www.researchgate.net/publication/316671933_
INFORMACIONNOTO_MODELIRANE_NA_SGRADI_V_BLGARIA, accessed 27
December 2017.

Kouteva-Guentcheva, M. and Boshnakov, K. (2018). *Engineering reading of the
international BIM Standards*, Annual of the UACEG Sofia Vol 51 2018.

Kouteva-Guentcheva, M. (2017). *First challenges related to the introduction and
implementation of BIM in Bulgaria*, Annual of the UACEG Sofia Vol 50 4/2017. http://
uacg.bg/UserFiles/File/UACEG_Annual/2017/%D0%91%D1%80%D0%BE%D0%B9_4/
17-S.pdf, accessed 26 December 2017.

Moskova, N. (2013). Framework Contracts, PhD Dissertation at the University for National and International Economy, 2013.

Spassova, A. (2014). *Is the Bulgarian Building Control regime more bureaucratic than the regimes in the leading European countries?* (Dissertation submitted in 2014 for a MSc degree in Construction Law & Dispute Resolution, King's College London). http://bscl .bg/en/is-the-bulgarian-building-regime-more-bureaucratic.

The Parliamentary Assembly of the Council of Europe PACE (2017). Session 12 October 2017. http://assembly.coe.int/nw/xml/News/News-View-EN.asp?newsid=6828& lang=2&cat=8, accessed 16 December 2017.

Tormanov, Z. (2012). *The contract in construction* (SIBI 2012 2 Ed ISBN 978–954–730-759-9).

Zheleva, M. and Ivanova, A. (2016). BIM and the benefits for the construction sector, *Stroitel* 29 May 2016. http://vestnikstroitel.bg/news/131457_bim-i-polzata-za-stroitelniya-sektor, accessed 26 December 2017.

Chapter 23 (Germany)

Agthe, V., Löchner, S. and Schmitt, S. (2016). *Intelligente Vergabestrategien bei Großprojekten: Ein Überblick*. Wiesbaden: Springer Verlag.

ARGE BIM Guide (2013). *BIM-Leitfaden für Deutschland* AEC3 & OPB, 2013.

Boldt, A. (2015). SimsalaBIM – Kann der Zauber der Technik auch unser Vertragsrecht verzaubern? *NZBau* 4: 216–224.

Bosch, G. and Zühlke-Robinet, K. (2000). *Der Bauarbeitsmarkt*. Campus Verlag.

Busche, J. (2018). *Münchener Kommentar zum BGB*, 7e. C.H. Beck.

Diekmann, F., Kröger, M. and Reimann, A. (2013). A History of Political Deception in Germany. *Spiegel Online* 10 January 2013. http://www.spiegel.de/international/business/disastrous-public-works-projects-in-germany-a-876856.html

Drömann, D. (2008). Wann sind Betreibermodelle so komplex, dass sie im wettbewerblichen Dialog zu vergeben sind? *IBR* 2.

Emmerich, V. (2016). *Münchener Kommentar zum BGB*, 7e. C.H. Beck.

Eschenbruch, K. and Grüner, J. (2014). *BIM – Building Information Modeling*. NZBau.

Eschenbruch, K. (2016). *Building Information Modelling*. BauR.

Eschenbruch, K. and Leupertz, S. (2016). *BIM und Recht*. Werner Verlag.

Euronews (2017). Berlin Airport Faces New Delays. *Euronews* (online) 23 November 2017. http://www.euronews.com/2017/11/23/berlin-airport-opening-faces-new-delays

Federal Ministry of Transport and Digital Infrastructure (2015). Stufenplan für digitales Planen und Bauen 2015. https://www.bmvi.de/SharedDocs/DE/Publikationen/DG/stufenplan-digitales-bauen.pdf

Gardiner, J. (2014). Latham's Report: Did it Change Us? *Building* (online) (27 June 2014). www.building.co.uk/lathams-report-did-it-change-us?/5069333.article

Heiermann, W. (2005). *Der wettbewerbliche Dialog*. ZfBR.

Heilfort, T. and Strich, A. (2003). Neue Chancen mit alternativen Geschäftsmodellen. *Baumarkt + Bauwirtschaft* 2.

Kapellmann, K.D. and Messerschmidt, D. (2018). *VOB-Kommentar*, 6e. C.H. Beck.

Kemper (2016). *BIM und HOAI*. BauR.

Kleine-Möller, N., Merl, H. and Glöckner, J. (2014). *Handbuch des privaten Baurechts*, 5e. C.H. Beck.

Knight, B. (2017). Stuttgart 21 – Germany's Other Engineering Fiasco Goes Off the Rails. *Deutsche Welle* (online), 13 December 2017. http://www.dw.com/en/stuttgart-21-germanys-other-engineering-fiasco-goes-off-the-rails/a-41782621.

Langen in Nomos Kommentar BGB (n.d.). Appendix IV to Section 631–651: *Vertragstypen im Baurecht*, para 46, 49.

Messerschmidt, D. and Voit, W. (2012). *Privates Baurecht*, 2e. C.H. Beck.

Ritter, N. (2017). *Mehrparteienverträge mit BIM*. Bauwirtschaft.

Schröder, H. (2007). Voraussetzungen, Strukturen und Verfahrensabläufe des Wettbewerblichen Dialogs in der Vergabepraxis. NZBau, 222.

Stadler, A. (2016). *Kommentar zum BGB*, 16e. C.H. Beck.

Stuttgarter Zeitung (2016). Neckarsteg wird Abgerissen. *Stuttgarter Zeitung* 13 July 2016. https://www.stuttgarter-zeitung.de/inhalt.stuttgart-21-neckarsteg-wird-abgerissen.16044498-8c0c-4b98-8550-9af9e1ba75c4.html

Ward, J. (2010). A Four Billion Euro Makeover. *Spiegel Online* 13 August 2010. http://www.spiegel.de/international/germany/stuttgart-21-a-four-billion-euro-makeover-a-710388.html

Zeit Online (2017). S21 wird später Fertig. *Zeit Online* 29 November 2017. http://www.zeit.de/mobilitaet/2017-11/stuttgart-21-verzoegerung-2024-externe-pruefung-deutsche-bahn

Chapter 24 (Italy)

Angeloni, F. (2002). La patrimonialità della prestazione. In: *Trattato di diritto privato* (ed. P. Rescnigno), 41. Turin: UTET.

Barcellona, M. (1987), Attribuzione normativa e mercato nella teoria dei beni giuridici, Quadrimestre, 61.

Cabiddu, M.A. and Colombo, M.C. (2016). Commentario al nuovo codice dei contratti pubblici. Appalti e concessioni, Pescara, CEL.

Camardi, C. (1997). *Economie individuali e connessioni contrattuali*. Milan: Giuffrè.

Cannata, C.A. (1999). Le obbligazioni in generale. In: *Obbligazioni e contratti, Trattato di diritto privato* (ed. P. Rescnigno). Turin: UTET.

Carullo, A. and Iudica, G. (2011). *Commentario breve alla legislazione sugli appalti pubblici e privati*. CEDAM: Padua.

Cian, G. (1968). Interesse del creditore e patrimonialità della prestazione (valore normativo dell'articolo 1174 c.c.), Rivista di diritto civile, 200.

Cianflone, A. and Giovannini, G. (2002). *L'appalto di opere pubbliche*. Milano: Giuffrè.

Ciribini, A. (2013). *L'information modeling e il settore delle costruzioni: IIM e BIM*. Santarcangelo di Romagna: Maggioli.

Ciribini, A. (2016). *BIM e digitalizzazione dell'ambiente costruito*. Palermo: Grafill.

Clarich, M. (2010). *Commentario al codice dei contratti pubblici*. Turin: Giappichelli.

Costanza, M. (2000). *L'appalto privato*. Turin: UTET.

Cuffaro, V. (2011). I contratti di appalto privato. In: *Trattato dei contratti* (ed. P. Rescnigno and E. Gabrielli). Turin: UTET.

De Nova, G. (2010). *Il contratto alieno*. Turin: Giappichelli.

De Nova, G. (2014). *Il tipo contrattuale.* Naples: Edizioni scientifiche italiane.

De Vincentiis, Q. (1906). Della patrimonialità della prestazione nelle obbligazioni contrattuali. In: *Studi giuridici in onore di C. Fadda, IV*. Naples: Luigi Pierro.

De Tilla, M. (2007). *L'appalto privato.* Milan: Giuffrè.

Di Gregorio, V. (2013). *L'appalto privato e pubblico.* Turin: UTET.

Donisi, C. (1980). Verso la depatrimonializzazione del diritto privato, *Rass. Dir. Civ.*, 644.

Eastman, C., Teicholz, P., Sacks, R. and Liston, K. (2016). *Il BIM. Guida completa al Building Information Modeling per committenti, architetti, ingegneri, gestori immobiliari e imprese,* (ed. G.M Giuda and V. Villa). Milan: Hoepli.

Ermolli, S.R. and De Toro, P. (2017). Innovazione di processo per la digitalizzazione degli appalti pubblici: sinergie tra BIM e analisi multicriterio. *Techne* 313.

Ferrari, G.F. and Morbidelli, G. (2013). *Commentario al codice dei contratti pubblici.* Milan: EGEA.

Galgano, F. (2011). *Il contratto.* Padua: Cedam.

Garella, F. and Mariani, M. (2016). Il codice dei contratti pubblici. Commento al decreto legislativo 18 Aprile 2016, n. 50, Turin, Giappichelli.

Giorgianni, M. (1951). *L'obbligazione.* Milan: Giuffrè.

La Rocca, D. (2006). *Diritto e denaro. Il valore della patrimonialità.* Milan: Giuffrè.

La Chimia, A. and Valaguzza, S. (2017). A new approach in implementing the 2014 public procurement directives in Italy: ANAC's soft law regulatory powers amidst uncertainty and need for clarity. *Public Procurement Law Review* 26 (4): 165–186.

Lasalvia, M. (2017). *Commentario al nuovo codice dei contratti pubblici e dei contratti di concessione.* Rome: EPC.

Luminoso, A. (2010). *Il Codice dell'appalto privato.* Milan: Giuffrè.

Mangini, V. and Iacuaniello Bruggi, M. (1997). *L'appalto privato.* Turin: UTET.

Miglietta, M. and Miglietta, A. (2006). *L'appalto privato.* Turin: UTET.

Monateri, P.G. and Somma, A. (2016). *Il modello di civil law.* Turin: Giappichelli.

Panetta, R. (2016). *Il contratto di appalto.* Turin: Giappichelli.

Racca, G.M. (2012). The electronic award and execution of public procurement. *Ius Publicum Network Review.*

Racca, G.M., Cavallo Perin, R. and Albano, G.L. (2011). Competition in the execution phase of public procurement. *Public Contract Law Journal* 41 (1): 89.

Racca, G.M. and Yukins, C.R. (2014). *Integrity and Efficiency in Sustainable Public Contracts.* Brussels: Bruylant.

Ramajoli, M. and Galli, L. (2017). Il ruolo della reputazione nel mercato dei contratti pubblici: il rating d'impresa. *Rivista della Regolazione dei mercati* 63–90.

Rolfi, F. (2006). La patrimonialità della prestazione e l'interesse del creditore. Brevi note sull'art. 1174 c.c. *Jus* 485.

Russo, E. and Criaco, C. (2005). *L'appalto privato.* Turin: UTET.

Sacco, R. and De Nova, G. (2004). *Il contratto, Le fonti delle obbligazioni.* Turin: UTET.

Torchia, L. (2016). La regolazione del mercato dei contratti pubblici. *Rivista della Regolazione dei mercati* 2: 72–79.

Valaguzza, S. (2016a). *Sustainable Development in Public Contracts. An Example of Strategic Regulation.* Naples: Editoriale Scientifica.

Valaguzza, S. (2016b). La regolazione strategica dell'Autorità Nazionale Anticorruzione. *Rivista della Regolazione dei Mercati* 1: 9–47.

Valaguzza, S. (2018). *Governare Per Contratto.* Naples: Editoriale Scientifica.

Chapter 25 (USA)

Cheng, R. et al. (2012). *IPD Case Studies*. American Institute of Architects, AIA Minnesota, School of Architecture University of Minnesota.

Cheng, R. et al (2015). *IPD: Performance Expectations and Future Use*, Univ. of Minn. 2015. http://arch.design.umn.edu/directory/chengr/documents/IPDSurvey_2015.pdf

Cheng, R. et al (2016). *Motivation and Means: How and Why IPD and Lean Lead to Success*, Univ. of Minn., Integrated Project Delivery Alliance, Lean Construction Institute, 2016. http://arch.design.umn.edu/directory/chengr/documents/motivation_means2016.pdf

Hurtado, K. (2015). BIM comes of age: the new ConsensusDocs BIM addendum (2015) for lifecycle building information modeling. *The Construction Lawyer* 36 (4): 37–45.

Integrated Project Delivery 2010 Case Studies, AIA/AIACC 2010. http://www.ipda.ca/site/assets/files/1111/aia-2010-ipd-case-studies.pdf

IPD: A Guide (2007). Integrated Project Delivery: A Guide, preface, AIA/AIACC 2007.

IPD: An Updated Working Definition (2014). AIACC http://www.aiacc.org/wp-content/uploads/2014/07/AIACC_IPD.pdf

Jung, W. and Lee, G. (2015). The status of BIM adoption on six continents. *International Journal of Civil, Structural, Construction and Architectural Engineering* 9 (5): 433.

Lean Construction Institute (2016). *Owner Satisfaction Survey*, Dodge Date and Analytics.

Molenaar, K. et al. (2014). *Examining the Role of Integration in the Success of Building Construction Projects*. Charles Pankow Foundation/Construction Industry Institute.

National Aeronautics and Space Administration (2011). *Building Information Modeling Scope of Services and Requires for Construction Contractor in a Design-Bid-Build Process*. National Aeronautics and Space Administration, 2011. https://www.hq.nasa.gov/office/codej/codejx/Assets/Docs/2-2011-0316%20Final%20DBB%20Construction%20BIM%20RFP.pdf

US Dept of Veterans Affairs (2017). *BIM Manual v. 2.2*. US Dept. of Veterans Affairs. www.cfm.va.gov/til/bim/BIM-Manual.pdf

World Economic Forum (2016). *Shaping the Future of Construction: A Breakthrough in Mindset and Technology*. World Economic Forum 2016. https://www.weforum.org/reports/shaping-the-future-of-construction-a-breakthrough-in-mindset-and-technology

Index